# THE GREAT LOS ANGELES SWINDLE

# THE GREAT
# LOS ANGELES
# SWINDLE

## Oil, Stocks, and Scandal
## During the Roaring Twenties

JULES TYGIEL

New York   Oxford
**OXFORD UNIVERSITY PRESS**
1994

Oxford University Press

Oxford   New York   Toronto
Delhi   Bombay   Calcutta   Madras   Karachi
Kuala Lumpur   Singapore   Hong Kong   Tokyo
Nairobi   Dar es Salaam   Cape Town
Melbourne   Auckland   Madrid

and associated companies in
Berlin   Ibadan

Published by Oxford University Press, Inc.
200 Madison Avenue, New York, New York 10016

Oxford is a registered trademark of Oxford University Press

Library of Congress Cataloging-in-Publication Data
Tygiel, Jules.
The great Los Angeles swindle : oil, stocks, and scandal
during the roaring twenties  / Jules Tygiel.
p. cm.   Includes bibliographical references and index.
ISBN 0-19-505489-X
1. Julian Petroleum Corporation.
2. Petroleum industry and trade—California—Los Angeles—Corrupt practices.
I. Title.   HD9569.J8T93   1994   364.1'68'0979493—dc20   98-19552

9 8 7 6 5 4 3 2 1

Printed in the United States of America
on acid-free paper

For Luise,
no one else can ever know

# ACKNOWLEDGMENTS

"Brave is he who would attempt to write the true story of the notorious Julian Petroleum overissue fiasco," observed *The Critic of Critics*, a Los Angeles weekly, in 1930. More than sixty years later, it takes far less bravery, and far more diligence and determination, to unravel the threads of the tale. I first chanced upon the Julian Petroleum swindle in Carey McWilliams's classic 1946 book, *Southern California: An Island on the Land*. It seemed a simple saga of greed, chicanery, and corruption set against the backdrop of the 1920s Los Angeles boom. Furthermore, it featured many of the elements I look for in a historical project: a strong narrative storyline; bold, vibrant characters; a window into the essence of a bygone era; and a mirror reflecting aspects of our current age as well.

At first glance, it appeared that the Julian story could be researched and written within a reasonably brief period of time. Indeed, my primary fear was that there would not be enough information to flesh it out. Yet, as so often happens in historical research, the more closely I probed, the more complex the subject became. The cast of characters ran first into the dozens, then the scores, then the hundreds. The available sources proved seemingly limitless. Newspaper accounts were supplemented by voluminous advertising, myriad city and federal court cases, and assorted clues and pieces to the puzzle hidden in personal papers, FBI records, and other archival resources.

Perhaps more perplexing was the challenge that the Julian Petroleum scandal presented to my conception of historical truth. The swindle offered not just what *The Critic* called "addenda of charge and countercharge," but it immersed me in a world of confidence men, promoters, and boosters who made their livings concocting convincing lies and artful exaggerations. Arrayed against them was a body of business leaders, politicians, and public servants, ensnared in the con artists' webs, who invented defenses out of whole or partial cloth and newspapers skilled at twisting the truth for promotional and political purposes. I felt at one with the the grand jury foreman who complained that "about the time one group of witnesses had succeeded in painting a well-rounded picture and the jury was preparing to vote an indictment or two, along came another witness with an entirely different 'set-up' of the whole situation. . . . Then would come another picture"; or the defense attorney who challenged the jury to "picture Diogenes looking for an honest man among these . . . witnesses."

Piecing together this "simple saga" thus took the better part of a decade, and even so I do not claim to have written "the true story" of the Julian Petroleum swindle. I have, however, attempted to weigh the evidence and sift through the conflicting testimony to arrive at the likeliest version of events. In the end, of course, that is all that any historian can hope for.

As is true of any work whose completion spans so long a period of time, I have incurred debts of gratitude to many individuals. If there is anyone I forget to mention, rest assured it is due to absentmindedness, rather than neglect or lack of appreciation.

My first debt is to the late Carey McWilliams for introducing me to the Julian Petroleum swindle. All historians of southern California follow in his formidable footsteps. While my conclusions do not always agree with his, the example McWilliams set of lively, lucid, and iconoclastic interpretive writing, underpinned by a strong moral conscience, remains an inspiration. I also must thank two men who stand at the opposite end of the political spectrum from McWilliams: Ronald Reagan and George Bush. Their economic policies, and the consequences thereof, continually breathed fresh life and relevance into the Julian Petroleum saga.

As always, my colleagues in the San Francisco State History Department offered support, encouragement, and that oft-talked-about but rarely achieved commodity, collegiality. The School of Behavioral and Social Sciences and the Professional Research and Development Committee at San Francisco State provided released time and research funds that greatly facilitated the completion of this project. Many people provided invaluable research assistance. They include Eleanor Stardom, who traced Julian's life in Canada, also Don Jordan, Bob Giroux, Tom Clark, Cynthia Taylor, Nancy Quam-Wickham, Jeff Quam-Wickham, Stuart McElderry, and Frances DeNisco.

I am also deeply indebted to the many librarians who offered assistance, often above and beyond the call of duty: Dace Traube, who makes working at the Regional History Center at the University of Southern California an always-pleasant experience; Kay Briegel, at the California State University Long Beach Oral History Project; Dick, Nadine, and Will Hathaway, who have created and operate the wonderful Hathaway Ranch Museum in Santa Fe Springs; Joe Da Rold, the former curator of the Hathaway Ranch Museum; Robert Freeman of the Los Angeles City Archives; William Frank at the Huntington Library; and Laura McCarthy at the Pacific Southwest Regional Branch of the National Archives.

Gary Hoffman of the Texaco Legal Department gave me access to company files related to the Julian Petroleum Corporation and offered valuable clues to unearthing the fate of the Julian properties; Rockwell Hereford and the late Dick Hathaway provided me with their enlightening recollections of the Julian scandal; Tom MacDonald allowed me to read his manuscript on the history of the Los

Angeles District attorney's office; Bill Rintoul offered constructive criticism on the oil industry sections; William Gardiner Hutson generously shared his research materials with me; and the late Gerald White gave valued encouragement.

Peter Carroll, Mike Pincus, and Richard Zitrin, all of whom had painstakingly critiqued my first book, consented to repeat the chore for this one, plowing through the lengthy original draft. I cannot thank them enough for their perseverance and friendship. Luise Custer, Bill Issel, and Howard Roitman also read and critiqued the first draft. Robert Cherny, Naomi Weinstein, and Jim Gregory read parts of the book and offered helpful suggestions. Gail Cooper ably copy-edited the book at Oxford University Press, while Karen Wolny shepherded the art through the production process. Amy Hosa devoted her considerable creative talents to the maps and graphics.

Two other people consistently demonstrated their patience, faith, and forebearance throughout the lengthy evolution of this project: Peter Ginsberg, my agent at Curtis-Brown, always believed that this unfamiliar saga would yield a worthwhile book; Sheldon Meyer, my editor at Oxford University Press, offered his counsel, support, and most of all, his friendship.

Research for this project frequently brought me to Los Angeles, where I invariably enjoyed the hospitality of Tom, Mary, Tommy, and Caitlin Pilla, my southern California family. I cannot imagine more loyal and loving friends.

Finally, I would like to thank my wife, Luise Custer, who has shared me for the past ten years with those two talented con artists and taskmasters, C. C. Julian and C. C. Tygiel. I respect and treasure her contributions to my work, our children, and our life together above all else. One difficult decade later, she continues to make my world a far better place to live in.

# CONTENTS

Cast of Characters,  *xiii*

## I Oil

1. The Seminal Lunacy,  3
2. C. C. Julian Breaks Into Santa Fe Springs,  17
3. Widows and Orphans, This Is No Investment for You,  36
4. The Dividend Payer of the Age,  56
5. *The Truth*,  74
6. When a Feller Needs Some Friends,  90

## II Stocks

7. Wouldst Thou Make Money?,  107
8. Death Valley's Hidden Treasure,  130
9. A Thoroughgoing Businessman,  157
10. The Million Dollar Pool,  174
11. It Appears That There Is an Overissue,  192

## III Scandal

12. The Greatest Swindle Ever Perpetrated in America,  213
13. The Institutions Themselves Must Not Be Tried,  229
14. You Cannot Convict a Million Dollars,  243
15. Well, What of It?,  258
16. When Justice Fails,  276
17. New Vistas of Rottenness,  292
18. What Price Fugitive?,  310

Notes,  329
Bibliography,  381
Index,  389

# ILLUSTRATIONS AND TABLES

Map, Location of Major Oil Fields, Los Angeles Basin, 1920–22,   *15*

Map, Santa Fe Springs, ca. October 1922,   *28*

"Widows and Orphans" advertisement (facsimile),   *41*

C. C. Julian advertising collage (facsimiles),   *44*

Map, Downtown Los Angeles 1920s,   *121*

Western Lead advertising collage (facsimiles),   *138*

Julian Petroleum stock certificate (facsimile),   *175*

Table 1. Julian Petroleum Chronology and Stock Prices,   *115*

Table 2. Julian Petroleum Stock Overissue,   *117*

Table 3. Western Lead Stock Prices,   *137*

Photo section follows page 224.

# CAST OF CHARACTERS

**Adkisson, Arthur**
vice-president, Merchants National Bank; arranges loans for Julian Petroleum

**Allen, Walter B.**
former president, Los Angeles Harbor Commission; Julian Pete investor

**Barber, John E.**
president, First Securities Company

**Barneson, Harold J.**
stockbroker, Barneson and Co.

**Bauer, Harry J.**
chairman, board of directors, Southern California Edison and California-Eastern

**Bell, Herbert A.**
vice-president, Pacific Southwest Bank; member, Bankers' Pool No. 1

**Bennett, Jack**
see Jacob Berman

**Berger, Jacob**
boyhood friend of C. C. Julian's; director, Julian Petroleum, Western Lead

**Berman, Jacob (Jack Bennett)**
associate of S. C. Lewis; "bright youngster" of Julian Petroleum

**Berman, Louis**
brother of Jacob Berman

**Campbell, Henry F.**
vice-president, Julian Petroleum

**Cannon, David**
former U. S. Special Attorney; lawyer for C. C. Julian

**Carnahan, H. L.**
former corporation commissioner; receiver for Julian Petroleum

**Chandler, Harry**
publisher, *Los Angeles Times*

**Chaplin, Charlie**
movie star

**Chessher, H. B.**
stockbroker, A. C. Wagy and Company

**Clark, David**
attorney; candidate for Municipal Court

**Conroy, T. P.**
secretary, Julian Petroleum

**Crawford, Charles**
leader of Los Angeles underworld

**Cryer, Edward**
mayor of Los Angeles, 1921–1929

**Dabney, Joseph**
director, California-Eastern

**Daugherty, Edwin M. "Mike"**
California Corporation Commissioner, 1921–1926

**Davis, Harold L. "Buddy"**
assistant district attorney

**DeMille, Cecil B.**
Hollywood director; Julian Petroleum investor

**Doran, William C.**
Superior Court Justice

**Durbrow, C. W.**
official, Southern Pacific Railroad; director, California Eastern

**Fitts, Buron**
Los Angeles District Attorney, 1929–1940

**Flint, Frank**
attorney, former United States Senator

**Flint, Motley**
vice-president, Pacific Southwest Bank; member, Bankers' Pool No. 1

**Frank, Alvin**
stockbroker; member, Bankers' Pool No. 1

**Friedlander, Jack**
California Corporation Commissioner, 1927–1929

**Gans, Robert**
slot machine king of Los Angeles

**Getzoff, Ben**
owner of tailor shop

**Greer, Perry**
auto dealer, politician; Julian Pete investor

**Hackel, A. W.**
Julian Pete pool member

**Haldeman, Harry M.**
businessman; head of Better American Federation; member, Bankers' Pool No. 1

**Hervey, William Rhodes**
vice-president, Pacific Southwest Trust and Savings; member, Bankers' Pool No. 1

**Hill, Liberty**
county clerk

**Hollingsworth, W. I.**
realtor; member, Bankers' Pool No. 1

**Jardine, John Earl**
president, Los Angeles Stock Exchange

**Johnson, Archibald**
San Francisco attorney; son of Senator Hiram Johnson; pool member

**Johnson, Leontine**
secretary to S. C. Lewis

**Julian, C. C.**
founder, Julian Petroleum, Western Lead

**Julian, Mary Olive**
wife of C. C. Julian

**Keaton, Frank**
murderer of Motley Flint

**Keyes, Asa**
District Attorney of Los Angeles County, 1923–1929

**King, L. J.**
board of directors, California-Eastern

**King, William**
U. S. Senator from Utah

**Kottemann, William C.**
accountant

**Krause, Louis**
jury-fixer

**Lavine, Morris**
reporter, *Los Angeles Examiner*

**Levy, Leonora**
secretary for C. C. Julian in Shanghai

**Lewis, S. C.**
president, Julian Petroleum

**Lickeley, E. J.**
Los Angeles City Prosecutor, 1927–1929

**Loeb, Arthur**
Julian Petroleum stockholder

**Mayer, Louis B.**
president, Metro-Goldwyn-Mayer; member, Bankers' Pool No. 1

**McComb, Marshall**
Municipal court judge

**McCormick, Paul J.**
U. S. District Court justice in charge of Julian Petroleum receivership

**McKay, Henry S., Jr.**
attorney; member, Bankers' Pool No.1

**McPherson, Aimee Semple**
evangelist

**Miller, J. H.**
U. S. Department of Justice agent; comptroller, Julian Petroleum

**Morris, A. W.**
stockbroker

**Morrisey, Thomas**
vice-president, cashier, Merchants National Bank

**Nix, Lloyd**
Los Angeles City Prosecutor, 1929–1930

**Packard, Fred**
Julian Petroleum transfer department employee

**Parrot, Kent**
Los Angeles political boss

**Pettingell, Frank**
president, Los Angeles Stock Exchange

**Pike, Milton**
tailor in Ben Getzoff's shop

**Porter, John**
mayor of Los Angeles, 1929–1933

**Ramish, Adolph**
theater owner; member, Bankers' Pool No. 1

**Reese, Charles E.**
vice-president, A. C. Wagy and Co.

**Reese, Richard**
president, A. C. Wagy and Co.

**Richardson, Friend**
governor of California, 1923–1927

**Robinson, Henry M.**
president, First National Bank

**Rolapp, Frank**
receiver, Sunset-Pacific

**Rosenberg, Ed**
associate of Jacob Berman's

**Roth, Jules H. "Jack"**
vice-president, Julian Petroleum

**Rouse, I. Linden**
vice-president, Pacific Southwest Trust and Savings; member, Bankers' Pools

**Rule, O. Rey**
president, Pacific Finance; director, California-Eastern

**Sartori, Joseph**
president, Security Trust and Savings

**Scott, Joseph**
receiver, Julian Petroleum

**Shipp, Pat**
Julian Petroleum transfer department

**Shuler, Robert**
evangelist

**Silberberg, Mendel**
attorney; member, Tia Juana Pool

**Smith, John A., Jr**
California oil man

**Stanley, Hubert A.**
Price-Waterhouse accountant

**Stern, Charles F.**
president; Pacific Southwest Trust and Savings

**Streeter, C. C.**
stockbroker, H. J. Barneson and Company

**Toplitsky, Joe**
realtor; member, Bankers' Pool No. 1

**Wagy, A. C.**
stockbroker, Wagy and Co.

**White, Leslie**
detective, district attorney's office

**Young, C. C.**
governor of California, 1927–1931

# PART I   OIL

# 1

# THE SEMINAL LUNACY

I

Frank Keaton could well have served as a prototype for the original Homer Simpson, the nightmarish caricature of the transplanted Midwesterner driven mad by the failure of the Los Angeles dream in Nathanael West's *Day of the Locust.* Keaton was born in 1872 in Titusville, Pennsylvania, the site thirteen years earlier of the first great American oil discovery. He grew to manhood in the Midwest, married, fathered two sons, and divorced. In 1902, Keaton moved to Los Angeles, where he worked as an expert machinist. On one job in 1910, a swinging block and tackle crashed into his head, temporarily disabling him. By World War I he had recovered sufficiently to earn high wages at the San Pedro shipyards. During these years he married and divorced again.

Like many southern Californians of modest means, Keaton invested in stocks and real estate. By 1921, when he married his third wife, Miriam, he had accumulated what she described as "a little fortune" of about $35,000 to $40,000. Keaton owned a home in Los Angeles and property in the town of Lennox in addition to banking and other stocks. During the next few years, he and Miriam lived a comfortable life as Frank continued to work as a machinist and speculate successfully. In 1925, Miriam gave birth to a daughter.

But two years later, things began to go horribly wrong. Since 1922, Keaton's investment portfolio had included forty-five shares of First National Bank stock. In May 1927, the Julian Petroleum Corporation, a company underwritten by a First National subsidiary, collapsed amidst charges of fraud and corrruption. A panicky Keaton sold his First National stock for $4,500 less than he might have before the Julian Petroleum scandal. The setback marked the beginning of a series of misfortunes. Keaton suffered a second major industrial accident when a high-power wire burned him severely. In 1929, he incurred further injuries in an automobile accident leaving him with a recurrent pain in his head.

Keaton continued to invest in the stock market, with diminishing returns. The Great Crash of October 1929 dropped the value of his remaining stock portfolio to $1,000. In 1928, Miriam Keaton had given birth to a second daughter, and Frank's losses forced the family to leave their home in Los Angeles and move to a smaller home in Inglewood. Unable to find work, Keaton opened a small real estate office, where he unsuccessfully tried to dispose of his remaining property. He made but one minor sale in two years.

After the stock market crash, said Miriam Keaton, "my husband seemed to have lost his reason altogether. . . . I remember during last November when the stock crash came. Frank would lie in bed and cry for hours at a time. . . . My husband has had a terrible time getting food for those two babies. Where he has got the money to do it I don't know. Possibly he borrowed it and possibly he mortgaged what little he had left."

During these trying times, Keaton grew obsessed by the calamities that had befallen him and the nation. He came to believe that they all stemmed from the collapse of Julian Petroleum. According to Miriam, "Frank blamed the Julian crash for every reaction on the stock market and every security in which he was interested. . . . Frank explained it to me this way. If you take a hatchet and cut off your foot it affects the head and rest of the body." In his growing dementia, Keaton identified a scapegoat for his plight. It was not C. C. Julian, the flamboyant promoter who had founded Julian Petroleum; nor was it S. C. Lewis, the smooth-talking oil executive who had caused the stock overissue that destroyed the company. Keaton fixed the blame on First National banker Motley H. Flint.

If the dour, nondescript Frank Keaton symbolized the hundreds of thousands of ordinary Americans lured by the promise of Los Angeles, Motley Flint, the well-groomed, flashily dressed bon vivant, personified the world of Los Angeles high finance. Flint liked to boast that for more than thirty years he had reigned as "a Santa Claus to Los Angeles." Indeed, to many, Motley Flint and his brother, Frank, epitomized the marriage of business, boosterism, politics, and public service that characterized the finest features of the Los Angeles élite.

Descended from colonial forebears, the Flints were born in Massachusetts and raised in San Francisco. They moved to Los Angeles in their early twenties, lured by the great southern California land boom of the 1880s. Together they had plunged into local Republican politics, which in Los Angeles meant collusion with the political machine financed by the Southern Pacific Railroad. The arrangement proved beneficial for both Flints. During the 1880s and 1890s, each received a succession of federal patronage appointments. Frank rose through the ranks of the United States District Attorney's office to become the United States Attorney for the Southern District of California in 1897. Motley worked as a postal inspector until 1898, when he was named Los Angeles Postmaster.

In 1905, the Flints abandoned government service to enter banking. They organized the Metropolitan Bank and Trust Company, and Frank became the institution's first president. Shortly thereafter, the state legislature, dominated by the Southern Pacific, elected Frank Flint to the United States Senate. Senator Flint turned over his bank presidency to his younger brother. Four years later, when Los Angeles Trust and Savings absorbed the Metropolitan Bank, Motley Flint became a vice-president and director of the new organization. In 1911, political and banking developments reunited the brothers. Frank, confronted with the Progressive rebellion against the Southern Pacific and the first direct election of senators, chose to retire from the Senate. At the same time, First National Bank merged with Los Angeles Trust and Savings. Both Motley and Frank became directors at First National. Frank became the bank's legal counsel; Motley remained as a vice-president of Los Angeles (later Pacific Southwest) Trust and Savings.

By this time, the tall, distinguished Motley Flint had developed a reputation as one of the city's leading boosters. "He was a genuine enthusiast about Los Angeles," commented the *Los Angeles Examiner*. "He predicted its future; he helped make his predictions actualities." Long active in numerous fraternal and social organizations, most notably the local Shriners and Elks, Flint was often referred to as "the best known lodge man in the United States." He organized and chaired the Los Angeles Convention League, which encouraged groups to hold their national meetings in the city. He became an unofficial "greeter" for Los Angeles, always present when visiting dignitaries appeared. The home he shared with his wife, Gertrude, became the center of lavish social gatherings.

His wealth and influence notwithstanding, Motley Flint took even greater pride in his charitable work and fund-raising activities. "Under an exterior that was sometimes gruff and a manner that was often misunderstood, there beat a great heart full of tenderness and sympathy," stated his friend William Rhodes Hervey. Flint served for twenty years as president of the local board of relief, finding jobs for the unemployed and offering assistance to the downtrodden. His efforts on behalf of impoverished children earned Flint his "Santa Claus" reputation.

With the rise of Hollywood, Motley Flint's life took another fortuitous turn. Flint was the first member of the local banking establishment to recognize the importance of the movie industry to Los Angeles. He organized the Cinema Finance Company, which made First National a leading financier for Hollywood. After 1917, when he separated from his wife, Flint's downtown hotel suite became a favorite gathering place for the Hollywood crowd. Top stars and studio executives regularly frequented his lavish luncheon and dinner parties. When the fledgling Warner Brothers Corporation faced bankruptcy in 1920, Flint encouraged First National to back the struggling studio. Thereafter Flint formed a close personal friendship with Jack Warner, reassuring the innovative motion picture producer, "I never worry about your debts. You and Sam are going in the right direction and I know you'll make it." At Flint's urging, First National became a key backer in Warner Brothers' revolutionary experiments with talking films.

Brother Frank, meanwhile, had also prospered. Historian Kevin Starr describes Frank Flint (in terms that could just as easily fit Motley) as "one of the most important players in the drama of making Los Angeles happen . . . a comforting symbol of public and private ambitions working together in creative synergy." Frank Flint's law partnership with Henry S. McKay was among the city's most successful. His diversified business interests, board directorships, and real estate developments flourished. In the 1920s he founded Flintridge, an exclusive residential community built into a hillside to resemble a Greek Mediterranean village. Motley erected a fourteen-room mansion at Flintridge as a retreat from his celebrated downtown apartment.

To Frank Keaton, however, Motley Flint resembled, not Santa Claus, but Shylock. While Keaton and thousands of others had lost millions of dollars as a result of the Julian Petroleum scandal, Flint and his fellow bankers and businessmen had organized a "Million Dollar Pool" backed by the company's stock. The pool delivered a 57 percent profit for its investors. One month after its liquidation, Julian Petroleum collapsed. Many believed that Flint and his fellow millionaires had knowingly bled Julian Petroleum while thousands of smaller speculators lost their life's savings. Frank Flint had served as legal counsel and director for the discredited firm. Devastated by the charges emanating from the scandal, Motley Flint had returned his profits and withdrawn from his myriad business, social, and charitable activities. "They have broken my heart in the city I love," Flint told his secretary; "I never want to see Los Angeles again." Flint moved to France. As the Frank Keatons of Los Angeles struggled to recover from their losses, Flint lived in Nice, from which he sent back reports of life on the Riviera and in Paris.

In the months after the October 1929 stock market crash, Keaton became increasingly obsessed with the unfolding Julian Petroleum scandal. "He followed carefully all the exposures in the Julian failure," related his wife. He expressed sympathy for the tens of thousands of investors who had lost money, blaming their misfortune on Flint and his fellow bankers. Keaton drew encouragement from the radio broadcasts of fundamentalist preacher Robert P. Shuler who regularly attacked the Los Angeles "banking crowd" for its role in the Julian affair. Shuler's diatribes fueled Keaton's paranoia. In the end, explained Miriam, "he finally came to the belief that Mr. Flint was the cause of all his losses in stock deals. I believe that his mind became so full of this that it snapped."

Despite his disillusionment, Keaton never lost faith in the curative powers of the stock market. Every morning he would leave home before five in order to arrive at the Los Angeles Stock Exchange when it opened. He would visit his stockbrokers and, despite his steadily declining resources and mounting debts, place new buy and sell orders. When finished he would haunt the myriad trials emanating from the Julian Petroleum swindle.

On the morning of July 14, 1930, Keaton left his home before dawn as usual. He had not slept well the past two nights, and the pain in his head had intensified. In his pocket was a .38 caliber nickel-plated hammerless revolver he had owned for forty years. He said that he planned to have it cleaned and polished. There were bullets in three of the chambers. Keaton reached downtown Los Angeles and stopped at his broker's office, where he placed an order for 100 shares of motion picture stock. He overdrew his bank account to cover the margin.

Keaton wandered the downtown streets until an article in the newspaper caught his attention. At ten o'clock that morning, read the item, Motley Flint would appear as a witness in a Los Angeles courtroom. Keaton walked to City Hall and asked for directions. He entered the courtroom quietly as the hearing began and took a seat.

The trial pitted David O. Selznick against First National Bank. Selznick's greatest successes (most notably the 1939 production of *Gone with the Wind*) lay in the future. But at age twenty-eight he already reigned as one of the brightest young producers in Hollywood. Two months earlier he had married Irene Mayer, the daughter of M-G-M mogul Louis B. Mayer, assuring his position in the local firmament. His suit involved the disposition of 1,000 shares of United Studios stock worth between $150,000 and $250,000, which Selznick and his family had placed with Pacific Southwest Bank, a First National affiliate, as security for a loan in 1922. Flint, a bank vice-president at the time the loan was made, had recently returned from France to face criminal charges stemming from the Julian Petroleum scandal. Attorneys hoped that he might shed some light on the disputed transaction.

As he seated himself in the witness chair, the sixty-five-year-old Flint appeared as debonaire and distinguished as ever. His silver mane and mustache offset his blue serge suit and blue shirt with a striped collar. Expensive jewelry bedecked his hands and wrists. A Panama hat rested on the table in front of him. Flint sat facing the bench directly opposite and below Judge Frank C. Collier. Attorneys for Selznick occupied the long table to Flint's left; the bank's lawyers flanked him on the right. To Flint's rear, behind the rail that demarcated the seating area, Selznick's brother Myron and his parents, Lewis and Florence, occupied the first three seats of the front row. Only a handful of people were present in the small courtroom; most of them court officials, attorneys, or parties to the case. Frank Keaton, seated in the second row directly behind Florence Selznick, was the sole uninvited spectator. At one point he leaned forward to ask Mrs. Selznick if this was the Selznick case. Keaton later asked another woman to point out Motley Flint.

Keaton watched silently, but intently, as Flint identified letters and telegrams exchanged by Pacific Southwest and the Selznicks in 1922. Flint failed to recall the details of most transactions. "You have no definite recollection as to how the sum was disposed of, have you, Mr. Flint?" asked the plaintiff's attorney after twenty-five minutes of testimony. "I couldn't positively state, no, sir," responded Flint. Following this exchange, Judge Collier dismissed the retired banker from the stand. Flint picked up his hat and turned to leave. A slight, relaxed smile came over his face.

In the spectator section, Keaton felt a twinge in his arm and the pain in his head grew sharper. "The gun that I held concealed in my pocket seemed to be as big as a cannon," he said later.

Flint opened the enclosure gate and stepped into the spectator section. He bent forward as if to talk to Mrs. Selznick. Suddenly Keaton leapt to his feet behind her. His hat hid the gun in his right hand as he fired directly over Florence Selznick's shoulder, just a foot or two away from the leaning Flint. Flame leapt from the muzzle of the old revolver. The shot shattered the banker's neck and lodged in his collarbone. Flint fell to his knees without uttering a sound. Mrs. Selznick began to scream. Keaton fired two more shots, one which pierced Flint's heart and the other his lung and liver. The impact knocked Flint backward, leaving his body lying full length, angling from the rail, his feet pointed at the door. Flint was dead before he hit the floor.

Keaton tossed his gun in the direction of Flint's body where it landed between the arm and body of the corpse. He sat down, squeezed his eyes shut and clenched his hands. "Oh, my God, why did I do it?" Keaton repeated over and over again. Judge Collier, upon hearing the reports of the gun, had jumped from the bench and boldly ran toward the shots. Keaton offered no resistance as the judge seized and restrained him.

When police deputies arrived, they searched both the assassin and his victim. In Flint's pockets they found $63,000 in cash. In Keaton's they found ten cents, a pack of cigarettes, and a well-thumbed sixty-four-page pamphlet written by the Reverend Robert P. Shuler. The treatise was entitled "Julian Thieves." It purported to unravel the tangled web of affairs known as the Julian Petroleum scandal.

## II

In Los Angeles in the early years of the Great Depression, C.C. Julian and the Julian Petroleum Corporation were household words. They symbolized, not merely what President Franklin D. Roosevelt would later deplore as "a decade of debauchery of group selfishness," but the failed hopes and dreams of the great boom of the 1920s. Indeed, no single story captures the essence of the 1920s in America — its booster optimism and rampant speculation, its entrepreneurial mania for mergers, its overlap of business and politics, its application of new communications technology, and its cast of oilmen, stock promoters, Hollywood stars, cinema moguls, banking executives, Prohibition era-gangsters, and evangelists — quite so well as the Julian Petroleum swindle.

During the 1920s the American people, chides economist John Kenneth Galbraith, "display[ed] an inordinate desire to get rich quickly with a minimum of effort." No one engineered this speculation or led investors to the slaughter, writes Galbraith. It was "the product of free choice and decision by hundreds of thousands of individuals . . . impelled to it by the seminal lunacy which has always seized people who are seized in turn by the notion that they can become very rich."

The "seminal lunacy" appeared not just on Wall Street or in Los Angeles, but in virtually every section of the nation. The Florida land boom remains the most celebrated of the mid-'twenties speculative excesses, but suburbanization and high-rise building construction triggered smaller bubbles in urban areas throughout the United States. Oil discoveries in Oklahoma, Texas, and Louisiana attracted speculators from around the nation. Mining prospects in the Far West and Canada excited the interest of hundreds of thousands of investors. Ivar Krueger, the Swedish "Match King," sold almost $150,000,000 in International Match securities; Samuel

Insull constructed a half-billion dollar nationwide utilities pyramid on the dreams of those seeking instant riches.

American culture encouraged the acquisitive mentality. In New York City an electric sign advised passersby: "You should have $10,000 at the age of 30; $25,000 at the age of 40; $50,000 at the age of 50." Self-made millionaire and National Democratic chairman John J. Raskob entitled a 1929 *Ladies' Home Journal* article "Everybody Ought to Be Rich."

In many respects Italian immigrant Charles Ponzi stands as the patron saint of the 1920s. In late 1919, the Boston-based Ponzi created the Old Colony Foreign Exchange Company. Ponzi claimed that by buying foreign currencies at low market prices and selling them at higher official rates he could guarantee investors quick and easy profits. He promised to return $15 for every $10 entrusted to him after ninety days. Miraculously, Ponzi consistently made good on his pledge. Within nine months he had become rich and famous. More than $1 million poured into Old Colony Foreign Exchange every week. Ponzi announced the opening of branch offices and bought a controlling interest in the Hanover Trust Company. Investors continued to receive their 50 percent profit every ninety days. But in August 1920, the money ceased to flow. Ponzi, it turned out, had purchased few foreign currencies. The only exchange that Ponzi had performed was to pay off old investors with the ever-increasing money collected from the new. This strategy could only last as long as the investment pool continued to grow. Ultimately it had to collapse, with the most recent arrivals footing the bill. Millions of dollars disappeared. Ponzi spent several years in prison before being deported, but his missing millions were never found.

It is not surprising, as Robert Sobel has argued, that the decade that began with the development of the "Ponzi scheme," as these rotating investments are still known, ended with the collapse of Wall Street's great bull market. Both reflected the speculative mania that had seized the American public in these "ballyhoo years." According to historian Ellis Hawley, "the 1920s are best understood not as the Indian summer of an outmoded order . . . but rather as the premature spring of the kind of modern capitalism that would take shape in the America of the 1940s and 1950s." Warren I. Susman identifies the culmination of "a crucial change in the nature of the capitalist order and its culture" occurring in the 1920s.

The "ballyhoo" aspects of this decade were part and parcel of the emergent "capitalist order." Confronted by an "extraordinarily rapid accumulation of both new knowledge and new experiences," Americans, argued Susman, "were made constantly and fully aware that they lived in a new era." The key to this embryonic society was a revolution in communications. The automobile, motion pictures, and radio were the most dramatic innovations. But tabloid newspapers with their greater emphasis on photography, the increasingly visual nature of modern advertising, and the popularity of cartoons and comics all added to the transformation. Together, argued Susman, these developments marked the "emergence of new ways of knowing that stood in sharp contrast with the old ways of knowing available in the book and the printed word." Through newsreels, photographs, and radio broadcasts,

Americans vicariously experienced and participated in the exploits of Charles Lindbergh and Babe Ruth or events like the Scopes trial or the Leopold and Loeb murder case on a level of intimacy never before possible.

Moreover, according to Susman, the communications revolution offered "newer and more effective techniques in the manipulation of men." Many commentators during the 1920s recognized the rising tide of consumerism and the emergence of a mass market made possible by these strategems. The expanded volume of advertising offered ample evidence of this phenomenon. Before World War I, American business had spent an estimated $300 to $400 million a year in advertising. By 1927, the figure had jumped to $1.5 billion; by decade's end it would exceed $1.8 billion. Newspaper advertising increased by 600 percent, claiming the lion's share of these figures. Magazines, billboard, and direct mail campaigns also mushroomed. Radio represented the fastest growing segment of the advertising market.

The techniques of mass marketing found a welcome among those seeking to squeeze investment dollars from an eager populace. Promoters of all stripes took advantage of the increased flow of advertising. The introduction of machines like the addressograph and the multigraph revolutionized mail-order solicitation, allowing con men like Frederick Cook, the polar explorer turned Texas oil man, to circulate tens of thousands of come-ons each day. Some experimented with radio and movie promotions. Others adapted the tried and true methods of the real estate and mining promoters of the past, substituting bus and auto excursions for the train and streetcar tours of an earlier era.

Federal tax policies also nourished the speculative urge. At the behest of Treasury Secretary Andrew Mellon, the scion of a leading banking family, Congress enacted a series of cuts primarily designed to reduce the taxes paid by America's wealthiest individuals. Freed from the onerous burdens of taxation, argued Mellon, the rich would invest in commerce and industry, providing jobs and prosperity for all. In 1921, Congress repealed the wartime excess profits tax, eliminated many luxury taxes, lowered corporate taxes, and slashed the maximum personal income tax from 65 percent to 32 percent. Mellon successfully advocated further reductions in 1924, 1926, and 1928, all of which offered windfalls for the wealthy. Mellon's personal tax reduction exceeded that of the entire population of Nebraska.

That this type of supply-side economics frees money for investment and may stimulate short-term growth is beyond question; whether the beneficiaries of this federal largesse invest wisely is another matter. Many have blamed the infusion of new money and inexperienced investors for the excessive speculation of the 1920s. But one is also struck by how readily the nation's old wealth and leading businessmen succumbed to lures of short-term profits and get-rich quick schemes, diverting capital from constructive projects into high-turnover, speculative ventures and risky securities.

The benefits of these tax cuts may or may not have "trickled down" to the masses, but the speculative fever that they helped release definitely did. Many have dismissed the misleading hyperbole that "every corner bootblack was speculating in the stock market" during the 1920s. Yet, in a nation of 120 million people, an estimated four million Americans owned stock. The vast majority were undoubtedly small investors. The size of their stakes only tempered, but rarely extinguished, their enthusiasm. More significantly, the stock market represented just one destination for risk capital. When one considers the full panoply of investment opportunities — from oil, to mining, to real estate, to industrial opportunities, both realistic and bizarre, offered on and off local and regional stock exchanges — a considerable segment of the American populace took at least one plunge, and probably several, into the speculative swim.

Nowhere were the waters more inviting than in southern California. That Los Angeles should succumb to the lures of speculation in the 1920s seemed as natural as the climate. Indeed, the city's entire history resembled a Ponzi scheme of fantastic dimensions. Since the 1880s, as one banker noted, "the perfectly prodigious business of growth" had been the city's leading industry, its primary reason for existence. "The steady, speedy growth is the one important thing to understand about Los Angeles," wrote Bruce Bliven during the 1920s. "It creates an easy optimism, a lax prosperity, which dominates people's lives. . . . The first comers if they can just get their fingers on a little property, are sure to grow rich with the unearned increment."

"Los Angeles," explained Morrow Mayo in the 1930s, "is not a mere city. On the contrary, it is, and has been since 1888, a commodity; something to be advertised and sold to the people of the United States like automobiles, cigarettes and mouthwashes." In 1888, city business leaders, reeling from the collapse of the first great land boom, declared a "new beginning." They centered this rebirth upon a reconceptualized Chamber of Commerce. In Los Angeles the Chamber would function, not simply as an informational agency, but rather as a high-powered municipal promotional vehicle. The Chamber launched an ambitious nationwide advertising onslaught to extol the virtues of Los Angeles. It distributed more than two million pieces of literature in its first three years and dispatched a "California on Wheels" train, laden with photos and exhibits, to the major cities of the South and Midwest, to boost the promise of Los Angeles. The campaign elevated Los Angeles from a modest community of 50,000 in 1888 to a metropolis of over half a million in 1920.

The booster spirit permeated Los Angeles. "These square miles are peopled with nothing but boosters. . . . They've been psyched into a hoorah frame of mind which knowns no scoffing," praised Paul Augsberg in "Advertising Did It!" his 1922 paean to the city's success. "From the highest to the humblest of its citizens," wrote Guy Finney in 1929, "their chief desire seemed to be to contribute, with fine enthusiasm, all they could in material and vocal acceleration to the movement that was sending the city ahead at a dazzling pace."

Louis Adamic, among others, dismissed this enthusiasm as crude Babbittry. The big businessmen, wrote Adamic in the mid-'twenties, were "Super-Babbitts"

> . . . possessed by a mad and powerful drive . . . grim, rather inhuman, individuals with a terrifying singleness of intention: they see a tremendous opportunity to enrich themselves beyond anything they could have hoped for . . . and they mean to make the most of it.

Thousands of lesser Babbitts, "all motivated by the same motives for wealth, power, and personal glory, and a greater Los Angeles," trailed in their wake. Together they preached a gospel of growth and adhered to what the *Los Angeles Record* called "the psychology of 'boost-don't knock.'" The darker side of Los Angeles—the epidemic arrests for vagrancy, the rabid repression of dissent and organized labor, the extraordinary rates of divorce, insanity, and drug addiction—could not be mentioned, wrote Adamic, "for the tourists must not get the idea that anything is wrong with Los Angeles."

The contribution of the Los Angeles élites, however, went beyond advertising and boosterism. Los Angeles in the 1880s had lacked not just recognition but significant harbor facilities and adequate water supplies to sustain growth. By the second decade of the twentieth century, the concerted efforts of Los Angeles business and political leaders had created a man-made harbor at San Pedro and procured (or, as many would argue, purloined) water from the Owens Valley via a 200-mile aqueduct.

With this infrastructure in place, the visions of the "Super Babbitts" and the "lesser Babbitts" converged in the 1920s in a reality that fulfilled the dreams of even the most optimistic boosters. Los Angeles boomed as never before. Between 1920 and 1924, over 100,000 people a year settled permanently in the city. The rate of growth slowed in the latter part of the decade, but by 1930, the city's population had soared from 577,000 in 1920 to over 1.2 million.

The Chamber of Commerce effectively promoted the municipality as the nation's "white spot," one of the few areas to escape the recession of 1920-1921. It organized a photo news service that distributed thousands of portraits "to exploit the nearby bathing beaches and give the sun-backed damsels a chance to reveal their charms." In 1921, the All-Year Club, a Chamber subsidiary financed by both public and private funds, embarked on a multimillion-dollar campaign to convince potential tourists that Los Angeles had "a year-round appeal and not merely one that lasts through the winter months." Millions of advertisements in newspapers and magazines convinced the nation's tourists that southern California was more than a retreat from cold weather. Within three years, summer visitors exceeded the traditional winter influx. Thousands of tourists became permanent residents.

No commentator captured Los Angeles in the 1920s better than Albert Atwood, who wrote a series of articles about the city for the *Saturday Evening Post* in 1923. Los Angeles alternately bemused, shocked, astounded, and awed Atwood with its vibrant energy, shameless boosterism, astonishing growth, and speculative excesses. In an article entitled "Money from Everywhere," Atwood commented on the "extraordinary and almost unprecedented pouring of population, money and prosperity into one section of the country, and more particularly into one city." What struck Atwood was the unusual affluence of the new arrivals: "There gravitates toward this region a constant succession of crops as it were of retiring persons, farmers from the Middle West, and manufacturers, merchants and business and professional people from all arts of the country. People retire at the most liquid moment of their lives . . . in the aggregate they make this a very rich country."

The influx of retirees represented a massive transfer of wealth and capital to southern California. Lucrative investment opportunities awaited these newcomers. In the past, agriculture, commerce, tourism, and service had dominated the local economy, but with the rise of Hollywood, the discovery of oil, and the expansion of the tire and other industries, Los Angeles developed an industrial base as well. Manufacturing output tripled during the 1920s, jumping the city from twenty-sixth to ninth place among the nation's manufacturing centers. The harbor at San Pedro became the busiest port on the Pacific Coast.

Southern Californians, recent arrivals and established settlers alike, sought to share in the prosperity. Many invested conservatively, taking advantage of the legitimate opportunities generated by industrial growth, but the get-rich-quick mentality pervaded the region. "Speculation is in the air," stated one banker, "because the people think of it as the land of easy money, and because for so many it has proved these things." In *The Great Los Angeles Bubble*, local journalist Guy Finney described a city "particularly ripe for heedless financial adventure." The "emotional hasheesh" of the get-rich-quick message "scattered whatever horse sense" people possessed.

> They knew their city was galloping along at dizzying speed. . . . The easy-money carnival spirit gripped them. . . . They sang it from every real estate and stock peddling platform, in every glittering cafe and club, in banks and at newstands. . . . The dollar sign was on parade. Why marvel then that like epidemic measles among the young, it spread from banker to broker to merchant to clerk to stenographer to scrubwoman to office boy, to the man who carried his dinner pail. . . . The mass craving wanted its honey on the table when the banquet was on. It simply couldn't wait a soberer day.

Ambitious hustlers prowled hotel lobbies courting recent arrivals. "Hopeful young men with something to sell penned them cordial notes, called them by telephone, invited them to come and take a look around," wrote Finney. Salesmen offering sightseeing tours and opportunities "to see the movie ladies in their cozy bungalows on the big 'lot,'" instantly became "friend[s] for life" before they sprang their investment schemes. Bankers played the same game on a more lavish scale. Junior vice-presidents, and even bank presidents, entertained wealthy prospects who had dropped in to transfer funds from the East with tickets to the Hollywood Bowl or other attractions.

An "insistent element of speculation . . . permeates all walks of life," wrote Atwood. But the real estate subdivision continued to symbolize the speculative soul of Los Angeles. Real estate in southern California, noted Atwood, resembled "a stock market craze of the first magnitude." During a twelve-month period starting in November 1921, realtors in Los Angeles County proposed 631 new subdivisions. Enough housing was built in 1922 to supply a city the size of Albany, New York.

The legitimate demand for housing for new residents kindled the real estate boom, but the speculative dreams of the masses fueled the bonfire. "One is struck, not altogether pleasantly by the great number of people who have sold, or expect to sell their homes at a profit," observed Atwood. "Time and again people have bought land at prices that were highway robbery at the time, only to sell at an advance in a few years. People have bought into subdivisions and found themselves unable to sell. . . . Compelled to build and live on the property, they have sold in a few years at a large profit." One real estate leader estimated that nearly 65 percent of those buying houses in the subdivisions anticipated reselling at an increased price.

At the peak of the boom in 1923, a veritable army of 12,000 realtors trooped across the subdivided terrain. Unregulated by legal or moral restraints, the lords of real estate gave new meaning to the terms *laissez-faire* and *caveat emptor.* "One is disgusted by the grotesque circus character of the subdivision development," wrote Atwood. "High pressure sales methods have passed their hitherto known peak in an orgy of fantastic blah." Robert Cleland found the aesthetics and the architecture of the boom as limited as its ethics. Cleland described Sunday afternoons at the Palos Verdes peninsula featuring Hollywood stars, Spanish dancing, stunt flying, athletic contests, aquaplaning, yacht racing, and a "Kiddies Tent" complete with playground teachers, physicians, and free toys. "Like the enterprising philistine that he was," criticized Cleland, "the Los Angeles real estate promoter, who thought primarily of quick and easy profits, subordinated beauty to utility and ridiculed the idea of letting aesthetic considerations interfere with the maximum gain to be derived from the subdivision and sale of the good earth."

Real estate remained a primary focus of speculative dreams throughout the 1920s, but in 1920 and 1921, astonishing events just south of Los Angeles added a new, more exciting dimension to the boom. Oil was discovered in southern California. Oil, wrote Samuel Blythe, was "the flame in which the small moneyed moths of this country have burned their wings for more than 60 years." And, he reported, the already speculation-crazed population of southern California went "stark, staring, oil mad."

The discoveries occurred along the coastal corridor where Los Angeles and Orange counties converged. The first strike came in the quiet resort town of Huntington Beach. Standard Oil of California had drilled several wildcat wells at Huntington Beach in 1920. The first produced a modest forty-barrel-a-day well on Reservoir Hill. A second, on the edge of the "Gospel Swamp" peat bogs, roared to life with an explosion heard fifteen miles away. The initial flow of 20,000 barrels of high gravity oil per day announced the birth of a great new oil field.

Five months later and less than fifteen miles distant, the Royal Dutch Shell Company sank a well on the east slope of Signal Hill, a 345-foot dome protruding above the small city of Long Beach. Shell had already spent $3 million in the California oil fields, to no avail. Industry experts predicted further disappointment on the Long Beach promontory. A Shell crew labored for three months atop Signal Hill. On May 23, they found oil standing in the hole. One month later, as the night crew worked under the derrick lights, they heard an underground rumble, which grew louder and louder until the black liquid spouted from the earth, shooting 114 feet into the air. Hundreds of local residents came running to share the excitement. Signal Hill would prove to be one of the world's richest oil fields.

Nearby, in Santa Fe Springs, Union Oil awaited results on a wildcat well on

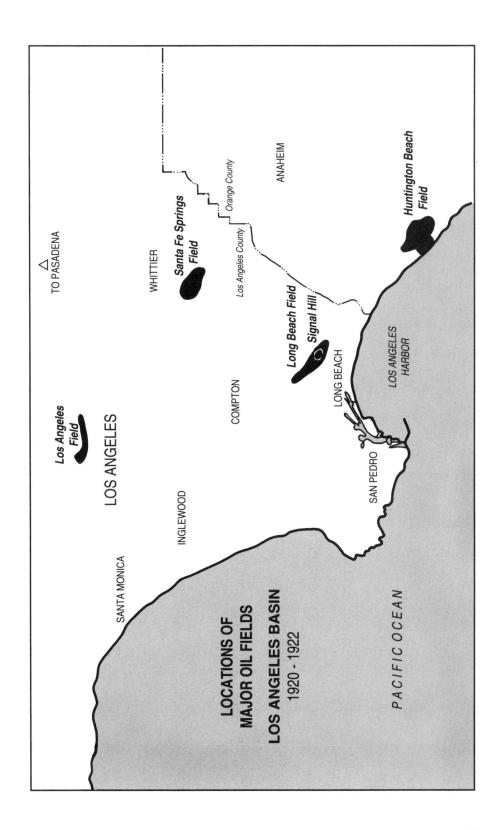

LOCATIONS OF
MAJOR OIL FIELDS

LOS ANGELES BASIN
1920 - 1922

SANTA MONICA

INGLEWOOD

LOS ANGELES

Los Angeles
Field

COMPTON

SAN PEDRO

LONG BEACH

Long Beach Field
Signal Hill

LOS ANGELES
HARBOR

PACIFIC OCEAN

WHITTIER

Santa Fe Springs
Field

TO PASADENA

Los Angeles County

Orange County

ANAHEIM

Huntington Beach
Field

15

land owned by former tennis champion Alphonzo Bell. On the morning of October 30, 1921, the earth roared and mud and water came rushing out the hole, followed by a rich stream of black liquid. Bell scrambled out of bed and donned a corduroy suit and sombrero, both of which were soon covered with oil.

Thus, wrote Atwood, "before the effects of Huntington Beach had a chance to wear off, Signal Hill came in; and when the flood from both was just coming on, the third and largest of all, Santa Fe Springs came rushing in." Within a fifteen-mile triangle, there materialized what one geologist has called "the greatest out-pouring of mineral wealth the world has ever known." Never before had three prolific fields in such close proximity appeared within such a short span of time. Individually, each was a sensation; collectively, they rent the fabric of southern California society.

"For all the high tides in the affairs of men," wrote Atwood, "none is at once so productive and destructive of . . . wealth, and none combines such beneficent effects as one of those floods of liquid gold . . . with its accompaniment of sheer frenzy." The Los Angeles oil boom, he noted, replicated conditions of earlier strikes: "the same optimism in the air, the same enrichment of landowners, the same ener-gizing and stimulating of the communities directly affected . . . the same exagger-ated expectation of riches for all."

But several major differences distinguished the new mania. Although most pre-vious discoveries had occurred in rugged, isolated terrain, the new strikes heralded the first oil boom to unfold within reach of a great metropolis. In addition, the three oil fields were not located on vast wilderness estates or lands in the public domain but on previously developed real estate subdivisions. Signal Hill contained almost one thousand individual lots. Huntington Beach and Santa Fe Springs added hun-dreds more. Huntington Beach had won renown for its "Encyclopedia lots." In 1914, a publishing house had marketed its encyclopedias by offering Huntington Beach land as premiums. Each purchaser received a 25-by-100-foot parcel. Many recipients failed to pay taxes and their ownership lapsed. Others throughout the nation retained their interests. Suddenly the Encyclopedia lots sat atop of one of the nation's richest oil fields. Those who had kept their "premiums" received healthy monthly royalties, symbols of the oil fortunes available to anyone.

Los Angeles thus offered, in Atwood's words, "a wonderful, superb, unprece-dented situation from the oil promoter's standpoint. The wells are there, the oil is there and the investors are there." These conditions, he wrote, "afforded an oppor-tunity for human greed to operate on a scale which has overreached itself."

C. C. Julian strode onto this stage in 1922. The events he set in motion transfixed the attention of southern Californians and transformed their lives throughout the 1920s and for many years beyond.

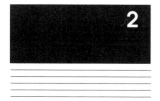

# 2 C.C. JULIAN BREAKS
# INTO SANTA FE SPRINGS

**I**

Any description of Courtney Chauncey Julian's early life must be viewed with skepticism. Available accounts of his formative years derive from his own often self-serving saga of hard work, ambition, and fortunes gained and lost. While public records support the broad outlines of his tale, many details escape confirmation; others fall into the realm of fabrication.

C. C. Julian portrayed himself as one part Horatio Alger and one part frog prince, born a generation removed from a life of wealth and leisure. He was, he claimed, the grandson of a prominent Irish attorney, the Crown solicitor for King's County in Ireland, who at his death in 1896 had amassed a $5,000,000 estate, and British documents do list the appointment in 1844 of a John Julian to the solicitor post. According to C. C., his father, Francis, had thus been born into a world of comfort and culture, "a man of extensive education, a graduate of Oxford, and master of seven languages." Francis Julian served as a captain in the British army, but "reared as a gentleman without any training along commercial lines . . . he had no notion in the world of what work was."

Francis, recounted C. C., fell from grace at age twenty-eight when he dared to love and marry an Irish Catholic girl. John Julian, an Orangeman and a Mason, bequeathed his son $15,000 and severed all relations. In the early 1880s Francis and his young bride left Ireland to seek a new life in the Canadian colonies. With the money received from his father, Francis envisoned becoming "a colonial on the grand scale, in the best British manner." He bought 600 acres of land near the town of Morris, Manitoba, halfway between Winnipeg and the United States border.

C. C. recalled his father as a man whose comfortable upbringing "made him forever afterward incapable of understanding the true value of money or taking care of it." As he had sunk most of his capital into land, "ready cash" became "too scarce an article with a man who didn't know how to get along on anything else." The land itself yielded only an unprofitable mustard seed crop.

Seven births were registered in Morris, Manitoba, in 1885. Among them was Courtney Chauncey Julian, born on October 10 on what he liked to call "one of the poorest farms of North America." By 1894, C. C. had four younger brothers and sisters. He spent his early years performing the arduous tasks of a farmboy: milking cows, herding sheep, chopping firewood, and harvesting the mustard crop. The habits fostered by the rigorous seventeen-hour work day, he later boasted, never left him.

Morris, Manitoba, marked the first of Julian's repeated exposures to the boom-and-bust mentality of the industrial age. Julian doubtless took note of the unabashed promotional activities of the townspeople. A local history, using language that might apply to Julian's later career, described the era as one of "activity, courage, determination, and an overly generous vision of expenditure." But the city's modest boom ended in the 1890s, and Francis Julian ranked among the victims. C. C. remembered how "the refractory land suck[ed] the life blood" out of his father, "engaging him in a fight in which he seemed unable to win." In 1898, the family sold its possessions, surrendered the mortgage on the land, and moved to Winnipeg.

A contemporary described the Winnipeg of C. C.'s youth as "a lusty, gutsy, bawdy frontier boomtown roaring through an unequalled economic debauch." A fledgling city of 40,000 at the turn of the century, Winnipeg more than tripled in size during the next decade, emerging as the metropolis of the Canadian West. Twenty-four railway lines radiated from the city, dispersing grain and cattle from Canadian prairies to national and international markets. By 1902, Winnipeg had surpassed Chicago as a center of the wheat trade. Hordes of young, single men seeking employment on the farms, factories, and railroads passed through Winnipeg, the gateway to the prairies, while an influx of "undesirable" immigrants from Russia and the Slavic countries unleashed a torrent of ethnic tensions.

Julian later depicted his life in Winnipeg in heroic terms. His family arrived, he related, virtually penniless. He himself sported a cut-down suit from his father's wardrobe that made him the butt of cruel neighborhood jokes, leaving an indelible imprint on his young mind. His father, claimed C. C., had died in 1898, precipitating the abandonment of the farm and leaving C. C., the oldest son at thirteen, as family head. His mother rented a ten-room house and took in boarders to support her large brood, while C. C. worked a series of odd jobs to supplement the family income and allow his siblings to attend school.

Winnipeg city directories, however, show Francis Julian alive as late as 1904, describing him first as a "farmer" and later as "retired." As a widow and boarding house-keeper, C. C.'s mother would normally be listed, but neither she nor her enterprise appear. In addition, the family moved three times in five years, unusual behavior for boardinghouse-keepers. More revealingly, the directories include a second Francis Julian, living at the same address and working for the Canadian Northern Railroad. C. C., at one time, spoke of an older brother born and left in

England and raised by his mother's family, who never joined the Canadian exiles. But this tale, like the premature death of his father, appears to have been fabricated to enhance the Algeresque imagery.

C. C. found work as a newsboy, "bucking the snowdrifts, peddling papers early and late . . . trying to grab enough nickels and pennies each day to supply bread and butter for a lot of little hungry mouths at home." The city boys, wrote Julian, "were 'plenty tough.' But I soon got just as tough as they were." One evening "with the mercury hanging around forty below zero," he spotted two newsboys "pounding the knuckles" to a third. Although he did not know the victim, and "I expected to get the whey beat out of me, I dived in with feet, teeth and fists [and] pryed the underneath kid loose." C. C. and his comrade stood "toe to toe, swapping punch for punch with two of the toughest kids I ever mixed with." In the end, C. C. had acquired a "little Yiddish friend," Jake Berger, who would join him in his later ventures.

At age fifteen, C. C. graduated from peddling newspapers to driving a milk wagon from 3:00 a.m. to noon each day. For two years he braved the blustery Winnipeg mornings before landing his first indoor job, clerking at a drugstore. His eager, engaging personality proved ideal for sales work. He soon secured a job for higher wages at a clothing store where he "carried forward his education in the art of meeting people and influencing their decisions."

Through this succession of dead-end, low-paying jobs, C. C. dreamed of becoming a lawyer, like his estranged grandfather. But because he lacked funds or formal education, law school was beyond his reach. He enrolled instead in night classes at a Winnipeg business college, where he learned shorthand, typing, and stenography. In 1903, armed with these rudimentary weapons of commerce and his first tailor-made suit, seventeen-year-old C. C. Julian moved 300 miles westward to Regina in the Saskatchewan territory. A frontier community of 2,500 people, Regina was primed for growth. Within two years Canada officially recognized Saskatchewan as a province, and Regina became the capital city. New railroad connections and a flood of westward migration inflated hopes and property values.

This latest Canadian boomtown exposed Julian to the great industry of the American West: real estate speculation. He became personal secretary to the head of a real estate firm, taking dictation, writing letters, and handling insurance, deeds, and records. He earned $75 a month, more than doubling his previous income, which allowed him to provide for his family in Winnipeg and accumulate modest savings. Claiming to have watched the wild escalation of real estate prices for a year before yielding to the temptation to speculate, C. C. bought a lot for $200, selling it six months later for $250. He quickly invested in five more land parcels, turning them over after three months for an $800 profit. During the next year, by his own account, C. C. Julian quickly acquired and disposed of thousands of dollars worth of empty lots.

In 1906, a land boom erupted in Edmonton, Alberta, several hundred miles farther west. Julian liquidated his Regina holdings, with the exception of one unsalable lot, and headed for Edmonton. He was twenty years old and worth $6,000. He signed on with an Edmonton real estate firm and, attempting to duplicate his Regina success, invested his profits in Edmonton properties. This time he had arrived at the wrong end of the boom. Within months the Edmonton bubble burst, and C. C.'s first fortune evapor. ted into the cold Alberta air.

In later life Julian reported that, at that moment, ashamed to face his mother, he returned to Regina, sold his one remaining lot, and headed west for Vancouver to recoup his losses. The city directory, however, reveals that Courtney lived, at least briefly, in Winnipeg in 1906, employed as a stenographer. The following year, he appeared in the Regina directory as the manager of a men's clothing and tailoring establishment, a position he omitted in later accounts of his life.

Julian's brief return to Winnipeg may have been motivated by the presence of Mary O'Donohue. He had met Mary, the "dark and attractive" schoolmate of his sister Violet, before originally departing for Regina in 1903. Her picture, he told a biographer, "kept him from running around with the girls and spending his pocket money on them." The pair later claimed to have exchanged daily letters during his three-year absence, a practice they continued throughout their unusual courtship.

In 1907, Julian left Regina for Vancouver, but the recession had already reached the western metropolis. Unable to find work, Julian and a friend saw an advertisement for laborers in the oil fields of Bakersfield, California. The two men abandoned their native Canada for the lure of California's black gold.

Bakersfield, in the heart of the rugged Kern River oil field, introduced Julian to a new type of boom town, one based, not on agriculture, railroads, or land speculation, but on crude oil pumped from the earth. The American industrial revolution of the late nineteenth century had relied heavily on the availability of two natural resources: coal for propulsion and heating, and oil for lubrication and illumination. California had seemingly lacked both. As early as 1864, Yale chemistry professor Benjamin Silliman, Jr., had predicted, "California will have more oil in its soil than all the whales in the Pacific Ocean." But initial explorations inspired by Silliman's optimism had yielded few positive results. Wildcat drilling brought forth either dry holes or modest flows below commercially viable levels. Furthermore, California crude produced inferior lubricants and kerosene that generated more smoke and smell than light.

Silliman's prophecy had foundered, not on the absence or quality of oil, but on the difficulty of its discovery. "California geology," writes Hartzell Spence, "is a study in surprises." The scenic, low, rolling hills that traverse southern California covered what Spence described as "eccentric rocks and crevices," forged by earthquakes and underground eruptions that "twisted and distorted rock folds to great depths." Within these "stratigraphic traps . . . locked in by hard rock cages," lay vast reservoirs of oil. But the unfamiliarity of the terrain and primitive nature of exploration and drilling technology protected the earth's secrets.

By 1890, the Pico Canyon in Ventura County was California's sole commercial oil field. As late as 1895, a correspondent reported from Los Angeles, "Fuel is so scarce and high here that it is the practice of local manufacturers to utilize for their furnaces the dried manure of the streets, and even office sweepings, carted for the purpose to their door."

Edward L. Doheny, an attorney turned mining prospector, became the first to tap the riches hidden beneath the southern California soil. Doheny arrived in Los Angeles in 1892. Gazing out his boardinghouse window, he noticed a wagon laden with chunks of tar. Mexicans had used the *brea* to waterproof roofs and fenceposts. Americans lubricated machinery with it and, on rare occasions, burned the tar for fuel and heat. Doheny knew nothing of oil and had never seen a drilling rig. But he sensed intuitively that the tar might be oil seeping from the ground and congealing on contact with the air. Doheny and his former silver mining partner, Charles Canfield, leased a lot in a residential neighborhood just east of downtown Los Angeles. Using picks and shovels they dug a 155-foot mine shaft, then sharpened the trunk of a eucalyptus tree into a makeshift drill. On November 4, 1892, at the modest depth of 460 feet, they struck oil, which they bailed from the well with a bucket.

The Doheny discovery inaugurated an oil boom in the heart of downtown Los Angeles. During the next five years derricks "as thick as the holes in a pepper box" sprouted along a narrow two-mile residential strip. Gardens, palm trees, and even homes disappeared as steam-powered rigs and primitive hand-driven drills probed the earth. Most found oil at shallow depths and produced a paltry ten daily barrels. But by the turn of the century, thousands of wells in the Los Angeles field yielded almost two million barrels a year.

The California oil strikes coincided with a growing nationwide demand for petroleum. The nineteenth century had been the "age of illumination." The oil industry, centered in Pennsylvania, Ohio, and Indiana and dominated by John D. Rockefeller's Standard Oil, refined kerosene for gas lamps as its primary product. Coal provided the energy that propelled locomotives and heated homes. Gasoline, a minor byproduct of the refining process, was used exclusively for cleaning, solvents, and some cooking stoves. By the turn of the century, however, Western railroads had converted to oil-burning engines, and fuel oil warmed a growing number of homes and businesses. More significantly, the automobile had captured the imagination of America. In 1899, only 8,000 motor vehicles were registered in the United States; twenty years later, automobile registrations numbered 7.6 million. By 1919, automobiles accounted for 85 percent of all gasoline use, and gasoline had long since replaced kerosene as the industry's major revenue producer.

The increased demand for petroleum transformed the oil business. Widespread exploration opened up new fields in California, Texas, Oklahoma, and Louisiana, creating opportunities for new companies and loosening Standard Oil's stranglehold on the industry. Beginning in 1900, the heart of the California oil boom shifted to Kern County, 120 miles north of Los Angeles. Discoveries at Kern River, McKittrick, and Midway-Sunset made Bakersfield and nearby towns "the center of the oil universe."

C. C. Julian arrived in Bakersfield in 1907 with "just enough money to rent a room and buy a square meal." He quickly procured pick-and-shovel jobs digging ditches for pipelines, tasks that required "only muscular arms and a knowledge of which side of the shovel is the top side." The arduous, unskilled work paid $1.50 a day. During the next three years Julian received his training as a "roughneck," the common term for the men who worked in drilling crews. He learned to fire boilers, dress tools, fish out dropped casing and tools, cement and complete wells, and handle and store oil. Wages were high and expenses low.

In late 1909, Julian returned to Canada. Personal considerations may have motivated his decision. He was twenty-four years old. For seven long years he had courted Mary O'Donohue from afar, writing and receiving daily letters. The time for a commitment had arrived, and the life of a roughneck held few attractions for a young city woman. In addition, his mother required care, which he could not supply amidst the oil camps.

Accounts of Julian's life diverge at this point. Julian claimed that he had saved $3,200 in the oil fields before returning to Vancouver where, after working several months in a jewelry store, he secured a job with Champion and White, the city's largest supplier of building materials, heating appliances, pipe, and hardware. Julian told later biographers that he quickly rose to the front rank of Champion and White's thirty-five traveling salesmen "by a wide margin." Within ten months, he states, the company named him sales manager at a salary of $500 a month. With the security of a well-paying job, Julian moved his mother to Vancouver and on December 1, 1910, finally married Mary O'Donohue.

Mary Julian presented a less flattering account of their marriage. In 1910, she later recounted, C. C. was employed in a plumber's shop in Winnipeg, with little means of support other than his meager earnings. He borrowed "various large sums of money" from her, which she had obtained from her father. Available records could support either account, but even if Julian exaggerated his sales success and adjusted the chronology of events, it is clear that by 1911 he was employed by Champion and White in Vancouver.

Since his birth in Morris, Julian had been nurtured on booms in Winnipeg, Regina, and Bakersfield, but the pre–World War excitement in Vancouver dwarfed these earlier experiences. Plans for a new transcontinental railroad line in 1909 spurred an influx of foreign capital from the United States, England, Germany, and other European countries. Population increased by 1,000 residents a month and, by 1911, exceeded 110,000. Infectious optimism engulfed Vancouver and its surrounding communities.

Once again Julian found himself amidst a raging real estate boom. Land values quadrupled in 1911 as apartment houses, office buildings, and hotels sprouted throughout the city and new subdivisions ringed the suburbs. Realty offices outnumbered grocery stores in Vancouver by three to one. Julian plunged into the market, registering, by his own count, a $30,000 profit in the first few months. In 1912, he left Champion and White and joined a real estate firm. But competition in Vancouver, where there was a "real estate salesman for every telephone poll" proved forbidding. By 1914, Julian had moved to Victoria on nearby Vancouver Island and launched his own realty concern.

On Vancouver Island in these years, "hammering could be heard everywhere." Julian remembered the heady days of the Victoria boom with great relish. "Everything I touched seemed to turn to money," he recalled. "I made it more rapidly than I had ever done before. I cleared $80,000 net profit on one subdivision of 56 acres. . . . I had my holdings concentrated in $200,000 of city business property." But beyond confirming that Julian was engaged in the land market, first in Vancouver and later in Victoria, it is impossible to verify his claims of success.

Whatever the extent of Julian's holdings, his fortune tottered on an unstable base. While real estate speculation flourished, mining and agriculture, the two basic industries of British Columbia, had stagnated. Reports of land swindles circulated through the province. As World War I approached, foreign investors withdrew their capital, and the economy sagged. When hostilities erupted in Europe, the real estate market collapsed.

"Soon you couldn't give anything away in Victoria," recalled Julian. "All my property was clear, but the taxes were devilish. Without any money coming in I paid the first year's taxes, but I couldn't pay the second." Ultimately, he stated, the city of Victoria seized his properties.

In late 1914, Julian left Victoria, bankrupt. After a decade as office clerk, oil field roughneck, hardware salesman, and real estate speculator, C. C. found himself back where he had started, selling men's clothing in Winnipeg. According to Julian he opened his own retail store, but "war conditions doomed it to failure," and he found employment with another company. By 1917, Julian, now thirty-one, had two daughters, Lois and Frances. The income of a clothing salesman barely supported his family. The lifestyle did not appeal to him. In California, meanwhile, American entry into the war had stimulated demand for oil and oil workers. Julian, although absent from the petroleum fields for almost eight years, moved his family to Taft, California, an oil town southwest of Bakersfield.

C. C. secured a job with Tecumseh Oil and worked twelve-hour shifts and received a healthy $22 daily wage. Although employed as a driller, Julian boasted, typically, that he "acted practically as a superintendent of operations." When Tecumseh Oil opened new fields in Fullerton, south of Los Angeles, Julian moved Mary and the girls to the Los Angeles beach community of Ocean Park and regularly made the triangular two-hundred-mile commute from Taft to Ocean Park to Fullerton.

Within a year of arriving in California, Julian set his sights on larger flows of oil. The great discoveries of the World War I era had occurred, not on the Pacific Slope, but in Texas and Oklahoma. In mid-1918, Julian left his family in Los Angeles and with three other drillers drove his Ford touring car to Texas. For several months Julian worked for independent contractors, who paid to drill by the foot, drove their workers mercilessly, but compensated them well.

Skilled roughnecks received extraordinary wages for that era, but the real profits in the petroleum industry came from leasing and developing oil lands. In September 1918, Julian met Edward St. Albans, a California oil man seeking to expand his operations into Texas. "He was forming the Texas Holding Company," Julian recalled, "and wanted an oil man to look over the holdings. I got a geologist friend of mine and we made a complete inspection for St. Albans. He made me a partnership proposition if I would stay in the field and take charge of the operating end." Julian quickly seized the opportunity to produce oil on his own.

St. Albans returned to Los Angeles to "push his financing" for the Texas Holding Company. Julian joined him there in December 1918. Over the next two months, C. C. gained his initial experience in selling oil stock. By March 1919, St. Albans and Julian had fully subscribed the Texas Holding issue and secured 600 acres of oil leases on the outskirts of productive Texas fields. Julian took command of the drilling operations, his first real effort as drilling supervisor.

Financial experts later estimated that between 1918 and 1921, over 270 similar Texas oil schemes took root in Los Angeles. Ninety percent proved worthless; none returned significant profits. Texas Holding, to the chagrin of its investors, produced four dry holes. According to Julian, the company withdrew from the Texas fields, and he resigned his position. He formed his own contract drilling business and over the next year and a half sank twenty-four wells. His clients included Union Oil and a former governor of Oklahoma. Despite his earlier failures, he recounted, most of these efforts struck oil. Still anxious to join the ranks of independent producers, Julian invested his earnings in oil leases. He drilled three wildcat wells on these properties, each one a "duster."

In early 1921, Julian returned to his family in Los Angeles. He had spent two and a half years in the Texas oil fields and, he later claimed, accumulated just one dollar per month for his efforts. In addition, according to his wife, Julian had also acquired a mistress. While in Texas, C. C. met "a young girl" named Maybelle Smith, who soon followed him to California. With a wife, two daughters, and a mistress to support, Julian was flat broke.

Events thirty miles south of his Ocean Park home, however, encouraged new hope. In November 1920, Standard Oil discovered its spectacular Huntington Beach bonanza. Since Huntington Beach had already been subdivided, the dispersed land ownership opened opportunities for small operators to obtain leases. Julian, fresh from his Texas failures, sought to recoup his losses. He recruited former Texas Holding Company backers to stake him in a new venture. Persuaded by Julian's energetic appeal and the tales of wealth emerging from the new field, they laid out money for a lease and drilling syndicate.

The only surviving accounts of Julian's exploit at Huntington Beach are based on his own reminiscences. He spent the year raising funds and supervising drilling, "nurs[ing] his oil well along like a sickly Boston Terrier." His busy schedule, he claimed, allowed him few nights at home. But the apartment he had established for Maybelle Smith may explain his frequent absences. At the well, Julian later lamented, "about everything happened that could happen." The well "contracted all the diseases oil holes are heir to—from swallowing tools that had to be fished

for to malnutrition due to lack of funds." On one occasion Julian traveled to northern California to raise money when he received a desperate telegram. A string of
tools had dropped down the well, and repeated efforts to remove them had dispatched more gear down the shaft.

"Have four strings of tools in the hole," wired his foreman. "What'll I do?"

"Dump in the boiler and call it a day," shot back Julian. He quickly returned
to Huntington Beach and supervised operations that successfully removed the tools.

On a cold, gray dawn in May 1922, Julian and his crew prepared for the decisive
thrust into the oil sands. Julian had spent $140,000 to drill this well. Thirty of his
backers and friends had shared his vigil through the night. Julian had not slept for
two days. At 5:00 a.m. the drill bit into the sand and liquid came pouring out. But
the flow was water, not oil. Julian had hit another "dry" hole.

Julian, according to his later account, stood alongside the derrick, "tombstone
at the grave of many dollars of cash and hours of effort," chewing on a toothpick.
The long sleepless nights had left "his eyes and cheeks . . . pretty hollow and his
face . . . kind of grey." As the morning sun appeared in the east, he stepped into his
automobile and turned the wheel, not toward home, but in the direction of southern
California's newest oil field, Santa Fe Springs.

## II

Twelve miles southeast of Los Angeles, Santa Fe Springs was, according to one
resident, "a gentle, quiet, serene place, too small to be a town." The community
traced its origins to the 1880s, when a local doctor discovered a sulphur spring
beneath his land. Claiming great curative powers for these waters, the doctor built
a lavish resort where hundreds of "excessively portly matrons of society, gouty old
gentlemen, rheumatics of all classes, and conformed 'soaks'" sought a cure for a
variety of ailments. Sanitariums appeared in surrounding areas promising treatments
"to dissolve and remove all morbid accumulations . . . restore the vital powers of
nature . . . [and] strengthen and harden the whole organism." This harbinger of
California health fads proved short-lived, however, and in the early twentieth century the area began to attract wealthier residents who purchased substantial acreage
and constructed lavish homes.

Prominent among these settlers was Alphonzo Bell. Bell was a true native son
of southern California nurtured by the orthodoxies of Protestant theology, irrigation,
and real estate speculation. His father, James Bell, had acquired and subdivided
thousands of acres of undeveloped Los Angeles land, and Alphonzo, born in 1875,
grew up amidst the heady euphoria of the 1880s land boom. In 1888 he entered
Occidental Academy, the austere Presbyterian forerunner of Occidental College
and, upon graduation in 1894, enrolled at the San Anselmo Theological Seminary,
to pursue the ministry. But within two years Bell inherited 110 acres and promptly
left to seek his fortune in real estate. Bell irrigated the land and subdivided it into
homesites, which sold briskly, giving him his first business success.

In 1902, Bell reinvested his profits in land in Santa Fe Springs. He irrigated the property and moved his family there in 1908. At its peak, Bell's Santa Fe Springs estate boasted 200 acres of prime agricultural land, producing alfalfa, oats, hay, barley, lemons, and oranges. Bell drove an expensive Hudson automobile and employed the services of two servants and as many as twenty-five hired hands. The most unusual feature of Bell's home was his tennis court, a rarity in southern California and a testimony to his athletic prowess. From 1900 to 1910, Bell gained fame as the outstanding tennis player on the Pacific coast, three times winning the California singles' championship and ranking briefly among the nation's top ten players.

But by 1910, a new passion had replaced Bell's devotion to tennis and farming. While drilling for water, Bell repeatedly struck pockets of gas that shot streams of water over several acres. Although he had no background in geology, Bell surmised that where gas appeared, oil must follow. Scattered pieces of tar in the diggings fed his conviction. By 1910, Bell was certain that his land sat above a vast reservoir of petroleum.

Alphonzo Bell was not the first to suspect the presence of petroleum in Santa Fe Springs. As early as 1865, Pioneer Oil had explored the land for petroleum. In 1907, Marius Meyer, a shrewd Basque sheepherder, leased his property to Union Oil, which immediately began drilling efforts. Although gas escaped from the hole in copious amounts, sand, gravel, and water blocked the path to oil, leading Union to abandon the well. In 1908, Union made a second unsuccessful attempt to probe Meyer's land.

The tribulations on the Meyer property bedeviled Bell's efforts to interest oil companies in his estate. For years Bell pestered officials at Standard Oil and Union Oil, but his entreaties evoked only vague promises of future exploration. World War I brightened Bell's prospects. In April 1917, he leased his land to Union Oil, but the company decided to first test Marius Meyer's claim one more time. For two years and eight months Union's roughnecks battled the precarious sand, gravel, and water of the Meyer land. In October 1919, their efforts produced a spectacular 3,000-barrel-per-day stream of precious high gravity oil. The capricious Meyer soils, however, had one last trump to play. Within hours, water broke into the well, interrupting the flow of oil.

The depth, difficulties, and ultimate disappointment of Meyer No. 3 caused most companies to shy away from Santa Fe Springs. Nonetheless, Union Oil agreed to sink the long-awaited test well on the Bell property. For Alphonzo Bell the completion of a successful well had acquired a new urgency. The once prosperous country squire now sat perched on the edge of a financial abyss. The mortgage on his ranch had fallen into default, as had a second short-term loan. His remaining workers received property in lieu of wages, and for several years the town grocer had provided Bell with food and supplies on credit. Bell was clearly approaching the limits of his resources.

The former tennis champion waited anxiously for two years as the drill plunged deeper into his land. By the morning of October 30, 1921, Bell No. 1 had reached a depth of 3,763 feet. As Bell slept, workers resumed the downward drilling. After five more feet of progress, the earth roared, and mud and water came rushing out of the hole, followed by a rich stream of black liquid. Bell No. 1 was the greatest well in the West, and two-thirds of Bell's Santa Fe Springs property sat atop a vast pool of oil. More than a decade after he had first smelled sulphurous gas at a water well, Alphonzo Bell had become incredibly rich, a yardstick to measure success by in southern California.

But the excitement had barely begun. On the evening of January 4, 1922, two months after the Bell strike, an oil crew resumed drilling on Alexander No. 1, a few hundred yards from the Bell discovery well. Mud suddenly began to boil from the hole and the head driller warned his men to run. Nine hundred feet of drill pipe blew high into the air and landed 200 yards away. Mud and debris covered the surrounding area, and windows rattled in Whittier, several miles to the northeast. Within hours a crater 100 feet wide and 30 feet deep had formed, swallowing the derrick, water tanks, and two automobiles. Terrified local residents took flight. Gas, mud, and rocks spewed continuously for over a month before Alexander No. 1 ceased its eruption, choked off by its own mud.

Before the devastation from the great Alexander blast had subsided, on February 11, the drill bit on nearby Bell No. 2 pierced a subterranean gas cap. A section of pipe arced skyward, then plunged rapidly, imbedding itself in the earth. Workmen struggled to control the flow of gas as rocks shot 300 feet into the air. After thirty hours a spark ignited the sky. The blast destroyed telephone poles within a 100-yard radius and threatened the Bell ranch house. "When the paint in our kitchen began to blister," recalled Alphonzo Bell, Jr., "we knew we'd better get out."

The *Los Angeles Times* reported flames reaching 500-foot heights. Illumination could be seen for twenty miles. Thousands of sightseers flocked to the scene. "We drove to view it and its heat kept us many blocks away and its roar made conversation impossible," reported one observer. Hundreds of workers battled the blaze, to no avail. After ten days the fire extinguished itself, only to erupt again the following day. The flames, raging as hard as ever, burned across the base of a fifty-foot crater created by the initial explosion. Fireballs soared high above the surface. After several more days sand and mud again plugged up the well, bringing the spectacular con-flagration to a halt.

The Alexander blowout and Bell fire made Santa Fe Springs the best-known oil field in the nation and enhanced the romance of the oil industry. Here, within a short car ride of Los Angeles, was adventure, danger, and spectacle, "the greatest thrill of recent years," said the *Times*. What better way to spend a weekend than to pack the family into the car to watch the roughnecks at work. If lucky, one could witness the arrival of a gusher or a grand conflagration. Los Angeles already offered real estate booms, Hollywood, and a wondrous climate, but could even these com-pare to the excitement of Santa Fe Springs?

Throughout southern California, Santa Fe Springs became a beacon for those seeking instant fortune. Overnight, a resident complained, the community became an "ugly place. . . . All the lovely trees were gone. Most of the houses were torn

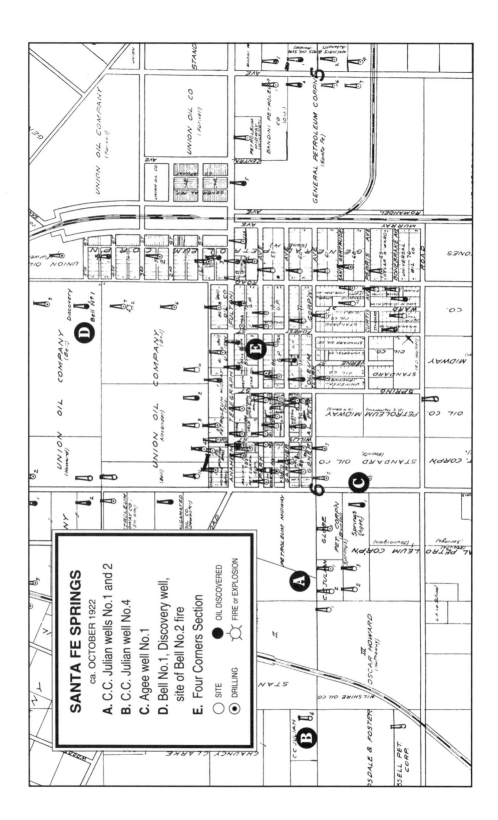

**SANTA FE SPRINGS**
ca. OCTOBER 1922

**A.** C.C. Julian wells No.1 and 2
**B.** C.C. Julian well No.4
**C.** Agee well No.1
**D.** Bell No.1, Discovery well, site of Bell No.2 fire
**E.** Four Corners Section

○ SITE
◉ DRILLING
● OIL DISCOVERED
☼ FIRE or EXPLOSION

down." Trucks bearing lumber and drilling equipment clogged the roads. Along the main street at Telegraph Road a "great carnival type midway . . . just like a circus" materialized with red and white tents and "big marquees with signs all over the place." Estate mansions became oil company headquarters. The Bell ranch house, transplanted from his property, resurfaced as a saloon.

Tent cities called "Gum Grove" and "Spring Slums" sprang up along Norwalk Boulevard to shelter and service the roughneck population. "California-style buildings," narrow boards bolted together into front poolrooms and rear chambers for poker and slot machines, with a "quick escape back door," appeared amidst the canvas homes. Sandwich and coffee stands, complete with counters for rolling dice, lined the streets. Young girls stood in front of tents, recalled driller Clifford Davis, with "an ironing board and shirt to iron. They kept ironing that same shirt while waiting for business customers." Other tents housed "magazine-selling" girls. "They got our money, but we never got no magazines," laughed Davis.

Establishments like May's Place and the Baker Winery Barn offered full-service entertainment. "If you wanted to," related Davis, "you could give your paycheck to May, and this entitled you to everything for a month—eats, gambling, girls and all." At the Baker Winery Barn, "The girls were as bare as the walls. Once you got in you could see the girls shimmy for nothing, that is if you didn't shoot craps or play the blackjack table or draw poker or roulette wheels or slot machines." Prohibition ruled the land, but Santa Fe Springs offered a free-flowing oasis. Among the rugged roughnecks, disputes were settled by brass knuckles, knives, hatchets, and crowbars.

Santa Fe Springs contributed its own unique features to the boom. The Bell discovery, like others before it, triggered a wild rush for oil land and leases. California's "Big Five" oil companies—Union, Standard, Shell, General Petroleum, and Amalgamated—rushed to secure drilling rights on the large estates surrounding the Bell properties. In the downtown and outlying areas, however, prospective operators frantically competed for control of farms, cemeteries, schoolyards, church grounds, and insane asylums. As in Huntington Beach, much of this land was already subdivided into "town lots," and independent oil men found a golden opportunity to gain a foothold.

In the United States the "rule of capture," or as historian Hartzell Spence succinctly defines it, the "right to steal oil underlying a neighbor's land," governs oil production. While most other countries consider sub-surface wealth part of the national domain to be developed in the public interest, the United States grants landowners exclusive mineral rights. Since petroleum reserves recognize no property lines, pools of oil may extend beneath land owned by many individuals. By the "rule of capture" the oil belongs to the person who pumps it from the ground.

Under these conditions, observed journalist Albert Atwood, "the idea is to get there first and draw out not so much the oil that lies under one's own land, as that which lies under the property of one's neighbors." Previous experience in California oil fields had established the practice of drilling a well for every five acres of oil land. This amount of land in the new fields, however, might require the merger of as many as seventy-five separate holdings. Many landowners united to sign "community leases" with the major oil companies. Others accepted the more lucrative inducements offered by independent oil men and stock promoters.

The five-acre minimum became the first casualty of the "town lot" war. The *Standard Oil Bulletin* reported in the fall of 1922 that one eighty-acre plot in the heart of Santa Fe Springs hosted forty-six derricks, an average of one well to every 1.74 acres. Six months later the more congested half of that tract held sixty wells. Rigs owned by Standard and Getty appeared only sixty feet apart, and mud passed through the formations from one hole to another. To prevent adjoining drillers from draining off their oil, companies sank "offset" wells, additional rigs that speeded their own production and added to the crowding.

C. C. Julian arrived amidst this chaotic hub of activity in May 1922. Exhausted and bankrupt from his Huntington Beach failure, he spent his first days in Santa Fe Springs "prowling around," assessing the likely direction of the field and evaluating lease possibilities. An oil lease is an agreement between a landowner and an oil company or individual granting drilling and exploration rights for a given property. Long-established industry standards rewarded landowners with a "royalty" amounting to one-eighth of the oil pumped from their property. But conditions in southern California rendered this figure obsolete. According to the industry journal *Oil Age*, cash bonuses that "a few years ago would have been considered beyond the pale of possibility" became the rule. Royalty rates spiraled upward to one-quarter, one-third, and even one-half. "Leasehounds," men who speculated in oil leases, buying them for a few thousand dollars one day and reselling them for five times that the next, fueled the inflation.

Julian, his lack of funds notwithstanding, threw himself into the lease-hunting scramble. The boundaries of the Santa Fe Springs field still remained uncertain. Indeed, since the spectacular fire, no new strikes had occurred. Some began to speculate that this would be a "one-well field." As late as July, the *Times* reported that Santa Fe Springs "has not proven the wonderful field it looked like six months ago." Nonetheless, most landholders had long since leased their land. Julian had to deal with the "lease-hounds" or independent companies seeking to sell off a portion of their holdings to finance drilling on the remainder. After talking to friends, Julian decided that the "Four Corners" intersection of Telegraph Road and Norwalk Boulevard, less than a mile distant from proven territory, was the most affordable and likely oil-producing area. Julian was not alone in this judgement. George Getty and his son Jean Paul had also reached this conclusion. In November 1921, they leased four lots from the Nordstrom family. This lease would mark the beginning of the Getty empire in California.

Julian probed the lease market for two weeks, attempting to trade promises for drilling rights in the "Four Corners" region. In late May he approached Globe Petroleum, developer of Huntington Beach's famed "Encyclopedia lots," about a lease it had acquired near the Getty holdings. Unable to compete with the major companies for the larger Santa Fe Springs estates, Globe had sought out smaller landholders, like Lem and Clara Brunson. For most of his fifty years, Lem Brunson had traveled the West as a shirt salesman. In the 1910s the Brunsons bought a small, rundown ten-acre farm at Santa Fe Springs, which they gradually improved. Ten days after Bell No. 1 came in, they granted exploration and drilling rights to Globe Petroleum.

Globe Petroleum, however, had been in no hurry to begin drilling. Santa Fe Springs remained an unproven field, and the Brunson lease stood three-quarters of a mile from the discovery well. A few hundred feet south, General Petroleum had commenced work on Agee No. 1, which would test the area. Globe moved a rig to the property and waited for the results of the Agee well. In May 1922, with the fate of Agee No. 1 uncertain, Globe decided to take profits on a western portion of the Brunson lease to finance drilling on the more promising eastern sector. C. C. Julian stepped forward as the buyer.

On June 1, according to Julian, he offered Globe Petroleum a cash bonus of $30,000 and a whopping 30 percent royalty for four acres of the Brunson lease. Globe agreed to his terms, but only if Julian could produce the cash by the following day. For the deeply indebted Julian, this became a "now or never" proposition. With no personal assets, he launched a round-the-clock fund-raising effort, returning once again to the people who had backed him unsuccessfully in Texas and Huntington Beach.

"Julian went home that night for dinner," reported his authorized *Los Angeles Record* biographer, "but spent his time telephoning instead of eating. . . . Telephoning found unsatisfactory, he hopped into his machine and started out to make a series of calls. At midnight the goal was as distant as ever. . . . Then he buckled down to work in earnest. He drove from house to house, friend to friend, acquaintance to acquaintance." At 3:00 a.m. he arrived at the home of a "well known Los Angeles physician" and "took some money away with him." At 4:00 a.m. he awakened a lawyer who had lost money in Julian's earlier ventures. The lawyer expressed interest but refused to commit himself until he could consult with his sleeping wife, whom he refused to rouse until morning.

"When noon came," related Julian, "I was still $10,000 short of my goal. I had been dodging in and out of my office all morning. I was getting ready to go out again on a last ditch appeal when I looked out and saw my lawyer friend coming in the door. . . . [He] planked down a check for $10,000 on my desk [and] said, 'I'm shooting with you again, Julian.'"

Julian's tale of his fund-raising marathon characteristically stretches the truth and omits key facts. The twenty-four-hour limit and all-night odyssey may well be a dramatic fabrication. Escrow records show that Julian deposited $2,500 on June 3, an additional $15,000 two days later, and a final payment of $12,500 on June 26. A 1924 FBI investigation listed fourteen people who loaned Julian a total of $27,300 in June 1922. The loans ranged from $300 to $5,000. Six people gave Julian $2,000 or more. No "lawyer friend" (in a subsequent account, Julian claimed it was a physician) contributed $10,000. To persuade these people to "shoot" with him again, Julian offered lucrative terms. He guaranteed to repay all participants double the amount of their investment. In addition he granted each one royalties in the first two wells, to be paid before other investors received their shares.

Whatever means Julian had employed to acquire the down payment, on June 3 he placed the money in escrow and assumed drilling rights on the Brunson lease. But Julian's financial worries had not ended. The contract required him to begin drilling operations within ten days and "be spudded in to commence the actual drilling of the well . . . within forty days." To do so, Julian would need tens of thousands of dollars, but he had exhausted most of his resources.

"I had to have the money," he later explained, "and I'd badgered to death every prospect I knew. The only place I had to go was to the people. I knew my ground was in the path of production. I knew there was oil there. All I needed was the money to get it out. I knew if I could just make the public see it as I saw it they'd go into partnership with me."

To make the public enter this partnership, Julian turned to a peculiar business structure, the "common law trust." Evolved from English common law and sometimes known as the "Massachusetts trust," this investment strategy had come into widespread use in the Texas oil fields. Virtually unknown in California before the 1920s, it became the economic foundation for the "town lot" drilling boom. A common law trust offered promoters wishing to sell shares to the public an alternative to incorporation. To establish this entity, three trustees simply made a declaration of trust with a bank or trust company. No investment was necessary. The bank became the depository for all funds accumulated by the company. Unlike a corporation, which sold shares of stock, the common law trust sold "units." A unit usually cost $100 and, depending on the number sold in each well or syndicate of wells, entitled the owner to a percentage of the the oil produced. If 1,750 units were sold, the purchaser received payments equal to $1/1750$th of the value of a percentage of the oil pumped. These units, like stock, could be bought and sold on the open market. William Emmens, who later became mayor of Santa Fe Springs, recalled his first visit to the town. "I never saw anything like it in my life," he exclaimed. "They were selling units. I didn't know what a unit was—I still don't—but we bought some."

Oil promoters, playing on American antipathy to big business and large corporations like Standard Oil, repeatedly proclaimed the advantages of units over stock. Unit-holders, they claimed, had no liability for the debts of the company. Since funds were deposited in a designated bank, they could not be plundered by unscrupulous businessmen. Unit-holders would be paid directly from oil sales rather than company profits and would receive full reimbursement for their investment before the trustees shared in the proceeds.

In reality, the unit-holders had fewer rights and exercised less control than the corporate stock owner. The extent of their liability awaited final determination in the courts. They had no vote in the selection of trustees or management of the concern. Although the funds had to be placed in a specific repository, trustees, like any depositors, retained full rights to spend the money as they saw fit. Indeed, the trustees possessed unlimited power. The declarations of trust typically allowed them to "drill for and produce oil," "market and refine crude oil and its products," and do "any and all things the trustee herein may deem wise and necessary for the general good of the estate." A writer in the *Los Angeles Record* queried, "Is there anything there that [they] cannot do?"

Local newspapers and groups like the National Vigilance Committee of the Better Business Bureau publicized the disadvantages of investing in common law trusts, but southern Californians displayed a decided preference for unit buying. "It is difficult to understand why there exists locally such a prejudice against a 'corporation' and 'stock,'" protested *Oil Age* in March 1923. While units in some instances produced more spectacular short-run returns, investments in the major oil companies returned far greater long-term profits. Albert Atwood claimed that he had "never run across such an unintelligible and unreasoning bit of psychology as this preference for units in a business trust," and suggested, "Perhaps the investor feels that a unit in a trust is closer to the land than a share of stock, more personal and individual."

"If [the oil promoter] is an honest man," advised an "expert" in the *Record*, "a trust agreement is as good an instrument from your point of view as a corporation. If he is dishonest, all the laws to protect corporation stockholders cannot protect you." C. C. Julian had to convince potential investors that he fell within the first category.

To project a substantial image, Julian operated on a lavish scale. He rented a "sumptuous suite" of offices in the Loew's State Theater building in downtown Los Angeles. He bought thousands of dollars of carpets, rugs, and office furniture, "serviceable, but in spots verging on elegance," instructing the salesman to "send me the bill when the stuff's installed." He engaged a rig-builder to construct two derricks at a cost of $11,000, "to be paid upon completion in eight days." And he began to recruit his office and sales staff.

As Julian proceeded with his whirlwind preparations, a remarkable stroke of fortune occurred in Santa Fe Springs. On June 11, General Petroleum's Agee well on the boundaries of his property came in with a 4,000-barrel-a-day flow. Julian no longer was drilling in unproven territory but several hundred feet from a bona fide gusher.

In light of this development, Julian's sales strategy became all the more important. How could he best get his message to the public? Many small operators had pitched tents in the oil fields and hired buses to transport potential investors. Others placed daily advertisements in the Los Angeles papers. Julian chose the latter course. But with newspapers flooded with investment opportunities, Julian had to distinguish his ads from the rest. He later claimed to have employed two advertising agencies, presented them with the facts, and asked them to prepare copy. When the agency representatives returned, Julian read the first ad, swiveled to his secretary, and barked sarcastically, "You shoot him, Pedro." He read the second version and tore it up.

"You see," he told the agency men, "the purpose of these ads is to get money from the public. These ads wouldn't charm a dollar out of my pocket. . . . If you're going after other people's money, you've got to talk to them in a different tone of voice." Julian dismissed the ad men and, with a publishing deadline growing closer, began to write his own copy.

Julian often liked to recount his first venture in ad writing. He sat at his desk with a sharpened pile of pencils and worked all night. The opening heading took three minutes to write. The remainder he "ground . . . out slowly and painfully, using pencil after pencil," rewriting it thirty-five times. As always in the Julian saga, he hurriedly scraped together the $700 publishing fee and barely beat the deadline for Sunday advertising.

On June 18, 1922, Julian's masterpiece, occupying two-thirds of a page, appeared in the *Los Angeles Times* and *Examiner*. It boldly proclaimed:

### C. C. Julian Breaks into
### SANTA FE SPRINGS
### The Gusher Oil Field of America

Three weeks ago I laid **$30,000 cash** on the barrelhead to the Globe Petroleum Corporation for **4 acres** of their holdings there. That day it was no cinch. I have since refused $100,000 for my lease. One week ago today the General Petroleum Corporation drilled in their **Agee Well** flowing 4000 barrels a day **250 feet** from the South line of my property. I am going to drill a well on one acre of this myself, but a well costs real money and I shot my bankroll when I bought this tract. I want you with me and I am going to make you the cleanest offer ever tendered the public in this or any other oil field.

I need $175,000 to drill this well. I am assigning 70% of my total net production and one acre to you with the

### Citizens Trust and Savings Bank

. . . for every $100.00 you invest you will receive 1-1750th part of this production. On a 4000 barrel well at $2.35 a barrel figure it out yourself . . .

I have **fifteen picked men** to start at once on this well, all drawing top wages and a cash bonus to them of $2,000.00 to have our well on production in 90 days. My rig is being rushed to completion now . . . [and] in two weeks we will be ballin' the jack and there will be a lot of them trailing when I hit that pay.

I am trying to offer you the squarest and surest opportunity for big returns it is humanly possible to make.

I have **no Salesmen** to call on you or stenographers to answer your correspondence so if you like my offer just mail your check to me for the amount you want to shoot . . . or come in and see me personally. All I ask is **"do it now"** for this advertisement will not appear many times.

In this first effort at ad writing, Julian demonstrated the keen understanding of the concerns of the small investor that would typify his career. "I have no board of

directors at fat salaries to juggle your profits or build up a surplus for future development," he declared. At the same time he stressed his own expertise as a veteran
of "15 years in the oil fields" who had "drilled 137 wells and . . . never failed to
finish a job." The details of his offer were plainly defined. Emblazoned at the bottom
of the ad in bold, straightforward capitals, was his name, **C. C. JULIAN**.

Now Julian awaited the verdict on whether he had "charmed the dollars out of
the people's pockets."

# 3  WIDOWS AND ORPHANS, THIS IS NO INVESTMENT FOR YOU

## I

If C. C. Julian's initial appeal possessed a distinct tone of urgency, it was because by June 1922, competition for the southern California oil dollar had grown intense. The numbers of people speculating in oil defied exaggeration. There were "thousands of oil lunatics rushing thither to throw their dollars into the oleaginous maw of the new field," wrote Samuel Blythe. In the promise and possibility of sudden wealth, oil had a universal appeal. As one oil promoter explained, "People of small means want to invest and don't know how. What they need is a definite specific proposal in words of one syllable."

To meet this, an incessant flow of promoters, both honest and dishonest, descended upon the new oil fields. Department of Justice investigators estimated that by 1923, southern California oil stock swindlers were netting $100,000 a week. The variety and ingenuity of their schemes to exploit the speculative mania seemed unlimited. Oil investment solicitations flooded the mails and telegraph lines. Entrepreneurs roamed among the crowds visiting the oil fields, selling worthless stock. Others formed their own drilling companies and hired themselves to complete their wells at inflated costs. Several companies specialized in names, compiling "sucker" lists to sell to promoters.

Experienced promoters knew myriad tricks to stimulate sales. They could take a well with minimal oil and turn it into a gusher by inserting a small pipe connected to an air compressor. They advertised astonishing oil-finding devices that guaranteed against dry holes. One man invented a "radioscope":

> a mechanical invention designed for using the wireless method of turning into ethereal waves or vibrations from crude oil and the amplification and measuring of the wave energy so as to scientifically locate and define the oil bearing strata and the probable volume, depth, and extent of petroleum deposits in the earth.

These gadgets generated thousands of dollars in stock sales but located only dry holes.

The most visible and colorful aspect of the speculative boom, however, involved a promotion unique to Los Angeles: the sightseeing tour to the oil fields. The 1920s' oil boom merged two traditions of speculative enterprise: the mid-continent sale of oil stocks and the southern California real estate promotion. Since most previous oil discoveries had occurred in isolated areas, oil men seeking investors usually relied on high-pressure newspaper and mail campaigns. Southern Californians, on the other hand, had perfected the strategy of the tent show and free lunch during the great land boom of the 1880s. Realtors sponsored trips to nearby subdivisions, entertaining their guests with brass bands, lectures, and hot meals. In Los Angeles, where the oil fields were but a short bus ride away, these approaches merged, producing what Albert Atwood described as "one of the most peculiar, bizarre and picturesque pieces of sheer business practice that has developed in many a year."

Every morning the buses laid siege to Pershing Square, the plaza in the heart of Los Angeles. Around the square and along the surrounding blocks, imposing passenger coaches destined for the oil fields lined the streets as far as the eye could see. Streaming pennants and banners brightly festooned the buses, promising "Delicious Beef Barbecue" and "Band Concert Entertainment." Throughout the downtown area, solicitors haunted intersections and doorways, insistently accosting passersby with free tickets to the fields. In residential neighborhoods female auxiliaries trooped door to door distributing passes. By 11:00 a.m., hundreds and sometimes thousands of people departed for the oil fields. The buses, which also originated in Long Beach, Pasadena, and other locales, had become the vehicles of what Atwood called "the most complete method for attracting money from the public that has yet been devised."

Many of the buses still serviced real estate subdivisions. But in 1922, the oil fields at Signal Hill, Huntington Beach, and, especially, Santa Fe Springs became the prime destinations. Newspaper ads invited readers to "SEE THE GUSHERS AT SANTA FE SPRINGS" and "realize what fortunes are pouring out of the ground." More than fifty concerns ran daily buses to the former health resort, some as many as fifteen each Sunday.

The buses took circuitous routes, artfully contrived "to thrill the people with the idea of wealth, to hammer in the desire for money, to make us as avaricious as possible," wrote Atwood. They motored past the mansions of business magnates, movie stars, and especially oil moguls. One reporter heard the tale of a man who arrived in California two years earlier "with only $400 and a sick wife," but became a "millionaire many times over." "And all," intoned the guide, "from oil, all from oil. . . . He had the same chance that you have today."

At Santa Fe Springs the tours drove past the legendary Bell properties and the ruins of derricks destroyed by blowouts and fires. Some buses stopped at the mammoth tents along the midway on Telegraph Road, but others proceeded into areas where "the derricks thinned, [and] the oil tanks became fewer," beyond proven territory. "Here we are, friends," the barker would announce, "right in the heart of the field."

The tents resembled the tabernacles of traveling evangelists. Red, white, and blue flags fluttered along the top and sides, while huge marquees welcomed the prospective investors. Inside, a blackboard rested on a platform and maps of the oil fields covered the walls. Placards offered epigrammatic advice: "Buy these lots and enjoy independence for life." Row after row of dining tables and benches filled the tent. Atwood found the atmosphere "cheap" and "hectic," worsened by heat, flies, and the general din of hundreds of people crowded into a small space. Local residents dubbed them "sucker tents."

The buses arrived at noon, and lunch was the first order of business. Numerous concerns advertised hot meals, but after several excursions Atwood complained, "Always I had a cold and clammy lunch." Coffee, "coarse sandwiches," and cookies made up the usual menu. Reporter Paul Wilson dined on a "seven course luncheon consisting of hot dogs and coffee." Atwood and Wilson may have been fortunate. One investigator claimed that the famed hot servings more often consisted of seagull than chicken.

With the luncheon complete, the festivities began. Salesmen opened the flaps of the tent so the well-fed throng could see the roughnecks at work and hear the sound of the steam engines driving the drill. At times the lecturers had to shout over the din of the work. But sometimes the sounds became mysteriously muted. "On the earlier wells that I worked," recalled one driller, "we had to slow down the operation because . . . the noise of the drilling would interrupt the sale of the stock."

Enterprising promoters bussed in church and veterans' groups, women's clubs, black railroad porters, and members of religious sects. Special guests and lavish events often supplemented the performance. On Sundays, when the crowds numbered in the thousands, tent shows included parachute jumps and brass bands, as well as movie stars like Norma Talmadge. Some promoters hired clergymen, judges, politicians, and other figures of prominence to meet the crowds, reportedly paying them as much as $1,000 a week. In one tent a man dressed like a preacher praised the gusher they were about to bring forth and piously proclaimed, "We should bow our heads and thank the Great Lord for this opportunity we have. Amen."

When the lecture ended, "the slaughter of the lambs" commenced. Salespeople descended into the crowd to complete the day's work. Working for commissions, they moved rapidly from customer to customer. Audiences, primed by the afternoon's promotion, responded avidly. Long lines formed to purchase units with cash, checks, and Liberty bonds. Buyers received units, which normally sold for $100 each, secured by a small down payment. Most promoters required a five-dollar minimum; some accepted as little as one or two dollars. Fur coats and gold watches also served as collateral. Wilson estimated that 250 of the 300 people at the tent show he visited committed themselves for investments totaling more than $40,000.

The tent shows, concluded Atwood, were "the last word in salesmanship, the ultimate perception of sheer mesmerism or hypnotism." The people acted less like investors, he claimed, than "an audience at some form of seance. Intense emotion rather than cold calculation was the predominant note." As the afternoon progressed, the pressure escalated. Mail solicitations and newspaper advertising allowed investors to weigh the risks in the privacy of their homes. But in the tabernacles, wrote Atwood, "the prospect is cornered. He has no mental escape . . . insistence, repetition, and brutal hammering [gave him] the courage to act." One observer incredulously discovered that people "buy on the very day they go to the fields without any consideration even overnight of the truth or untruth of the statements made."

Atwood described the people who visited oil tent show promotions as "a sad aggregation of humanity" united by curiosity and greed. They were, he wrote, "people who could not be driven too hard mentally," who bore "unmistakable signs of ignorance and poverty . . . [and] a monotonous and mediocre life," elderly men and women in "the last stages of doddering decrepitude," domestic servants and young women with children, all people "too childish mentally to possess any such thing as business judgment."

Atwood's description reflected a common trend among 1920s journalists and intellectuals to belittle southern Californians. Louis Adamic disparaged "the retired farmers, grocers, Ford agents, petty hardware and shoe merchants from the Middle West . . . who . . . [having] made their little pile, grown old and rheumatic . . . came to Sunny California to rest, regain their vigor, enjoy climate." Adamic found these "Folks" to be "unwell, vacuous, biologically finished men and women," but "good simple honest people as people go." H. L. Mencken was less sympathetic. "The supply of morons from the Midwest is inexhaustible," wrote Mencken, and Los Angeles was "the southern paradise of morons."

Mencken, Adamic, and their literary descendants, like Morrow Mayo and Nathanael West, helped create the image of the transplanted Midwesterner whose naïveté made Los Angeles a mecca for fraudulent promotions. "People here are especially gullible," Atwood quoted a banker. "There are so many suckers here," said another businessmen, "I know of no other place where it is easier to put out a fake security based on a good thing." But these stereotypes overestimated both the distinctiveness of Los Angeles and the importance of the retired Midwesterners in the local speculative frenzy.

Southern Californians were no more susceptible to get-rich quick schemes in the 1920s than New Yorkers, Floridians, or any other Americans. The magnitude of the boom simply offered greater opportunities. Even within Los Angeles itself, there seems no reason to believe that the retirees were more prone to risky investments than other residents with surplus capital. Atwood admitted that the "more intelligent people," many of whom had prospered in real estate speculation, could not resist gambling in oil. One former oil promoter named doctors as his prime target. "Doctors take the lead any place, anytime," he explained. "Stock sellers know the average physician will fall quicker than anyone else." Prune growers ranked second on his list. Those who ultimately invested with C. C. Julian included large numbers of Jews and Japanese. If Midwestern Protestants predominated among oil speculators, it reflected their preponderance in the region, not their inherent gullibility.

It is a tribute to C. C. Julian's stature as a purveyor of dreams that several oldtimers remember him as "the king of the sucker tents." Yet Julian relied on neither bus rides nor tent shows to entice people. His talents lay in writing advertising copy that encapsulated the hopes and dispelled the fears of the small investor. When his office opened on June 19, 1922, the Monday morning after his first ad appeared, a line of people stretched down the hallway. Customers arrived throughout the day, and the afternoon mail brought additional orders. In the evening when Julian totaled his receipts, he counted $19,500! Operating without any outside salesmen and based upon just one ad (which reappeared once in the Monday papers), Julian had subscribed almost one-eighth of his issue on the first day.

Buoyed by his initial success, Julian churned out daily ads to woo the public. "*SANTA FE SPRINGS*," stated his June 20 effort, "*NO STOCK! NO UNITS! No Directors at Fat Salary!*" Over the next few days Julian began to hit his stride. "**Never Before Perhaps Never Again**," he advised his readers on June 21, as he introduced a theme that ran through many of his ads. "If you were given a tip on a race horse that was a strong favorite at odds of thirty to one, would you place a $50.00 or $100.00 bet on him to win? I am that race horse. . . ." On Sunday, June 25, Julian produced the classic advertisement that he later boasted elicited the greatest response of his career:

> ### Julian Refuses to Accept Your Money
> ### Unless You Can Afford to Lose!
> ### Widows and Orphans, This Is No Investment for
> ### You!
>
> I cannot conscientiously tell you that there is no element of chance in my project. . . .
>
> I don't tell you that I am offering you a "**Gold Bond**" backed by a ten-million-dollar deposit in the bank, and if you are drawing out your little savings account that you cannot afford to lose, to support me, my advice to you is, **leave it in the Bank.**
>
> My appeal is addressed to people who can legitimately afford to take a chance. "**And Mr. Thoroughbred**" I have tried to offset this element of risk in my offer to you, by laying my cards face up on the table. . . .
>
> I am not asking you to place a bet on something that I have not already bet mine on. For I want to tell you that when I stepped up and laid $30,000.00 cash on the "**Come Line**" . . . it took every dime that **I could beg, borrow, or scrape together.**
>
> . . .
>
> If you feel that I am worthy of your support I want to promise you that until my well is on production, I will extend every ounce of **energy and ability** that I possess to **protect your investment, and will stick by you** until the last dog is hung.

With one masterful stroke Julian had separated himself from other oil promoters. Unlike the promoters investors had been repeatedly warned against, Julian did not prey on "widows and orphans." He accepted only bona fide investors.

"**SIXTY THOUSAND DOLLARS IN SIX DAYS!**" read Julian's ad headline on June 28. "**SEVENTY THOUSAND DOLLARS IN SEVEN DAYS!**" he proclaimed the following day. In his first ad Julian had boasted that he had no salesmen,

but within two weeks he had enlisted representatives in Glendale, San Pedro, Long Beach, Eagle Rock City, and Hollywood.

Julian's lease allowed for the drilling of three wells on the Brunson property. With money pouring in and achieving his goal of $175,000 for the first well imminent, Julian subtly shifted his appeal to include two additional wells. "Do you know that I am going to drill three wells on this tract and that you have the privilege of participating with me on each one?" he asked on June 28. Four days later, Julian offered those investing in his first well "the privilege of participating for the same amount in my second and third wells." He had thus tripled the ante; his goal was now $525,000. Even at this elevated level, Julian could announce on July 2: "**13 DAYS MORE And the Opportunity to Place a Bet on Julian at Santa Fe Springs Will Have Passed You By.**"

For the next two weeks the headlines on Julian's ads counted down the days until his offer would be *"closed forever."* "'For the love of Mike,'" he exclaimed, "don't keep putting it off, make a noise like a check." "**Only 5 Days Left to Buy a Ticket on the Favorite in the Big Race at Santa Fe Springs**"; "Talk about being snowed under my offices have been jammed all day long. . . . I'm all in my shoelaces tonight, no pep to write you much of a message"; "'**GOOD-NIGHT NURSE' 'Just One Day Left!'**" and I will have 'kissed you good-night forevermore'. . . ." On July 15 Julian reduced his usual verbosity to a simple: "**GOING GOING GONE** . . . Tonight at ten o'clock my offer to you will be ancient history and 'you' will have slipped a 'cog.'"

Two days later Julian reported the final tally. "'**IT'S A GRAND OLD WORLD' AT THAT**," he wrote. "I Am $67,000 Oversubscribed on My Project at 'Santa Fe Springs.'" In less than a month C. C. Julian had collected almost $600,000.

Julian's stock offer had closed, but his daily ads continued, the reigning sensation of the Los Angeles press. The public anticipated Julian's messages as with the same avidity with which they awaited "the Sunday comic page, or the newest Hollywood scandal," wrote Robert Cleland. Julian obliged them even when he no longer had units to sell. Reading Julian's prose today one marvels at the appeal it held for southern Californians in the 1920s. "I wrote just as if I was talking to them," Julian explained. But even in a Los Angeles overflowing with retired Midwestern farmers, no one spoke in the corny, homespun manner that Julian projected. "Do you realize that my offer to you is nothing short of the '**OLD CAT'S TONSILS**'?" he asked in one ad. On other occasions he described his offers as the "Kitten's Cuffs," the "Pigeon's Pajamas," and "Finer than the Fuzz on a Frog's Face." Bad puns proliferated:

> "You tell 'em 'Bradstreet,' what I have 'Dun'";
> "YOU TELL 'EM RUBBER I'M THERE IN THE STRETCH";
> "YOU TELL 'EM 'WELLS FARGO' IT'S HARD TO EXPRESS."

"Written by anyone else but Julian," observed journalist Guy Finney, "without the friendly go-gettem personality behind it [the ads] would have been as flat as stale beer dregs."

Finney and other contemporaries struggled to explain the Julian phenomenon. "He demonstrated to the satisfaction of the population of a great American state," concluded Finney, "that he knew exactly how a speculation-daffy crowd reasons . . . that it doesn't reason at all . . . that it moves quickly toward the get rich quick bubble by its emotional urge alone." Julian, suggested journalist W. V. Woehlke, tapped "the gambling instinct of the crowd, holding out the lure of the big profits if he should win." The quest for oil became a horse race with Julian as the "Pole Horse . . . Setting A Dizzy Pace . . . leading the field and riding easy," paying off on odds of "30 to 1."

Julian's appeal, however, went beyond pungent prose and race-course rhetoric. Nurtured since birth in the heady euphoria of oil and real estate boom towns, Julian had acquired an uncanny perception of the apprehensions, doubts, and uncertainties of the small investor. His ads played upon public antipathy to big business, widespread fears of unscrupulous or inexperienced operators. "I have no millionaires standing behind me," exclaimed Julian. "I'm the guts behind this offer me and me alone. . . . Every word of every message I ever sent you, I sat down and wrote myself, and you must believe me when I tell you they express my sincere emotions."

But if Julian as not a corporate mogul, neither, he boasted, was he a fly-by-night operator. "Folks," he wrote, "I don't want you to class me with these 'wild-eyed promoters' that never miss a chance to clean up a 'wad of dough' where the going looks easy. . . . I do not ask you to come in and buy units in a well that is capitalized at two million dollars or to drill one on a town lot." Julian expressed sympathy for those trying to "distinguish between the fellow that is trying to be on the square and 'the every day hot air artist.'" But, he pledged, "if I can only win your confidence, just long enough to get a chance to show you, I'm sure you will never regret having been shown."

To gain their confidence, Julian reassured investors about his operations. He repeatedly bragged of his "handpicked Expert Crew" of fifteen "Bearcat" drillers. He exaggerated his own experience in the old fields, boasting "that for the past fifteen years I have never earned a dollar any other way than drilling holes in the ground." Julian cultivated the image of "an ordinary, every-day driller," who had "worked 'dog-on' hard in the fields for every dollar I've earned" and would continue to do so on behalf of his clients. "Action is my slogan," he trumpeted in his second ad. "Just watch me go." "While you are snuggled up in the feathers every night, do you know where I am?" he later asked. "I'm out at my wells, watching every move, grabbing a few winks now and then."

Above all, Julian stressed the trustworthiness and honesty that distinguished him from the promoters they had been warned against. "Do you know that I have pledged my word of honor to you that there is no incentive in the world could make me betray the confidence that you place in me?" he asked. "Folks, if you would

only realize how truly sincere I am trying to be with you, this would be the last message I would have to send." "A promise made is a debt unpaid," Julian preached in a later ad. "The day I break one of mine, '**I want you to walk in and shoot me.**'" He hoped to "fall headfirst into the deepest well in Santa Fe Springs" or let the "Buzzards pick my bones and the wind whistle Star Spangled Banner through my ribs" should he betray the "royal confidence" of his investors.

Confronting an audience anxious to invest but surrounded by dire warnings about con men and oil schemes, Julian directly countered the charges leveled against oil promoters and turned them to his advantage. Even before he struck oil, competitors aped his approach and boasted of their proximity to his lease. But none could match C. C. Julian as the articulator of the no-risk, get-rich-quick speculative dream.

## II

C. C. Julian had mesmerized his audience, but he had yet to strike oil, and, despite his confident assurances, delivering one well, much less three, was far from a sure thing. The Brunson lease, while no longer distant from the main Santa Fe Springs field, remained on the fringes of proven territory. To reach oil Julian would have to drill to a depth of almost 4,000 feet, a process fraught with peril. Even in the richest field, warned Samuel Blythe, "an oil well is a tricky and malignant object."

Under the terms of Julian's lease, he had to launch preparations within ten days and commence drilling within forty. Workers immediately began constructing the wooden derrick, which would tower over 120 feet in the air, bedecked with electric lights for round-the-clock work. Huge boilers to generate steam to drive the engine had to be set in place, along with a solid cement block created to support the large stationary engine that would power the drill. On June 30, the rotary drilling equipment arrived. It included a "rotary table" for the center of the derrick floor, which, when turned, would rotate the drill bit into the earth. The table was connected to the engine by a steel chain of the type once described by Upton Sinclair as "exactly like the sprocket of a bicycle, except that the links were as big as your fist." A square hole graced the center of the table and a corresponding opening penetrated the derrick floor. The top drill stem, equipped with an eighteen-inch bit, fit through these holes.

With rig built, boilers set, and rotary table in place, Julian held a gala ceremony to launch Julian No. 1. "Come down to the field Sunday and look it all over from cellar to Crown block and see me spud this baby in," he wrote on Friday, July 7. "Seeing is believing." Two days later, five days before the forty stipulated on the lease expired, a large, festive crowd gathered near the corners of Clark Avenue and Norwalk Boulevard in Santa Fe Springs. With a flourish Julian christened the derrick with champagne, which flowed so freely, according to one participant, that the local ducks and geese staggered drunkenly through the town for the next day. The drill pierced the earth with the familar "Spud! Spud!" sound. **"JULIAN 'BALLIN' THE JACK' AT SANTA FE SPRINGS,"** crowed the ad in Sunday's papers.

"The basic principle of rotary drilling," writes William Rintoul, "is the same as that for a carpenter's auger. A hole is made—in wood or the earth—by turning a cutting tool while weight is put on the tool." Underground, the "blunt steel teeth" of the drill bit operated "like a nutmeg grater," grinding dirt and rock into powder. The drill bit was attached to a steel pipe, or drill stem, through which passed a special fluid, called drilling mud. The fluid escaped through holes in the bit, picked up crushed pieces of rock, and was forced up to the surface, where it sluiced through a screen that separated the rock cuttings, and into a "sump hole" from which it circulated back into the well. The process also created a mud cake along the sides of the hole, which protected against the loss of the fluid and shored up the walls against cave-ins. Competent mudding prevented fluids, whether oil, gas, or water, from passing in or out of the hole. A porous mudding job doomed the well.

As the well grew deeper, additional lengths of steel pipe had to be added to the drill stem. Each thirty-foot-long 380-pound cylinder had to be carefully slid down the hole and screwed to the next. A 4,000-foot hole required forty tons of steel pipe. When the bit became dull, or when a smaller bit was necessary, the roughnecks hoisted the entire drill stem from the hole with a block and tackle system, stacking the pipe vertically in stands on the derrick. They attached the new bit and then lowered the thousands of feet of pipe back into the hole to resume drilling.

Throughout the process, care had to be taken to stop water, "the foe and conqueror of oil fields," from killing the well. The drill stem regularly passed through underground streams and pools. If this water mixed with the oil, it would destroy, not only the well, but possibly the entire field. To prevent this, oil workers would "cement off" the well, forcing vast quantities of pure cement down the well under heavy pressure so that it rose outside the casing and tightly sealed off the sides of the hole, filling every crevice. The entire job had to be completed within an hour, before the cement began to set. Work would cease for two to three weeks while the cement hardened. In the end a giant plug would remain, sealing off the well. The drillers would then bore a hole through the block and resume their quest. If these efforts failed to stem the flow of water, as Blythe wrote, "there's an oil well that has gone to glory."

Julian's ads enthusiastically chronicled the "downward plunge" in detail. "Do you stop to remember that my offer has been before you for three weeks and I am now drilling at 900 feet?" he asked on July 14. Ten days later Julian No. 1 reached 1,500 feet and his crew successfully engineered the first cementing task. By July 30 it passed the 2,000-foot mark.

The troublefree journey ended abruptly on August 10. With No. 1 at 2,804 feet, 150 feet of drill stem twisted off and blocked the hole, necessitating, as Julian described in his ad, "A Mean Fishing Job." To save the well, thousands of feet of pipe had to be hoisted and unscrewed until the break was discovered. The drillers then began "fishing," lowering a heavy "overshot" down the hole with a cable, attempting to hook onto the damaged drill stem and pull it up. Fishing jobs could take days, weeks, or months. Sometimes the junk could not be moved, in which case it became necessary to drill around or "sidetrack" the broken equipment, or worse, drill through it in a lengthy, expensive "milling" process. In many cases the well had to be abandoned.

Maintaining his straight-shooting image, Julian reported the setback in his August 11 ad. Instead of the vague promises offered by tent show orators, he "[laid his] cards face up on the table." "When you lay your 'Jack' on the line with me, you know exactly what you get and how you get it," Julian promised. "I am here to give you the straight dope. Whether it is good or bad, I'll give you the lowdown always."

Julian reassured his investors, "You may bet your spare change I will have the fish out within 24 hours. 'Are we downhearted?' Not on your old straw hat." Julian proved as good as his word. "FOLKS," he crowed on August 12, "At three o'clock this morning we laid the 'old fish' on the walk and my 'Bearcats' are again rambling to that 'Gusher' oil sand. . . . I'm a little groggy but had to scribble you the word before hitting the hay."

Three days later, on August 15, however, Julian again bore bad tidings. "Ain't it Awful?" he asked. "We no sooner got out of one fishing job than we stepped into another. 'Folks Hold Your Breath.'" This task proved more challenging. On August 17, Julian wrote, "'Yes Indeed' Still Fishing . . . but don't you lose any sleep. I'm doing that for you." Julian decided to cease fishing and instead sidetracked the drill bit and collar lodged in the hole. "HI-BALLIN THROUGH," announced Julian on August 18. "We're Sitting Pretty Again."

In the aftermath of these difficulties, Julian dramatically offered his investors an opportunity to sell back their units. "I'm Calling the Roll Today," he announced on August 23. "Does anyone think I'm wrong? Does anyone doubt my sincerity? If there is anyone among my 'Thoroughbred supporters' who have 'cold feet,' step up to my office today and I will cheerfully hand you back your money." For Julian, this represented a no-lose proposition. With other wells at Santa Fe Springs now coming in and demand for his units running high, he could easily resell them at a profit. By his own account, the "roll call" found his backers " '99' per cent standing behind me like a 'Stone Wall.' "

Julian's willingness to satisfy disgruntled customers, his forthrightness on the "fishing" difficulties, and the alacrity with which he and his drillers had handled that challenge, all reinforced his reputation for honesty and ability. These lessons occurred at an opportune time; Julian was about to launch another offer to the public.

On July 26, Julian had started to prime his readers for a new investment opportunity. "KEEP YOUR HAND ON THAT BANKROLL FOLKS," he advised. "I have another sweet deal brewing at Santa Fe Springs on the most wonderful piece of oil land you ever laid your eyes on." His offer, he promised, would be "the Eagle's Hips and you will be on the ground floor and I guarantee there will be no basement."

The new property sat 1,320 feet west of the Brunson lease, far beyond proven territory. But Julian was not alone in his interest in this sector of Santa Fe Springs. Standard Oil had leased the land on three sides of Julian's claim. George Getty occupied his northern border. Throughout August, along with progress reports on wells no. 1 and 2, which had spudded in at the start of the month, Julian whetted investor appetites for his new proposition. "Are you saving up dough for my new offer?" he asked. "It's going to 'knock you dead.'" Finally, on September 3, Julian tossed his pitch:

"IF YOU LOVE ME SHOW ME NOW"

Here it is, Folks, short, sweet and simple, and take it from me as clean and keen a gamble as was ever submitted anywhere in the world.

. . . .

Folks, I am mighty tired of piling up dough from the big fellows. There is lots of money to be made in producing oil, providing you operate legitimately, and with your assurance I know I can make us all some real dough.

I don't want to hog the game and become a millionaire over night, and in return for your support I will always give you the "lion's share."

**"With me, your interests come first—mine come second."**

In this offer, unlike his first, Julian pooled the proceeds of two wells, Nos. 4 and 5, into one syndicate. Each $100 invested netted $\frac{1}{3500}$th of 60 percent of the combined production of the wells, after the landowners and Julian's backers received their shares. Furthermore, promised Julian, "If I drill the first well on this tract dry, I certainly won't drill the second one on the same property. . . . I'll step out and select the surest spot in California that I can find and drill the second of these two wells there."

For his new campaign Julian had honed his technique and intensified his message of opportunity. "**FORTUNE KNOCKS BUT ONCE**," warned Julian. On the fourth day he beckoned:

**"FAINT HEART NE'ER**
**WON A LADY FAIR"**
**"YOU MUST TAKE CHANCES TO BECOME A**
**MILLIONAIRE"**

Life is a gamble after all, we are here today and gone tomorrow, it's just a collar, just a coffin when you die, don't you know. Every well drilled is more or less a gamble . . . the gamble I'm offering you is a "Royal One."

In another ad he quoted Shakespeare, who advised investors, "OUR DOUBTS ARE TRAITORS AND DO MAKE US LOSE THE GOOD WE OFT MIGHT WIN BY FEARING TO ATTEMPT." "He must have been thinking of me and Santa Fe Springs," concluded Julian, "when he said that mouthful."

With two wells drilling, No. 4 rigging up, and a syndicate to subscribe, Julian found himself "busier than a one-armed paper-hanger with the itch." In later years Julian liked to boast of and perhaps exaggerate his arduous work regimen. He claimed he would arrive at his downtown offices at 8:00 a.m. and spend the day selling units, supervising the sales staff, meeting customers, personally answering all letters, and buying supplies. After a brief break for dinner, Julian would return to the office to compose the next day's advertisement, then curl up in the back seat of his chauffeur-driven automobile for a quick nap on the way to the oil fields. At Santa Fe Springs he would monitor the progress of his wells and consult with his roughnecks. If possible he would return home by four to catch a few hours' sleep. More often he returned to his office and bedded down on the floor, before renewing his grueling routine. His top assistant claimed to have lost forty pounds chasing Julian. "Nobody could keep up with Julian while he had his syndicates on," stated an associate. "We all ran ourselves ragged trying to do it."

Julian had hoped to deliver Well No. 1 to coincide with his new offer, but the repeated fishing episodes made this impossible. In addition to the two bottlenecks reported in his ads, the drilling operation's log reveals two smaller fishing jobs. Meanwhile, No. 2 experienced a remarkably smooth trip. By September 12, the depth of the second well had surpassed that of the first. On September 24, No. 2 hit the oil sands, the coveted strata that lay atop the black reservoir, at 3,853 feet. Julian's crew cemented the well and waited while the plug hardened before beginning the final thrust. To celebrate, Julian "called the roll" again, offering to buy back units in either well for $140, a 40 percent profit after just sixty days. "No one ever went broke taking profits," he advised.

Meanwhile, according to Julian's probably apocryphal recollections, events built to a dramatic climax. His spartan regimen had taken a toll. Chills and fever plagued his nightly visits to the field. In early October, as No. 2 sat cemented and No. 4 prepared to spud in, a phone call from Julian's doctor confirmed his worst fears. An intermittent form of malaria he had contracted in Texas had resurfaced. The doctor advised him to spend the next month recuperating. Simultaneously, on another line, his drilling supervisor told Julian, "We've just dropped 3,000 feet of drill stem 600 feet" on No. 1. Julian sat shocked at his desk. "It was surely a one-two punch to the chin," he later recalled. "There I was, sitting in the mud, with my first well full of dropped stem and my blood full of malaria."

To his readers, Julian described the problem as "another little fishing job." If his later accounts can be believed, however, he feared the sunken drill stem meant "the certain loss of the hole." Miraculously, he recounted, "We managed to tie onto it and lift it out without dropping a single link—one chance in thousands." The drilling log confirms the fishing job but also reports sidetracking at least some of the junk. Whatever the truth, by October 13, No. 1 had joined its counterpart cemented in the oil sand. As for the malaria, Julian "went on working 18 hours a day, on my easy days, and the germs must have got discouraged, for they quickly left my system."

Meanwhile, Julian's new issue sold rapidly. "Only Four More Days," Julian trumpeted on October 19. By October 21, Julian announced, "Instead of Two Days More, I will shut her off tonight, because I will be absolutely sold out before 9 o'clock this evening." In less than two months, Julian, who had yet to deliver a single well, had raised $350,000 for his new project.

As if to punish those who had not supported him, on October 24, Julian's drillers began to bore through the cement on No. 1. Two days later, at five in the morning, Julian No. 1 "went over the top . . . pouring 'a wicked stream' into our storage tanks." While the 1,600-barrel-a-day rate did not match the "4,000-barrel pay" that Julian had predicted (or the 3,600-barrel flow he claimed in his ads), he had delivered what the *Times* called "one of the star producers of the Santa Fe Springs field." "If you happen to hold a ticket on me to win on 'this baby' you are in the right church and in the right pew," exulted Julian. Three weeks later, with "Daddy's Sweet Baby" (No. 1) 'struttin her stuff,' " her sister, Julian No. 2, came to life on November 15, with an initial flow of 2,400 barrels.

Less than five months had passed since an obscure C. C. Julian had published his first advertisment. In that brief span, his sales revenues approached the $1 million mark, his units had almost doubled in value, and his two wells poured 3,500 barrels of oil into his storage tanks each day. Julian reigned as one of southern California's great success stories, an inspiration to oil promoters and investors alike. "Julian No. 1 investors, We congratulate you!" advertised a rival company. "You were brave enough to scoff at those who deride oil investments. . . . You were awake while others were asleep." *Oil Age*, the respected industry journal, saluted him in an editorial entitled "Julian Makes Good." The *Los Angeles Examiner* hailed him as an "experienced operator, not a promoter of the fly-by-night type," concluding, "People who have invested in him will get a run for their money."

On December 19, the first dividend checks for unit-holders in wells no. 1 and 2 came due. "The Bank . . . [has] kindly consented to allow me to pass them out to you this month," wrote a gleeful Julian, "and I can promise you 'I am going to get a kick out of doing it.' " He had pledged to investors that he would "be your Santa Claus next Christmas Eve." As everyone now knew, C. C. Julian was as good as his word.


## III


At the dawn of 1923, C. C. Julian was the prince of oil promoters. His advertisements had become a southern California institution. Although he had no new units to sell, his prose flowed as regularly as the oil from Julian wells no. 1 and 2. C. C. entertained his public with a never-ending mélange of drilling reports, dividend notices, poetry, cartoons, sage advice, and homespun philosophy. Cartoons depicted "A Happy Family of Tomorrow" — a man, his wife (a Julian unit), and their offspring (scores of stock dividends); or a poker player demonstrating "A Winning Hand" — five cards depicting Julian wells number 1 through 5. Supporters composed poetry extolling his virtues. ("For "**Honesty's**" his middle name, / And "**action**" is his first; / Just take a tip and shoot your wad / If for dividends you thirst"), to which Julian added his own handiwork:

"Tho you have tried and lost
And tried and lost again,
My most sincere advice to you is
Reach for your check book, grab your pen."

Once sold to the public, Julian units assumed a life of their own, their prices fluctuating with the fortunes of the wells. By late 1922, a lively market in the buying and selling of Julian units had surfaced on Spring Street, the financial center of Los Angeles. With wells No. 1 and No. 2 both on production, units originally priced at $100 sold for as much as $190. Rumors disparaging Julian's success and deflating unit values coursed through the financial district. "I notice they are trying to pound down the price of my units on the market," he admonished on December 31, when prices on wells no. 3, 4, and 5 actually dropped below the $100 level, "but don't let that concern you." Julian blamed the drop on "sharks . . . concoct[ing] measly stories . . . to try and dampen your confidence in my security," and "knockers" who "seem to delight in circulating the meanest kind of stories about my different wells." When his other wells came in, he boasted, he would "present all the knockers with 'Barbed Wire Tooth Brushes' and a mouth wash of TNT."

In reality, the price manipulation of his units worked little hardship on Julian. Assuming the role of protector of his flock, C. C. reaffirmed his hold on the faithful and won converts to his cause. By offering to buy back units and periodically "calling the roll," he helped break price slides. At the same time, like other brokers, Julian purchased units at reduced values and resold them at a profit, pocketing additional sales commissions. Indeed, some charged that Julian himself was responsible for the rumors that created the roller-coaster market in his wells.

In the waning months of 1922, large corporations, independent producers, and wild-eyed promoters delivered gushers at Santa Fe Springs with uncanny regularity. Over 200 wells appeared over a vast stretch of terrain. More significantly, Standard Oil made a staggering discovery. In the early months of the boom, drillers had drawn oil from the "Bell" sands, a 300-foot-thick stratum of continuous oil-bearing sand and shale, approximately 3,500 feet below the surface. Disappointed with the potential of the Bell sands in the eastern portion of the field, Standard had driven one of its rigs deeper. On October 23, the drill bit pierced a second, even richer, reservoir of oil at a depth of 4,644 feet. On the first day, 4,250 barrels of petroleum gushed from the well. During the next week the flow unexpectedly increased by 50 percent to a spectacular 6,345 barrels a day.

The Standard well was hailed as the "Wonder Well of the West," a title it held for less than three weeks. On November 12, a small company operating a few hundred yards away reached the new "Meyer" sands, and delivered a well at 8,000 barrels. The latest well, reported the *Times*, "stands without an equal in the world today for the amount of oil that is being produced and the high gravity of the product." Overnight, the phrase "Meyer Sands" entered the everyday lexicon of southern Californians.

This development came not a moment too soon for C. C. Julian. Julian No. 1

produced almost 40,000 barrels in its first month, but production had dropped off and water had begun to seep into the well. In the southwestern portion of the field, Julian No. 4 had already passed 3,000 feet and showed little promise of producing in the Bell sands. Whether or not the Meyer sands extended to either of these tracts remained uncertain, but Julian unhesitatingly seized the initiative.

"Folks," he informed his readers on November 27, "I have closed Well No. 1 and am getting ready to drill her deeper." On Julian No. 4 and Julian No. 3 (which was just rigging up), he decided to bypass the Bell sands entirely and head for the big "Meyer Pay." Only Julian No. 2 would continue production from the higher stratum. "As long as we can pull 2000 barrels a day each from this upper stuff we had better do so, but if [it] ever shows any signs of weakening on me, I'll kick it down to the big pay," vowed Julian.

As his drillers probed the limits of the Meyer sands, C. C. scanned the horizon for new leases. With the discovery of the more prolific reserves, bonuses and royalty rates had again skyrocketed. On January 31, Julian acquired drilling rights to five more acres at Santa Fe Springs. Lease-hounds had driven the price of the new tract, adjoining the Getty interests just southeast of the main town-lot drilling area, to unprecedented levels. Since the end of September the land had been leased and subleased three times. Julian paid a staggering $80,000 and a 50 percent royalty for the acreage. "This new issue of mine is going to be a knockout," he boasted two days later, converting his high overhead into an asset. "Just let me tell you I'll be drilling on the highest priced piece of oil land in the world."

Within a week Julian gained rights to a second five-acre parcel, this one in the southwestern edge of the field, far from any proven production. The cost, though not as exorbitant as that of the other lease, was still substantial. Julian paid $50,000 and granted a 30 percent royalty. Julian planned to include both tracts in a new four-well syndicate. To sweeten the offer, however, he awaited the imminent arrival of Julian No. 4 in the Meyer sands.

Julian had originally "scheduled" No. 4 as a "Christmas present" to his backers. But oil wells, particularly those drilling to 4,600 feet, rarely adhere to schedule. On December 27, with the rig "cutting ditch pretty fast" at 4,455 feet, Julian beseeched his followers, "Don't weaken any on this child." Two weeks later, he delivered "bad news." "We twisted off the drill stem . . . yesterday leaving seven joints in the hole, and right now we have a brute of a fishing job, and being so deep makes it rather mean." His crew "speared the old fish" the next day and continued downward.

By mid-January 1923, Julian had dubbed his recalcitrant well "the prodigal son of the family." But, he promised, "we'll kill the fatted calf very soon now to celebrate the incoming of this wayward child." On January 23, Julian's men cemented well No. 4 at 4,460 feet. Within days, Julian's original well, No. 1, also reached the Meyer sands. But the battle had not yet ended. "Here I am fighting my way foot by foot deeper and deeper into the Meyer pay on both No. 1 and No. 4," note Julian on February 11, "and the deeper we go, the harder it is to control them. The gas pressure on both of them is terrific."

By this time, as *Oil Age* reported, "all eyes" in the oil industry were "being directed to Julian No. 4." to prove the Meyer sands in the southwestern reaches of the field. On February 18 at 4:00 a.m. on a Sunday morning, "that darling No. 4" roared to life, the crude oil flowing at a rate of 6,200 barrels a day.

The *Los Angeles Times* lauded No. 4 as "a real discovery." *Oil Age* hailed the arrival of the Julian well as a development "of tremendous importance" that extended the producing area of the Meyer sands. For Julian, the well spelled vindication and opportunity. "Where are the birds that promised to drink all the oil I ever got on that six-acre lease?" he crowed. With wells no. 1, 2, 3, and 5 now "a cinch for still bigger production," he asked, "do you wonder that I am wearing 'a smile that won't come off?'" Furthermore, noted Julian, "I feel that the results I produced for you entitle me to submit my new offer."

One week later, on February 25, Julian inaugurated his latest campaign, harking back to his most successful theme:

### "If You Can't Afford to Take a Chance"
### "You Can't Afford to Play With Me"

And while I have reduced the element of risk of losing your money to the minimum, every oil investment, no matter how sure it looks is more or less speculative . . . no matter how carefully or skillfully a man may operate, there is always that element of chance.

. . . I am going to strain every nerve in my body to make your investment . . . the most profitable one of your life, but at the same time "I don't want any widow's mite." I don't want any life savings that you have sweat blood to put away to take care of you in your old age; I not only don't want it but I positively refuse here and now to accept one dollar from anyone if by losing it there will be a hardship worked on them.

The offer was straightforward. Julian would drill two new wells on each of his new Santa Fe Springs leases. Investors would share in the total production of all four wells. "There are 2500 assignments in each well or 10,000 in all, at $100 each," wrote Julian. The capitalization thus totaled $1,000,000. In addition, Julian offered two overriding guarantees. While he retained a 12 ½ percent interest in the production, he would defer his share until all assignment holders "received back 100% on their investment." Second, Julian promised "to finish all of these four wells regardless of any condition that may arise. What I mean by that is should I be unfortunate enough to lose a hole before going to the necessary depth I will skid the rig and drill the well over at my own expense without ever calling on you for another dime."

Over the next three weeks, Julian conjured up his old magic and added several new tricks. "I'm not appealing to the unintelligent," he explained. "I'm talking to you Mr. Man who makes the investments on the actual merit of his project." Julian's appeal was addressed, not only to those unfamiliar with the oil business, but "to every oil operator and every driller" in California. Among his backers, he boasted, were "men who have been engaged in producing oil all their lives," "representative men" from all the major companies. Nonetheless, he reassured the small investor, "you and your little hundred dollars means as much to me as a man who walks in and throws me five, ten, or twenty thousand." If he failed to deliver on his four wells, Julian would "pretty near promise to eat 4000 feet of drill stem."

With No. 4 pumping thousands of barrels of oil each day, customers needed little prodding. Six hundred thousand dollars arrived in the first twelve days. "It Will Soon Be Curtains For You," warned C. C., "and the old saying 'He who hesitates is lost,' will have run true to form again." On March 25, exactly one month after the issue had opened, Julian issued "THE LAST CALL" on "the greatest money making opportunity ever submitted to any community." The following day Julian officially halted sales, having attracted $1,000,000 in twenty-two working days. Hundreds of thousands of dollars, worth of orders continued to pour in.

March 26 gave Julian and his backers additional cause for celebration. For over a month his crew had battled Julian No. 1, which had become "a little unruly" at the lower depths. "Her gas pressure is so terribly strong," he reported on March 3, "that it's a little difficult to make the cement set." It took three tries before the state Mining Bureau approved the cement job. Finally, at 11:00 a.m. on the day his "four-well entry" closed, Julian No. 1 began "Struttin Her Stuff" with an 8,100 barrel-flow.

"Julian No. 1 is today one of the 'greatest wells' in the greatest oil field in the world," boasted Julian. Not even his harshest critics could disagree. Julian units in nos. 1, 4, and 5 now commanded as much as $225 on the open market, and in April a "cartload" of dividend checks went out to unit-holders.

With his star in the ascendancy, Julian wasted little time organizing another syndicate. Ironically, while Julian had found fortune on the western portion of the Brunson lease, Globe Petroleum had only met with disappointment on the five acres it had retained for itself. Thus, when in March, Julian offered a $250,000 bonus and a 50 percent royalty for drilling rights to the remainder of the Brunson lease, reportedly the highest price ever paid for an oil lease, Globe readily accepted. In addition, for the first time, Julian ventured away from Santa Fe Springs, to the nearby Compton oil field in a "wildcat" enterprise. For a modest $5000 bonus and a 30 percent royalty for twenty acres of land, Julian joined corporate giants Union Oil and General Petroleum in sinking test wells in the unproven new field.

On April 22, 1923, Julian linked these leases into a new proposition. Capitalized at $750,000 (7,500 $100 units) this syndicate offered interests in two wells, Julian no. 11 and 12, drilled in Santa Fe Springs; and a third, a wildcat named "Pico," at Compton. This will be, announced Julian, "My Last Oil Syndicate This Year." "Play Ball," he cried with a seasonal flourish, "This is the 'last game of the season' and the 'Grandstand seats' are going fast. I'm at bat now and due to knock out another home run." By May 8, Julian's three-well $750,000 "Issue Deluxe" had sold out.

Since June 1922, in less than a year of operation, Julian had collected $2.5 million from the investing public. His wells had produced tens of thousands of gallons of oil. Now, at last, he began to enjoy his rewards. "How is Julian taking his success?" he had mimicked street talk in February. "Is he 'all swelled up' or is he 'holding up his head?'" He was not, he guaranteed his followers at that time, "the kind of 'hair-pin' that a little success could make 'nigger-rich.' . . . You won't find me living in any swell home . . . in 'Beverly Hills'; I'm still parked in the little old furnished flat. I'm not spending any of my profits on Rolls Royce cars—I still bump along in my same old fifteen hundred dollar boat." After all, "even with 'two million dollars' a man can 'wear only one collar' 'one tie' and occupy 'one coffin when he dies.'"

By May, however, a great transformation had begun. Clothes-conscious since his youthful embarrassment in Winnipeg and his years of work in Canadian men's stores, Julian now sported many collars and ties—a new wardrobe of expensive suits, candy-striped shirts, polkadot ties, and white spats. A rakish derby and gold-headed cane completed the ensemble. Julian still disdained the Rolls-Royce but purchased instead a "great grey" Pierce-Arrow. "Oh! by the way folks," he gushed in his April 5 ad, "do you remember me telling you about the '$1500 Hack' I've been driving, well its gone now, just traded it in on a 'real wagon' thanks to good old number one."

As Julian approached the first anniversary of his arrival in Santa Fe Springs, he moved his wife and two daughters from the "little old furnished flat" into a $150,000 home on Los Feliz Boulevard on four and a half acres overlooking the lights of Hollywood. The ten-room house boasted a gymnasium and a thirty-by-forty foot living room "panelled in beautiful soft woods." Over the next few years, Julian would sink a reported $100,000 into his estate, adding nine new rooms, hidden pools and waterfalls, a heart-shaped swimming pool, a gold-lined bathtub, a large library of rare books, and "an Ali-Baba like array of rare rugs, furniture, tapestries, art objects and paintings." Julian had attained the wealth he had pursued throughout his Canadian-American odyssey. But success had emboldened his dreams and grander visions loomed before him.

# 4 THE DIVIDEND PAYER OF THE AGE

## I

The 1920s oil boom offers a compact, accelerated demonstration of the free market at work. Fears of petroleum shortages after World War I had driven oil rates to almost $3 a barrel. High prices precipitated expanded exploration and technological innovation. By 1923, these efforts had proven successful, not only in southern California, but in other parts of the world, creating a sudden and unexpected glut of oil. Supply quickly outstripped both demand and the existing capacity of pipelines and storage tanks. The price of high gravity crude oil plummeted from almost $2.50 a barrel in 1922 to as low as .76¢ in October 1923. In addition, the fevered drilling of the southern California oil boom led to massive waste. "Town lot" and offset drilling drained oil fields that should have flourished for decades in a matter of years. Overproduction replaced fuel shortages as the greatest threat to the oil industry.

The new crisis pitted large corporations against promoters and small producers. Major companies and their industry experts laid the blame for the surplus at the doorsteps of greedy landholders who included rapid drilling clauses in their leases and "town lot" promoters who encouraged wasteful practices with large bonuses and royalties. In an attempt to stabilize the industry, they counseled limiting production to sustain yields, eliminate waste, and maintain prices.

The small independents, however, while hurt by declining prices and recognizing the long-term wisdom of this approach, lacked the resources to survive a cutback. Standard, Union, and the other giants had vast tank farms to store crude oil while awaiting higher prices. Smaller companies had to vend their oil as it rose from the ground, accepting the going rate. In addition, attempts to "pinch back" wells and limit their flow often permanently damaged production, an outcome that large corporations could absorb, but that was fatal to the firm with only one, two, or even ten wells.

Thus, oil promoters resisted the efforts of industry leaders to coordinate production, denouncing them as ploys to monopolize the market. "There is a lot of propaganda in the oil business," admitted the conservative *Los Angeles Times* in September 1922, as fears of a serious overproduction crisis gained credence, "and it isn't always confined to the stock, lease, and unit promoters. . . . The near panic induced by the recent agitation may be slightly overdone." A visit to the oil fields, argued the *Times*, would dispel any notion that the industry feared overproduction, since the large companies were "doing more to increase their production, any one of them, than a half dozen of the little fellows."

C. C. Julian, whose ads gave him a broad forum, entered the fray even before striking oil. Designating himself the spokesman for the small producers in the crusade against Standard Oil, he repeatedly belittled the idea of impending crisis. "Big Boy Don't Kid Me," he responded on August 2. "HOW DO THEY GET THAT WAY ON THIS OVERPRODUCTION GUSH?" he asked on September 1, reproducing articles about the demand for oil in Japan and South America. As he later explained, "Naturally the 'Big Boys' don't like our competition in this 'world market' nor do they like such 'birds' as me paying legitimate prices for leases on their property."

As the frenzy at Santa Fe Springs, Huntington Beach, and Signal Hill showed no signs of abating and prices continued to fall, the advocates of "pro-rationing," a voluntary curtailment of production, gained strength. In this ongoing struggle, the large companies held one trump card. They alone controlled the refineries necessary to convert crude oil into marketable gasoline. By threatening to refuse to accept excess production, they could force the smaller oil men to cut back.

The pro-rationing campaign reached a climax in the spring of 1923 as oil prices plunged below the dollar-a-barrel mark. On April 27, Julian joined 400 oil producers "from the big fish to the small fry" at a meeting to resolve the issue. Industry leaders introduced a resolution calling for all producers and marketers to effect a system of pro-rationing by reducing the flow from all wells by a minimum of 30 percent. They sought unanimous approval, but one lone voice dissented. C. C. Julian demanded a vote, and then, standing atop his table, cast a loud, lonely "No."

Julian defended his opposition on both idealistic and practical grounds. Pro-rationing, he argued, injured the public by short-circuiting the laws of supply and demand and helped only large companies. While the "big boys could honeyfugle the little fellows into this course on the grounds of patriotism and good business," charged Julian, large quantities of oil continued to flow into their storage tanks. From a purely selfish standpoint, he admitted, the "pinch back" could not have come at a worse time. If he were unable to market the increasing flow of oil from his wells, he would be driven out of business.

One day after the industry-wide meeting, General Petroleum, which refined Julian's oil, informed him that they would take only 70 percent of his production. These events, claimed Julian, drove home his ultimate dependence on the oil barons. The great fortunes in oil, like those of John D. Rockefeller, had been made, not in production, but in refining. "The refining of crude petroleum is the balance wheel and stabilizer of the oil industry from an earnings standpoint," Julian later explained. "When production increases . . . the price of crude automatically decreases while the prices of refined products vary but little. . . . The loss of the producer is the gain of the refiner." To break the stranglehold of Standard Oil, General Petroleum, and their ilk, he had to create his own vertically integrated company, which would provide all operations, from production, to pipelines, to storage, to refinery, to distribution. In this burst of inspiration, the Julian Petroleum Corporation was born.

Julian's recollection of these events, as always, embellished the truth. The "pinch back" decision undoubtedly accelerated his plans, but Julian had long incubated the idea for his own corporation. C. C. epitomized the small producers described by his contemporary Samuel Blythe, who, "having achieved one or two or three maybe good producing profitable wells . . . get delusions of grandeur. . . . They see themselves putting the Standard out of business." As early as March 30, 1923, Julian had informed his followers, "My great ambition in life today is to build an oil corporation second to none in our "**Golden State.**"

On May 15, 1923, less than three weeks after the oil industry meeting, Julian incorporated Julian Petroleum. The corporate charter, isssued in Delaware, called for a capitalization of $10 million, including 200,000 shares of preferred stock, valued at $50 each, and an additional 200,000 shares of common stock without nominal value. On May 18, he hinted to his readers, "I know I can make six to eight dollars a barrel . . . and all it takes to accomplish this result is a '**Real Wad of Money.**' '**You have that**' and 'You'll tell 'em Jamaica, I've got the ginger to carry it all through.'"

Two days later, on May 20, Julian unveiled his grand scheme:

**REFINERIES**
**How Does This**
**Lineup Appeal**
**To You?**

1—A settled production of from 30,000 to 40,000 barrels a day, entirely owned by ourselves, and with no possibility of ever having to step out in the market and buy our production in competition with the "**BIG BOYS.**"

2—A tank farm of over 100 acres at Santa Fe Springs . . .

3—Pipe lines from each of my properties direct to my steel storage tank farm.

4—Wharfage facilities at the Harbor, where our tankers can steam in under their own power and load out for points on the "**Atlantic Coast,**" "**the Orient,**" "**South America,**" "**Great Britain,**" "**Canada**" and "**France.**"

5—A refinery site adjoining the Harbor and right in beside all the plants of the "**REAL BIG BOYS.**"

. . . .

8—A Corporation with no **"high and mighty"** board of directors always waiting like a **"flock of vultures"** to swoop down and take the **"cream"** while you get the **"crumbs that fall off the table."**

The following month Julian embarked on his customary whirlwind preparations. "My policy has always been to be well under way with my program before I invite you to participate," he advised. "**Do You Hear Me Calling You to the Tune of Five Million Dollars?**" he asked on May 23. "I'm here to tell you I'm sure spending it like a **'drunken sailor'** these days." On June 17, Julian reported astonishing progress: the purchase of an eighty-acre tank farm site; the acquisition of "what I consider to be the finest refinery site of 100 acres in Southern California"; and over $1.5 million worth of material on order for immediate delivery. To expedite matters, Julian advanced his personal funds to the corporation.

Simultaneously he primed his investors with promises of extravagant profits. " **'One lonely share'** of Henry Ford's original stock, which sold for $100, is worth over $47,000 today," he wrote. "'**One lonely share'** of Atlantic Refining Company, which sold for $100 paid over 900% in one year. '**One lonely share'** of Standard Oil Co., which sold for a few dollars, is today worth a fortune." Standard, he noted, made 87 percent of its profits from refining and marketing, "and they make **some** profits too." In response to alleged whispers that the large oil companies viewed him "**as a menace that should be rightfully stamped out,** and were going to 'clip my wings,'" Julian predicted, "**It can't be did**. . . . I'm losing a lot of sleep about that angle, '**in a Pig's Nightie**,' I am."

According to Julian, an "avalanche" of letters, telegrams, telephone calls, and visitors descended on his office requesting reservations for stock in Julian Petroleum. By May 31, he reported, prospective investors had pledged nearly a million dollars in amounts ranging from $100 to $50,000. "Were I to live to be thousand years old," he enthused, "and accumulated more money than Henry Ford, 'I will never again get the thrill that 'your thousands of messages of encouragement gave me.'"

On June 25, Julian proclaimed the birth of "**THE DIVIDEND PAYER OF THE AGE**," and outlined his coming offer. While the ultimate capitalization of the project would be $10 million, the initial subscription would be for half that amount. One hundred thousand shares of preferred stock would be placed on the market for $50 per share. Each two shares of preferred stock purchased would include a bonus of one share of common stock. With the preferred, investors would receive an annual 8 percent dividend. Earnings above this would be divided among common stockholders three times a year. The common stock held the voting power in the company. Although Julian conveniently omitted it from his advertising, the corporate charter also rewarded Julian with one share of common stock for every two shares of preferred sold, guaranteeing him control of the company.

During the last week of June, Julian No. 3, the last of his original syndicate wells, began producing from the Meyer sands. Julian gleefully dispatched over

$150,000 worth of dividend checks. With all components of Julian Petroleum now in place, only one obstacle stood between Julian and the start of his greatest stock-selling campaign: the Corporation Commissioner of California, "Iron Mike" Daugherty.

## II

On the surface, Edward Michael Daugherty and C. C. Julian had much in common. Like Julian, Mike Daugherty was in his late thirties and father to two children. A tall, muscular man, he shared with the oil promoter a reputation as "a most likable man . . . the kind of hail-fellow-well-met, who inspires confidence and amiability." He arrived in California from his native Illinois as a young man and plied a number of trades, first as a newsman with the *Los Angeles Times*, then a house builder, and later a publicity man. Active in Republican politics, Daugherty worked in George Cryer's 1921 mayoralty campaign. A group of Los Angeles Republicans and businessmen prevailed upon Governor William Stephens to place Daugherty in charge of the state Corporations Department. In 1922, just as Julian began drilling his first Santa Fe Springs well, Daugherty assumed the post of corporation commissioner.

The department Daugherty inherited represented one of the great experiments of the Progressive Era. Since the turn of the century, public officials and businessmen had sought a means to eliminate fraudulent promoters who promised their clients everything, including the "blue sky" above. Progressive politicians called for a regulatory agency to protect small investors. They received support from a wide array of business groups that feared that speculation in illegitimate endeavors would injure the reputation of all enterprises raising capital from the public. The campaign for "blue sky" laws reached national proportions. California adopted the Investment Companies Act in 1913, one of thirteen states to enact this type of legislation that year.

California's blue sky law required that all corporations selling securities in the state procure a permit from the newly created Corporations Department. The act gave broad discretionary powers to a corporation commissioner, appointed by the governor, who could deny or revoke permits from the enterprises he deemed "unfair, unjust, or inequitable," or whose methods would "work a fraud on the public." The commissioner could attach conditions to the granting of a permit and had the "power to establish such rules and regulations as may be reasonable to carry out" his mandate.

While most businessmen praised the legislation, its revolutionary nature generated fierce opposition from the start. Many real estate and investment brokers, both legitimate and otherwise, feared they would be forced out of business. Others raised philosophical objections to the creation of a regulatory agency that intervened in the free market and put arbitrary authority in the hands of a single unelected official. A 1914 referendum to repeal the blue sky law failed when the Los Angeles Investment Corporation, the leading advocate of repeal, collapsed shortly before the election amid revelations of widespread fraud. The scandal convinced voters of the need for a regulatory agency.

In 1917, the legislature passed the Corporate Securities Act, strengthening the Corporations Department and giving California the strictest blue sky law in the nation. By 1918, Corporation Commissioner H. L. Carnahan boasted that corporate financing had achieved a high "degree of sanity and stability."

Nonetheless, upon assuming office in 1922, Daugherty discovered that the oil boom had reintroduced insanity and instability at a previously undreamed-of level. Blue sky laws always reflect the inherent conflict between the need to curb fraud and the avoidance of unwarranted interference in honest business activities. Conditions in the oil industry, highly speculative at best, aggravated this dilemma. Fraud ran rampant through the oil fields. Hundreds of companies seeking permits had swamped the meager resources of the department. Many promoters misrepresented their permits as endorsements or certifications of probity. Others bypassed the agency entirely, claiming their sales fell outside its jurisdiction. Daugherty charged that these "outlaw" or "coyote" companies, operating without a permit, were often "hopelessly overcapitalized."

The critical distinction between legitimate oil man and stock swindler often lay in the level of capitalization—the amount of money a well had to produce before returning a profit to its investors. Unit-holders, like those in Julian's syndicates, bought not $\frac{1}{1750}$th of total oil production, but that fraction of the oil after promoters deducted bonuses, royalties, drilling, sales, and office outlays before paying dividends. For example, a well leased for a 50 percent royalty and capitalized over $200,000 an acre held scant hope of returning a profit to individual investors receiving $\frac{1}{1750}$th of the remaining production.

While most promoters retained a percentage of the oil for themselves, arguing that this guaranteed they would profit only when their clients did, they relied primarily on sales commissions for their income. Daugherty discovered that many firms "pocketed enormous profits as selling costs," running as high as ninety cents on every dollar invested. Lavish promotional outlays, including advertisements, buses, free lunches, and healthy sales commissions, meant "the money of one sucker was being used largely to catch another." According to Daugherty, these enterprises "were purely promotional schemes which dissipate the funds of the investing public." The *Record* reported capitalization in the Santa Fe Springs field running as high as $1 million an acre, a figure well beyond the potential return of all but the greatest wells.

In these instances, a dry hole could be preferable to a gusher. "One of the mangiest of the fakes," explained Samuel Blythe, "was to sell $200,000 worth of units in a well that cost $150,000 to drill whereby the promoters made $50,000 whether they struck or not." These wells might be found on the outskirts of the field, well beyond proven territory. Santa Fe Springs mayor William Emmens recalled that several promoters "sold more percents than they had. They could not afford to bring the well in so they muddied them and just let them go."

Not even a gusher, however, guaranteed profits. Dividend amounts were often paltry and failed to match the initial investment. Although promoters predicted four- and five-thousand barrel wells and promised long-term income, flush production typically lasted only a few days or weeks at most before leveling off. The greatest yield occurred during the first year, after which the flow ceased or fell to a relative trickle. Thus, while hundreds of wells produced oil and some syndicates returned large profits to their investors, state officials estimated that two-thirds to three-quarters of the money deposited in promotional oil ventures would be lost.

In the summer of 1922, Commissioner Daugherty declared war on the coyote operators and unleashed a new "field investigating division." "All oil companies selling securities in any form without a permit," warned Daugherty, "will be considered as resorting to subterfuge and will be required to submit themselves to supervision." The wily unit-selling coyotes, however, were not so easily trapped. For one thing, Daugherty's heralded "division" had only two overworked operatives. Furthermore, most promoters had organized common law trusts, which they contended exempted them from the Corporate Securities Act. Units, they claimed, represented "interests" rather than stock, removing them from the jurisdiction of the statute. In October 1922, the State Supreme Court ruled in favor of the corporation commissioner. The court held that units were securities and therefore fell within the purview of the blue sky law.

This decision failed to resolve the issue. In March 1923, a federal district court dealt Daugherty a sharp setback. The judge ruled that a promoter who had sold units as an individual, not as a trustee of his syndicate, was not subject to the Corporate Securities Act. This ruling, protested the *Record*, "handcuff[ed]" Daugherty and sent a message to promoters to: "Go to it! The corporation commissioner is powerless to touch you! Broadcast your lurid advertisements. Entice 'em into your free buses . . . fill 'em up with free dinners and rob 'em right and left." Confronted by a loophole that threatened his authority, Daugherty turned to the legislature, which quickly amended the Corporate Securities Act to specifically include bonds, mining stocks, and oil units sold by individuals.

Daugherty faced yet another obstacle. The new governor, Friend Richardson, a fiscal conservative hostile to Progressive reform, had launched a round of spending cuts. While he reappointed Daugherty and endorsed his efforts, he slashed the already inadequate Corporations Department budget. Daugherty put on a bold face. "As long as I am in office," he declared, "there will be no letting down. . . . There is plenty of money left us to keep an eye on the oil sharks."

On April 20, Daugherty reopened his oil campaign. "The reduction in the price of oil makes the condition of these unit holders even more precarious than before," proclaimed Daugherty. "Drastic action is warranted on our part. . . . The fellows we will try to curb are those that use the public's money." With his funding curtailed, however, Daugherty needed a prominent example to bolster his authority. Increasingly, his attention came to focus on the promoter who had gathered more of "the public's money" than any other. In the vast menagerie of honest oil men, promoters, and swindlers, Daugherty wondered, echoing the question asked by many, what breed was C. C. Julian?

The answer proved elusive. Julian had undeniably mastered the tricks of his trade. His ads promised "the cleanest offer[s] ever tendered the public," but, on close examination, they obscured as much as they revealed. His repeated predictions of 100 percent returns each month and 30 to 1 for each unit fell well beyond the range of even the greatest wells at peak prices. "I am riding dollar for dollar right down the line for you," he asserted, "and until this well is flowing into the tanks I will never pull down a solitary 'thin' dime." however, Julian, like all promoters, had assured his profits once he had fully subscribed the issue. He had capitalized his first five wells for $175,000 each, for a total of $875,000. Julian claimed that this reflected the actual drilling and operating costs. But this figure, while arguably correct, gave him a comfortable margin of error. His total expenses on these five wells, including drilling, operations, advertising, and the payment of loans and lease bonuses totaled less than $700,000, leaving Julian with over a $175,000 surplus.

In addition, Julian had secretly subdivided his personal oil interests in the first three wells and sold them as "Special" units, bringing him an additional $162,500 in clear profit. On these wells Julian would realize no additional gain by striking oil. Thus, Julian received a guaranteed $312,500 profit, even without sales commissions, long before any oil flowed from his wells. These commissions (for the oil fields, a modest ten percent) amounted to $103,000. Much of this accrued to Julian. Julian also made money by buying and reselling his units.

Julian's next two syndicates brought him additional profits. His $1,000,000 package on wells 6 through 9 produced a $400,000 surplus, without commissions or oil. On his final three-well offer, high drilling costs ultimately reduced his guaranteed earnings to a still-respectable $175,000. Thus, in less than a year, Julian had garnered a minimum income of $875,000.

Although Julian had undeniably played loose with the truth and pocketed large profits, had he actually violated the Corporate Securities Act? Julian had operated without a permit in his initial venture, but the legal position of unit sellers was still in doubt at that time. On all subsequent offers, he acquired the necessary permit. Since Julian frequently "called the roll" and offered full repayment to dissatisfied customers, the Corporations Department had received few complaints against him.

More important, had Julian committed the cardinal sin by overcapitalizing his syndicates? "For every $100 you invest you will receive $1/1750$th part of 70% of my total net production," Julian had advertised for wells 1, 2, and 3. The language implied that the investor's share would be based on the total flow of oil after the landowner had received his 30 percent royalty. Actually, unit-holders were to divide 70 percent of the remaining 70 percent of production, or only 49 percent of the value of the oil. Twenty -one percent was reserved for Julian and his original backers. Nonetheless, Julian had definitely not overcapitalized. His total of $525,000 for five acres fell far below the $200,000-per-acre limit accepted by Corporation Commissioner Daugherty. Wells No. 4 and 5, while farther from proven territory, offered an even better proposition. Investors would divide 60 percent of the total production. By the end of May 1923, Well No. 1 had already paid back 87 percent of its investment, while No. 2 and the 4-5 syndicate had returned 32 and 44 percent, respectively.

The Julian syndicates for wells no. 6 through 9 and 11, 12, and Pico present a less clear verdict. The 50 percent royalties and $80,000 and $250,000 bonuses on the more promising tracts strained the prescribed $200,000-per-acre limit. Furthermore, Julian divided his offer into 2,500 rather than 1,750 shares per well, capitalizing each at $250,000. Nonetheless, given the acreage involved, Julian had not exceeded the boundaries delineated by the corporation commissioner.

In early May 1923, Julian and his attorneys held a preliminary meeting with a deputy corporation commissioner and state engineers to outline the proposed Julian Petroleum Corporation. According to Julian, the government representatives approved his plan as "fair, just, and equitable," and encouraged him to proceed. They promised to issue the company a permit to sell $5 million worth of stock as soon as incorporation was complete. In late May, however, when a formal permit request reached Daugherty's desk, it encountered a far less hospitable reception.

Recent developments had reinforced Daugherty's doubts about Julian. In late May, falling gas prices had prompted thirty-five companies to file plans for publicly funded refineries. This meant almost $100 million dollars in stock and a doubling of the state's refining capacity. Daugherty sensed a new oil scam afoot. Most of the plans, including Julian's, allowed up to 20 percent in selling expenses and issued a majority of the voting stock to the promoter. Under these conditions, the refineries might be overcapitalized. "That's a heavy burden," warned Daugherty, "even if the load is efficiently managed."

On June 1, Daugherty announced that the processing of all applications would be delayed pending thorough investigation. The commissioner stipulated that applicants must show sufficient supplies of oil, enough capital to finance their projects, efficient refining processes, and long-term contracts for output before permits would be issued. For Julian, with materials arriving daily and payments due immediately, Daugherty's decision proved devastating. "Ruin was just around the corner," he later lamented. "It looked as if my first bright idea of ordering materials in advance was to be my downfall and business burial."

Julian unsuccessfully sought an audience with Daugherty. On June 15, after two weeks of frustration, Julian composed a long letter explaining his plight. "Your Department has backed me into a corner and I am fighting for my very existence at this moment," wrote Julian. "I am only requesting that you . . . extend to me what I consider I am entitled to, a white man's break." Despite financial obligations "piling up on my shoulders," Julian refused to accept the $50,000 to $75,000 a day offered by his investors. "I have already spent every dollar I possess in the world to further this enterprise," he protested, "all this without a dollar from the people I expect to finance this project." Julian specifically outlined his preparations and expenditures, warning, "The material is rolling in here now by both boat and rail with hundreds of thousands in payments on same falling due almost immediately." Without additional funds he would also lose his options on the wharfage and refinery sites. "I am at the end of my rope," he told Daugherty.

Julian's plea went unanswered. In desperation, he later recounted, he turned to Harry Chandler, the publisher of the *Los Angeles Times*. Chandler, Daugherty's former employer and the most powerful figure in southern California, had readily accepted thousands of dollars' worth of Julian's advertising. Julian later claimed that Chandler promised to help and even met with Daugherty. But the following morning Daugherty called Julian's attorneys and, according to Julian, vowed that if the promoter lived to be ninety, he would never get another permit from the corporation department. Daugherty finally agreed to see Julian, but when asked when the permit might be issued, he replied, said Julian, "It might be a week, a month, a year or ten years."

Legally, Julian's sole remedy for a perceived abuse of power by the commissioner was to file a writ of mandate in court, compelling Daugherty to show cause why a permit should not be issued. This procedure, however, would take weeks, exacerbating Julian's financial straits. Hemmed in, Julian wondered if he were "Houdini enough to figure a way out of the hopeless jam." The escape artistry ultimately came from his attorney, Dean Goodwin. Goodwin, a former Corporations Department deputy, suggested that despite recent changes in the law, Julian Petroleum could sell the stock to Julian himself outside of California and Julian could then resell it as his personal stock without obtaining a permit.

On June 22, Julian and the four other members of the Julian Petroleum Board of Directors boarded a train to Las Vegas. The directors approved the sale of 100,000 shares of preferred stock and 50,000 shares of common stock to Julian for the sum of $5 million as a "preparedness move for use only as a last resort." Julian would pay $200,000 down (toward which they credited the monies he had already advanced) and $200,000 every fifteen days until repayment was complete. The board then returned to Los Angeles.

Julian waited five more days in hopes of obtaining a permit. Finally, on June 28, his ads announced:

### "FOLKS"
#### "I'm Calling on You"

Yes, and I'm calling like no human being ever called before, and I've spent a goodly gob of cash to make my refining project one hundred percent legitimate before I invited you to participate. . . .

I have used all my own money up to date, to accomplish in ninety days time, what I fully believe would take the average company a year to accomplish. . . .

I've made you **"a thousand sacred promises"** during the past year and I'm calling on the world to ask, have I ever violated one, even in the minutest detail?

. . . .

I feel you believe me when I stand up and tell you on my **"word of honor"** that I would sooner be on the corner of Seventh and Broadway in overalls, with all my pockets turned inside out, and without a dime in the world to eat on, **"than to stray from the straight and narrow path"**

. . . .

Folks, time is the essence of this deal and your dollars mean more to me today than I can begin to tell you, so if you believe in me, **"my great urgent appeal"** is **"show me now"** and I'll **"die like a dog"** in a ditch before I will ever give you any cause to say I betrayed your confidence.

That day, working from his own offices without the assistance of outside brokers, Julian sold over $200,000 worth of stock.

Commissioner Daugherty responded swiftly and decisively. "California investors are warned that the stock offered in [Julian Petroleum] . . . may be determined by the courts to be invalid," Daugherty proclaimed in a formal statement on June 28. "This Julian Petroleum Corporation has no permit to sell stock in the State of California. . . . It is therefore the request of the Corporation Department that all persons . . . withhold or rescind their subscriptions until the legality of said stock is determined."

Daugherty charged that the company had failed to answer questions regarding its supplies of oil, transportation facilities, refining processes, and distribution. Julian's attorneys, he stated, offered "generalities where specific answers could have been made." Daugherty also alleged that Julian had attempted to "bludgeon a permit out of the department." An unnamed friend of the promoter's had called Daugherty and advised him to compromise with Julian. If he refused, the caller threatened, Julian would try to ruin Daugherty with rumors that he had acted at the bidding of the oil trust.

On June 29, Daugherty dispatched a deputy commissioner to Julian Petroleum offices to confiscate all books and records. After a telephone call, presumably to Julian, company vice-president Jack Roth refused to release the books until Julian returned that afternoon. Daugherty immediately suspended Julian's broker's license and asked the district attorney for a warrant to seize the books and arrest Julian. Spring Street brokerage houses instantly felt the effects of these actions. Larger-than-usual crowds of spectators and speculators appeared downtown as a brisk trade ensued in both Julian Petroleum stock and Julian's syndicate units.

Los Angeles newspapers refused to carry Julian's June 29 ad soliciting investors, but the following day they allowed him to publish a Rudyard Kipling poem:

> If you can hold your head when all about you
>      Are losing theirs, and blaming it on you;
> If you can trust yourself when all men doubt you,
>      But make allowance for their doubting too;
> If you can wait and not be tired by waiting,
>      Or being lied about don't deal in lies,
> Or being hated don't give way to hating,
>      And yet don't look too good, nor talk too wise;
>
>                         . . . .
>
> Yours is the earth and everything that's in it,
>      And — what is more — you'll be a Man, my son.

"Mr. Daugherty appears to have the floor just now," concluded Julian. "'RIGHT IS RIGHT' and 'WRONG IS WRONG' I'll stand or fall on your verdict."

That morning Julian, dressed in a "natty Palm Beach suit," appeared at the district attorney's office. Informed that a complaint was being prepared, he waited to surrender himself and held an impromptu press conference. Asked if he would plead guilty, Julian replied, "Not very readily." He warned that he had a lot of high-caliber ammunition to fire. "It will be '75's," he noted laconically.

The show moved to the courtroom. Bail was set at $7,000, $3,000 in cash. "We can furnish $3,000 cash, can't we Mr. Julian?" asked his attorney. "I think so," responded Julian. He then, reported the *Times*, "dipped his neatly manicured fingers into his breast pocket and extracted a fat roll of currency. He leisurely peeled three crisp $1000 bills and tossed them to the clerk."

Outside the court Julian accused Daugherty of a "grandstand play." "I am at a loss to understand why he has singled me out and accorded me the treatment that has been my portion," exclaimed Julian. "I do not yet know what it is all about or wherein I have committed an offense." Julian denied any attempt to defy the commissioner. "As far as my books are concerned," he stated, "they are not only open to the state corporation commissioner, but they are open to inspection by the whole world at a moment's notice."

At his offices, Julian found three deputy sheriffs guarding the door. A delegation from the Corporations Department greeted him within. After a lengthy conference, the officials confiscated the records and books and removed them for investigation. "Come again—more often," quipped Julian as they left.

On Sunday, July 1, Julian presented his case in full page ads in his favorite medium.

"Friends, Foes and Countrymen"
"Lend Me Your Ears"
"I'm Here to Criticize, Not to Praise"
"The Corporation Department of
This Golden State"

The state agency, charged Julian, "has endeavored to perpetrate on me, Folks, one of the most severe injustices that was ever dealt to a citizen of California." His permit, he alleged, had been "written up a few days after my application went in and placed on [Daugherty's] desk for [his] approval." Why, asked Julian, "was it not signed after being prepared by your deputies?" The "most extreme tactics imaginable" had been used to discredit him. "Why," he wondered, "should the State Corporation Department try to retard one of the very few operators that has come through with flying colors?" Julian reproduced his lengthy June 15 letter to Daugherty, so the people could **"Be My Judge and Jury."** "Am I playing the game like a man?" he challenged.

Julian's plea struck a responsive chord. Supporting letters and telegrams flooded his offices. In Sacramento, Governor Richardson reported, "I am receiving a barrage of telegrams regarding one C. C. Julian of Los Angeles. . . . I know nothing about this matter." Richardson expressed full confidence in Daugherty and refused to review his decision.

Julian ceased his sales and adopted a conciliatory stance. He submitted additional data to the Corporations Department. "Patience Is a Virtue," he counseled his public. On July 5, his ad acknowledged that the large number of applications for refining projects had delayed procedures. Four days later, he reported, "The whole world looks bright and cheerful. Mr. Daugherty has four of his lieutenants working on my permit and unless someone drops a monkey wrench in the machinery, I say we will soon get some action."

The permit process, however, continued to drag on. The financial respite provided by the one-day stock sale soon ended. Checks totaling over $200,000 had arrived instructing Julian to "use them as [he] saw fit," but the commissioner's edict had rendered them worthless. "Hundreds of the kindest letters imaginable" from supporters arrived each day. In them, Julian discovered "the stroke that saved the bacon . . . and turned on the light when things were darkest."

Several correspondents asked Julian why he didn't raise funds not by selling stock, but by borrowing against his personal notes. On July 10, Julian began quietly accepting money in exchange for his personal note at 10 percent interest. But Julian's ads made no mention of the borrowing campaign until July 17, when he wrote cryptically "I threw my arm out signing personal notes for subscriptions from people that still believe 'I will' . . . build one of the greatest refineries in the State of California. It all goes to prove there are more ways to kill a cat than choking it to death with hot butter."

On July 22, Julian unveiled his campaign:

**"One Million
Dollars in Cash"
"IN 12 DAYS"**

That's the amount that has been advanced to me on my personal note up to last evening . . .

Owing to my not having received any action on my application for a permit to offer my refinery security for sale . . . I have resorted to the only legal means I know to carry this project of ours through to a successful conclusion. . . .

**"I Need 4 Million
Dollars More"**

But all I can offer you until my permit has been issued is my personal note, for any amount you feel inclined to advance me. This note will bear interest at the rate of 10% per annum . . .

My note is backed by every possible asset I possess, and on top of all this for each $100 you advance me I will put up as further security two shares of Preferred and one share of Common, which I now personally own, or will legally purchase in the Julian Petroleum Corporation.

Bolstered by the newspaper ads, money arrived at the rate of $100,000 a day. **"Time and Tide and C. C. Julian Wait for No Man,"** he admonished hesitant investors. With a ready cash flow, Julian finally completed the purchase of his refinery site and stepped up work on the tank farm. This bold action enhanced Julian's reputation as the champion of the bureaucratically oppressed. A $1,000 contest to name his new gasoline attracted 17,000 entries, and Julian grandly adopted the label he claimed was suggested by 30 percent of the respondents— "Defiance Gasoline."

In one swift master stroke, Julian had outflanked the Corporations Department. If Daugherty refused to issue a permit, Julian would raise his entire funding on personal notes. Within days of his July 22 ad Julian and the corporation department began negotiations to end their stalemate. Rumors flowed that the permit would be issued only if Julian severed all connections with the company. "When I resign, you will see grass growing on Main Street," boasted Julian.

On July 30, the Corporations Department ended two months of contention and issued a permit to Julian Petroleum authorizing the sale of $5 million worth of stock. The action represented a compromise. All stock certificates would bear the message "THIS ISSUE IS HIGHLY SPECULATIVE," and all funds would be placed in escrow until released by written order of the commissioner. Julian's personal broker's permit remained suspended, and criminal charges filed against him remained pending. Four days later, in compliance with another provision of the agreement, Julian resigned as president of Julian Petroleum. O. S. Witherell, a longtime associate, replaced him. But Julian still received one share of common stock for each one sold, effectively retaining control of the company.

Press accounts completely ignored Julian's resignation, and C. C. continued to personify his namesake company, which the public now affectionately dubbed "Julian Pete." To the people of Los Angeles there was little doubt who had won round one in the Julian-Daugherty conflict.

### III

"'**Come On, Folks' 'Let's Go!**'" crowed a gleeful C. C. Julian, on August 1, 1923, permit in hand. "'I'm a bear' on making hay while the sun shines, and it's sure pretty weather now."

"**What More Could You Ask?**'" he queried in his new sales campaign, "'**Than to positively know**' that you are investing in an enterprise that is generally conceded to be one of the greatest money makers of the age, and to know on top of that, **the man behind the gun** knows the oil industry from A to Z." The oil business, Julian assured investors, "can be carried on as safely as the grocery business if one has learned how." Julian Petroleum, backed by thousands of investors, possessed a built-in safeguard. "With 40,000 partners to buy my gasoline and motor oil," he predicted, the company would have "some ready market," for his projected sixty service stations selling Defiance gasoline.

"Did you ever hear of a Refining Company in the United States with ten million dollars to carry on their business going broke?" wondered Julian. "I never have." General Petroleum and Pan-American Oil had raised millions to build new refineries. "Who puts up the money for them to do all this? 'The smartest Money sharks off Wall Street.'" Julian predicted, that "in 24 months . . . when you talk of the

'Union Oil Company,' 'General Petroleum Corporation,' and 'Associated Oil Company,' you will talk of the Julian Petroleum Corporation in the same breath."

Throughout August and September, orders for Julian Pete stock poured in at a rate of $50,000 to $75,000 a day. "Folks, I'm busier these days than a one-eyed man looking through a knot-hole at the Ziegfeld Follies," he protested. "HOLY SMOKE," he reported on September 18, "HOW YOU STAMPEDED ME YES-TERDAY." Two days later Julian began his final countdown. "Crack Your Whip Now," he admonished. The stock sale ended at midnight, October 4. In fifty-six days, Julian had sold 100,000 shares of preferred stock for $5 million. Whatever reservations Commissioner Daugherty and others might have held about Julian and his venture, tens of thousands of investors had accepted his gospel.

The stock sales completed, Julian Petroleum sprang into high gear. Hundreds of employees laid miles of pipelines, constructed thousands of gallons of storage facilities, and prepared the refinery and wharfage sites. Julian negotiated oil leases in both new and old fields, and drilling began on several wells. Corporate officials laid the groundwork for a vast chain of gas stations, procuring $2,000,000 worth of gasoline from General Petroleum to tide the company over until its own refinery was completed. An agreement to purchase more than 2,000 gas pumps over five years demonstrated the ambitious scope of future plans. Ads promised the first fifteen service stations would be dispensing Defiance gasoline by late November.

But even as Julian envisioned the day when his efforts would make him "to the masses of the people the best loved man in the state," storm clouds were gathering. "'**I Wonder Who's Framing Me Now?**'" asked Julian on October 25. "Folks, there's sure something in the air again . . . but I'm curious to know what the new line of attack will be." Julian claimed that he had received "mysterious phone calls" and "anonymous letters" warning of a plot against him. "I never leave my office night or day that I'm not followed every place I go," he alleged. But, Julian warned, "If they come at me too strong, they better come shooting, because I'm not so slow on the trigger myself."

Julian might have been responding to rumors of a federal investigation, or he may have been bracing his followers for two upcoming hearings stemming from his earlier confrontation with Daugherty. On November 1, a judge heard testimony to determine whether Julian had violated the Corporate Securities Act in June when he sold personally owned stock. The following day, Daugherty convened a second, nonjudicial proceeding concerning the revocation of Julian's broker's permit.

Daugherty's hearing had an inquisitional atmosphere. Corporations Department deputies presented the evidence while Daugherty presided and would render the verdict. For three days, state officials, aided and abetted by Daugherty from the bench, attempted to prove that Julian had misled the public and collected excessive profits.

At the opening session an immaculately dressed Julian sat amidst three attorneys and gazed "moodily" with the "air of a 'hurt man.'" A deputy corporation commissioner testified that Julian had invested little of his own money in his syndicates. The chief engineer contested Julian's advertised claim that he was the "largest independent oil producer in the state." (Julian's attorneys responded somewhat ingenuously that this referred to the number of stockholders, not the volume of oil.) A state accountant challenged Julian's assertion that a $100 investment in Union Oil had earned $40,000, but admitted that an accurate profit figure was not available. None of these points seemed particularly damaging.

On November 5, the tempo accelerated. The Corporations Department went to great lengths to prove that the fleet of tankers loaded with oil and bound for the Orient, which Julian had depicted in his ads, did not exist, making several "unkind insinuations" about the phantom fleet and calling representatives from other oil companies to testify that no such shipments had occurred. Julian's attorney objected. The advertisement, he argued, did not describe an actual shipment, "but a mythical conversation on the street," relating to future corporate activities. "I don't see anything mythical about the ad," concluded a humorless Daugherty ruling against the defense.

Julian buried his face in a newspaper, pretending not to hear or care. Suddenly, late in the morning session, he came to life. The oil promoter rose and requested permission to personally examine the witnesses. Daugherty, noting Julian's more than adequate legal representation, denied the request and ordered Julian to be seated. The oil man complied but moments later threatened to withdraw from the hearing if he could not question the witnesses. Daugherty said he could do as he pleased. Julian departed defiantly, ending the morning session.

That afternoon, Daugherty responded to Julian's challenge. He announced that if Julian walked out again, he would be cited for contempt. Julian, however, was not so easily thwarted. His attorneys informed Daugherty that, in order to avail himself of his right to conduct his own defense, Julian would dismiss his counsel. The commissioner relented. Julian, he allowed, could question witnesses. Late in the afternoon, Julian took the floor, cross-examining witnesses, according to the *Times*, "with remarkable vigor and directness." He challenged Corporations Department figures and cost estimates, attempting to demonstrate that he had not raised unreasonable amounts of money for his syndicates.

The following day, another Daugherty aide testified that an advertisement describing R. W. Cowan, whom Julian had lured from Union Oil to manage his refinery, had blatantly misled the public. Cowan, Julian had claimed, had invested $50,000 in Julian Petroleum "on the same basis as you have or will invest yours." The Corporations Department could find no record of this sale on Julian's books. The transaction, protested defense attorneys, had transpired with no money changing hands. If so, countered the prosecution, winning one of its few points, this violated the permit, which allowed only cash exchanges.

The hearing concluded with the appearance of the deputy commissioner who had attempted to seize the Julian Petroleum books in July. The inability to acquire the records, he asserted, "seriously delayed" the functioning of his department. Julian's counsel challenged this assertion. Did three hours constitute an unreasonable delay? they asked. Was the bookkeeping not exceptionally good? The deputy admitted he could find no fault with the accounting.

The outcome of the proceedings had never been in doubt. "Inasmuch as the Corporation Commissioner, some months ago, saw fit to have a complaint issued against me for violation of our Corporate Securities Act," Julian had predicted on November 5, "it would be reasonable to believe that he will undoubtedly feel justified in making the revocation of the broker's license permanent." Daugherty, despite the often trivial content of the prosecution's case, fulfilled this prophecy. Even one month later, when a judge dismissed all criminal charges against Julian, Daugherty refused to reissue the permit.

In the public mind the impression that remained was one of a persecuted, yet gallant, Julian striding forth into battle against arbitrary authority, outwitting and outclassing his opponents once again. Daugherty clearly recognized this perception. On November 7, the day after the hearings ended, he contacted the United States Department of Justice and requested that the agency examine Julian's affairs.

Julian's battles with the corporation commissioner had elicited a groundswell of popular support. Had people paid more attention to the dwindling fortunes of his syndicates, however, they might have tempered their approval. Wells No. 1, 3, and 4 continued heavy production despite the "pinch back." But Julian, unable to find a buyer for additional oil, kept Well No. 2 idle until September. No. 5 did not arrive until November and delivered a far smaller flow than its predecessors.

The 6 through 9 syndicate wells, heralded in July as "monster producers," had all experienced difficulties. On August 26, Julian reported "flowers and soft music at 5000 feet" for Well No. 8, his first dry hole at Santa Fe Springs, forcing him to abandon the adjoining site for No. 9. Disaster also struck Well No. 6. Drilling was completed in September with an initial flow of 1,500 barrels a day. But three weeks later, Julian dolefully reported, "An outfit on an adjoining lease put so much pressure on their cement job that they forced their drilling mud clear over into our well and killed number 6 deader than a doornail." When cleaned out and put back on production, No. 6 yielded only 350 barrels a day, a figure that declined steadily. Its running mate, No. 7, also proved disappointing on delivery.

Julian's final three-well syndicate fared no better. Wells No. 11 and 12, which adjoined his initial Brunson properties, each produced a paltry few hundred gallons a day. The Pico wildcat in the unproved Compton field encountered one difficulty after another. "Has a sick headache, with the stomach pump not doing so well," Julian revealed on October 26. "We may have to order in a coffin."

Those not distracted by the more flamboyant affairs of Julian Petroleum began to question these failures. Julian had capitalized the later syndicates more heavily than the first and rumors of excess sales continued to circulate. Had Julian deliberately scuttled his wells to mask an oversubscription? Had Daugherty's suspicions been justified?

Ample alternative explanations existed to allay the fears of Julian's supporters. The Santa Fe Springs field, crippled by overdrilling and weakened by the "pinch back," had reached its peak in September. A precipitous decline ensued. New wells became fewer and, like Julian's, produced far less. Within a few months *Oil Age* would call Santa Fe Springs "a pleasant memory for the fortunate investors and a nightmare to the unfortunate ones." In Compton, optimism generated by Union Oil's discovery well in August had turned to "considerable anxiety" in December. Like Julian, General Petroleum had failed to strike oil with its wildcat.

Nor did dry holes automatically doom Julian's syndicates. "This is my funeral, not yours," he had promised on the Pico well, "because even if I lose the hole I'll skid the rig and drill the well without a dime of cost to anyone but myself." If Julian could relocate wells 8 and 9 and the Compton wildcat to more productive soils, he might still redeem his promises.

Meanwhile, Julian Pete celebrated the opening of its first gas stations in December. Bright red, white, and green "Julian" signs bedecked the sites. On opening day, brass bands christened each location, playing "For He's a Jolly Good Fellow" as Julian, clad in white overalls, toured the stations dispensing the first tankful of Defiance gasoline at each one. But these festive scenes played against a backdrop of growing crisis. Commissioner Daugherty had again set his sights on Julian and enlisted the powerful *Los Angeles Times* as his ally.

*THE TRUTH*

For C. C. Julian, 1924 began with the subtlety of an avalanche. Within one week he had published his own newspaper, incurred the wrath of the *Los Angeles Times*, sued the corporation commissioner, and dodged three bullets aimed at his head.

The origins of these turbulent events lay in Julian's 1923 summer stock sale. C. C. had not distributed stock certificates, but had provided temporary receipts that could later be redeemed for stock. By late 1923, these receipts flooded Spring Street. Their value fluctuated wildly as if some person or syndicate was conspiring to drive down prices and buy shares at artificially low levels. Was Julian himself manipulating the stock to absorb and cover up an overissue?

In later years Julian readily admitted that he had collected funds well in excess of the $5 million allowed. Since all monies were to be placed in escrow and controlled by Corporation Commissioner Daugherty, Julian explained that he wished to amass a hidden reserve to cover immediate expenditures. He therefore oversubscribed the stock issue by $1.69 million, planning to redeem these funds as part of a second $5 million offering. Julian applied to the Corporations Department for a second permit in September 1923, but when Daugherty only authorized an additional $1.5 million sale and imposed several restrictive conditions, Julian angrily withdrew his application.

Thus Julian found himself in a financial vise. Each time he desired access to his legally acquired funds, he had to petition the corporation commissioner. Two or three weeks might pass before a hearing could be held. Julian claimed that state officials once took twenty-two days to approve the release of $500,000 to purchase property in a new oil field. In the interim his option expired, and the land was sold twice at a $500,000 profit. "The public," complained Julian, "was depending on

[me] to make money for it, and the state was really running [my] business." Julian charged that Daugherty had become the "dictator" of Julian Petroleum.

For several months Julian skirted these restrictions by tapping the oversubscription. But by December 1923, this reserve had dissolved. With 1,100 workers and a $15,000 monthly payroll, Julian had already advanced the company $1 million dollars of his personal funds and mortgaged his home to raise more money. Creditors, acting as "pawns of some powerful opposition," according to Julian, had begun to demand that Julian Pete be placed in receivership. The corporation commissioner nonetheless refused to release the $1.8 million he held in escrow.

Meanwhile, Julian Petroleum had begun to issue stock certificates to replace the temporary receipts, but it still had to resolve the remaining oversubscription. While Julian never admitted manipulating his stock, he apparently took advantage of the sinking prices to buy up the receipts and retire them from circulation. These activities did not escape the notice of Daugherty and his deputies.

To resolve his financial dilemma, Julian again attempted to circumvent Daugherty's authority. At a December 10 meeting of the Julian Pete board of directors, Julian officially reclaimed the company presidency and accepted $3.5 million worth of stock. One and a half million dollars represented money that the corporation owed to Julian. The remainder he hoped to resell to the public as personally owned shares.

Five days later, on December 15, the *Los Angeles Times* unexpectedly refused Julian's daily advertisement. The *Examiner, Herald,* and *Express* quickly followed suit. On the eve of a new stock-selling campaign, Julian had no public medium to peddle his wares or plead his case.

On December 17, Julian formally applied for the release of the $1.8 million still held in escrow by the corporation commissioner. Daugherty responded by demanding a complete examination and investigation of Julian Petroleum books and records. Julian Petroleum defiantly refused to cooperate. The Corporations Department had already audited these books and records "on a half a dozen occasions," protested a company official. In addition, Julian Petroleum had provided a complete list of shareholders, "which in our opinion should be considered secret and confidential . . . not because we believe you have a legal right to demand it" but because "we have at all times endeavored to cooperate with your department." The compilation of this list had entailed "unnecessary expense" and several weeks of work. Since the Corporations Department would not issue a new permit in California, added the official, Julian Petroleum had undertaken a nationwide sales effort and removed all stock certificates, subscription blanks, and other sales materials to its Delaware office.

Daugherty viewed this as a declaration of war. He ordered Julian Petroleum to make all sales materials available and instructed twenty Spring Street brokers to surrender their records of Julian Pete transactions. All but one, Roy West, a former Julian salesman who dealt exclusively in Julian issues, complied.

On December 29, Daugherty attacked Julian in a lengthy statement to the *Times.* The Corporations Department, he alleged, had received information that

Julian Petroleum had sold hundreds of thousands of dollars' worth of stock without its permission and was attempting to determine which sales were legal and which violated the law. Julian Petroleum had stymied these efforts by removing its books outside the state. As a result, Daugherty announced, he had impounded the $1.8 million remaining in escrow "pending the disclosure of the true facts."

Daugherty charged that "a large amount of money has been used by certain persons in manipulating the market for their own benefit by running up and down the market prices at different times." His agency wished to "find out who profited by such manipulations." The commissioner warned against buying Julian's personally owned stock, as additional financing "is not warranted until the officers of the corporation show sufficient evidence of putting the corporation on a businesslike basis, instead of making the sale of stock [its] principal business." Two days later, police arrested Roy West on charges of violating the Corporate Securities Act.

Julian raised his now-familiar cry of conspiracy. He denounced Daugherty's statement as "the most cruel attack ever perpetrated to my knowledge in the State of California" and threatened to sue the commissioner and the *Times* for $1 million dollars each. But with his ads barred from local newspapers, the beleaguered oil man lacked a public forum. With "powerful unseen forces" refusing to print his replies or accept his advertising, charged Julian, "the injustice . . . is practically the same as tying my hands and feet and then clubbing me over the head."

If the existing dailies could distort reality and deny him access to the public, who he was sure "resent[ed] these contemptible tactics as much as myself," the ever-resourceful Julian could find another way to convey his message. Julian vowed to publish his own newspaper. He entitled his journal *The Truth.*

"I locked myself into my office on the Saturday night before New Year's and started to prepare the copy," Julian later recalled. "I didn't take my clothes off or lie down in a bed until 10 a.m. the next Wednesday." Julian's lawyers attempted to temper his rhetoric and accusations, but Julian ultimately "sent them home and took the entire responsibility on his own shoulders." On Monday, December 31, 1923, at 2:00 a.m., Julian had completed enough copy and illustrations to fill a four-page newspaper, and arranged for distribution. The *Shopping News*, an advertising periodical owned by local department stores, rented him its presses.

By Monday morning the printers had prepared the plates and started the machinery. Copies of *The Truth* began to roll off the presses, when suddenly the *Shopping News* manager ordered all work halted. The owners of the paper had changed their minds and would not allow *The Truth* to be printed in their plant. Julian argued and pleaded and threatened lawsuits, all to no avail.

Julian contacted another printing firm, which agreed to complete the job if Julian could supply workers. Julian hunted for printers, offering triple time in wages, but discovered that most were gone for the New Year's holiday. Julian finally located several journeymen at an offshore island resort and rented a launch to transport them in. The *Shopping News* had used rotary presses; the new printer ran flatbed presses. All forms thus had to be remade. By dinnertime all finally seemed ready. Julian and two associates returned to his home and were preparing to eat when the telephone rang. Attorneys had informed the printing firm that *The Truth* might be libelous and advised against cooperating with Julian.

When C. C. arrived at the shop, he found it locked and barred, with all gone save the night watchman. He sent for two Pinkerton men, and he and his colleagues, "though ignorant of the printing business, removed the forms with our own hands and stacked them on a truck under the Pinkerton's guard." Julian hoped to rush the plates to San Francisco and have them printed there.

Before departing, Julian played one last card in Los Angeles. He sought out Cornelius Vanderbilt, Jr., the twenty-five-year old maverick publisher of the city's newest periodical, the *Illustrated Daily News*. Vanderbilt, the great-great grandson of the famed railroad magnate, had been raised in one of America's wealthiest families. He had fought in World War I and worked as a "millionaire reporter" for the *New York Herald* and the *New York Times*, before arriving in Los Angeles in the fall of 1923 to start the first newspaper of a projected national chain.

Vanderbilt saw the *Daily News* as a liberal alternative to the conservative Los Angeles media oligopoly. Whereas his famous forebear had once uttered "the public be damned," Vanderbilt adopted the motto "The Public Be Served." In the bitterly anti-union climate of southern California, Vanderbilt became a voice for the working-man. He sold his tabloid-style, picture-filled newspaper for an affordable one penny and repeatedly attacked the Los Angeles power structure, symbolized by Harry Chandler and the *Times*. To finance his fledgling empire, Vanderbilt planned a public stock sale aimed at the same clientele as Julian's. Julian hoped that in Vanderbilt he had found a kindred spirit and ally.

Julian located the young publisher at a New Year's Eve celebration. For Vanderbilt, struggling to establish his newspaper, the Julian situation offered a unique opportunity. By siding with the popular, flamboyant oil man, he might increase circulation and attract investors. In addition, Vanderbilt, who had recently started a second paper in San Francisco, needed cash. Vanderbilt agreed to print *The Truth* and to run Julian's advertisements in exchange for a $100,000 loan.

Julian delivered the printing forms to the *Daily News* plant, one of the few union shops in Los Angeles, only to suffer another setback. The foreman balked at using plates prepared by non-union labor. After much debate and, according to Julian, money spent freely, the foreman allowed his men to redo the forms.

*The Truth* finally won out. On the morning of January 2 the four-page newspaper rolled off the presses, and an army of 350 men and boys distributed copies throughout southern California. Truckloads of papers, protected by armed guards, traveled north to San Francisco and south to San Diego. In Los Angeles, Julian later recounted, obstacles continued to appear. Two armed men waylaid a delivery boy and stole thousands of papers. At least ten of Julian's workers landed in jail.

Julian's allegations of strong-arm tactics might be readily dismissed if it were not for a persistent pattern of intimidation in the Los Angeles newspaper wars. Harry Chandler had first gained influence on the *Times* in the 1880s by acquiring a circulation monopoly and using it as leverage with publisher Harrison Gray Otis. Chandler later married Otis's daughter and succeeded his father-in-law at the paper's helm. He never forgot the importance of distribution. In 1912, after voters had established the city-subsidized *Municipal News*, its deliverymen were physically attacked in residential neighborhoods, and, under pressure from the *Times*, the paper lost its subsidy and died.

Well into the 1920s, according to one former employee, the *Times* employed "thugs" to protect and expand its newsstand sites and delivery routes. Vanderbilt had received warnings from Chandler and others not to start the *Daily News*, and confronted efforts to sabotage his presses and cripple his distribution. Copies of the tabloid were thrown into ditches and sprayed with oil. Delivery boys faced "intimidation and bribery, then fists and clubs." To protect his trucks and newspaper boys, Vanderbilt maintained a fleet of Ford runabouts for his men to troubleshoot the streets.

Despite the efforts of Julian's opponents, one million copies of *The Truth* found their way into the public's hands, according to Julian. In the newspaper, he denied all charges levelled against him and attacked Daugherty and others for conspiring to destroy Julian Pete. *The Truth*, he pledged, would become a "permanent feature, issued at regular intervals." But no subsequent issue appeared. Julian had worked nearly eighty continuous hours and spent thousands of dollars, leading him to the rueful conclusion that, "It was cheaper to let the Commissioner say anything he wished than to combat his statements with such expensive publicity." With *The Truth* finally on the streets, Julian went home and slept for two days.

## II

C. C. Julian returned to his offices on January 3, 1924. News that a judge had dismissed Daugherty's case against Roy West buoyed his spirits but a death threat found among his messages tempered his mood. That night, at 10:30, as Julian worked with Julian Petroleum vice-president Jack Roth, an anonymous caller warned he would be killed before morning. Julian and Roth left the office and drove to Julian's Los Feliz mansion. His chauffeur detected a taxi following them. Shortly before 1:00 a.m., as Julian and Roth sat on a davenport, two .38 caliber bullets pierced the screen and French windows at the front of the house, barely missing Julian's head. As Julian and Roth dove for cover, a third shot crashed into the wall. The closely grouped bullets, fired from thirty feet away, formed a small circle of holes in the window.

Julian charged that the people who had tried to block publication of *The Truth* had planned the shooting. "They were trying to scare me, rather than kill me," he claimed years later. "I just failed to scare." Others speculated that Julian himself had arranged the incident to win sympathy for his cause. Police investigators found the handgun used the next day, but could shed no further light on the attack.

Throughout the week, Los Angeles newspapers accorded the Julian saga banner treatment. The *Record*, which had never accepted Julian's advertisements, ran page-one editorials on the "Julian tangle" for three consecutive days. Since 40,000 people "with bended backs and calloused hands," had handed over "their savings in response to his steam calliope advertising," argued the *Record*, the conflict between Daugherty and Julian concerned the entire community. "In justice to Julian," the *Record* admitted, "it must be conceded that he has performed some astute gymnastics. . . . Unfortunately, melodramatics will not straighten out the Julian promotional tangle."

The *Times* responded to Julian's onslaught with an editorial protesting Julian's "picturesque abuse," "petty malice," "trivial rumors," and "preposterous intimations." It had ceased accepting Julian's ads when it received information "indicating the possibility, if not the probability, of some of the investments failing to develop." As the publisher of "a larger volume of advertising . . . than any other newspaper in the world," the *Times* printed "only advertising that can be relied upon," shielding readers "from misrepresentation, fraud, and irresponsibility," sparing "no effort or expense to protect itself and its family of readers."

This sudden discovery of advertising scruples doubtless amused long-time Los Angeles residents. For decades the *Times* had amassed its vaunted volume of ads by soliciting and printing the blandishments of a wide variety of promoters and con artists. Patent medicines and health treatments shared the pages with countless opportunities in business, oil, and real estate speculation. The *Record* estimated that the *Times* had earned in excess of $1 million from advertising in each of the previous two years with "no small share of that profit [coming] from oil promoters." Julian alone had paid approximately $40,000 to the *Times*, and similar amounts to other newspapers that now refused to accept his copy.

On January 5, attorneys for Julian filed a writ of mandamus in Superior Court demanding release of the funds impounded by Corporation Commissioner Daugherty. Julian appealed for moral and financial support in his first ad in Vanderbilt's *Daily News*:

### "RED BLOODED
### AMERICANS"
### I'M CALLING ON YOU
### TODAY

**"The Powers aligned against me,"** after months of vicious attack from every possible angle, have finally thrown every reserve power they control into the breach, in a last desperate attempt to annihilate the **"Julian Petroleum Corporation."**

We have fought them toe to toe every inch of the way, always knowing that sooner or later the showdown must come and always betting all ours that in the final battle, "public opinion" and the financial support of the California people would be the necessary weight swung into the balance to defeat one of the most diabolical political conspiracies that has ever been hatched in the State of California . . .

I want you to know that at this moment we are "barely holding" the "Powers" that are assailing us, and that the "line" is swaying back and forth, and to win this battle we must have your further financial support now.

Julian offered $2 million of personally held stock for sale. "I am calling on every man and woman in this 'Golden State' who is a believer in seeing 'FAIR PLAY,' to lend me their financial assistance," he wrote, "and in return I will promise you you'll see the gamest bunch of fighters this side of the 'Mason-Dixon Line' go over the top and make an overwhelming success of our great enterprise."

Julian's charges of "unseen powers" struck a resonant note. Many people in Los Angeles firmly believed that a conspiracy, emanating from the offices of the *Los Angeles Times*, controlled their lives. Harry Chandler, the *Times* publisher, reigned as southern California's wealthiest and most influential figure. He served on over fifty boards of directors, established countless dummy corporations and secret trusts, and dispensed financial backing to numerous undertakings. "He is mixed up in so many ventures," commented the *Saturday Evening Post*, "that nobody, with the possible exception of himself, has ever been able to count them."

Speculative real estate syndicates involving the most influential people in the community lay at the heart of Chandler's empire. "It's not what you go into," he once said, "it's whom you go into a venture with." Leading bankers, businessmen, streetcar and utility owners, and political figures found their way into Chandler's investment alliances. In the most famous and controversial of these syndicates, Chandler and other community leaders had acquired land in the arid San Fernando Valley in 1904 with foreknowledge of a secret plan to construct an aqueduct from the Owens Valley to provide water for Los Angeles. Using their considerable influence to make the aqueduct a reality, the group garnered over $100 million in profits.

The belief in the existence of a "newspaper conspiracy" was an integral part of the local landscape. Many observers assailed what state official William G. Bonelli later called a "blackout on truth" in southern California. The political perspectives of the *Express, Examiner,* and *Herald,* once-liberal rivals of the *Times,* now differed little from that of Chandler's journal. Publishers E. T. Earl of the *Express* and Max Imhsen of the *Examiner* participated in Chandler real estate syndicates. Critics charged that, to ensure continuous growth and drive labor costs down and real estate values up, the *Times* and its crony newspapers carefully controlled the images dispensed to city and nation. Reports of traffic problems, labor unions, and political repression never appeared in the local print media. The fact that all four papers had simultaneously canceled Julian's ads, denying him an opportunity to present his version of the controversy, and that Daugherty had once worked for the *Times,* offered further evidence for collusion-minded observers.

"Who and what is this sinister and menacing power that controls newspapers and public officials?" asked one citizen in a letter to the *Daily News.* Other correspondents warned of "a sinister, subtle conspiracy, incapable perhaps of direct proof . . . to destroy an enterprise in the interest of old established privilege," and of "the seeming evil with which our very citizens have been tied hand and foot." "No man is free, evidently, from the influence of this crowd who strike in the dark," wrote one Julian supporter. Julian and Vanderbilt had "stood up against the crowd that runs the town," lauded another.

These sentiments reflected the vast outpouring of public support for Julian.

During the first two months of 1924, the *Daily News* published over 130 letters endorsing its stand on Julian, as many as thirteen in a single day. "Mr. Julian may not be all that he claims," asserted one man, "but as a citizen of these United States he is at least entitled to tell his side of the story." Julian Pete supporters expressed common fears about Progressive-era regulations. "We are fast losing all the liberty that we have and that our fathers fought for," wrote one Jeremiah. Some challenged the rationale behind regulatory commissioners. "Who is this man who is trying to put us in the class of serfs of old?" wondered a letter writer. "It will not be safe to invest if one man holds such power," argued another. "When we need protection we will ask for it," a Julian backer wrote indignantly. "Are we children not to be trusted with our money?"

Increasingly, Julian appeared in these *Daily News* missives in an almost Christ-like guise. "What faith they have in this wonderful man," observed one letter writer. Supporters described him as "a square man . . . who towers above the ordinary;" "a man of the first water [who] has been sorely tried . . . honestly trying to help the poor and ordinary men and women to feed and clothe their children better and get a few luxuries for themselves." One real estate promoter anointed his subdivision adjoining the proposed refinery site "Julian City," a promised land of "rich, fertile, black loam soil."

This deification reached new heights when the trial in the Julian-Daugherty mandamus suit began on January 17. With a new courthouse under construction, the tiny Sunday school room of the Broadway Christian Church served as a make-shift court chamber. The seventy-five seats filled rapidly with twice that number of people cramming the rear of the courtroom—"old and middle aged men and women, working people many of them daguerrotypes . . . [standing] tirelessly through the slow, snailing minutes." They laughed and applauded when the judge ruled in Julian's favor, their "blurry, lampless faces, transformed for a brief, sublime moment." At the end of each day they surged forward to greet Julian, shake his hand, or just touch him. "God is on your side, Mr. Julian," proclaimed an aged women with "withered and fissured face," "for you know, I pray to Him for you every night."

*The Record* reporter sensed the larger drama being played out:

> Strange . . . strange . . . somehow they seemed to be HIS people, people who bought HIS oil stock, people who believe in HIM with faith hard as granite. They seemed strangely attuned to HIM.
>
> . . . the romance of it all tingled in me . . .
>
> C. C. Julian, with his raucous, quixotic ads, has made high adventure out of the drabness of selling stock. He has brought romance . . . to thousands . . . . He has made it appear that the C. C. Julian corporation is fighting a gigantic desperate battle against immense, disgracefully immense odds.
>
> And all these middle-aged, tired, or old clerks, or laborers, or small merchants see in him the embodiment of their highest dreams. Once they've bought C. C. Julian's oil stock, they're C. C. Julian's soldiers, kindling to a great cause—great to them.

And the quaint and wistful monotone of their days is for the sublime hour
dispatched.

Julian played to the crowd, "with the amused inscrutability peculiar to matinee
idols in small cities." His colorful wardrobe dazzled the onlookers. "There is something of the fighter about him," concluded the *Record*, "something of a bantam
rooster . . . something peaked, yet doughty, and in profile there is a handsome
greyhound quality to the oil genius visage." Commissioner Daugherty, on the other
hand, despite the judge's demand that he appear, remained in San Francisco.

Julian's attorneys argued that Julian Petroleum had not violated the company
permit but instead was victim of a long history of mistreatment by Daugherty. Witnesses revealed that the Corporations Department had had complete access to the
Julian books for several months and had made repeated "test checks" before the
company had transferred them to Delaware. A Julian Pete official confirmed that
Julian had loaned the company $1.5 million, but had yet to receive any salary for
his efforts.

After two days of testimony, the Corporations Department offered terms for
surrender. Julian Pete would immediately receive all impounded funds in exchange
for a promise to return its books to California no later than April 18. The crowded
courtroom burst into cheers at the announcement. Julian's triumph offered his
followers further evidence of a dark conspiracy to persecute him. "We'll show them
Standard and Shell fellows yet," crowed a Julian supporter. "Isn't it strange that
Julian has won every case he has ever fought and yet they keep fighting him?"
remarked another.

For Daugherty the outcome was a humiliating defeat. His failure resulted, in
part, from the novelty of his role. State regulatory agencies were still an experiment
and lacked the resources, authority, and public acceptance to successfully challenge
a determined businessman. In addition, Daugherty himself had often acted in
heavy-handed fashion, nurturing the rumors of collusion and conspiracy. A February 1924 grand jury investigation of his office, perhaps triggered by the Julian
affair, supported these charges. "The department is under- and inefficiently manned
and its policy is vacillating and undignified," reported the grand jury. While no
specific instances of graft were proven, "favoritism was undoubtledly shown" and
"conditions of permits in some cases were unbusinesslike." The Corporations
Department would not easily overcome the public damage done by the Julian affair.


## III


The Corporations Department released the impounded funds at 2:40 on Friday
afternoon, January 18, 1924, inspiring a weekend of wild celebration. The following
evening, as the clock approached midnight, Julian, his wife, and several uninvited
comrades including Jack Roth, Julian's right-hand man, appeared at a party hosted

by fellow oil man Lewis B. Chase. Among the guests was Julian's physician, Dr. Ira Tower. Suffering from lumbago and leaning heavily on a cane, Tower extended his hand to Roth, who inexplicably spun angrily away. "I turned to talk to Julian," Tower later explained, "and the next moment, before I could gather myself together, saw Roth swing on me." Roth's punch broke Tower's nose and fractured his jaw. Julian made futile attempts to smooth matters over. Mrs. Chase, the hostess, pronounced, "I am not willing to have them as guests in my home again."

The following night the Julian caravan traveled to Hollywood's posh Café Petroushka, the domain of Russian émigré princess Dagmara Saricheva. In the kitchen, a former chef to Czar Nicholas II cooked borscht for a celebrity crowd. At one table sat violinist Jascha Heifetz. Nearby, Prince and Princess Narichkin entertained Russian author, actor, and strongman Nicholas Dunaev, who reputedly could bend a dime between his fingers. Across the dance floor, Charlie Chaplin, Hollywood's biggest star, dined with prominent scenarist Carey Wilson and controversial actress Mary Miles Minter. Minter had reigned as one of the screen's leading ladies until she became implicated in the mysterious 1922 murder of director William Desmond Taylor. Since then her career had foundered.

As these luminaries listened to the music of a Moscow violinist, Julian and his boisterous party arrived. Julian's group included his brother, C. A. Julian; and Roth. They were accompanied by an unidentified woman, possibly Julian's mistress Gladys Smith, described as "very attentive to [C. C.], very pretty," with "beautiful clothes . . . and too many diamonds to be a poor working girl in the movies," and actresses Peggy Browne and Mildred Harris, Chaplin's former wife. Julian, obviously drunk, flashed a wad of thousand-dollar bills. He cursed loudly as he accidentally kicked over and broke a floor spotlight. At one point he attempted to take to the dance floor, ignoring a Sunday night ban on dancing. The restaurant management forcibly restrained him.

Julian approached Chaplin's table and brushed against Mary Miles Minter. Peggy Browne later stated that Julian had promised Harris that he would "get" her former husband. Chaplin asked Julian to leave his guests alone, but C. C. sneered and took a swing at Chaplin. The "Little Tramp," demonstrating the dexterity he had displayed in his movies, ducked the punch and then decked Julian. A general free-for-all ensued. C. A. Julian struck Chaplin from behind before the nimble actor dispatched him as well. Nicolas Dunaev entered the fray to knock down C. C. a second time. "If Mr. Julian had been a dime, I'd have bent him in two," boasted Dunaev. Princess Saricheva rushed in to restrain Chaplin. According to witnesses, Chaplin, his nose bloodied, retreated to the kitchen, and then returned to announce, "Here I am again. Anybody who wants to fight me step up. Only one at a time will be accommodated."

The "Battle of the Petroushka" made front-page headlines. Chaplin pled self-defense. Mildred Harris expressed regret that Chaplin had "experienced humiliation." The café management hastened to guarantee that they had not violated Prohibition liquor laws, while a local fight promoter offered to arrange a rematch on his Saturday night boxing card. For his part, Julian adamantly denied the whole affair, claiming to have been hundreds of miles away in San Francisco. Nonetheless, he instructed Roth to pay the Petroushka $595 to cover damages.

Julian responded to the adverse publicity in a statement addressed to "the people of California." Willingly accepting the mantle of a fighter, C. C. warned his followers not to take reports of his nightlife too seriously:

> I only request that you bear in mind that the four Los Angeles daily newspapers that have been distorting the truth during the past few days have refused to accept my advertising. . . . They claim we are all fighters. But I'm here to tell you if we had not been, we would be in the discard now; Keep in mind that the truth distorted is the dish we are eating out of these days, but we all have strong stomachs.

C. C. reiterated his charges of biased reporting in his January 26 ad in the *Daily News*:

### "Time Is Money To Me"

But evidently it does not mean much to my good friends of the "**Los Angeles Press,**" for I hear Mr. Chandler of the Los Angeles Times has issued orders to his staff of reporters to spare no pains or expense in digging up dope on all the fistic battles of my career, so that his paper may publish them to the world.

"**Harry, Dear,**" you don't need to waste your good elegant reporters' time, hunting up that information, because, "**old kid,**" if you drop into my offices any day, I'll supply you all the dope on fifty different fistic encounters I have participated in. You know in the oil fields where I got my education we usually settle our differences that way . . .

And by the way, "**Harry,**" any time you are short of a "headline" for your "**Sunday paper,**" I might oblige you by staging a special bout to serve your purpose.

Even though you refuse to accept my advertising in the "**Times**" you're doing pretty well on my publicity end.

The battle royal with Charlie Chaplin added a new element to the Julian legend: the nightclub-hopping bon vivant, who cavorted with glamorous actresses, flashed thousand-dollar dollar bills, spent money lavishly, and bedecked his women in diamonds. Although Julian vehemently rejected this image, protesting that he could never have accomplished all he had while leading a wild nightlife, the incident did little to diminish his allure.

Even as Julian's legal and nocturnal exploits filled the news pages, the major dailies still refused his ads and paid scant attention to Julian Pete activities. This proved unfortunate for Julian, as several developments seemed to be bearing out his optimistic predictions. The prolonged downswing in oil prices had abated, and crude oil prices had jumped 25 cents a barrel. Julian hailed this as "the forerunner of a series of price advances," justifying his program of storing low-priced oil for future sales. In addition, several new leases, which Daugherty had attacked as speculative, had already paid off. In the Torrance field, Julian Pete brought in four wells in two weeks, all flowing 1,000 barrels a day. At Whittier a wildcat well adjoining the Julian properties delivered a 3,000-barrel gusher. Julian had also relocated the abortive No. 8 syndicate well to Huntington Beach, where a modest producer at least promised to bring some return to investors. In early February the first tanker loaded with oil at the new Julian Pete wharf headed for the East Coast.

More visible to Los Angeles residents were the red, white, and green Julian Petroleum service stations. Ten outlets had opened dispensing Defiance gasoline. "**'Smoke,' 'It doesn't make any,'**" boasted an ad in the *Daily News*. "**'Power,'** nothing else but." Although his stations might be "a little hard to find today, it won't be long before they are as prominent as lamp posts," predicted Julian.

Despite this undeniable progress, Julian found it impossible to sell his personally owned shares of Julian Pete stock. Uncertainties about his venture had driven prices down on the open market. Investors could purchase previously issued stock more cheaply from brokers than they could new shares from Julian. As stock prices continued to plummet, Julian searched desperately for a way to stem the tide.

At this moment Julian turned to an unlikely ally, the *Los Angeles Record*. Part of the national Scripps-Howard chain, the *Record* was one of the weaker entries in the city's journalistic wars. The iconoclastic Scripps had created a syndicate "to serve the working class," denouncing capitalists and big advertisers as "the mortal foe[s] of honest journalism." The *Record*, according to reporter Rueben Borough, "was very progressive in its whole outlook, pro-labor, which was something in Los Angeles." Like all Scripps papers, the *Record* specialized in muckraking exposés, dramatized, in Borough's words, with "big headlines and short stories with short sentences." For several years the newspaper had waged a vehement campaign against all oil promoters, refusing to publish their advertisements and revealing their unscrupulous practices in front-page exclusives. In early January it had presented Julian in a less-than-favorable light. But the *Record*, perhaps impressed by circulation gains made by the *Daily News* after its pro-Julian stand, suddenly united with C. C. to rehabilitate his image and his enterprise.

"**STOP JULIAN PANIC TALK!** SAVE 43,000 INVESTORS!" read the banner headline to an editorial that occupied the *Record*'s entire front page on February 9. "Vultures have been circling overhead for a chance to snatch Julian's assets at the expense of the stockholders," warned the *Record*. "If they are victimized, the whole community suffers with them." The newspaper, resuscitating a suggestion made a month earlier by both Julian and the *Times*, called for a group of "men of unquestioned business character and ability" to "review Julian's operations, analyze his status and help guide his future moves."

The speed with which Julian and others responded to the *Record*'s call suggests a carefully orchestrated scenario. "**JULIAN ACCEPTS OFFER OF HELP!**" the *Record* reported on February 11. "Your newspaper has sounded the first constructive note that has sung out after a period of months," wrote Julian. "It will be entirely satisfactory and agreeable to me to have a committee of leading, capable, disinterested citizens come into the affairs of the Julian Petroleum Corporation as counsellors and reviewers."

That afternoon Julian appeared at the offices of the Los Angeles Chamber of Commerce. Charles H. Treat, a former oil and real estate man who had recently resigned as president of the Los Angeles Board of Public Works, accompanied Julian. Five prominent businessmen awaited them inside. After one and a half hours, Julian emerged from the meeting and made a prepared statement. Denying any financial difficulties, but noting "an undercurrent of propaganda well spread," Julian announced that he had requested that a committee investigate his affairs "from beginning to end." The panel members, stated Julian, "are prompted by the most unselfish motives and are men of unquestionable character for whom I have the highest respect." He called upon them to "make an impartial report to the world on my company and its affairs as they find them," but predicted that they would discover "the whole enterprise . . . [was] as clean as a hound's tooth."

No one in the Los Angeles business community could question the qualifications of Julian's citizens' committee. The six men ranked among the most respected figures in the city. In addition to Treat, the group included; William Lacy, a prominent steel and metal manufacturer and the president of the Chamber of Commerce; William T. Bishop, a former Chamber president and head of Bishop and Company, makers of cookies and crackers; banker Irving H. Hellman, a former city engineer and member of one of the city's leading banking families; former Superior Court justice William Rhodes Hervey, a vice-president of the Pacific Southwest Trust and Savings Bank; and R. C. Gillis, president of both the Los Angeles Union Terminal and Iron Chief Mining Companies. Like most members of the Los Angeles business establishment, these men served on the boards of directors of several local companies, belonged to the city's leading country clubs, and actively participated in Republican politics. The conventional wisdom held that one or more of these men would be named directors of Julian Pete in the aftermath of the investigation.

The *Record* hailed the committee as the "Answer to The Julian Tangle." They "will not stand for any whitewash of Julian if he is black," asserted the *Record*, or "any blackwash of him if he is white." Taking full credit for this turn of events, the paper asked these "gentlemen," all of whom had been "successful in accumulating money," to "place yourselves for the moment in the position of these 43,000 people who have saved and skimped and finally handed their dollars to Julian. YOU GENTLEMEN HAVE BEEN HIGHLY HONORED. YOU HAVE BEEN ASKED TO PROTECT THESE PEOPLE."

Losing no time, on February 14, Julian guided the committee on a well-publicized tour of his facilities. Led by Julian's large, gray Pierce Arrow, "a small cavalcade of automobiles" inspected the company's properties. The parade stopped first at the planned site of a supply depot for filling stations. The group then visited a "wide expanse of orange groves" in Whittier, where Julian had leased 170 acres of prospective oil lands. Next Julian showed off his famous Santa Fe Springs holdings, and then directed the odyssey to his tank farm in Norwalk. Committee members climbed a long spiral stairway to the pinnacle of one of a dozen steel tanks. Julian lifted the cap to show the oil beneath "glistening in the dark depths." From Norwalk, the fleet of cars sped to Huntington Beach, where Julian displayed three producing wells and the origins of a vast pipeline system. Throughout the tour, Julian regaled his fellow travelers with statistical information about the corporation.

The procession halted while Julian distributed beef sandwiches, ginger ale, and milk. "Help yourselves, there's plenty," offered the congenial host, reasoning that millionaires, like "sucker tent" habitués, would be impressed by a free lunch. With the meal completed, Julian delivered his judges to the wharf at San Pedro. Four steel tanks sat behind a high concrete firewall on four acres leased from the city. Large pipes led to a wooden dock where tankers could load. The afternoon progressed with stops at Julian wells in the Lomita and Torrance fields, and then continued on to a barren 231-acre refinery site. Julian promised to have the refinery completed in 120 days. The final stop was at the Compton oil fields, where Julian had sunk a wildcat well on a 400-acre lease. Treat, the oil man in the group, volunteered that it looked like good territory.

"The committee doesn't say much but it is evidently impressed," concluded the *Record*. Investigators and reporters, "hustle[d] back" behind Julian's Pierce Arrow. "We do not hustle as fast as Julian," commented the *Record* reporter. "We can't keep up with him."

Meanwhile, The *Record* began another phase in the Julian defense. On February 13, the first installment of a week-long front-page biography of Julian commenced under a heading familiar to Julian followers, "The Truth." Disguised as a hard-hitting muckraking exposé, the lengthy articles, relying primarily on Julian's own recollections, presented an all-American rags-to-riches tale. Julian strode through these accounts as a hardworking, devoted family man, who, after years of failure in Canada, California, and Texas, had finally found well-earned success. With Julian Pete, he wished to share his good fortune with the people who had backed him so faithfully. At the conclusion of the series, the *Record*, while reminding people of its stand against oil promoters, genially confessed, "The truth about Julian proved to be not so unfavorable as the *Record* expected it to be."

For the next two months, as southern California awaited the report of the citizens' committee, the Julian saga temporarily faded from the headlines. Even Julian's ads in the *Daily News* grew relatively infrequent. Behind the scenes, however, Julian found himself besieged by a horde of examiners clamoring for the corporation's books, including federal Justice Department agents invited into the fray by Daugherty. "I have not used the general books, but a very few days," reported the federal agent in charge. "The fact is C. C. Julian has had Internal Revenue Agents working on his books, a special accountant making up his 1923 income tax return, and a Citizens Committee investigating his methods of doing business . . . with several Price-Waterhouse accountants who have had the books most of the time since January first."

On March 21, Julian informed the Justice Department that he would no longer cooperate with its efforts. Julian noted that his attorneys had advised him that the federal agents "are not quite within their jurisdiction in this matter." Nonetheless, he expressed regret, perhaps with a trace of sarcasm, "that we have to part with such congenial and gentlemanly company," assuring the government "that the only reason of making this request" was to "facilitate the handling of the clerical work with a greater account of efficiency." The agent vacated Julian's offices but decided to continue his investigation "on some outside work," examining the books and records of stockbrokers to determine whether Julian had formed "some conspiracy to manipulate the market." These efforts failed to unveil conclusive evidence of illegality, and the agent ultimately recommended discontinuing the Julian probe.

On April 8, the citizens' committee issued its long awaited report on Julian Petroleum. Bolstered by "a small army of experts and auditors," the five-man panel (William Lacy had withdrawn earlier) offered a stunning vindication of Julian and his controversial enterprise. "The committee was impressed by the vast amount of work accomplished by the corporation since its organization eight months ago," stated the report. It praised Julian for "devot[ing] his best efforts in good faith to the service of the corporation."

Evaluations of Julian Pete's assets prepared for the committee painted a highly flattering portrait. A former president of the Los Angeles Realty Board appraised the corporation's property holdings at $888,000, exclusive of oil possibilities and locational considerations. In reality, Julian Pete real estate was worth two to three times this amount. A petroleum engineer valued the operational facilities at almost $7 million. The committee, noting that this represented the current market value of these assets while the company had recorded only the purchase price on its books, arbitrarily deducted $1.25 million dollars from these estimates. Even so, they established a $49.83 value for each share of preferred stock, just pennies below its sale price.

The highly respected accounting firm of Price-Waterhouse, in a report prepared by the director of its Los Angeles office, verified the inventory of crude oil, accounts receivable, and cash on hand. While noting that the corporation lacked the usual reserves against taxes, bad accounts, and depreciation, Price-Waterhouse found only modest liabilities and made no criticisms of bookkeeping methods or the values at which properties were carried on the books.

"Respecting Mr. Julian," the citizens' committee concluded, "it is fair to state that he has served the corporation from its inception to this date without salary or other compensation except the common stock issues." Julian had advanced the company over $1 million and had not collected an additional $970,000 in commissions owed to him. He had charged the company no interest on these sums. For his services, Julian had received common stock, the value of which depended on the fate of the corporation.

While generally laudatory in its conclusions, the committee did express several reservations. It noted that, while the company planned "a complete unit from production through all the steps . . . to distribution," it still lacked a refinery. The businessmen also expressed concern about adequate supplies of crude oil. Julian Pete wells generated 23,000 barrels a day, but this was "wholly insufficient for the economic utilization of the plant." Crude oil purchased under contract represented "a source of supply [that] may not always be dependable." Finally, to "round out"

Julian Petroleum "into an earning business of value commensurate to its cost will require good management and considerable additional money." The committee recommended that Julian add several well known local business leaders to his board of directors.

Julian welcomed the report with characteristic hyperbole, hailing it "as the greatest vindication and approval ever extended by a group of hard-headed, conservative, capable, banking and business experts." The committee members, he pointed out, were "not the type of men calculated to handle the Julian Petroleum Corporation gently." Their findings put "to an end all uncertainty . . . and [clear] the track for completion of its program and the realization of the hopes of its stockholders." Julian noted the conservative nature of the report. Had the original assessments been accepted, the value of each share would stand at $58. This, stated Julian, was even more remarkable when the usual 20 percent sales' commission was taken into account. A pair of preferred shares purchased for $100 "have no reason to have a book value of more than $80." The audit showed that since he personally had borne all sales expenses of the stock, these shares merited a $116 valuation.

Julian defended the current management of the firm but readily agreed to seek new directors and additional financing. As to the source of these funds, Julian had few doubts. "I have every confidence of finding the rest of the money needed just where the money already forthcoming was found," he pronounced; "among the people of Southern California."

Commissioner Daugherty made no comment, but the Corporations Department ceased its investigations and issued Julian Petroleum a new stock-selling permit. The *Times* and other local papers gave Julian a green light to resume advertising. Julian Pete received additional legitimacy when the Los Angeles Stock Exchange agreed to list its preferred and common stocks. Exchange president Frank Pettingell proclaimed it a "public service" and a "civic duty" to "provide an open and legitimate market . . . thereby relieving the stockholders of the burden of being subjected to the wide spreads so prevalent in dealing with the shares on the street." Doubters and detractors remained, but for the moment Julian had routed his opponents, and his triumph seemed complete.

# 6 WHEN A FELLER NEEDS SOME FRIENDS

**"WELL,
LOOK WHO'S HERE"
"FOLKS"
"IT'S NOBODY ELSE BUT"**

And whether that old saying "you can't keep a good man down" prevails or whether
it is a case of "a bad penny being sure to always turn up again" is for you to decide.
However, here I am breaking back into print with that same old line of chatter that
I've peddled to you so long.

With that cheeky greeting, C. C. Julian returned to the advertising columns of the
*Los Angeles Times* and other newspapers on April 19, 1924, after a five-month
absence. In the preceding ten days, Julian had moved decisively to capitalize on
the citizens' committee report. On April 15, Julian Petroleum had received a new
permit from the Corporations Department to sell an additional $5 million worth of
stock. Julian also added three prestigious corporate vice-presidents to bolster his
board of directors: Charles H. Treat, the erstwhile Los Angeles Harbor Commissioner, who had headed the investigative panel; Parley M. Johnson, recently
resigned as Los Angeles Police Commissioner; and most impressively, William D.
Stephens, former governor of California.

Julian hoped that the addition of a former governor and two other respected
businessmen would give greater legitimacy to his embattled firm. Stephens, Treat,
and Johnson declared that, after personally inspecting Julian Pete properties and
thoroughly studying the citizens' committee report, they had confidence in Julian's
leadership and were ready to serve the 43,000 stockholders. Later accounts by Julian
attributed other motives to their participation. Upon their appointment, he claimed,

each man had demanded a salary of $1,000 a month. When Julian informed them that the corporation could not afford this expenditure, Stephens and Johnson accepted a reduction to $500, but Treat held out for, and received, the full amount.

One final slot remained open on the Julian Pete board. On May 1, Julian filled this position with Jacob Berger. Corporate press releases described Berger as a "pioneer Alaska banker," formerly the head of the Bank of Nome, and "one of the outstanding figures of the gold rush in Alaska," who had also won fame as the owner of championship dog-sled teams. Now settled in San Francisco, Berger was associated in the oil business with that city's prominent financiers, according to these dispatches. In reality, Jake Berger was none other than Julian's "little Yiddish friend" of Winnipeg newsboy days. Julian later admitted that by 1924 Berger had "dumped off" whatever, if any, fortune he had made in Alaska "in failure after failure in the oil business." His last remaining stake was an unlikely 4,000-acre lease at Half Moon Bay, just south of San Francisco, where, concluded Julian, "there was no oil within a hundred miles." Nonetheless, Julian agreed to drill two wells on the property and brought Berger into his firm.

Julian spent the latter part of April priming the public for his new sale. His advertisements recapitulated the citizens' committee findings and revealed the existence of a new 10,000-acre lease near Durango, Colorado, which had not been included in their report. On Sunday, May 4, Julian sprang his offer:

### "When A Feller Needs Some Friends"
### "FOLKS"
### "IT'S JUST LIKE THIS"

Suppose you went tramping out through the hills and ran into a young mountain of solid gold, but before you could pry off any of the treasure you must secure enough money to purchase the necessary machinery and supposing you could not get the money, your mountain of gold would slumber on perhaps for another million years, regardless of the great wealth almost within your grasp.

Well "People of California" that's just my position today . . .

I'm barking at my door with a "mighty healthy baby" in my arms, that you will never make me believe is not going to develop into . . . the greatest dividend paying oil Company in the US.

"I haven't weakened. I don't intend to weaken."

"I want $5 million and I want it now."

But Julian's hope that the citizens' committee report would stimulate demand for his stock proved false. In reality, as he later realized, the report "was the beginning of the end for me." While waiting for the committee to complete its work, Julian stock had plummeted to less than $60 a unit. Julian blamed this on "the forces opposing the corporation's success" and local brokers who had "feverishly sold short." With stock available so cheaply on the open market, how could even Julian hope to persuade people to pay $100 for his issue?

"'This Will Never Do' 'Folks,'" he complained on May 13; "She's not rolling in like it should." Three days later, Julian appealed to those who had previously found his terms prohibitive. He offered stock on the installment plan: $20 down and eight dollars a month for ten months. But his best efforts failed to move the stock at a sufficiently brisk pace.

In addition, despite 109 stations now selling Defiance gasoline, Julian Petroleum production efforts continued to falter. After a promising start, Julian Pete holdings in the Torrance field had proved disappointing, and *Oil Age* reported that Julian had erected a "nine-foot bird-proof board fence" around his Pico syndicate well to hide "everything from stockholders and curious oil and paper scouts." Shortly thereafter he shut down the Pico wildcat. More ominously, Julian No. 5 at Santa Fe Springs became the first of his famed early wells to be abandoned. Supplies had become so deficient that Julian had to purchase oil at a loss from other producers to fulfill his contracts.

Increasingly Julian relied upon the Colorado lease to reverse his declining fortunes. "The Biggest Thing in the World," he called it, predicting that a gusher on this plot "could easily make our holdings there worth '**twenty five million dollars.**'" By late May Julian had upped the prize to "**forty million dollars.**"

These inducements now fell largely on deaf ears, and on May 28, Julian attempted to divert blame for his lagging sales and document his assertions of conspiracy by filing a $100,000 damage suit against federal officials. Julian charged that Post Office inspector W. I. Madeira, with the foreknowledge and approval of two special assistants to the U.S. Attorney General, had sent a "false and defamatory letter . . . with expressed ill-will and malice and hatred" in a "deliberate attempt to damage [Julian] personally and [his] business." The special assistants had arrived in California one year earlier to spearhead a federal investigation of oil industry fraud. Madeira, working in conjunction with them, had sent out letters to Julian Pete stockholders inquiring into their dealings with Julian. When a northern California stockholder requested additional information, Madeira, despite official policy to issue no statement regarding a pending investigation, responded with a lengthy attack on Julian.

Madeira claimed that the oil promoter had paid dividends on his first five wells "for a purpose." The next seven wells had not returned money, "nor does it appear to be his intention to do so." Julian, said Madeira, had accepted thousands of dollars in oversubscriptions, and "it is public knowledge" that he had bought up stock receipts at lowered prices to avoid paying them back at full price. "He has cleaned up quite a sum on this racket," charged Madeira. The inspector also asserted that Julian Petroleum, which had collected $7 million from the public, had only $2 million worth of assets and that Julian, who had watered the stock, was now trying to dispose of his interests. "There is no large company which has in any way hampered him," concluded Madeira, "although he may use that as an excuse. You may judge for yourself in this. All of which I have personally verified."

The Madeira correspondence infuriated Julian. Madeira's letter, he stormed, "cannot by the widest stretch of the imagination be considered part of an investigation in good faith." Although this missive, as federal officials pointed out, was never made public until Julian himself exposed it, it nevertheless constituted an extraordinary breach of investigative ethics. Furthermore, since the citizens' committee report had assessed the company's value in excess of $8 million and praised Julian's management, Madeira's charges seemed baseless.

At last, Julian felt he had evidence of a campaign to mislead potential and actual stockholders, but the controversy did nothing to rescue Julian's offer. On June 4, he announced that he would withdraw the issue in six days. One week later he advertised, with a show of bravado:

She's
Closed
To
You
Forever

Miraculously, given the availability of cheaper shares, Julian had sold almost $1 million worth of stock. But $4 million of the issue remained unclaimed. "I decided that a stock which would go that well here against such odds," Julian later recounted, "ought to sell anywhere on the basis of the Citizens' Committee report. I now decided to go [to New York City], where there was the most money, in an attempt to clean up the issue quickly."

En route to New York, Julian stopped at Colorado to be on hand when his wildcat "went over the top." On June 16, he advertised, "Folks, I'm promising you all an 'interesting piece of news' real soon. . . . Hold everything for it won't be long now." Two days later, he published just two words: "**SOON NOW.**" But with that message the reports from Colorado ceased and after June 26, Julian's ads once again disappeared from the newspapers. Something had gone terribly wrong in Durango.

In late June, Julian arrived in New York City for a last-gasp fund-raising campaign. According to Julian's grandiose account, Jack Roth preceded him to Manhattan, where he opened corporate offices and arranged for several New York newspapers, including the Hearst dailies, to accept Julian Pete advertising. Hearst's business manager wired Julian "that they would keep my copy out of his paper only over his dead body." Julian sent ahead a week's worth of ads.

Julian detrained in New York with a sales force of twenty men, or as he related, "you might say 22 men, only two of them were not employed by me." Julian claimed that two Burns Agency detectives had accompanied him from Los Angeles: "I was never out of sight of both of them at once as long as I was in the East." Upon arrival, Julian learned that only the Hearst papers had carried his first advertisement, and they had refused to print any additional ads, upon orders from William Randolph Hearst himself. Accusing his old enemies of following him to Manhattan, Julian alleged that "I know personally, definitely, and am willing to make an affidavit to the fact, that a prominent San Francisco representative of the powers that be went to New York to see an influential publisher and block my advertising."

Julian claimed to have "laid siege" to the Hearst offices and after two weeks gained an audience with the powerful publisher. According to Julian, after listening to him for an hour and a half, Hearst assembled his subordinates to reach a final decision. For two more hours Julian pleaded with the newsmen. "I saw some of them in tears," he recalled, "for I knew I was talking of my financial life and I made them see the pitiful aspects of my case." In the end, however, they turned him down.

Blocked in New York City, Julian lined up eighty-five newspapers in the rest of New York, Connecticut, Pennsylvania, and Massachusetts and began running his ads. "The copy of the first eight days was introductory, telling the history of my proposition," said Julian. "My ninth ad was the first selling ad." But, according to Julian, Burns agents had visited each of the home cities of the newspapers and mobilized the chambers of commerce, banks, and bond houses against him. These forces influenced local publishers to drop his ads. On the ninth day Julian sat in his New York offices, "opening telegrams as fast as I could slit the envelopes. . . . The telegrams were monotonously the same. They could not accept any more of my advertising."

"The steam roller had rolled and I had been sitting right where it passed when it was rolling," he lamented. Effectively barred from advertising, Julian arranged a $1.5 million advance from an English syndicate for oil shipments over the next year, only to have other companies undersell his offer. He negotiated a $2.9 million bond issue, but it was revoked on the day papers were to be signed. According to one assessment, after two months in the East, Julian had sold $30,000 worth of stock at a cost of $100,000.

Like most of Julian's tales, his New York chronicle mixes fact with gross hyperbole. Newspapers most likely accepted, then rejected his ads, not necessarily due to concerted opposition but upon learning of his controversial California career. Detectives working for one of any number of people may well have dogged his path. Nonetheless, so vast a conspiracy seems difficult to believe.

Reports filtered back to Los Angeles of a more flamboyant side to Julian's New York adventure. According to these accounts, Julian rented a $3,000-a-month Park Avenue apartment and purchased the most expensive Cadillac he could find. He toured New York's nightspots, lavishing $100 tips on cab drivers, waiters, and hatcheck girls. He quickly acquired new girlfriends. He gave the Cadillac to one of them; he gave another a $10,000 trinket from Tiffany's. Journalist Carey McWilliams reported that one evening Julian hailed a taxi, offered the driver $1,500, and told him, "You be the fare—I'll drive." With Julian driving, "the cab careened around corners, shot through traffic signals, jumped over curbs, and finally smashed into an automat." Damage estimates from this spree reportedly approached $25,000.

Rumors circulated that Julian had fled to Egypt with the Julian Pete bankroll and that the company had ceased operations. Stock prices dropped as low as $11 for preferred and $7 for common. His corporate managers urged Julian to return to California to dispel the rumors. In late August 1924, Julian departed New York and headed for home.

## II

Upon arriving in Los Angeles, C. C. Julian confronted the monumental task of winning back investor confidence. Julian moved to reorganize Julian Petroleum, accepting the resignations of businessmen Treat, Johnson, and Stephens. "These men proved to be of no value whatever in the actual management of the corporation," he reported. Their departure "cut a useless expense . . . from the salary list." For the first time, Julian openly discussed the "tough luck" in Colorado that had hastened his disappearance. "(We) got down to 3600 feet and figured we had only another hundred feet to go . . . when we got into a mean fishing job and lost the hole completely," he explained. Nonetheless, Julian still offered to "pretty near stake my reputation" that this lease would some day be worth $20 million dollars or more. "Cheer Up, Old Timers," he comforted; "one of these days our ship will sail into port and when it does she'll be loaded to the guards."

Daily ads attempted to counter "adverse propaganda" from "the same old sources," but Julian decided that "the best means of nailing the false rumors of my flight" was to hold a mass meeting of shareholders. With tens of thousands of people needing to be accommodated, Julian reserved the spectacular new Hollywood Bowl. "I am personally going to tell you folks stuff about the inner workings of the oil industry that will make Tea Pot Dome look like an ant hill," he promised; "I will answer every imaginable question relating to 'Julian Pete' . . . and [show] how the 'Octopus' operates in restraint of trade." Julian offered free parking, special buses, and weather reports. "Don't forget your wraps," he warned, "because it may be chilly. . . . Everybody and his dog are welcome."

On September 5, an estimated 30,000 people overflowed the Hollywood Bowl's vast outdoor seating capacity to attend a financial revival meeting worthy of Julian's evangelical contemporary, Aimee Semple MacPherson. Introduced by a thirty-five-piece band, Julian reviewed the troubled history of his struggling corporation. Julian Pete, he charged, "had been investigated to death." Stockbrokers had "hammered down" corporation stock. The large oil companies had deliberately attempted to misinform stockholders. Julian described his New York experiences and the obstacles he had faced. He reminded the crowd that he had not taken any money for his efforts, and pledged, "I am not going to until every one of my stockholders has received 100 cents for every dollar invested."

Julian then unveiled his latest scheme to save the company. In order to brake skidding stock prices and stabilize the market for Julian Pete securities, Julian proposed removing the stock from the public arena. Julian called upon his followers to place their stock in escrow under his control for two years. As he explained in an advertisement the following week:

> Now here's what I'm doing. I'm getting the "JULIAN" shareholders to deliver all of the stock they hold into my hands immediately. I take it and place it in a vault in the Bank and deliver it back to them just as I received it, as soon as the Corporation is on a dividend paying basis when our security cannot be hammered down.

In addition Julian asked his supporters to send him whatever available funds they had to establish a pool "to pick up all the cheap stock on the market." Here, he advised investors, was an opportunity for "quick money." "Sometime within the next 30 days," he predicted, "I expect to have enough money in this pool to clean the market of 'Julian Pete' and the result must be a big jump in the market price." As Julian later elaborated, his plan would restore the market value of the stock and make possible additional sales to finance the refinery and put the corporation on a sound operating basis.

Over the next several weeks Julian ardently peddled his new panacea. He ran daily buses from downtown Los Angeles to the oil fields so stockholders could inspect corporate properties and presented nightly showings of a motion picture depicting his struggle. He promised to appear personally at these screenings, to "tell my audience things that I cannot write . . . because the papers would dare not print what I have to say." On opening night, a fistfight between Jack Roth and a heckler, allegedly armed with brass knuckles, enlivened the festivities. Julian charged that the intruder was one of more than a score of "gangsters" hired to disrupt the meeting, leading Julian to hire his own strong-arm team to safeguard his enterprises. "When they get too tough for everyone else," he boasted, "they're just getting right for me."

According to Julian, $3 million worth of stock certificates poured into his office following the Hollywood Bowl extravaganza, placing more than half of the $6.5 million issue in escrow. Stock prices jumped sharply in heavy trading. But the burst proved short-lived. By the end of September the stocks dropped close to their earlier low. On October 1, Julian issued one final advertising blast:

> Every intelligent citizen realizes that I am fighting a very uneven battle, but even so I am going to win my fight, because the essence of it all is "RIGHT OVER WRONG," and I'm a staunch believer that "Right" must prevail.
>
> I am fighting with every ounce of energy I have to protect the investments of nearly 50,000 people, many of whom have trusted me with their life savings.
>
> . . . .
>
> I am fighting a powerful, merciless, and unscrupulous hidden force that has respect for nothing but their greed for money.

Julian's ads then ceased.

Two days later, the federal grand jury in Los Angeles subpoenaed Julian and thirty of his employees, demanding that they appear in court with the corporate books and records. An exasperated Julian arrived at the courthouse alongside employees carrying armfuls of ledgers; Julian's car stood at the curb, awash with documents. The oil man was "as debonair as ever." His bright yellow shirt stood out boldly against his dark blue suit. A red carnation adorned his buttonhole. Speaking in a voice rendered hoarse by addresses at stockholder meetings, Julian charged that the indictment was "spite work" and a "frameup" resulting from the damage suit filed in May against federal officials. He joked that government teams, working in rotation, regularly reviewed his records. "They've got nothing on me," he told reporters, "I'm clean and above board and they know it."

In the courtroom Julian protested this latest assault. Delivering the company's books would require four trucks, he declared. "It means," he told the judge, "that we will have to cease all office operations and throw our office force out of work until this matter is decided." Nonetheless, Julian pledged his cooperation. But when the court also demanded that he turn over documents regarding his original syndicates, Julian refused. These books, he asserted, were personal documents unrelated to the current probe. The judge cited him for contempt of court and ordered United States marshals to take possession of the Julian Pete offices.

The latest crisis passed within a week. On October 10, in exchange for the dropping of the contempt charges, Julian appeared before the grand jury and relinquished the disputed books. But the controversy effectively stifled the escrow drive and prompted J. Edgar Hoover, then acting director of the Justice Department's Bureau of Investigation, to dispatch accountant J. H. Miller to conduct a second federal inquiry. Miller's investigation stands as one of the most unorthodox in FBI annals. Within days of his arrival Julian began to "flatter" and woo the agent. He offered to provide Miller with "a verbal history of his business experiences." Arguing that Julian's "viewpoint . . . would aid us in checking the information secured from other sources," Miller agreed. He also decided to conduct the interview alone "without any of the official group," under the assumption that "Julian would probably talk more freely if there were no witnesses present."

On the night of October 24, at Julian's private offices, Miller listened for four hours to a rapid-fire account of Julian's career, from his childhood to his current difficulties with Julian Pete. Unable to keep pace and take notes, Miller then condensed this tale, "based almost entirely on [his] memory" into a 5,000-word narrative. Remarkably, he resubmitted the rough draft a week later for Julian's approval. Julian assured him "that it was the most accurate statement of the kind that he had ever seen, not one pertinent fact having been omitted." This "authorized" biography constituted the bulk of Miller's November 6 report. It featured Julian's now-familiar charges of conspiracy and a spirited defense of his activities. A subsequent agent described it as "nothing more than a self-serving declaration of C. C. Julian."

The local U.S. Attorney General's office hoped to present a case against Julian to the grand jury by mid-December, but Miller, attempting to buy time for the oil man, projected the need for a "detail audit" that would delay prosecution for eight to twelve months. In addition, Miller, although still employed by the FBI to examine Julian's books, began to supervise accounting activities in the Julian Pete offices.

Meanwhile, Julian Petroleum had suffered further setbacks. Julian suspended new drilling in the Athens fields, charging that Standard Oil planned to "send out a wrecking crew and junk all his wells." At Santa Fe Springs, wells no. 1 through 4 sat neglected and idle, needing cleanout work and repair to resume pumping. Total daily oil production had dropped to 500 barrels a day. By December all Julian Pete operations, save the service stations, had shut down due to lack of funds.

"I was like an animal caught in a cage, pacing back and forth and turning every way in an endeavor to find a loophole of escape," Julian later described these dark days. Looming ahead was bankruptcy and receivership, entailing losses for tens of thousands of stockholders. Through Agent Miller, Julian doubtless knew of an impending federal indictment. "There was only one way [out]," reasoned Julian. "That was to sell out on some basis to some outfit that had the finances."

During the summer of 1924, oil man Sheridan C. Lewis, owner and founder of the New York–based Lewis Oil Company, had expressed interest in the Julian properties. "I was approached by one of the major oil company officials, who wanted to know if I would be interested in buying the Julian Petroleum Corporation or making a loan to them of a substantial amount of money, as it would be very unfortunate if the company could not be financed," related Lewis. Lewis said that he examined the properties and then spoke with Julian.

The thirty-five-year-old Lewis impressed Julian, as he would others, with his considerable knowledge of the oil industry, mastery of minute facts and details, and imposing presence. "Dark of complexion, with penetrating brown eyes and a high voice," wrote newsman Morris Lavine, Lewis "stands out in any group of men, at all times, as the master of the situation. He is plausible and emphatic in everything he says." Lorin Baker depicted the thick-set, broad-shouldered, balding Lewis as a "senior senator type." His tall, erect bearing and soft Southern drawl earned Lewis the nickname "Judge," an appellation that he "carries well." "As a walking compendium of the oil business, [Lewis] had few equals," according to Baker. "His mind is chain lightning," wrote Lavine. "He can sit and watch you out of his piercing eyes, with wrinkled forehead, while he mops his pate and says little, but thinks plenty; or he can talk for weeks until he has gained his objective." "He compels confidence and to all apparent indications, he justifies it," concluded one journal. Unlike Julian, wrote the Reverend Robert Shuler, Lewis "never loses his head, never flies mad, never orders anyone out of his office."

Lewis presented an admirable résumé. He traced his birth to a plantation near Richmond, Virginia, which his ancestors had owned for several generations. A physician's son, he claimed to have graduated from the University of Virginia with a law degree at age eighteen. Lewis had practiced law in Texas and Alaska and also represented several companies in legal matters in Mexico. He ultimately abandoned the pursuit of law for the romance of oil, becoming, by one account, a "man who is known wherever oil is drilled from the Arctic circle to Tampico in Old Mexico." Lewis took credit for developing the Katella shallow oil fields in Alaska. In 1921, he formed Lewis Oil, absorbing smaller firms in Texas, Louisiana, and Ohio. Lewis had recently entered the California arena and acquired seventeen wells in three different fields.

Among his friends and associates, Lewis numbered William Gibbs McAdoo, son-in-law of the late President Wilson and leader of the California Democratic Party, and U.S. Senator William King of Utah. King served as legal counsel for Lewis Oil and used his influence to arrange export transactions with Japan. The Broadway Central Bank in New York City vouched for Lewis's financial stability. Lewis assured Julian that he had access to virtually unlimited capital from Eastern sources.

Their summer talks had produced no results, but in early December, Julian, "as a last resource," wired Lewis "to write his own ticket on acquiring control" of Julian Pete. Lewis hastened to Los Angeles, accompanied by Senator King and two associates, Fred Packard and Jack Bennett. The four men established headquarters at the Biltmore Hotel and opened negotiations with Julian.

During the talks Lewis reinforced his earlier image. He recruited an impressively diverse board of directors, which included Senator King, an Indianapolis automobile manufacturer, the president of the Broadway Central Bank, a Southern lumber baron, and several experienced oil men who listed extensive interests in the United States and Mexico. Senator King introduced Lewis to local financiers and businessmen and helped him win the approval of Joseph Sartori, the most influential banker in Los Angeles. Unlike Julian, who had repeatedly found local financing closed to him, Lewis quickly arranged a $300,000 loan from Sartori's Security Trust and Savings Bank.

Julian offered to turn over his own common stock, which carried with it control of Julian Petroleum, to Lewis Oil in exchange for a pledge of "fresh capital." He did this, he claimed, "without one thin dime of payment or consideration in any way shape or form." While technically this may have been true, Julian clearly profited from the transaction, or at least cut his losses. On December 10, shortly before concluding the deal with Lewis, the board voted to cancel Julian's obligation to purchase 10,000 shares of preferred stock at $50 a share, thereby forgiving a $500,000 debt. In addition, Lewis agreed to reimburse Julian for funds he had advanced to the corporation. Julian also agreed to remain with the corporation for one year, without compensation, "in any capacity in which the new board might care to utilize me."

The formal transfer of power occurred at the December 19 meeting of the Julian Petroleum board of directors. The final resolution of the outgoing board charged that "This corporation has been handicapped for various and sundry reasons in properly financing itself," and recognized the need for "some definite arrangements . . . insuring to this corporation additional capital immediately." Citing the agreement with Lewis Oil as the "most practical, feasible and desirable of all plans, means and methods which have been heretofore considered," the board approved the contract drafted by Julian and Lewis. According to this agreement, Lewis Oil would provide $1.8 million in capital: $300,000 in cash, seventeen wells worth $500,000, and an additional $1 million by January 1, 1926. In exchange, Lewis would assume effective control of the corporation, which would still bear the Julian Petroleum name, and be empowered to appoint his own board of directors.

One by one, the old directors tendered their resignations. In a final gesture, Julian resigned as president. The reconstituted executive panel unanimously elected Lewis to the post. Nonetheless, as Julian later explained, "Lewis and I decided that it was best for us not to let the stockholders know that control of the company had changed hands." Thus, for public consumption, Julian remained the firm's president.

On December 22, Julian and Lewis jointly announced the reorganization of Julian Petroleum. Julian Pete, they explained, had absorbed the Lewis Oil Company. "It is simply a question of adding our combined strength to that of the present organization" of Julian Petroleum, stated Lewis. Lewis announced plans to inaugurate a $5 million bond issue, underwritten by the new directors and their associates in the East, which would be used to complete the refinery, add a fleet of tankers, and extend the system of filling stations. Julian, posing as president, nonetheless made it clear that he had surrendered "a very substantial part of the control." "I feel," he proclaimed, "that the addition of these men to our board, the large help to our future management, and their resources and standing in the business world practically solve all of the troubles of the Julian Petroleum Corporation and give the brightest possible outlook for the future."

"The day I turned over control," Julian later stated, "was far from being one of the unhappiest of my life. I had a strong sense of relief with my regret. I figured I had saved the company."

## III

Julian's transfer of power to S. C. Lewis in December 1924 occurred a scant two and a half years after he had boldly heralded his arrival at Santa Fe Springs. During that period he had emerged as one of the most celebrated and spectacular figures in Los Angeles. "He has put his name on hundreds of thousands of lips," reported the *Record*, "and fastened hundreds of thousands of eyes on his meteor course." In a city besieged by the glamour of Hollywood stars and movie moguls, oil millionaires and financial barons, "the public citizen would rather read about [Julian] than any other fellow citizen," asserted the *Record*.

Yet southern Californians were no closer to a consensus about Julian in 1925 than they had been one, two, or three years earlier. Was he, asked the *Record*, "a gay cafe figure, throwing investors' money away on giddy affairs in Hollywood nightclubs, or a hard-working businessman always putting his investors' interests first?"

To journalist Walter Woelke, Julian appeared to be "the farmer boy . . . milking his Los Angeles dupes" and "a flat failure" in the affairs of Julian Pete. Others saw him in a more sympathetic light. "It was the same old story," reported the *Coast Investor and Industrial Review*, "of a man untrained in big business and incapable of meeting the responsibilities that attend the consummation of great projects, facing inevitable failure." Contemporary commentators Guy Finney and Carey McWilliams, rendering perhaps the most common verdict, remembered Julian as he had portrayed himself: a sincere, if unorthodox, businessman who had challenged "Big Business and Big Finance" but succumbed to the pressures they had employed to defeat him.

That this image prevailed was as much a consequence of the highhandedness and inefficiency of the relatively new state and federal regulatory agencies as it was of Julian's unsurpassed skills as promoter and confidence man. Time and again, investigations of Julian fell short of compiling a conclusive case. Neither the corporation commissioner nor federal agents ever presented enough evidence to support an indictment of Julian, much less a conviction. Moreover, the heavyhanded tactics of Commissioner Daugherty and the postal inspectors reinforced Julian's portrait of persecution, while the 1924 citizens' committee report convinced thousands that Julian Petroleum was an honest concern.

Julian skillfully turned the shortcomings of the predators to the advantage of the prey. He keenly appreciated, even amidst the incessant glorification of business in the 1920s, the lingering fears of monopoly power and the widespread belief in conspiracies. He understood the growing concern with emerging government bureaucracies and added them to his list of demons to be exorcised. Julian's charges and accusations always resonated with the echo of truth. In this manner he presented a defense that not only convinced his contemporaries but that, even in retrospect, maintains considerable credibility. Indeed, only the FBI reports on Julian, unavailable to the public in the 1920s, allow a clearer judgement to be reached.

In their examination of Julian's earliest syndicates, Justice Department agents uncovered no evidence proving fraudulent intent. Investments in wells no. 1 through 5 were highly speculative by definition and utilized common-law trust and other features of dubious oil promotions. Julian's early ads overstated potential returns and exaggerated other details. But in most respects, these syndicates stayed within the law. Julian returned all oversubscriptions or recycled them into "special units" drawn out of his personal share of production. FBI investigators found no evidence of a conspiracy to manipulate the market in these Julian syndicates. On wells no. 1 through 5, investigators found that Julian had repurchased few units, and that price fluctuations seemed timed quite normally with dividend payments. Most significantly, all five wells struck oil and while Julian profited handsomely, earning at least $350,000, so did his investors.

Julian's experience in these wells, however, drove home a truth that he already must have suspected. The quickest path to wealth lay, not in pumping oil, but in promoting units. The Julian wells no. 6 through 9 syndicate bore all the earmarks of a fraudulent entry. In Agent Miller's November 1924 report, Julian admitted oversubscribing this syndicate, confirming information already known to federal officials. As always, he offered a plausible explanation. On his earlier wells, he stated, he had discovered "that there were many people who always wanted their money back or failed to pay up in full." The surplus sales offered "a reserve to take care of these conditions." The oversubscription, however, amounted to a staggering $600,000, or 60 percent more than authorized sales. Throughout 1923, Julian engaged in significant manipulations of these units, buying large numbers of units at low prices to artificially prop up the market. In the end, none of these four wells produced enough oil to justify dividends.

Had Julian deliberately "junked" these wells to mask the oversubscription? Maps of the Santa Fe Springs oil field show that No. 8, Julian's first dry hole, stood just outside the limits of the productive region. This failure delayed the location of No. 9 for almost two years. Wells no. 6 and 7, on the other hand, stood in one of the most prolific quadrants of field. Both had struck oil, but neither yielded sustained flows. Julian cited water difficulties as the explanation. Wilshire Oil, the original leaseholder on this property, suspected foul play. In July 1925, Wilshire Oil filed suit against Julian, charging that "he failed, refused and neglected to continuously pump and or operate to their full capacity, or at all," either of the wells. In 1926, the court found these allegations true and ordered the leasehold returned to Wilshire Oil. Under its management Well No. 6 revived, but failed to become a major producer. Julian investors received no return on this syndicate; his profit approached $1 million.

The story of Julian's final syndicate is similar. Julian told Miller he oversold these wells by $200,000. While Julian no. 11 and 12 both struck oil, each experienced problems and stood idle after July 1924. The Pico wildcat was a dry hole. The FBI estimated Julian's profits at $375,849. Overall, Julian collected $3,759,500 for his syndicates (the report mentioned no overissue), but by June 1924, had paid out only a little over $1 million in dividends. Julian's personal resources had risen from $2,565 in July 1921 to $1.24 million in July 1923, but units in Julian's syndicate wells, which had once had sold as high as $290, became virtually worthless by 1924.

The affairs of Julian Petroleum proved more complex. As conceived, Julian Pete may well have represented the type of shady refinery proposition that Commissioner Daugherty had warned against. Julian had oversold his first $5 million issue by $1.7 million. "He promised the stockholders that a refinery would be built from the proceeds of the first $5 million worth of stock sold," wrote Agent Miller, but "these proceeds have been spent and the refinery is not yet constructed." The large profits that Julian had stated would be realized in a very short time had "never materialized."

Although the refinery site remained vacant, Julian had spent millions of dollars for pipelines, storage facilities, oil wells, and leases. Based on appraisals by prominent experts and an audit by the prestigious Price-Waterhouse accounting firm, the citizens' committee in April 1924 valued company assets at over $8 million dollars. This, more than any single element, became the cornerstone for Julian's claims of legitimacy.

Agent Miller never questioned this study. In January 1925, however, the FBI dispatched another accountant to replace him. This unnamed agent expressed severe misgivings about Price-Waterhouse. The firm, he noted in his June 1925 report, had been paid, not by the citizens' committee, but by Julian Petroleum. The Price-Waterhouse audit had reported that a surplus of $45,000 existed on January 31, 1924, but the agent argued that this was "false and misleading." On March 15, 1924, ten days before Price-Waterhouse submitted its findings, Julian Petroleum had filed a tax return showing a $28,000 deficit at the end of 1923. To reconcile these accounts, the company would have had to show a $73,000 profit during January and February. In fact, corporate records showed a $118,000 loss. Either the tax return or the Price-Waterhouse audit was fraudulent.

The FBI agent interviewed accountant Hubert Stanley, who had prepared the Price-Waterhouse report. Stanley, the son of a pioneer figure in the firm and a Price-Waterhouse employee since 1912, could offer no satisfactory explanation of the discrepancy. Reviewing Stanley's work papers, the federal offical discovered that Stanley had counted in the profit ledger $87,500 received from Sun Oil in January for the delivery of crude oil. Julian Pete had been unable to fill this order, however. Julian "had to go into the open market and buy the majority of crude oil at prices in excess of what he had agreed to sell it for." By April, when Stanley prepared his statement, Julian Petroleum had suffered a loss on the transaction. The FBI accountant also questioned the audit's treatment of unpaid sales' commissions owed Julian. The agent concluded, "I cannot help but feel that [Stanley] is preparing the most favorable statement he can possibly prepare for this corporation. . . . I feel it is not honest when he signs a statement which he knows to be false."

Nonetheless, if Julian and Stanley had adapted the record to show nonexistent profits, the broader imprimatur of the citzens' committee stood. Julian had not absconded with money or blatantly misused the funds he had raised. He had acquired properties worth millions of dollars. He began a successful chain of service stations. He deferred payment of over a million dollars in sales commissions owed him and served for almost two years without salary. His bank accounts dropped from a height of $1.24 million in July 1923 to only $100,000 in early 1924, supporting his claims of large personal outlays. By April 1924, his reserves had risen, but only to $417,000. Even after Julian received $500,000 for turning over his interests to Lewis, it seems likely, as Julian later claimed, that "the net financial result to me of my whole experience with the Julian Petroleum Corporation from first to last is that I am considerably . . . poorer than I would have been if I had never organized the company."

In a sense, Julian may have become the victim of his own rhetoric. While Julian Pete remained first and foremost a moneymaking scheme, C. C. came to believe that he could nurture it into a great oil company. To do so, he had to build the promised refinery. But the initial capitalization allowed by the corporation commissioner, even including the oversubscription, was inadequate for this task. Unable to sell more stock or tap conventional sources of finance, Julian resorted to the strategy he knew best: speculating in oil leases in hopes of duplicating his success at Santa Fe Springs. Increasingly the fate of Julian Petroleum came to rest on a bonanza strike at one of the new southern California fields or the high-risk wildcat in Colorado. When these ventures failed, or yielded inadequate flows of oil, or simply ran out of money for continued drilling, Julian Pete, under Julian's leadership, was doomed.

Remarkably, the failure of the later syndicates, the mismanagement of Julian Petroleum, the depressed condition of Julian Pete stocks, and the reports of Julian's spendthrift ways failed to significantly tarnish his reputation. Many continued to believe that, but for the persecution of his enemies in business and government, Julian would have prevailed in his quest to rival Standard Oil. Tens of thousands of people willingly escrowed their investments at his command. If and when he would again beckon for their trust, they would respond to his call.

# PART II  STOCKS

# 7 WOULDST THOU MAKE MONEY?

## I

On Saturday afternoon, January 17, 1925, the roads to the Hollywood Bowl overflowed with Julian Pete investors for the second time in five months. "This is the one time I refuse to accept apologies for your absence," chided Julian in announcing the meeting to introduce the company's new directors. More than 20,000 people obeyed his summons. "No stockholders' gathering in the world's history has ever attained these dimensions," marveled the *Los Angeles Record*.

Julian himself opened the festivities. His bold, checkered yellow tie, held in place by a diamond stickpin, stood out for all to see. "You've been disappointed, and I've been disappointed," he shouted to his followers. "But it's been a hard battle. Now we're going to WIN! I tell you frankly that considering what has happened I wouldn't be here today but for the 40,000 stockholders who've never shown the white feather despite our reverses. A man can't quit on people like that."

The crowd cheered as C. C. attacked the "high bidders and second story men, some of whom I see sitting up there," for undermining the company's stock. But he had found "men of integrity, high standing in the financial world and knowledge of the oil industry to help me make this corporation a success." To recruit these men, Julian acknowledged, "I had to relinquish a substantial part of control" of Julian Petroleum, but nonetheless, "You will find me during 1925 fighting with you, as I have done in the past."

Julian introduced S. C. Lewis, who presented a very different image from that of the flamboyant Julian, to preside over the meeting. Lewis emanated an aura of dull solidity. He stood before his vast audience and read the company's 1924 balance sheet—item by item. Julian Pete, he reported, had a gross operating profit in excess of $500,000. Depreciation and depletion, however, had left the company $39,000 in the red, "an amazingly small sum for a new producing and marketing organization started under the conditions that confronted the Julian company." Assets far exceeded liabilities and profits would soar in 1925, he predicted. Lewis drew loud applause with the announcement that he had completed negotiations for $2.5 million of additional capital. Thousands cheered as the meeting adjourned.

With the Hollywood Bowl gathering behind them, Lewis and Julian divided corporate duties. Julian retained responsibility for completing the stock escrow plan and composing ads. As general manager, Lewis supervised operations. He also campaigned to earn the company a measure of respect from Los Angeles business élites. To achieve this, Lewis continued his courtship of Joseph Sartori and other leading bankers and hired the prestigious William Gibbs McAdoo as the firm's legal counsel.

The procurement of adequate supplies for the company's flourishing service stations posed Lewis's most pressing problem. Julian Pete wells produced far less oil than the demands of the retail outlets. Lewis negotiated a contract with Sierra Refining for the bulk of the gasoline to be sold under the Julian Pete banner. He also resumed drilling and pumping operations at Athens and Santa Fe Springs, expanded procurement pacts with independent producers, and acquired three producing rigs at Huntington Beach. On April 13, he applied for a Julian Petroleum listing on the New York Curb Exchange.

Lewis's greatest coup in his early months of leadership was the introduction of "Lightning Gasoline," the first ethyl auto fuel to be sold on the Pacific coast. General Motors research engineers had developed ethyl fluid, which, when blended with gasoline, increased its efficency by eliminating carbon. The new combination improved the pick-up of the car, eliminated knocking, and enhanced gas mileage. Standard Oil had begun distributing ethyl gasoline in the East. But no one on the West Coast had yet refined or marketed the product, and the cost of shipping it by rail was considered prohibitive. To the surprise of the industry, Julian Petroleum began dispensing "Lightning" premium gasoline on February 9, announcing it with customary fanfare in Julian's daily ad:

"Lightning Strikes Today"
"The End of the World?"
"NO"
**"BUT THE BEST QUALITY GASOLINE"**
**"IN THE WORLD"**
And Folks, that don't mean maybe either.

The introduction of ethyl gasoline provoked concern among competitors. How had Julian Petroleum acquired the product? Where did it originate? Lewis claimed that Lewis Oil engineers had developed Lightning. *Oil Age* speculated that the company transported the fuel in tankers, rather than by rail, to cut costs. Others dismissed it as a "trick gasoline." The Los Angeles city chemist twice attempted to examine samples of Lightning, but company officials rebuffed his efforts, charging that there was a plot to force Julian Pete to disclose its secret formula.

The controversy, combined with Julian's pungent advertising, piqued public curiosity about the new product. If Julian's ads are to be believed, Julian Pete sales jumped by 400 percent. By March 6, he boasted daily sales of 64,000 gallons, or about one-fifth of the gasoline sold in Los Angeles and its suburbs.

Despite this progress, Julian Pete still lacked the single most important element of a major oil company—a refinery. In April, Lewis remedied this state of affairs. Bypassing Julian's original plan to build a new facility, Lewis leased a refinery in nearby Hynes, California. The plant, previously operated by the Richfield and Pauley Oil companies, had the capacity to produce between 2,000 and 3,000 barrels a day. This coup completed a convincing opening display of Lewis's abilities. The Los Angeles business community hailed him as a refreshing change from his controversial, erratic predecessor.

Not everyone in Los Angeles was impressed by Lewis's performance, however. The accountant dispatched by J. Edgar Hoover in January 1925 found much to question in the Julian Pete situation. The agent had been sent to Los Angeles "on account of the unsatisfactory manner in which [his predecessor] Special Accountant Miller was conducting the investigation of this company." After his highly unorthodox interview with Julian in October, Miller had submitted his resignation in order to accept a $10,000-a-year post as Julian Petroleum comptroller. His replacement arrived on January 14 and found Miller, though officially on the FBI payroll until March 1, directing the Julian Pete staff in its preparation of statements and schedules to be used by Julian and Lewis at the shareholders' meeting. Other employees confirmed that Miller had been working for Julian Pete "for some time." The new agent immediately telegraphed Justice Department officials in Washington, who dismissed Miller from his post.

The new accountant attended the January 17 stockholders' meeting, but, unlike most gathered at the Hollywood Bowl, he examined the financial statement more critically. He found Lewis's 1924 year-end report, prepared under Miller's direction, "false and misleading." Assets of the company had been artificially appreciated by $1.5 million. Thus the deficit, represented at the meeting at under $40,000, in reality stood at $1,540,000.

The agent also raised questions about the validity of the accounting done by Price-Waterhouse on both the April 1924 citizens' committee report and a subsequent audit in February 1925. Hubert Stanley, formerly head of the Los Angeles office of Price-Waterhouse and currently manager of the firm's California branch, conducted both analyses. Stanley underrepresented the Julian Pete deficit by including appreciation of fixed assets among the profits. In addition, Stanley omitted "a contingent liability of accumulated dividends amounting to $626,824." Price-Waterhouse had calculated a deficit of $918,000. The federal accountant placed it closer to $2.8 million. The agent also questioned the cost and form of the report. Julian Petroleum had paid Price-Waterhouse $5,000, but Stanley had provided only a simple balance sheet.

When confronted with these discrepancies, Stanley defended his work. He argued that the omission of contingent liabilities was his usual practice and that

Julian Petroleum, unwilling to pay the cost of a full report, had only requested a balance sheet. The FBI agent remained unconvinced. "I have examined a number of Price-Waterhouse . . . reports," he wrote, "and this is the first one wherein I have failed to find contingent liability shown on the signed statement." Regarding the absence of a complete report, which Stanley admitted would have cost only an additional $200, the agent concluded, "I am led to believe that the reason a complete report was not furnished was because there were so many assets shown on this statement . . . that were worthless or of doubtful value, that if he had written a complete report he would have to qualify his report as well as his financial statement."

The diligent investigator also had serious doubts about the conditions under which Lewis had gained control of Julian Petroleum. Lewis, he noted, had agreed to furnish $1.8 million in additional capital: $300,000 in cash, seventeen wells worth $500,000, and an additional $1 million within a year. In reality, the inital cash payment had come, not from Lewis, but in a loan to Julian Petroleum, against its own assets, from Sartori's Security Trust and Savings. Nor could the agent find any records to confirm the value of the wells. "I have been unable to get the cost of any of these wells, neither have I been able to get the production of these wells," he reported. At the same time, the Price-Waterhouse statement showed these mystery wells to have appreciated in value to $675,000 in two months. Thus, contrary to the claims of Julian and Lewis, Lewis had acquired control of Julian Pete without investing a single penny of additional capital.

Indeed, according to later testimony by Fred Packard, Lewis's personal secretary, when Lewis arrived in Los Angeles, he had virtually no cash and a string of debts in Texas. Packard, a former employee of Utah senator William King, claimed that Lewis had recruited him into his entourage to gain an acquaintanceship with King. Lewis promised Packard substantial remuneration and a good position for this service. In December 1924, Lewis established headquarters at the Biltmore Hotel, renting one room for himself and a second for Packard and King. Packard later recalled, "He was always wiring different places for a small amount of money, $500 or $1000, which looked ridiculous to me at the time, because of the representations that had been made to us as to the standing of Mr. Lewis and his company." After two weeks, Packard discovered that the hotel accounts remained unpaid. "These bills must be paid," Packard warned Lewis. "If Senator King knows the bill is not paid, he is going to be suspicious." Lewis reassured Packard that "in the course of a few days everything would turn out all right." Packard said that he paid the bills with his own funds and that only after securing control of Julian Petroleum and his first Los Angeles bank loan did Lewis's finances significantly improve. Lewis adamantly denied Packard's story.

"It is hard for me to determine just what the probable loss [of Julian Petroleum investors] will be," wrote the federal agent in his final report in June 1925, but he estimated that it would amount to over $8 million within eight months of his report. He listed Julian, Lewis, Miller, and Stanley as "possible defendants" in a criminal suit.

Neither the Department of Justice nor any other branch of the federal government, however, took any immediate action to prosecute the Julian Petroleum participants. This is particularly perplexing in light of the fact that the department's Bureau of Investigation already had compiled a damning dossier on Lewis's stewardship of Lewis Oil. In June 1923, the Bureau, then under the direction of William J. Burns, had received a complaint about Lewis Oil from one Alphonso Gales Johnson, who detailed his allegations in a second letter, dated August 17, 1923.

Lewis Oil, Johnson charged, had been organized in 1921 "for the purpose of taking over the Monarch Petroleum Company . . . and several other small companies [all of which] were in bad condition financially." The condition of these companies under Lewis's guidance not only failed to improve, but Lewis Oil itself "was not able to meet [its] obligations." In 1922, Lewis Oil moved its offices to New York "for the purpose of taking over other small companies by an exchange of stock," basing these acquisitions on "gross misrepresentations." In the spring of 1923, according to Johnson, "after realizing the serious financial condition they were in . . . they organized the Lewis Oil Market-Export Corporation." The new corporation sold stock through a mail campaign using statements that "they are unable to substantiate with facts."

Johnson claimed that Lewis Oil issued erroneous financial statements designed specifically to mislead stockholders. In addition, Johnson alleged that Lewis Oil had declared stock dividends on common stock despite a lack of funds, "for the sole purpose of pacifying the stockholders while further stock selling schemes are promoted." "I firmly believe," wrote Johnson, "that the methods of the Lewis Oil company are questionable and should be carefully investigated."

Director Burns turned over the Lewis Oil matter to Post Office inspectors in Texas who launched an investigation into possible mail fraud. Lewis, who had his own sources in either Washington or Texas, quickly learned of Johnson's complaint. When Johnson visited his office in September 1923, Lewis introduced him, according to a subsequent Johnson communication, as the person "who was causing so much trouble with the stockholders . . . and also with the Department of Justice." As Johnson left, he related, Lewis threatened him on account of these actions.

Subsequent Justice Department and Post Office investigations substantiated Johnson's charges and uncovered a different life story than the one Lewis had presented in Los Angeles. Lewis had been born, not on a Virginia plantation, but in the town of Gravelly Springs, Alabama. Except for a brief stint in Richmond, Virginia, Lewis had received his schooling in Texas and spent his teenage years not at the University of Virginia, nor in the Alaska oil fields as he alternately claimed, but in a variety of railroad clerking jobs in Texas and Colorado. He graduated from Dallas Law school in 1911 and worked for the legal claims department of the Texas and Pacific Railroad until 1913, when he was admitted to the Texas bar. He practiced law in Dallas for nine years before opening the New York offices of Lewis Oil in 1922.

Lewis Oil, much as Johnson had charged, resulted from the merger of nineteen financially strapped companies. From the start, according to FBI reports, "It was a one man concern. . . . The corporation was at all times under the complete influence of Lewis and the Officers and Directors were never considered in the activities except to ratify certain actions when he saw fit to disclose same." Fred Packard later recalled that, although he held the position of vice-president of one affiliate and endorsed many documents, he never attended any meetings. Lewis would hand him papers and say, "You saw these last night. Sign them."

As he would later demonstrate with Julian Petroleum, Lewis had mastered the art of the false financial statement. A federal audit found the 1923 Lewis Oil report overvalued by $1,256,000. Lewis had achieved this sleight-of-hand by juggling lease values. A lease purchased in September 1921 for $73,000 was reassessed just prior to the 1923 accounting at $325,000. Five months later, Lewis Oil sold it for $68,000. A second lease was appreciated by $400,000 in order to wipe out a $207,000 operating deficit for oil production. These and other transactions, reported Bureau accountants, "were placed on the books to build up a surplus balance to be reflected in the Financial Statement that was being made at that time. It appears that the entries, although bearing the date of March 31, 1923, were not set up until a later date." A similar pattern appeared in the 1924 financial statements.

In November 1923, Lewis began to raise funds by issuing $500,000 worth of five-year "Gold Notes." To sell these notes, Lewis's mailings and salesmen depicted Lewis Oil as "in good financial condition," promising that it "could and would pay dividends at an early date." The earnings of the corporation, they stated, "were more than enough to pay the interest three times over in any year." In reality, only one of the many companies listed as part of the Lewis Oil empire at this time was returning a profit.

In the Gold Note campaign, Lewis came to rely increasingly on the efforts of his top salesman, Jacob Berman. Little is known of Berman's background. Born to a family of Jewish immigrants in Brooklyn in 1896, Berman had previously worked as a salesman for a New York City brokerage firm. Some claim that he served time in a New Jersey prison. Berman entered the employ of Lewis Oil in January 1924, but it is possible that his relationship with Lewis preceded that date. Standing five feet six inches, with dark-brown receding hair, brown eyes, a dark complexion, and what the FBI later described as "hebraic features," Berman proved to be an expert salesman and stock manipulator. He traveled through the United States and Canada under a variety of aliases—J. Lamar, J. Sheppard, J. Wagner, J. Stewart, and, most often, Jack Bennett—purveying Lewis Oil Gold Notes or exchanging them for other, more valuable stocks.

The persuasive Berman used many ruses to stimulate sales. In Meridian, Mississippi, he offered an elderly mechanic control of the local Lewis Oil service station in exchange for a $1,000 investment. The mechanic turned over his life's savings, but no station materialized. Most often Berman guaranteed spectacular short-term profits. He promised to double a Montreal couple's money within ninety days. They exchanged $15,000 worth of American Telephone and Telegraph stock for Gold Notes with an equal face value. Berman immediately sold the A.T.&T. stock, but interest coupons on the Gold Notes were returned unpaid. Willie Baldwin, a wealthy Quebec merchant, lumberman, and member of the Canadian Parliament, invested $20,000. A department store owner in Burlington, Iowa, borrowed $24,000 when told that within three months he would receive $130 for every $100 he invested.

The acquisition of Julian Petroleum offered Lewis and Berman new opportunities for creative bookkeeping and misrepresentation. In Los Angeles, Lewis had represented Julian Petroleum as the parent company that now owned Lewis Oil's California properties. In the 1925 Lewis Oil financial statement, Lewis claimed the opposite. Lewis Oil, he reported, now had a controlling share of Julian Pete common stock. The assets of Julian Petroleum were listed as holdings of Lewis Oil. The duplicate accounting gave both companies the appearance of prosperity. In reality neither the transfers of property to Julian Petroleum nor ownership of common stock ever appeared on the Lewis Oil books. Lewis appropriated the common stock for his personal use.

Berman also capitalized on the Julian Pete acquisition. A Louisiana baker bought $1,500 in Gold Notes from Berman so that the money would be used to "take over and make more solid the Julian Company." Widow Anna Sharkey invested $1000 under the same assumption. Over a nineteen-month period, Lewis, Berman, and other salesmen disposed of $845,300 in Gold Notes, oversubscribing the authorized issue by 69 percent. Berman personally received over $100,000 worth of notes that were never accounted for in the company's records.

By late 1924, Jacob Berman had become Lewis's first lieutenant. "I trusted him implicitly. He had my utmost confidence," stated Lewis later. When Lewis arrived in Los Angeles to take over Julian Petroleum, Berman was at his side. Lewis introduced him as "Jack Bennett." The combination of Julian Petroleum and the speculative fervor of Los Angeles residents opened new vistas for their combined imagination.

## II

In the spring of 1925, employees in the Julian Pete stock transfer office noticed an unusual state of affairs. A corporate transfer office keeps track of all stock transactions. Whenever a share changes hands, the original stock certificate is endorsed by the owner and returned to the company. The transfer office mutilates and cancels the old certificate and prepares a new one, which is then sent to a corporate officer for a validating signature before it is issued. Normally, the secretary or another company official supervises stock transfer activities. At Julian Petroleum the transfer staff received instructions from the man they knew as Jack Bennett.

Bennett held no official position with Julian Petroleum, but his desk adjoined Lewis's in the corporate suite in the Pershing Square Building. Employees generally acknowledged him as "next in command" to Lewis. In the area of stocks, recalled secretary T. P. Conroy, Bennett had "the final word." Chief stock clerk C. T. Harris recalled, "Though Bennett was not an officer in the corporation, the stock transfer office accepted orders from him."

A great deal of confusion existed about the legal number of Julian shares outstanding. The original corporate charter had authorized the sale of 200,000 shares of preferred stock and an equal number of Julian common. In 1924, Julian had acquired a second charter, tripling the company's original $10,000,000 capitalization and calling for the issuance of 600,000 shares each of preferred and common stock. During Julian's regime, however, the Corporations Department had halted sales after the issuance of 159,064 shares of each type of stock. Barring a new public offer approved by the corporation commissioner, this would remain the total number of shares authorized for trading. Yet, a common perception existed among bankers, brokers, and investors that the 600,000-share figure represented the legal limit, a misconception that would plague Julian stock traders for the next two years.

Lewis inherited from Julian an overissue of approximately 10,000 shares above the 159,064 limit of preferred stock. On January 1, 1925, Lewis cancelled these excess shares, presumably from his personal holdings, temporarily restoring legality to the company's preferred stock issue. In the early months of 1925, according to transfer clerk Pat Shipp, Bennett began requesting new stock certificates, promising to produce older ones for cancellation within a few days. The original forms rarely appeared. Bennett would also instruct the clerks to sign or affix officers' signatures to the new issues. A corporate by-law enacted two days after Lewis had assumed command facilitated this practice. A rubber-stamped signature of one of the officers now sufficed to validate new stock certificates. "Bennett told me written authority would be forthcoming from T. P. Conroy, when I signed Conroy's name to the stock certificates," related Harris. "This authority never came."

In January and February of 1925, the transfer office issued 304 new shares of preferred stock and cancelled only two. In March and April, over 7,500 shares emanated from Julian Pete without a cancellation. In May the office released an additional 22,808 shares of preferred stock. Shortly thereafter, Julian Pete vice-president H. F. Campbell demanded that cancelled stock accompany the new certificates presented for his signature. Lewis objected that this would be too "cumbersome." When Campbell refused to sign additional issues, the transfer office subsequently bypassed him by using a rubber stamp. Within five months the new regime thus overissued the preferred stock by almost 20 percent. (Bennett and Lewis also overissued the common stock, but records of these transactions are unavailable.) Whatever their misgivings or suspicions, Julian Petroleum transfer clerks dutifully followed Bennett's orders.

Other irregularities also materialized. In his agreement with Julian, Lewis had received 51 percent of the common stock, giving him a controlling interest in the company. This stock was to be placed in escrow pending the acquisition of additional financing. Lewis, however, surreptitiously sold these shares within a few days, thereby raising funds for living and operating expenses in Los Angeles and to pay off debts in Texas. Since these transactions went unrecorded, no one realized that Lewis had forfeited voting control of Julian Petroleum.

TABLE 1. Julian Petroleum Chronology and Stock Prices
(December 1924 to May 1927)

| | | Stock Prices* | |
|---|---|---|---|
| *Date* | *Event* | *Preferred* | *Common* |
| *1924* | | | |
| December 22 | Lewis takes over Julian Pete | 19⅝ | 7½ |
| December 25 | | 24 | 11 |
| *1925* | | | |
| March 8 | | 21 | 9¾ |
| March 15 | Julian drive up common stock | 22 | 15⅝ |
| March 18 | | 23 | 18½ |
| March 29 | | 20¼ | 14 |
| April 26 | | 21 | 11⅝ |
| May 10 | 20,000 shares preferred issued | 17 | 7⅝ |
| May 17 | | 12⅝ | 7¼ |
| July | 120,022 shares preferred issued | | |
| July 26 | | 12 | 8 |
| August 16 | | 8¼ | 5¼ |
| September 15 | | 6⅛ | 3⅜ |
| September 18 | Lewis launches bull market | | |
| September 21 | | 9¾ | 3½ |
| October 1 | | 17 | 7 |
| October 10 | | 23⅞ | 8 |
| October 15 | | 30 | 9¾ |
| October 16 | Bull market collapses | 13⅞ | 5 |
| October 17 | | 12⅛ | 5¼ |
| *1926* | | | |
| January 16 | | 15 | 4⅛ |
| January 30 | | 18 | 5½ |
| February 11 | Stockholders' meeting | 20 | |
| February 18 | Second bull market begins | 22¼ | |
| February 28 | 29½ | | |
| March 2 | | 30 | 8 |
| March 12 | End of second bull market | 35 | 5⅛ |
| March 13 | | 28 | 4⅝ |
| March 18 | Lewis purchases A. C. Wagy & Co. | 20⅛ | 4 |
| March 27 | | 15¾ | 3¾ |
| April 27 | Rumors of merger | 20⅛ | 6⅝ |
| April 29 | Lewis announces merger plans | 19 | 5¾ |
| May 1 | Beginning of bear raid | 17⅝ | 5½ |
| May 13 | | 14 | 3½ |
| May 15 | | 11⅞ | 3 |
| May 29 | | 16 | 4 |
| June 3 | | 12⅝ | 3 |

TABLE 1. (continued)

| *Date* | *Event* | *Stock Prices** | |
|---|---|---|---|
| | | *Preferred* | *Common* |
| June 8 | | 12½ | 2½ |
| August 8 | Lewis announces Marine Oil merger | 15¼ | 2½ |
| September 17 | | 17 | 2⅞ |
| September 21 | Million-dollar pool created | | |
| September 24 | | 20¼ | 3¾ |
| October 6 | | 22⅝ | 3½ |
| October 7 | Market breaks | 13¼ | 2½ |
| *1927* | | | |
| February | Bennett overissues 558,056 shares | | |
| February 8 | | 13½ | 2¾ |
| February 9 | | 11½ | 2¼ |
| February 10 | | 7½ | 1⅞ |
| February 14 | | 10 | 2⅛ |
| February 18 | | 8 | 1⅞ |
| March 25 | California-Eastern financing announced | 11½ | |
| April 2 | | 8⅞ | |
| April 14 | Million-dollar pool liquidated | 7¼ | |
| April 19 | | 8¼ | |
| April 22 | California-Eastern financing collapses | 5 | 1 |
| May 2 | Jack Bennett flees Los Angeles | 4¼ | |
| May 4 | | 2⅛ | |
| May 6 | Julian Pete taken off market | 2⅞ | ¾ |

*Stock prices represent daily extremes rather than opening or closing prices.
(Source: Financial pages, *Los Angeles Times*)

How much C. C. Julian knew of the state of affairs at Lewis Oil is unclear, but within a few stormy months he had grown disillusioned with S. C. Lewis. In late January, despite three broken ribs suffered in an auto accident, Julian was putting in eighteen-hour days promoting the Julian Petroleum escrow scheme. "I've been working on this deal of escrowing your stock for 4 months now," he advertised. "It is essential to the success that we have every share of both Preferred and Common in the bank. . . ." Julian claimed that while 85 percent of the smaller stockholders had cooperated, unless he could persuade the shareholders who held $1,000 to $25,000 dollars' worth of stock to participate, "all my efforts must prove futile."

Throughout February, Julian's ads pounded away at reluctant stockholders. His persuasive powers, however, could not overcome the temptation of brokers and shareholders who hoped to capitalize on the price rise dictated by the anticipated stock shortage. On March 2, Julian threw in the towel, announcing his traditional six-day close-out sale. "I must have at least a million dollars more stock in our vault in the bank or my plan has blown up," he confessed. "[This] means better than a

TABLE 2. Julian Petroleum Stock Overissue

| Date | Shares Authorized: 159,604 | |
| | Preferred Shares Issued | Total Preferred Stock Outstanding* |
| --- | --- | --- |
| **1925** | | |
| January/February | 304 | 159,365 |
| March/April | 7,566 | 166,922 |
| May | 22,808 | 189,730 |
| June | 58 | 189,788 |
| July | 120,022 | 309,810 |
| August/September | 58 | 309,864 |
| October | 3,543 | 313,399 |
| November | 8,020 | 321,419 |
| December | 51,885 | 372,404 |
| | | |
| **1926** | | |
| January | 51,252 | 423,656 |
| February | 60,393 | 483,594 |
| March | 236,217 | 683,580 |
| April | 87,699 | 770,279 |
| May | 230,256 | 987,285 |
| June | 275,666 | 1,231,192 |
| July | 245,706 | 1,458,638 |
| August | 369,174 | 1,773,502 |
| September | 118,976 | 1,888,477 |
| October | 91,504 | 1,979,531 |
| November | 127,135 | 2,105,666 |
| December | 255,674 | 2,342,340 |
| | | |
| **1927** | | |
| January | 208,915 | 2,523,255 |
| February | 558,056 | 3,033,121 |
| March | 284,056 | 3,318,050 |
| April | 301,832 | 3,614,283 |

*After cancellations
(Source: Haskins & Sells Audit, FBI Reports)

hundred and fifty thousand a day and it certainly has not been coming at that rate lately." But deadlines on stock surrender had none of the urgency of earlier limited stock sales. Despite a last-minute rush that, according to Julian, "came in so fast and furious we were unable to keep tab on it," the escrow campaign closed on March 8, with a reported 80 percent of the stock in escrow.

The combined removal of substantial amounts of Julian stock from the market and the successful Lightning gasoline marketing campaign logically should have

driven up the price of both preferred and common shares of Julian Pete. Throughout January and February of 1925, however, stock prices hovered close to the levels
attained when Lewis had acquired control. The preferred stock, which had jumped
to 24 after the December announcement, held between 20 and 22 for the next two
months. The common stock, which had peaked at 11, had slipped as low as 8¼ and
had locked in under 10 throughout February.

Julian found this puzzling. He and Lewis had organized investment pools to
increase prices. "After Mr. Lewis took over control," he later revealed, "I told him
that a little money would put that market in a good substantial figure." Julian put
up $50,000 or $60,000 and raised an additional $150,000 for the Julian Petroleum
Trust Pool "to buy up stock on the market . . . and try to stabilize the price at a
higher figure." Lewis, Julian, and Bennett met nightly to "program their pool
schemes." These efforts failed, Julian discovered, because Lewis "had put out my
$200,000 worth of stock himself . . . that is, he had issued new stock. . . . I found
out I had been taken for whatever amount of money I had put in." According to
one account, Julian stormed into Lewis's office, "and when he got through his verbal
fireworks . . . everything was scorched for a hundred yards to the right and left."

Julian was determined to recoup these losses. He proposed a second pool to
Lewis and initiated a campaign to drive up the price of common stock.
"**WOULD'ST THOU MAKE MONEY**'" he advertised on March 13. "Well
here's the ticket. '**BUY JULIAN COMMON.**'" The stock, selling at $9, he pre-
dicted "was due for a snappy rise." Within two days, Julian common had jumped
to 15½. Stockbroker W. H. Durst, undoubtedly working in conjunction with Julian,
ran large ads in the March 15 papers, urging Julian Pete shareholders to "go into
the market and buy up small blocks of stock now floating around." If each of the
40,000 Julian investors would buy a single share, promised Durst in a remarkably
candid attempt at stock manipulation, "the price will shoot skyward overnight." By
Tuesday, March 17, the stock stood at 18⅓, double the price a week earlier. "We
can't help saying 'We told you so,'" boasted Durst. "Finish a job well done. All
together buy every share of Julian offered on the local market." Meanwhile, Julian
had "handed [Lewis] back the stock" and more than avenged his earlier losses.

Julian's ad of March 17 warned his supporters to "**WATCH YOUR STEP**"
and insist on immediate delivery of all stock. "Julian Preferred and Common are
both going plenty higher than they are quoted at today," he wrote. "'PREFERRED'
is a steal at the present quotations." Though Julian did not know it, these would be
the last lines of copy he would ever write for Julian Petroleum.

Three days earlier, on March 14, Julian had been driven to Yuma, California,
to inspect prospective oil land. Since some of the land was in Mexico, his chauffeur
had to cross the border at the junction of the towns of Calexico and Mexicali, where
customs officials, despite Julian's strenuous objections, searched the car. When an
officer discovered a bottle of liquor bearing a prescription label (a common decep-
tion during Prohibition), Julian ordered his chauffeur to drive off. The inspectors
halted his auto and turned the group over to the Calexico police.

At the police station Julian became abusive. He gave a false name and threatened to "spend a half-million dollars to have the entire police force fired." When he was confronted with papers bearing his name, Julian readily admitted his true identity. The police booked Julian, but released him on a $25 bond, which Julian forfeited when he left town.

The news of the Calexico incident reached Los Angeles on March 18, as Julian Pete issues rose on the stock market. Julian dismissed his arrest as the product of "some unnecessary and unjust action by some small officials at the border. . . . I refused to let them take [the prescription liquor] and that was my right. . . . I was arrested on general principles." Nonetheless, a common perception of the affair was that Julian had "jamboreed at Mexicali," gotten drunk, and "engaged in a fist fight." The incident, along with Julian's dumping of his common stock, dampened the buying fervor, plummeting stock prices precipitously.

Julian's escapade widened the breach with Lewis. "The new money that came in and my personal connections with the oil supply companies had started things ahead," Lewis later explained. The reaction to Julian's latest misconduct, however, "was very severe. . . . [The suppliers] felt that Julian was using me as a subterfuge and my money was a subterfuge to get the cooperation of the local bankers." Lewis rebuked Julian for his actions, and Julian "resented it very much that I would discuss it with him," Lewis recalled. As a result, the two men decided to sever Julian's remaining connections to the firm.

On April 6, the Julian Pete board of directors formalized this agreement. The directors approved Julian's expenditures for the escrow plan but disavowed any responsibility for the Julian Petroleum Trust Pool he had organized. The board also authorized the payment of $187,000 to Julian for monies owed him in interest and commissions. With these matters completed, the board accepted Julian's resignation. Nine days later, Lewis revealed the news to the public.

"JULIAN IS OUT OF OIL COMPANY OF HIS NAME," read the celebratory lead story headline in the *Los Angeles Times* on April 15. "C. C. Julian is no longer an officer or director of the Julian Petroleum Company and holds not one share of stock that carries a voice in the company's management." Lewis admitted that he had been president for four months. Julian confirmed this for the *Record*. "The reason everything was kept secret was because I did not want the stockholders to become alarmed," he stated. "If I had suddenly dropped out there would have been a panic. Now that the transfer is a proved success, however, I feel free to bare the transaction."

For the reporter from the *Times*, on the other hand, Julian refused to comment. When asked about his retirement, he "crossed his legs, cocked his feet on the desk and took a deep draft from his cigar."

"It is my opinion," he observed, "that when the Prince of Wales arrives in India, a determined effort will be made to stamp out these recurrent outbreaks of cholera. Don't you think?"

### III

With this characteristically flippant response to the *Los Angeles Times*, C. C. Julian turned over the public spotlight to S.C. Lewis. The days of daily Julian Pete ads and weekly controversies had ended. During the spring and summer of 1925, Lewis maintained a low profile, dividing his time between Los Angeles and New York, expanding operations and arranging financing. Under his guidance, the Julian Pete drama shifted from the oil fields and investigatory agencies to the floor of the Los Angeles Stock Exchange (LASE).

Founded in 1899 as the Los Angeles Oil Exchange, the LASE had grown slowly for two decades. At the dawn of the 1920s, the LASE, headquartered at Sixth and Spring streets in downtown Los Angeles, remained a modest operation. The same forces, however, that generated the massive expansion of securities transactions on Wall Street in the 1920s—general prosperity, low interest rates, growing public demand, and ever more generous loan and margin arrangements—energized the nation's regional exchanges. Nowhere did this process accelerate as rapidly as in Los Angeles. Oil and mining discoveries, the rise of Hollywood, and the increasing reliance upon local institutions to finance business and agriculture in the Southwest transformed the Los Angeles stock and bond market into one of the most important in the nation. "Ten years ago there were perhaps six bond houses in the city. Today there are at least fifty," reported Albert Atwood in 1923. "Probably all the office space occupied by bond houses ten years ago could be put into one of the larger offices today." Stock brokerages witnessed the same explosive growth.

The city's brokerage houses fell into two groups: the sixty to seventy "most reputable stockbrokers" represented on the LASE, and the "bucket shops," smaller, often unlicensed, non-members of the exchange. The LASE, governed by President Frank H. Pettingell and an elected board of directors, strove to bring efficiency and respectability to local securities trading. During the 1920s the Exchange steadily modernized its operations, installing fifty stock tickers in 1924 and expending $1 million in 1925–1926 to wire a direct connection to Wall Street.

The popular Pettingell presided over these developments. A native of Massachusetts and descendant of New England pioneer stock, Pettingell had forged a successful banking career in Colorado before arriving in Los Angeles in 1912 at age forty-four. Within a year he won election to the LASE presidency. In addition to his duties on the Exchange, Pettingell served on the board of directors of the Public Library and was an active participant in the local chapters of the Society of Colonial Wars and the Sons of the Revolution. Pettingell's regal, blueblood figure lent an aura of prestige to an environment that, unlike Wall Street, lacked a firmly entrenched trading establishment.

As in New York, LASE brokers in the 1920s increasingly relied on margin sales to expand their clientele and volume. Under the margin agreement, an investor could buy shares by depositing a fraction of their market value, usually 50 percent, in either cash or other securities, and borrowing the remainder from the broker at a fixed rate of interest. The balance of the purchase price would be paid in monthly

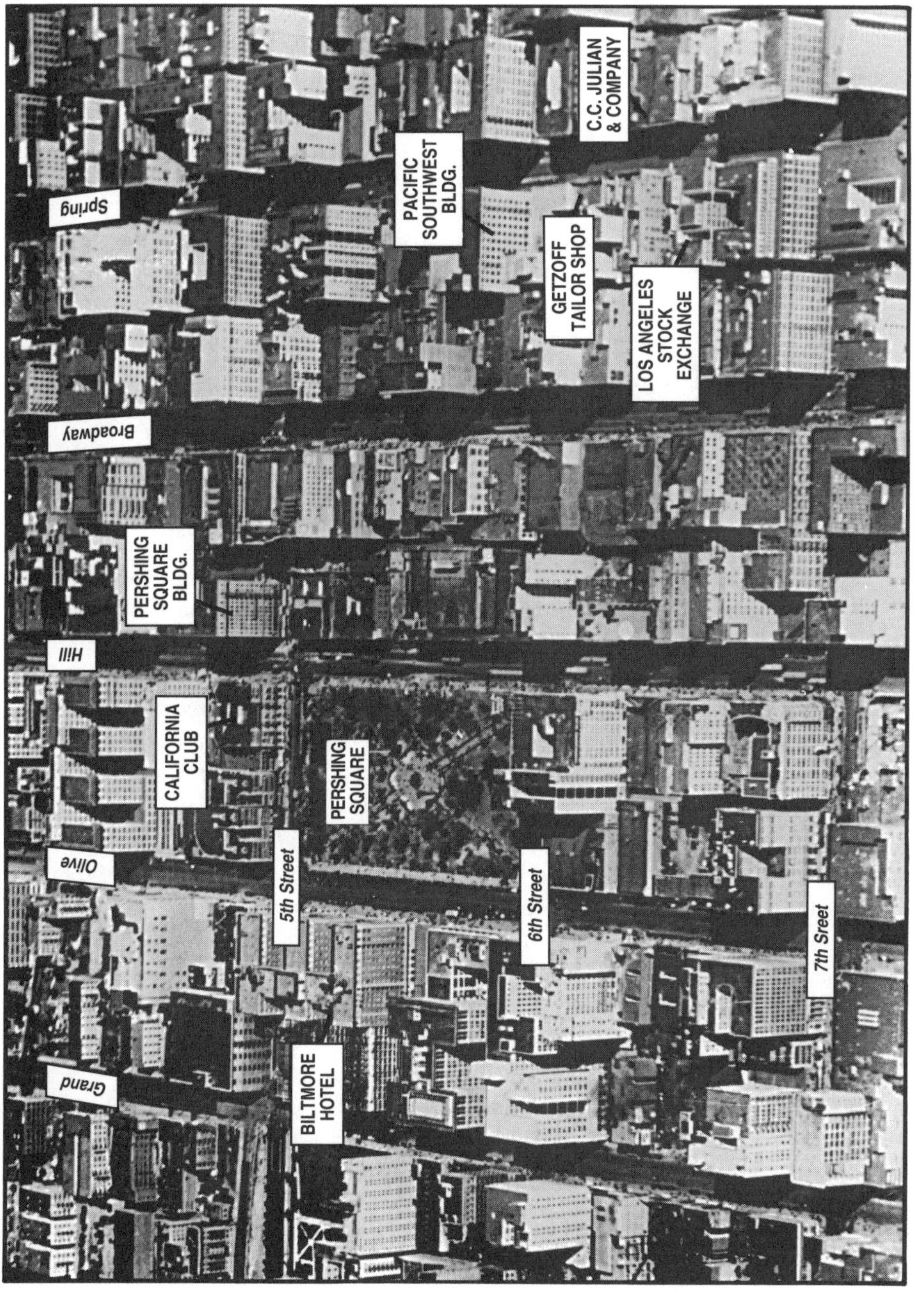
Spring
Broadway
Hill
Olive
Grand
PACIFIC SOUTHWEST BLDG.
C.C. JULIAN & COMPANY
GETZOFF TAILOR SHOP
LOS ANGELES STOCK EXCHANGE
PERSHING SQUARE BLDG.
CALIFORNIA CLUB
PERSHING SQUARE
5th Street
6th Street
7th Sreet
BILTMORE HOTEL

Downtown Los Angeles, 1920s

installments. Margin arrangements thus offered benefits to both buyer and broker. The margin plan allowed investors anticipating a rising market to expand their holdings at minimal cost. They retained the rights to all dividend payments. Brokers profited both from the commissions on increased sales and from the interest charged on margin loans.

A. C. Wagy and Company, a LASE member catering to the small investor, noting that the demand for margin trading had made it "almost obligatory" to extend this accommodation to its clients, offered a typical plan for the purchase of Julian Pete. "Lack of funds presents no handicap to those desiring to purchase Julian Petroleum shares at present prices," Wagy and Company advertised. The brokerage promised to sell "any reasonable number of shares of Julian equal to the shares you deposit to us as collateral . . . or you can deposit one-half cash." Buyers were obligated to maintain the margin at 50 percent. Thus, if the price of the stock dropped below this level, investors could be asked to provide additional collateral or be forced to sell the stock at a loss. Rules established by the LASE, based on those prevailing on the New York Stock Exchange, governed both the allowable margin and the interest rates which members could charge.

Those who found these terms too stringent could turn to the nether world of the bucket shop to broaden their investment opportunities. Bucket shops offered a wide variety of schemes. Some specialized in installment sales, "twenty percent down, the balance in eight equal monthly payments," and "short selling." The bucket shop operator played the law of averages, reported the *Record*. "He figures that before the eight months have elapsed and you make your final payment he can buy the stock from some other source . . . for about 25 cents on the dollar. Nine times out of ten he can."

Other operations included talented "switchers" and loan sharks. Switchers preyed on migrants from the East who brought with them securities, mortgages, and land contracts. Under California law, "foreign" securities were subject to taxation, while those purchased locally were not. Switchers persuaded new residents to exchange their Eastern holdings for locally issued, high-dividend stocks, most of which proved short-lived or were fraudulently issued. Loan sharks offered cash for sound securities, charging high rates of interest, then foreclosing to acquire control of the stock.

In New York during the 1920s, writes historian Robert Sobel, the rapid expansion of the securities markets meant that "uninitiated brokers were selling stocks to an uninformed public." The situation in Los Angeles was even more treacherous. In New York the firmly entrenched securities establishment at least attempted to temper the activities of shady and inexperienced operators both on and off the Exchange. In Los Angeles, on the other hand, where few brokerages could boast a long history, the line between the practices of the legitimate houses and the bucket shops was often dangerously thin.

This often chaotic and unstructured atmosphere offered both opportunity and challenge for S. C. Lewis and Jack Bennett. During the spring and early summer of 1925, an absence of information and the periodic overissuing of stock drove the price of Julian Petroleum securities steadily downward. From January to April, preferred shares had floated within a narrow range between 20 and 23. Common stock, influenced by Julian's pools, had moved more erratically, swinging from a low of 8⅜ to a high of 19, and, at the end of April, back to 11⅝. In May, however, Bennett's first major overissue of preferred stock caused a spectacular drop in prices. Preferred shares, which had sold at 21 as late as April 26, plummeted to 12⅝ three weeks later. Common stock also declined, slipping as low as 7¼ Prices held at this level through July.

In August the market for Julian securities suddenly collapsed. By mid-month, preferred shares could be bought as low as 8¼ and common stock moved under 5. Lewis blamed the slide on "'Sharpshooters' among the Los Angeles Brokerage fraternity, in their methods not unlike a pack of ravening wolves, [who] planted emissaries in the Julian Corporation offices" and floated rumors detrimental to the company. "The bolshevik element of the brokers," he later alleged, "were getting crowds of [stockholders] together to start attacks on the officers of the company." In addition, Lewis reported, high-pressure salesmen were attempting to persuade Julian Pete stockholders to exchange their securities for mining stock.

To combat the rumors and boost prices, Lewis resorted to one of Julian's old ploys, a public shareholders' meeting, scheduled for August 19, 1925, at the Philharmonic Auditorium. "Since taking over control of this company," Lewis advertised in prose that would never be mistaken for C. C. Julian's, "myself and my associates have extended our utmost efforts to accomplish . . . almost a herculean task, namely to put the CORPORATION on a dividend paying basis. We have made progress. We have made great progress. We are proud of our accomplishments."

Although Lewis lacked Julian's drawing power, a turnout of 8,000 people necessitated two back-to-back meetings. The participants cheered as Lewis outlined his achievements. Since January 1, he announced, Julian Pete had completed eight producing wells. Oil production now stood at $100,000 a month, and some 40,000 miles of pipelines had been built. Gasoline sales in August, he predicted, would surpass 300,000 gallons a day. Most important, Julian Petroleum now had controlling interest in three refineries—the Hynes facility, a second plant at National City near San Diego, and a third in Phoenix, Arizona.

The spirit of C. C. Julian hovered over the meeting. Prolonged cheers echoed through the hall at the first mention of his name. When Lewis reported that Julian had taken ill, both audiences responded with sympathy. Each group unanimously approved a motion to give the current officers authority over stock escrowed with Julian in 1924.

The enthusiastic meetings failed to stem the stock market tide. By September 6, preferred stock had fallen to 6⅛ and the common floundered at 3¼. As a result, on September 18, 1925, Los Angeles newspaper readers found an old friend on the financial pages. Julian Pete advertising had returned. "Nearly time for the Wise Investor to sit up and take notice!" warned the ad, which bore S. C. Lewis's signature at the bottom. Julian Petroleum, he stated, had $40 of appraised assets backing up every share of preferred stock. Yet these shares were selling on the LASE for only six dollars. At this price, noted Lewis, the entire issue could be had for $1,250,000. "Nearly time for the thinking stockholder who has purchased his at $50 a share to average down his holdings," suggested Lewis. "Figure it out: It should make you plenty of money if you don't wait too long."

The September 18 ad initiated a unique phenomenon—a corporation touting not its product, but its own stock. The fact that all shares authorized for sale were already outstanding and that Julian Petroleum could not sell additional stock made these advertisements all the more unusual. Yet neither the Los Angeles Times, nor other local newspapers, nor the LASE raised objections or seriously sought to curtail this practice.

During the next week, the volume of trading in Julian preferred increased, and the price tilted upward. "We advised you approximately a week ago to buy Julian Preferred," advertised Lewis. "Today we say, BUY JULIAN." The anticipation of continued price increases created a bull market, as investors rushed to take advantage of rising prices. Interest in Julian securities rose steadily. Then, on September 30, a sharp escalation in trading volume rocked the LASE.

"Pandemonium prevailed among brokers on the Los Angeles Stock Exchange yesterday afternoon," reported an A. C. Wagy advertisement. The Los Angeles Times attributed the activity to renewed rumors that Julian Pete stock would be listed on the New York Curb Exchange. The Wagy firm cited "obvious . . . insider buying" as the cause. "Many are predicting much higher prices principally because the insiders seemed so eager to purchase the preferred stock under $13.50 per share," noted the Wagy Company. This prognostication proved uncannily accurate. On October 1, a record turnover of almost 10,000 shares in Julian securities boosted both preferred and common as much as three points. At the week's end, on October 3, preferred shares had reached 17, and common, 7.

Lewis readily admitted to charges of inside buying. "This is true," he confessed on October 5, "if it means that officers, directors and eastern stockholders are buying in the preferred . . . Please remember that I gave you the facts and the same advice we have followed." Lewis repeated rumors of a New York Curb Exchange listing opening "in excess of $30 a share" and predicted that prices in Los Angeles would soon surpass that level. "You can still make plenty of money, if you don't wait too long," he advised.

The sudden emergence of a bull market generated panic among brokers who had "beared" or "shorted" the stock. Stock market "bears" thrive on pessimism. When they anticipate falling prices, they borrow shares from other brokers and immediately sell those shares at prevailing high prices. This temporarily leaves them "short" of the stock. When the time to repay the loan approaches, they purchase the necessary number of shares, optimally at far lower prices, and thereby profit from a declining market. If, on the other hand, their pessimism proves illfounded, and prices rise, they must purchase the requisite shares at higher cost, thereby suffering a loss. A variation on this practice, common in the 'twenties among bucket-shop brokers, was to sell shares they did not possess—i.e., were short of—for future delivery. If the market dropped, they could then buy these shares for a lower price than they had sold them for. Periodically investors who had shorted a stock would stage "bear raids," flooding the market and driving down prices before buying back the greater number of shares necessary to "cover their shorts." In the face of the ongoing surge, Julian Pete bears began to cut their losses, purchasing stock at market value, further fueling the rising market.

During the second week of October, Julian Petroleum dominated trading on the LASE, accounting for almost all activity in the oil group. Sixty-five thousand shares changed hands, almost a third of that amount bartered at constantly rising prices in a frantic Friday session. On Saturday, October 10, Julian Preferred closed at 23⅞ and common at a more modest 8. "THE STORY HAS JUST STARTED," advertised Lewis.

On October 11, Julian Petroleum released a favorable "Income, Profit and Loss" statement for August. In light of these figures, advertised one broker, "it is easy to understand why the INSIDE have been buying in such large quantities. . . . We see no reason for the Preferred to stop at $30 or the Common at $15 per share." For the next four days the bull market continued. More than 75,000 shares were traded. By Thursday, October 15, Julian preferred, which one month earlier had sold at 6⅛, closed at 30. Julian common, which had once languished at 3½, now brought 9¾.

An exultant Lewis celebrated the activity as "the greatest buying power that has ever been exhibited with an independent oil issue in the history of the exchange." Covering by the "shorts" alone, he predicted in his October 16 ad, would drive the stock above $35 within the week. That morning, however, the market for Julian Pete securities buckled and then snapped. "An avalanche of selling orders" sheared nine points off the price of preferred by the end of the morning session. Afternoon trading exacted an even deeper cut. By 2:00 p.m., Julian preferred could be bought at 10. Julian common had slipped from 9 to 4. The crash erased almost $2 million in stock values. Although a late rally raised prices modestly, the bull market in Julian Pete had ground to a decisive and shattering halt.

S. C. Lewis immediately cried foul play. The bucket-shop interests, "confined principally to unlisted and non-members" of the LASE, had engineered a "bear raid . . . to wipe out the margins of thousands of small investors," he charged. In many instances, buyers had never received their stock. Unscrupulous brokers had engaged in "cross sales"—"when there were bids of $29, stock was offered and crossed at $27 per share and on down the line"—and in many instances had trans-ferred stock back and forth to each other through artificial sales in which no actual stock changed hands. Lewis claimed to have spent $500,000 of his own money purchasing Julian securities at prices averaging more than $24 a share in an effort to stem the tide and protect Julian Pete investors. But the "desperate" efforts of the bears had overwhelmed him.

That afternoon, while Julian Pete stock plunged dizzily, Lewis declared war on the bucket shops. "The fight I propose to make is to clean out these so-called brokers and investment houses," he vowed. Julian shareholders, alleged Lewis, had not received "a run for [their] money" and had been "wiped out under the so-called margin system . . . [without which] the break would not have come." Lewis demanded a grand jury investigation of the LASE.

Not everyone shared Lewis's perspective on the Julian Pete stock collapse. The *Los Angeles Times* reported that "men connected with the operations of the exchange" put the blame on Lewis and others whose buying had generated the bull market. The consensus on Spring Street, according to the *Times*, "was that withdrawal of financial support behind the recent upward movement of the stock was responsible for the drop." LASE president Pettingell concluded that, the wild price swings notwithstanding, "there have been no developments on the exchange today to justify the board of directors in taking any action on this matter. So far as we know members of the exchange have been complying with the rules." Pettingell denied any knowledge of "cross sales" or illegitimate transactions and refused to allow the LASE to be drawn into what he termed "any controversy between individuals."

Other observers of the local financial circus found illumination in a curious sideshow starring Lewis and the Wagy brokerage. The Wagy firm, a member of the LASE, was headed by A. C. Wagy and H. B. Chessher, whom Lewis later described as "by far the shrewdest traders" on the LASE. It catered to small investors and served 4,000 customers, the largest clientele of any local brokerage. On October 11, in the midst of the Julian bull market, a Wagy advertisement claimed that "street gossip" had made the firm the "goat" of "rumors of a huge 'short' interest attempting to break the market," and offered a $100 reward for "absolute bona fide evidence that any person with visible assets in excess of $10,000" had stated that A. C. Wagy & Co. had lost money trading in Julian Pete or were short thousands of shares. Three days later, on October 15, Lewis countered with a $10,500 reward for A. C Wagy & Co. or a charity of its choice, if the firm could prove that it "and each individual member . . . is even or long as much as ONE share of Julian Preferred Stock as of . . . 10:00 A.M Thursday the 15th." "EASY MONEY!" responded Wagy and Chessher in a bold ad on October 16, the day the market in Julian securities crumbled.

During the next two weeks, while the market in Julian securities stabilized amidst continued heavy trading, Lewis and the Wagy brokerage patched up their differences. Lewis praised A. C. Wagy & Co. for its "spirit, promptness and execution in accepting the challenge," and admitted that the rumors about the brokerage had been "greatly exaggerated." Wagy & Co. proclaimed their "faith and belief that Julian Preferred would not remain very long at or around $10," and noted that despite the falling market, at the risk of thousands of dollars, they had "not closed out a single marginal account on Julian Petroleum." On October 26, an accounting firm reported that A. C. Wagy had been "long" 302 shares on October 15 with an additional 600 shares in a special joint account. "WAGY WINS," trumpeted the corporation's ad on October 28. "I LOSE," conceded Lewis in a conciliatory advertisement. Neither party, however, offered any explanation for the puzzling dispute.

The real story behind these events leaked out slowly over the following years. In 1930, S. C. Lewis presented his account of the mercurial market performance and his 1925 battle with Wagy & Co. According to Lewis, Chessher and Wagy approached him in the Summer of 1925 with a scheme to raise funds for Julian Petroleum. Lewis would transfer a large block of stock to Wagy & Co. in certificates made out in the names of various Wagy employees. None of these shares would be placed on the open market or sold to other brokers, but instead, they would be deposited into special accounts from which they could be purchased by Wagy customers on margin. Julian Petroleum would profit from the proceeds from these sales, while Wagy & Co. would receive $1.25 a share in commissions and bonuses. In exchange Lewis agreed to advise Julian Pete stockholders to direct all business to Wagy & Co. In July 1925, Lewis turned over 120,000 shares of Julian preferred, far and away the largest overissue to date, to Wagy and Chessher.

"From the very beginning," related Lewis, "the brokerage house began to doublecross me." By the terms of their agreement, Julian Petroleum should have garnered $1.2 million, and Wagy & Co., $190,000. In reality the oil company received only $700,000, and Wagy personally pocketed an equal amount. In addition Jack Bennett soon found himself buying shares on the open market that had been deposited with Wagy & Co. The summer plunge of Julian prices doubtless reflected the dumping of this stock. According to Lewis, he stormed into Wagy's private office and denounced him for wrecking Julian Petroleum. "I jerked two pistols out of my pocket," claimed Lewis, "and shoved one of them over to him and notified him that he would settle with me one way or another." Wagy fled the office.

The public dispute between Lewis and Wagy & Co. probably reflected either Lewis's effort to get revenge for the early betrayal or an attempt to assess how much of the overissue had found its way into the market. If the audit was accurate, only 902 shares of the 120,000-share overissue remained with Wagy & Co.

To recoup these losses, Lewis and Bennett launched the bull market. A series of lawsuits between Jack Bennett and stockbroker James Welch revealed how Bennett and Lewis orchestrated the stock climb. Welch charged that prior to October 16, 1925, Lewis and Bennett drove the price of Julian preferred up to $30 through "fictitious, pretended and fraudulent" transactions known in the brokerage trade as "wash sales." Under this arrangement, "stock was sold from one defendant to another or from one defendant to himself through different brokers" to create an artificially higher price. Several days before the bull market collapsed, Bennett instructed Welch to purchase large numbers of shares for him on margin. Lewis, who Welch claimed had advanced the money used by Bennett and who was "the true party in interest in the action," agreed to protect Welch against any losses. Then, alleged Welch, "the defendants ceased to support the market they had artificially created." On October 16, when the value of the stock plummeted by two-thirds, Welch demanded money from Bennett to cover the margin via "telephone, written and telegraphic messages." Bennett, according to Welch, "in order to avoid serving of notices upon him, concealed himself in the Biltmore Hotel."

When Bennett failed to meet the margin requirements, the stock was sold. Four days later, Bennett, who had actively reentered the market to buy up the depreciated stock, appeared with cash and demanded that Welch return his shares. Bennett knew, argued Welch, that these shares had already been sold, but made this request in order to force the brokerage to repurchase the stock, thereby "artificially" increasing the price of Julian preferred on the LASE.

The events of October 1925 ended ambiguously. The listing of Julian securities on Eastern exchanges, which had allegedly triggered the upward surge, never materialized. Lewis vowed that his "efforts to protect Julian stockholders and the investors against those who are admittedly 'bucketing'" would "go forward unabated," but nothing came of this pledge. Wagy & Co., its reputation enhanced by its victory over Lewis and refusal to close margin accounts, became even more active in the Julian market. By the end of the month, stock prices had levelled out in the mid-teens for Julian preferred and between 4 and 6 for the common. In all probability, both bulls and bears profited, while thousands of smaller investors absorbed the losses.

The tumultuous October trading in Julian securities should have raised eyebrows in the Los Angeles financial community. In addition to the blatant manipulation of stock values by both bulls and bears, the volume of trading offered ample evidence of an overissue. Over 245,000 shares changed hands during the month, almost 80 percent of the total authorized issue of preferred and common stock. Since from two-thirds to four-fifths of the stock had reportedly been placed in escrow and many of the original investors doubtless retained their stock and did not engage in the trading, the amount of activity should have been increasingly suspicious. Observers might also have questioned the role of Lewis's advertising. From the beginning the overt intrusion of Julian Petroleum in its own stock situation seemed unorthodox. Yet no one seriously investigated the matter.

November of 1925 restored an element of "normalcy" at Julian Petroleum. Activity in Julian securities remained heavy, but stock prices fluctuated only mildly. Lewis's advertisements, now weekly rather than daily, put greater emphasis on corporate operations than on stock market manipulations. Julian Pete roughnecks had delivered several wells in the Athens field, and the company had also purchased six producing rigs in the Torrance area. The discovery of a new oil field at Costa Mesa, where Julian Pete controlled 500 acres, seemed most promising. A Julian Petroleum test well yielded 200 barrels a day at a depth of only 600 feet. The shallowness of the pool meant that wells could be drilled in only ten days.

Business-as-usual had also returned to the Julian Petroleum stock transfer offices. While Jack Bennett had stopped creating new stock in the aftermath of the massive July overissue, he now resumed his earlier practices. At the end of October, Lewis began to advertise, urging those wishing to dabble in Julian securities to visit corporate offices. While Julian Petroleum had no new stock to sell, Lewis stated, its employees would provide "assistance and cooperation" in placing orders with local brokers. In reality, according to Pat Shipp, "we would get a list each day from Bennett of those purchasers. . . . [These] people were under the impression they were buying stock through some brokerage house and that we were merely acting as agents as a convenience for them in purchasing it. Instead of Bennett turning over [old] stock to the transfer office to transfer to these people, we merely issued them [new] stock." Bennett would also order large blocks of stock and issue them to names selected randomly from the telephone book.

During October and November an additional 12,000 shares of Julian preferred made their way into the marketplace. Since January, the transfer office had issued over 162,000 preferred shares, while cancelling only fourteen. The overissue alone now exceeded the legally authorized number of shares, more than doubling the amount of outstanding stock. Perhaps due to these growing irregularities, Lewis decided to move the company's books to the New York office. Lewis also transferred Pat Shipp to the Eastern headquarters. Shipp brought with him a little black book in which he had begun to enumerate the issuance of what he termed "Jack Bennett" shares.

As 1925 drew to a close, however, attention shifted from the Pershing Square offices of Julian Petroleum to an empty brokerage suite at 624 South Spring Street. Throughout November a sign on the door modestly, if ominously, proclaimed, "This space will be occupied by C. C. Julian on or about December 1."

# 8 DEATH VALLEY'S HIDDEN TREASURE

I

When C. C. Julian severed his ties to Julian Petroleum in April 1925, he had, perhaps, harbored dreams of at least a temporary retreat to his two homes—the palatial estate where his wife and children lived and the second residence he kept with Maybelle Smith—where he might enjoy his considerable wealth free from the concerns and controversies that had characterized his mercurial career. For several months, C. C. largely disappeared from public view. By late 1925, however, circumstances, both domestic and economic, beckoned him back to Spring Street for another foray into the byzantine Los Angeles financial maze.

During the early months of his "retirement," Julian set about resuscitating his original Santa Fe Springs syndicate wells. These wells, which he still controlled, had sat dormant and in disrepair since late 1924. While Julian No. 5 had been abandoned, the other four wells retained potential. Julian employed a production manager, who retooled nos. 1, 2, and 3, eliciting a combined monthly flow of 15,000 to 20,000 barrels. Efforts to restore No. 4, however, proved unsuccessful, forcing Julian to abandon the well.

Julian's private life proved far less sanguine. Prohibition notwithstanding, Mary Julian had launched on a "wild career" of drunkenness. According to C. C., Mary grew prone to "ungovernable fits of rage over trifles . . . using vile and abusive language toward her relatives, friends and strangers." Julian alleged that his daughters, "just budding into womanhood," were called upon to serve highballs to Mrs. Julian and her guests, "the scum of Hollywood."

For her part, Mary Julian had ample stimulus for her plunge into alcoholism. C. C.'s celebrated nightlife had embarrassed her deeply, and his continuing relationship with Maybelle Smith caused constant pain. Julian frequently traveled with

130

Smith and often presented her as his wife. In late 1925, two men appeared at the Julian estate and introduced themselves to Mary Julian as Maybelle Smith's father and brother. Maybelle had told her family that she was Mrs. C. C. Julian. The men had driven from Texas to visit her, but having lost the address, had simply looked up Julian in the telephone book. C. C. must have brought to bear all of his considerable talents of persuasion and obfuscation to extricate himself.

In August, Julian received two more unexpected and unwelcome visitors: the first, a return of his recurrent malaria; the second, agents of the Internal Revenue Service. The bout with malaria, which, according to one account, threatened "fatal consequences in three months," led Julian's doctors to prescribe a strict regimen of rest. Julian halted all business activities, turning over the trusteeship of escrowed Julian Petroleum stock to S. C. Lewis. Almost simultaneously, however, the IRS announced that Julian owed a staggering $792,000 in back taxes for the years 1922 through 1924. The IRS attached all of Julian's visible assets—an estimated $350,000 worth of bank accounts, properties, trust deeds and mortgages—pending disposition of the case.

As in 1924, at the height of his controversy with the corporation commissioner, Julian turned to the *Los Angeles Record* to defend himself. "Notoriety, The Great Broadcaster, is bugling C.C. Julian's name again to the four winds," reported the *Record* on August 27. "Uncle Sam is now pushing against him one of the largest tax suits ever filed." Since "the public citizen would rather read about [Julian] than any other fellow citizen," the *Record*, presumably as a public service, offered a revised and updated version of the Julian biography it had published earlier. For two weeks, accompanied by banner front-page headlines, the *Record* regaled its readers with the Julian saga.

The C. C. Julian who appeared in the *Record* seemed heroic, persevering, and unperturbed by the most recent turn of events. When confronted by IRS agents, Julian "never turned a hair." When a New York bank informed him that the government had attached his account, Julian grinned "like a mischievous schoolboy," and quipped, "I thought maybe they'd missed that one." To those who suspected him of dishonesty, Julian asked, "If I had been trying to beat the government out of income tax would I have carried all my property in my own name and kept it lying around inviting attachment?" The *Record* observed, "The facts and figures say he is down and out, bound and gagged. But he doesn't look it and he doesn't act it. He has taken a lot of punishment—seems to be a regular glutton for it—but he is just as decisive, as confident, and cheerful as ever."

Throughout the series, the *Record* dropped hints that Julian might soon resume his promotional activities. "Is C. C. Julian a dead meteor, exploded in the ether with only darkness and dust in his wake," wondered the newspaper, "or is he a comet of the business world, about to whirl; back again more brilliant than ever?" Julian himself expressed confidence in his ability to return. "Thousands know me through foul as well as fair weather," he explained. "What the name Julian stands for with them is nonetheless an asset of great value simply because it cannot be weighed in dollars and cents." Julian attributed his earlier failure to the "opposition of organized capital," but he avowed, "I'm converted. They've taught me my lesson and they never need teach it again. No more in my life will I go into an enterprise for full success where I have to buck an established monopoly." The *Record* concluded that Julian "is smart, sophisticated, and growing cynical." After a month "somewhere in the hills regaining his health . . . the public will hear from him again."

In reality, Julian appears to have devoted a minimum of time to recuperation. The IRS assault brought Julian's "retirement" to a premature end. He had already been searching for new promotional opportunities. If Julian had worn out his welcome in the oil industry, the earth held other minerals to exploit. Julian turned his gaze to California's eastern border, where a potential new lure for those seeking instant wealth had materialized.

In March 1924, an old "Desert Rat" named Ben Chambers had located a series of silver and lead mining claims in the Titus Canyon in the heart of the California quadrant of Death Valley. In the late nineteenth century, Death Valley had witnessed numerous gold, silver, and copper mining "excitements" and had supported a thriving borax mining industry. Since the turn of the century these activities had largely ceased. The Titus Canyon, hidden amidst the Grapevine Mountains along the California-Nevada border, had attracted the attention of two prospectors in 1905, but development problems had led them to abandon their claims. Two decades later Chambers rediscovered the canyon and saw evidence of silver and lead deposits. He staked fourteen claims over a 230-acre area.

Titus Canyon remains one of the most breathtaking locales in Death Valley, its beauty matched only by its inaccessibility. "I don't suppose anything ever walked, crawled or crept over it, but a lizard, a rabbit or a mountain goat," wrote Julian. One observer described Titus Canyon as an eight-mile "miniature grand canyon . . . an inspiring gorge with walls, in most places rising vertically, or actually leaning past the vertical for hundreds of feet . . . [which] will repay the seeker for the unusual in nature." Twenty miles distant from Beatty, Nevada, and only slightly closer to the former boomtowns of Bullfrog, Rhyolite, and Chloride Cliff, Titus Canyon posed daunting problems for profitable mineral extraction. No road provided access down its precipitous slopes, and no railroad ran anywhere near the site. The water supply was inadequate and unpredictable. To tap its potential riches, Chambers had to attract financial backing.

Chambers turned to John Salsberry, a former Alaska miner turned Nevada cattleman, who often "grubstaked," or financed, Chambers's prospecting ventures. Salsberry and Chambers incorporated the Western Lead Mines Company in Nevada in August 1925. The company issued two million shares of common stock with a par value of ten cents a share (the legal minimum) and a modest capitalization of $200,000. Salsberry contacted Jake Berger, an old acquaintance from the Alaska gold rush. Berger, Julian's boyhood friend and Julian Petroleum associate, reportedly went "raving wild" over the site. He suggested bringing in Julian as a source of development funds.

Julian knew nothing about mining, but he readily recognized a stock-promotion bonanza. In the fall of 1925, in the wake of the IRS seizure of his assets, he formed a partnership with Berger, Salsberry, and a fourth investor, W. E. Staunton. They bought out Chambers and reorganized Western Lead. Each of the partners received 225,000 shares, which they then placed in trust under the direction of Julian and Staunton. Salsberry and Berger agreed to supervise operations at the mine site, and Julian promised to lend all necessary capital needed to put the company in operation. These loans would not become payable until the company had paid a return to its shareholders of at least $1,000,000 in cash dividends. During the next few months, Julian proceeded to purchase an additional 800,000 shares at prices ranging from fifty cents to one dollar a share, giving him almost total control of the outstanding stock.

To sell this new venture to the public, Julian needed to display more than a series of undeveloped, inaccessible claims. During the waning months of 1925, Julian and his associates laid out the town of Leadfield along the slopes of the Titus Canyon; started two tunnels, the Bonanza and the Berger, into the rock face; and employed seventy-five men for the arduous task of building an auto road from the mine site to the railhead at Beatty, Nevada, almost twenty-two miles away. By January 1926, operations at Leadfield had reached the stage where it could be unveiled to investors.

Western Lead, however, was not Julian's sole bulwark against bankruptcy. On November 23, he incorporated C. C. Julian and Company, a real estate and brokerage business, in Nevada. Julian capitalized the firm at $7.5 million and petitioned the California Corporations Department for a permit to sell one million shares of stock.

While launching these new promotions, Julian also sought to resolve his lingering problems with federal agencies. Julian not only faced tax difficulties, but also earlier investigations by the Justice Department and Post Office inspectors into his oil promotions that were still pending. Julian eased the Internal Revenue situation by paying $254,024 in back taxes and appealing the remaining half-million-dollar levy. Short-circuiting an indictment for mail fraud proved more complicated, but not impossible. Upon completing their investigations, the FBI and Post Office had turned their findings over to Special Assistant U.S. Attorney General David H. Cannon. On February 5, 1926, Cannon advised his superiors that "he was of the opinion that the case did not justify submitting the facts to a grand jury." Relying on Cannon's opinion, J. Edgar Hoover advised his agents to drop the matter. Shortly thereafter, Cannon left the federal service and entered private practice in Los Angeles. C. C. Julian became one of his first and most prominent clients.

At the start of 1926, Julian began to whet the public appetite for his return. On January 7, an unsigned but unmistakable advertisement greeted readers of the Los Angeles financial pages:

"Somebody"
**Watch this space every day for the next 100**
**years.**
**$100,000 "TO YOU"**

For the next three days, the clues and promises of $100,000 continued. " 'Some-body' Wonder Who, Wonder When, and Wonder How," read the January 8 edition. " 'Somebody' Answer This:" asked the next day's ad, "Whose-it? Lighting up the street. Whose-it? You are going to meet. Whose-it? Going to Chatter to you now." For those who had not caught on, Julian broadened his hints on January 10:

**Well! Well! Well!**
**No! No! Marie Not Oil Wells**
**Well What?**
**Well just $100,000 in cash**
**"TO YOU"**
**"WHO" "WHEN" & "HOW"**
**"TOMORROW"**

On January 11, Julian formally announced his return with a lavish reminder of his past triumphs. "$100,000 in Cash 'TO YOU,' " he revealed, represented the latest dividends from his Santa Fe Springs syndicate wells (including, somewhat miraculously, a dividend on the Julian nos. 4 and 5 syndicate, neither of which had produced since October 1924), which would be paid on February 15. "Thousands and thousands of people have already made over 200 percent on their investment in these syndicates," he reminded his readers, "and most of the wells . . . should be [producing] for years to come." Julian also threw in a plug for Julian Petroleum, which, he remarked, is "showing all the earmarks of being a " 'whip.' " But, he observed, "Any old cat can be the cat's whiskers, but it takes a 'Tom Cat' to be the cat's Pa."

For the next ten days, Julian ads boosted what he called "Ze Grand Opening" of C. C. Julian and Co. on January 22. "Looks just like a bank," previewed Julian, "with carpet on the floor, brass cuspidors 'n' everything harmonizing." When opening day arrived, the party was vintage Julian. An estimated 15,000 people, many of them investors in his oil promotions, swarmed to his new offices. Julian received over 200 floral bouquets honoring his return. The celebration continued throughout the weekend as thousands more stopped by to offer support or just ogle. " 'Wonderful' 'Wonderful' 'Wonderful,' " Julian reported on January 26, describing the "tingling sensation" he felt "from the top of my head to the bottom of my feet . . . every time I received a new expression of kindness, or another basket of gorgeous flowers." Julian profusely thanked "those of you who have permitted me to drink a cup of kindness from your hands at life's most rare and beauteous fountain — 'The Fountain of Friendship.' "

A different sentiment flowed at the Department of Corporations. In early January, Julian's attorneys had advised him that his old adversary, Mike Daugherty, was "very doubtful" as to whether he would issue a broker's license to C. C. Julian and

Company. On January 10, Julian sent the commissioner a "strictly confidential" letter protesting this "very unfair stand." Julian reviewed their long history of contention, reminding Daugherty that his original permit had been revoked for alleged violations of the Corporate Securities Act but that the courts had acquitted him of these charges. Julian also cited his adherence to a "secret agreement," signed in the wake of the 1924 mandamus proceedings, that he would not "give any publicity to the fact that you [Daugherty] quit in the middle of the stream," despite the fact that "I could have blown you out of the water with newspaper publicity."

"Any consideration I have ever received from your department I have had to battle for," wrote Julian. Nonetheless, "I was in hopes that you would forget personalities and give my application an unbiased consideration." Julian closed with a challenge: "Los Angeles is my home and it is going to remain my home as long as I live and I intend to do business in this city for many years to come and your department or no other department is going to upset my plans."

Julian's behind-the-scenes efforts were to no avail. On January 29, the department opened a public hearing to consider his permit application. State officials, displaying the overzealousness that lent credibility to Julian's charges of persecution, subpoenaed records of, not only the new firm, but Julian Petroleum as well. When pressed, the Corporations Department could offer no reason for this request. S. C. Lewis protested that his company's activities were immaterial to the hearing and defied the order. The department also demanded an appearance by Julian, but Julian refused to comply, arguing that the agency already had ample documentation on the company's organization and plan of operation. The hearings were abruptly adjourned. On February 2, Daugherty denied the permit.

Julian appealed the decision in Superior Court, shrugging off the setback in his February 4 ad. "It Is To Laugh," read the headline. "I expected to receive no fair treatment from his department." By this time, however, the plan to sell stock in C. C. Julian and Company had diminished in importance. Julian's alternative fundraising venture had begun and, as he advised his readers, he had arranged this "BABY" so that it "in no way comes under the jurisdiction of the Corporation Commissioner, and thank goodness for that." Western Lead, Julian's spectacular mining promotion, had come to the fore.

## II

C. C. Julian's January 1926 advertising had heralded his return, revived the gospel of oil unit dividends, and introduced his lavish office suites. In the process, he had also offered clues to his impending investment opportunity. "I've got a 'WONDER' coming up, and Folks, she's not only warm, —she's 'RED HOT' right off the coals," he bellowed in circus barker fashion. "Just be prepared to 'Stop' 'Look' and 'Listen' when I holler 'All aboard.'" Julian stressed that he had put the oil industry behind him. "'IT IS NOT OIL,' no sirree," he asserted; "the 'BIG BOYS' converted me on oil. . . . I'm out of the oil business for keeps, and I don't care if I never see another drilling rig." His new venture, he promised, "should make the biggest oil project look like a punched out meal ticket." For the impatient, Julian composed his trademark puns to offer a glimpse at the future. "You can drive a horse to water, but a Pencil must be 'LEAD,'" he revealed on January 13. One week later he added, "It's Black and White and 'LEAD' all over."

With preparations in Death Valley proceeding apace, Julian schemed to evade Mike Daugherty's jurisdiction. A permit to sell stock in Western Lead was out of the question. Instead, Julian went directly to the LASE and requested a listing. If the recent manipulations of Julian Pete stock had sobered LASE directors about speculative ventures, they evinced no wariness now. On January 26, 1926, the LASE announced that Western Lead would be called four days later.

Subscriptions to Western Lead stock became available on January 29, one day before the official listing. Long lines of purchasers gathered at the LASE, necessitating four sets of salesmen to handle the demand. The following day Western Lead opened on the exchange. Mining stocks, unlike those for oil and other industrials, were listed, not in fractions, but at their exact price. Western Lead sold for $1.50. At the closing bell over 40,000 shares had been traded, and Western Lead had risen to $1.57.

Julian's Western Lead advertising campaign officially began on February 1. Full-page ads appeared in the major Los Angeles daily newspapers (as well as local Chinese newspapers) detailing Julian's first non-petroleum offering.

### "DEATH VALLEY" and "HER HIDDEN TREASURE" "THAT'S MY BABY NOW"

Some 5200 feet above sea level and peeping down into that great brown desert of Death Valley there stands a Mountain, austere and lordly, which embraces the holdings of the "WESTERN LEAD MINES COMPANY. . . ."

Our property comprises some 14 claims . . . just over the Nevada line in Inyo County, California . . . where so many hundreds of millions of dollars from gold silver and lead have been produced during the last 10 years.

The "MINE" in question, which is known as "WESTERN LEAD" is absolutely a new discovery . . .

True enough it's out in a desolate spot . . . I am convinced that it holds more hidden treasure, yet undiscovered, than any spot its size on the face of the map.

At all events it is out there where "MEN ARE MEN," "Where LIFE IS REAL. . . ." "WESTERN LEAD" property looks for all the world to me like a "HUNDRED MILLION DOLLAR SILVER-LEAD MINE."

"Eight of the foremost Engineers and Geologists of America," boasted Julian, had examined Western Lead, and "in every instance, their reports have been startling as to the magnitude of our Mine." J. W. Bandhauer, the man who Julian claimed had opened up two large copper mines in Arizona, asserted, "The surface showing is indeed very large, and has the largest outcrop of primary lead ore that I have ever seen. The ore will produce a high grade concentrate . . . [and] silver enough to pay the mining cost. . . ." A geologist made three trips to the Western

TABLE 3.    Western Lead Stock Prices, 1926

| Date | Event | Price* |
|------|-------|--------|
| | Listing Price | 1.50 |
| January 30 | LASE lists stock | 1.57 |
| February 1 | Julian ad campaign begins | 1.73 |
| February 20 | | 2.00 |
| March 2 | | 2.20 |
| March 10 | | 2.97 |
| March 13 | Leadville excursion | 2.59 |
| March 15 | | 3.30 |
| March 16 | Rumors of Corporations Dept. action | 1.88 |
| March 17 | Brokers summoned; ads stopped | 1.55 |
| March 18 | | .90 |
| March 23 | | 2.03 |
| March 26 | | 1.24 |
| March 28 | Corporations Dept. hearings open | 1.24 |
| April 1 | | 1.07 |
| April 6 | | .60 |
| April 13 | | 1.28 |
| April 29 | Hearings end | .58 |
| May 28 | Last day of trading | .70 |

*Prices reflect daily extremes rather than opening or closing prices.
(Source: Financial pages, *Los Angeles Times*)

Lead properties and called it "the largest lead discovery in the United States." Another engineer told Julian that he "would be dead and buried for 50 years and the 'WESTERN LEAD MINE' would still be producing millions of dollars a year."

Julian boasted of the twenty-mile road built at a cost of $37,000, the burgeoning town site, the seventy-five men already digging the two tunnels, and the six or eight other companies anticipating operations in the area. He also described a world shortage of lead, which would drive up prices, and revealed the presence of representatives from European countries "here now, trying to buy lead for future delivery."

In Julian's masterful hands, the case for Western Lead seemed compelling. During the next month Julian's advertisements, simultaneously extolling the virtues of caution and courage, prodded the market ever higher. Julian resuscitated his favorite sympathetic group, the "'widows and old folks' who have a little nest egg set away to provide the necessaries of life," and railed against the "many unscrupulous, blackhearted tricksters . . . these ulcers of humanity" who preyed and feasted on them. At the same time, Julian, describing life as "all a gamble after all," confessed, "I'll be tetotally 'COW-KIPPED' if I can figure how anyone with gambling blood in them can stay off 'Western Lead.'"

Julian regaled his readers with tales of mining stocks that had increased in value by 13,000 percent and of investors like the man who fourteen years earlier had

invested $25,000 in an Idaho lead mine "and every year since has drawn down a cool million in return." He repeatedly reassembled his band of experts, whose profit estimates he had to cut in half and was still "almost ashamed" to reveal. C. C. invited those who questioned these assessments to "Go hire yourself the best mining engineer in this country. . . . If he doesn't bring back a highly favorable report, I'll pay his fee and all his expenses." (In some instances, Julian allegedly took even greater liberties. His promotional literature for the East reportedly depicted ocean steamships laden with lead ore docked in Death Valley.) Julian promised a doubling of stock prices within 120 days and dividends by Christmas. "There ain't a-goin to be no grass to cut under my feet . . . because I'm going to have the throttle wide open, full speed ahead on developing my 'WESTERN LEAD MINE,' " wrote Julian. "She's a 'JAZZ BABY' and due for plenty of 'HIGH STEPPING.' "

The performance of Western Lead stock gave credibility to Julian's campaign. Throughout February 1926 the stock prices rose steadily as tens of thousands of shares changed hands daily. The February 15 disbursement of $100,000 in dividends from Julian's syndicate wells, much of which was redirected to the purchase of Western Lead, fueled the market. On February 20, Western Lead reached the two-dollar mark. In early March it topped $2.20. " 'WESTERN LEAD' is costing you more money every day you put off placing your order," warned Julian.

Three hundred miles away, in Death Valley, the lure of a rich new mineral discovery had attracted hundreds of people to Leadfield. Dave and Anna Poste migrated with their young son from the busted boom camp of Gilbert, Nevada. "As you came down the hill," recalled Anna, "first there was a couple of little shacks. Then a big garage. . . . Then the first thing on the main street, the only real street [dubbed Julian Avenue], was a saloon, and I think a grocery store on the same side. Across the street was another saloon, and a little farther, I think, another." Dave Poste got his first job building a schoolhouse. But, observed Mrs. Poste, "mostly it was tents."

Like many of the Leadfield pioneers, the Postes fervently supported C. C. Julian. Scotty Allan, who had migrated from Alaska, held Julian in "the highest regard." At one point, with the camp unable to feed its growing population, Allan "sent an S.O.S.," and Julian responded by shipping 500 turkeys. "No one came there he didn't give a job," asserted Anna Poste. "There was one fellow all crippled up—he carried water for the men working on the road." Julian "paid five dollars a day for a one-armed water carrier," recalled Dave Poste incredulously.

Reporter Bill Chalant took advantage of Julian's new road to visit Leadfield, where "it looked as if the rush had just begun." Automobiles were visible in numbers, loaded with camp impedimenta. Men were clearing spots of ground for the building of tents. Chalant reported the completion of a telephone line to Beatty and the imminent publication of a local newspaper. Local boosters promised construction of a forty-room hotel. "Levelling and building are always going on at a rate that promises a big place in a short time," observed Chalant. "It is sure that money is being spent, and not wastefully, to put [Leadfield] on the map and further its rapid development."

On the mining front, Chalant found that "Western Lead . . . is driving ahead with work and the operation of its machinery is heard upon entering camp. Tunnels in the mountainside are cutting into the ore, their sides glistening with galena." In the surrounding countryside, prospectors had staked additional claims, covering the landscape with monuments for six miles in every direction. "It is said that at least 1000 locations have already been made," reported Chalant, "and that the total will ultimately reach 5000." Chalant, however, remained skeptical about the region's potential. "So far no ore of Aurora richness has been taken out, nor is it likely that any will be," he concluded.

By March, Leadfield and the Western Lead mines had progressed far enough for a grand unveiling. From his earliest ads, Julian had promised to run trains to Leadfield and invite "the world along as my guests." C. C.'s March 7 ad announced the first tour:

"COME UP" or "SHUT UP"
"THAT CHOO CHOO LEAVES" for
"WESTERN LEAD"

"YES FOLKS," next Saturday, March 13th, at 4 o'clock in the afternoon my first "PRIVATE SPECIAL TRAIN" will leave the SP Depot for "LEADFIELD."

She will be a solid Pullman car with 10 sleepers, one observation car and 2 dining cars.

Leaving the SP Depot between 4 and 5 o'clock the afternoon of next Saturday, we will arrive in Beatty, Nevada, Sunday morning about 7 o'clock, where I will have 100 motor cars awaiting to drive us 20 miles to the Mine. We will give this "SILVER-LEAD" property the once over, indulge in a Barbecue lunch at the property, arriving back to our train about 6 P.M. Sunday and then on to Los Angeles by 7 o'clock Monday morning in time for business.

. . . You need bring nothing but your "nighties and pyjamas."

From the time you leave Los Angeles until we arrive back, there will be no charge to you of any description, the entire trip being complimentary by me.

"The only sad part about this trip," Julian noted, was his inability to satisfy the more than 1,000 people who had asked to take the tour. Only about 300 people could be accommodated. Julian extended "special invitations" to representatives from the Los Angeles Chamber of Commerce, Better Business Bureau, and Banker's Association, the city's six leading newspapers, and the California Chamber of Mines. "By no means wear any good clothes," advised Julian. "The country is rough and there'll be plenty of dust."

The impending Leadfield excursion injected an added stimulus to the market for Western Lead. Stock sales averaged better than 100,000 shares a day at steadily rising prices. On March 10, Western Lead closed at almost three dollars a share. Three days later, amidst rumors of a new Corporations Department crackdown, throngs of people gathered at the Southern Pacific station to catch the train to Beatty.

The new *Leadfield Chronicle* offered the most complete description of the events of March 14. "The Julian special train from Los Angeles," reported the *Chronicle*, "pulled into the Tonopah & Tidewater station at Beatty at 8:30 Sunday morning. It carried 340 passengers including 24 women." A fleet of ninety-four automobiles awaited the visitors to transport them to Leadfield. Over 800 people from Beatty and other nearby towns accompanied them. The vast motor cavalcade snaked its way through the desert along Julian's highway. "To the Californians the new road was a wonder," stated the *Chronicle*. "At some places along the road there is a sheer drop into the valley immediately below, . . . [and] a view of stirring mountain scenery. . . . From the summit the cars dropped rapidly down the western side of the Grapevines to the noise of squealing brakes and, as Leadfield came in sight, to the booming of dynamite in the hills surrounding the town."

Upon arrival, the travelers "milled around a bit and hoisted Julian on their shoulders, cheered for him repeatedly, [and] had their pictures taken by a moving picture camera." Nevada lieutenant governor Maurice Sullivan officially welcomed Julian: "Mr. Julian is a booster, and I myself am a booster . . . . I want to say that Mr. Julian is a sport and I greet him as a near neighbor, and we want him in our state." Julian told the crowd, "I didn't bring you here to buy Western Lead. If you don't buy it I will be better satisfied. No fooling. . . . This baby stands on her own feet. I want you to feel you are all one big family—a happy family. I have a few dollars, enough to care for you all." An estimated 1,120 people then settled down to a lavish dinner of turkey, pork, beer, salad, and "all the trimmings." A six-piece Negro ensemble, transported by Julian from Los Angeles, entertained the diners to celebrate Julian's Western Lead "Jazz Baby."

After lunch the party moved to a large open-air floor where an orchestra performed dancing music "amid blasts in the hills." Most of the visitors took time off from the festivities to tour the Western Lead tunnels and avail themselves of free ore specimens. The celebrants departed Leadfield in the late afternoon, and the southern California contingent caught the "Julian special" back to Los Angeles that evening.

In Beatty the revelry continued late into the night. One man reported, "The dance lasted until 3:30 and after that we fought. I lost my cap . . . I don't know where my overcoat is but I don't need one here anyway. My automobile? Oh, I don't know where that is." Julian added a final improbable note by announcing his candidacy for governor of California. "I promise you, if elected," he proclaimed, "we shall have a rip-snorting, two-fisted, stand-up, knock-down, take-it or leave-it, win-lose-or-draw government in California, and we shall be rid of the long hairs."

Julian's political career went no further, but the Leadfield excursion marked the apex of his promotional exploits. Of the many publicity stunts staged by 1920s oil, mining, and stock promoters, none matched the audacity or execution of C. C. Julian's 1926 Leadfield extravaganza. Julian borrowed the concepts of the oil field bus trips and free lunches and magnified them exponentially. As one historian of Death Valley commented, to "bring more than a thousand people into that canyon in one day, feed them, entertain them, and return them without incident to Beatty

. . . was a dazzling performance. And the thought of those 94 overloaded 1926 and earlier vintage autos and auto-stages transporting the throng down that precipitous grade into camp, and grinding back out with them, gives me the mild shakes."

## III

California corporation commissioner Mike Daugherty had watched the latest Julian carnival with growing alarm. Daugherty "knew" that the heavily promoted mining venture had to be fraudulent, but in going directly to the LASE, Julian had circumvented the commissioner's authority. Since all trading in Western Lead was handled by brokers, neither Julian nor his company needed a permit from the commissioner's office. The Department of Corporations did have jurisdiction over those who bought and sold stocks, but theoretically, stockbrokers merely executed the orders of others. They violated no laws by dealing in Western Lead. Daugherty would have to construct subterfuges and evasions of his own to thwart Julian's latest challenge.

On February 1, just two days after Western Lead appeared on the LASE, Daugherty sent letters to several stockbrokers requesting "a true and correct statement" concerning the company. The information he demanded included both readily available public knowledge—the location of the principal office, the names of the managing officers, and the assets, liabilities, and issued capital stock—and other data not normally possessed by brokers and more easily attainable from Western Lead itself. Daugherty asked the brokers to reveal "the gross income, expenses and fixed charges . . . the amount cost and description of development work done on the property . . . the condition of titles to the property . . . and the proposed plan for future development and estimated cost. . . . "

Daugherty's request was highly unorthodox. Stockbrokers would not be expected to possess this information for General Motors or Standard Oil. Why, then, should they be asked to produce it for Western Lead? As Julian later alleged, the February 1 letter was sent, not to elicit facts, but rather "with the full knowledge that the brokers could not supply the information requested . . . [and] for the sole purpose of giving [Daugherty] a pretense" for holding hearings regarding Western Lead. When Julian volunteered to supply the data asked for, Daugherty declined the offer, insisting that each broker be held accountable.

According to Julian, Daugherty also dispatched deputies and investigators to Spring Street to disrupt and discourage trading in Western Lead. Julian alleged that state agents would stand outside his offices, "advising, urging and coaxing stockholders and prospective purchasers" to refrain from buying Western Lead. They also audited brokers' books "for the sole purpose of securing names and addresses of the purchasers and owners" of Western Lead stock, after which they would make personal calls and "in direct and positive language or by innuendo and suggestion" downgrade its value. Julian later produced sworn affidavits from stockholders attesting to these activities.

Throughout the month of February, Daugherty amassed information casting doubt on Julian's claims for Western Lead. Bruce L. Clark, one of Julian's "experts," wrote to Daugherty protesting the use of his name in Western Lead ads. Clark stated that he was not a mining engineer or geologist but an associate professor of paleontology at the University of California. He had been approached by Ira Coe, another of Julian's experts, "to examine the geological structure of the mine." At no time had Clark been aware of Julian's participation in the mining venture. Clark contended the sections of the report that appeared in Julian's ads had been written by Coe alone and inaccurately interpeted by Julian.

On February 20, the *Mining Journal-Press*, a prominent trade magazine, also challenged the accuracy of Julian's ads. "The claims are nothing but prospects that are undeveloped," reported the journal. Furthermore, Julian's experts did not rank among the "foremost engineers and geologists in America." One had developed "gold and silver properties of minor importance." Two others could not be found in any engineering directories. Coe was a practicing civil engineer but had no connection to mining. "To represent these men as foremost engineers and geologists," concluded the journal, "is distinctly overstepping the boundaries of good sense and tact."

On the same day, Corporations Department mining engineers arrived at Leadfield to evaluate the claims. Walter Abel reported, "I have found nothing in my examinations to induce the belief that a profitable producing line will be developed by the company on its present holdings. . . . Geological conditions, structural formations and ore occurrences are decidedly against it." Julian's plans for erecting a 1,000-ton milling plant and promises of dividends by Christmas of 1926 "are absolute drivel," asserted Abel. Given the multimillion dollar capitalization of the firm, it would have had to earn $65,000 a month to return a profit. Yet, he predicted, the "actual ore in sight . . . will not exceed more than 200 tons," with a net value of $16,400. "Further comment is superfluous," concluded Abel.

L. C. Wyman, a second state mining engineer, concurred with these findings, charging that the cost of developing the property would exceed potential profits and that Julian's celebrated road would require substantial widening to handle heavy-duty trucks. "This road," stated Wyman, "is apparently being constructed for the main purpose of comfortably transporting by auto the many prospective investors."

Daugherty, Abel, and two other Corporations Department officials met with Harry Chandler at the *Los Angeles Times* building on March 9 to discuss Western Lead. Julian later alleged that the publisher had "summoned" Daugherty, his former employee, to plan strategies against Julian and Western Lead. Indeed, Chandler admitted to inviting Daugherty to his office. "When the claims set out in [Julian's] advertising were challenged in the columns of a national mining journal and elsewhere," Chandler explained in a subsequent editorial, "*The Times* itself began an inquiry. It sought information from persons presumed to be in a position to supply it, among others the State Corporation Commissioner." At the meeting, according to one account, state documents were spread across the floor and table and in the laps of the participants. Neither Daugherty nor Chandler saw any impropriety in a state official's sharing information in an ongoing investigation with a private citizen, much less with the press.

Three days later, on March 12, Julian launched a pre-emptive strike with an advertisement entitled "Public Interest":

> I am being asked many times a day how come none of the Los Angeles Brokers ever mentioned "WESTERN LEAD" in any of their daily advertisements, and right here is where I let you in on a dirty little secret.
>
> Immediately when "WESTERN LEAD" was listed . . . the Commissioner of Corporations . . . generally forbid [sic] the Brokers of this City to even mention its name in their daily quotations in the newspapers, thereby working a gross injustice on the Brokers.
>
> Ever since the listing of this security, the Corporation Commissioner has consistently endeavored to disrupt trade by using many unfair methods of procedure.
>
> Over a period of some 6 weeks, I guess, his deputies have called and examined the books of every Los Angeles Broker that might be trading in "WESTERN LEAD" — mind you, not their books generally, but only transactions relating to the above security . . . with always an air of mystery and whispering innuendo . . . and the sad part is that [he] plainly states that the penalty on them for failure to produce will be a "Hearing Before the 'Commissioner!'"

Julian challenged Daugherty that if he "can pick a flaw either in the property itself or the stock structure, I feel that I am the logical one for him to attack, and not some group of brokers."

When the "Julian special train" returned to Los Angeles on Monday morning, March 15, the conflict gained momentum. The junketers returned in time to watch 300,000 shares of Western Lead change hands at prices ranging from $2.60 to a closing high of $3.30. But the following day, rumors of imminent Corporations Department action dampened the surge and triggered a bear raid on the stock. Prices dropped as low as $1.88 before rallying back over $2.00 late in the afternoon.

On March 17, Daugherty and Chandler struck. The commissioner summoned eleven stockbrokers (later increased to twenty-four), including LASE president Frank Pettingell, to appear at hearings to "determine whether or not the sale or offer of sales of the capital stock of Western Lead Mines Company . . . will be unfair, unjust or inequitable to the purchaser thereof." These brokers, stated Daugherty, had been ordered to answer questions about Western Lead but had failed to comply. The *Times* discontinued Julian's advertising pending the outcome of the investigation. The market responded by driving Western Lead stock down to $1.55, a scant five cents above its original listing price.

Julian branded Daugherty's action "a disguised blow at me, the Western Lead mine and all the shareholders," and the brokerage community rallied behind Julian. "It is not our business," criticized Pettingell, "to determine whether a price is fair or unjust. We, as members of an organized market, furnish a place where buyers and sellers meet and only offer our services as brokers to complete transactions in securities regularly listed." Frank Gardner, a stockbroker and governor of the local exchange, defended the Western Lead listing. "We are not in a position to conduct an independent investigation of any mining claims," contended Gardner. "The Western Lead application showed a satisfactory amount of cash on hand and sufficient money to continue development of the mine."

The following morning thousands of Western Lead stockholders converged on Spring Street. Exchange officials hastily erected emergency barricades to restrain the crowd, which overflowed the LASE building and blocked traffic. Messengers from brokerage houses and telegraph offices waded through the mass of people to reach the traders. Several women reportedly fainted. On the exchange floor Western Lead opened at $1.50 a share and quickly tumbled to ninety cents. Perspiring brokers wildly shouted buy and sell orders. A flood of afternoon buying drove the stock up to $1.80, before it closed at $1.75. The *Times* labelled March 18 the "wildest day" in LASE history.

That evening Julian met with thirty-three stockbrokers to plan their defense against Daugherty. "I told the brokers that every resource at my command will be placed at their disposal in defending themselves against what I and the brokers consider unfair tactics by the Corporation Commissioner," announced Julian. "The brokers unanimously expressed their willingness to stand by me in this matter." Reports circulated that the brokers would question Daugherty's right to force them to supply information about listed securities or to make judgements as to their value.

On March 19, an even larger throng invaded downtown Los Angeles. Police had to be called to open and maintain a passageway for traders and messengers. The market for Western Lead opened at $1.75. Six minutes later it stood at $1.33. "Brokers strained against a brass rail" reported the *Record*, "veins standing out at their temples, hands waving madly over their heads." The exchange caller "yelled hoarsely through the din," trying to keep up with the latest bidding.

Several hundred yards away, surrounded by scores of supporters, Julian held court at his sumptuous suite of offices. "It will come back," he promised. At one point he picked up two chunks of ore, heavy with lead and silver, resting near his immaculately polished shoes. "They mean," he explained, "either than Western Lead is worth millions or it isn't worth a cent. It's a speculation, but the best I know of." Julian coordinated the efforts of friendly brokers to offset the bear raid. Together they succeeded in minimizing the day's losses to twenty-five cents a share as Western Lead closed at $1.50.

Los Angeles financial leaders reportedly asked Pettingell to calm the frenzy by cancelling the traditional Saturday morning session. Pettingell declined and remained at home claiming illness. Saturday sessions usually presented uneventful codas to the trading week, but on March 20, the tumultuous scenes of the preceding days recurred in "the biggest half day" in LASE annals. Over 100,000 shares of Western Lead changed hands as the price rose to $1.75. The Saturday trading brought the weekly volume of Western Lead shares to 1,230,872. The value of mining stocks exchanged during the week reached $2,287,075, a larger total than recorded in the category for all of 1925.

Julian continued to plead his case in his advertisements in the *Record*, which now became the major conduit for his messages. "WATCH THE FIREWORKS," he advised on March 25. "There's been something rotten in Denmark, and I've got it run to earth. All the noise you hear in the next 48 hours will be the echo of my heavy artillery."

That morning, Julian filed suit against not only Commissioner Daugherty but Harry Chandler as well, accusing them of conspiring "for the purpose of injuring and damaging [Western Lead] and destroying the market price and demand for the stock." Julian, who apparently had detectives trailing Daugherty, charged that Chandler had summoned the commissioner and his assistants to the *Times'* offices "to determine whether or not proceedings should go forward against the plaintiff and if so the ways and means which should be followed to carry out the design." According to Julian's complaint, the two men had decided on the following scheme: Daugherty would cite members of the LASE "with the pretended design and plan" of conducting a hearing to ascertain whether or not Western Lead offered a "fair, just or equitable" proposition. Brokers would be summoned under the implied threat that if they continued to deal in Western Lead their licenses would be revoked. The hearings would last for more than thirty days, during which time the "undue and unnecessary publicity" would force down the price of Western Lead.

Julian also alleged that the conspiracy included a ban on stockbrokers mentioning Western Lead in their ads, a "whispering campaign" by Daugherty's deputies downgrading Western Lead, and a negative report on the mining claims from Walter Abel, whom Julian described as "a young man who has had little or no experience in mining engineering or in metallurgy." The suit, recounting the history of Julian Petroleum, cited "motives of ill will and malice" toward Julian over a period of three years, during which the "defendants have at every opportunity harassed, annoyed and been guilty of other oppressive conduct" against him. In addition to Daugherty and Chandler, Julian named three Corporations Department employees and four officers of the *Times* as defendants. He asked for a total award of $350,000 in damages.

"**Pop Goes the Weasel**," crowed Julian's March 26 ad. "Now we'll find out just who is the investigator this trip, the Corporations Commissioner or the writer." Julian distributed thousands of copies of his complaint. The *Times* responded with an editorial admantly denying any collusion or conspiracy. The *Times* admitted the Daugherty-Chandler conclave, but "for no other purpose than to protect the interests of its readers and the integrity of its advertising columns." If purely selfish designs had motivated the journal, argued the editorial, it would have supported Julian so that he "would be able to advertise and pay money to the *Times*."

Corporations Department hearings on Western Lead began on March 28. Chief Deputy Commissioner Elmer Walther presided. Daugherty, choosing to downplay the popular notion of a personal feud with Julian, neither appeared at nor commented on the proceedings. Julian was called as the first witness and deliberately began to stall. He challenged the right of the commission to question him, provoking a lengthy delay while Walther searched for the appropriate law and read relevant sections into the record. Walther then asked Julian if he would "testify freely and of his own accord, waiving the rights to immunity from criminal prosecution." Julian invoked his constitutional rights and emphatically answered, "No." When a deputy commissioner explained to Julian that he could not be compelled to testify, Julian rose from the witness stand and took another seat. "Well, if I'm not compelled to testify, then I won't sit here," he explained.

The purpose of Julian's behavior became evident shortly thereafter when process servers from the United States District Court arrived with a temporary restraining order terminating the hearings. Julian's attorneys had filed suit seeking an injunction, alleging that the investigation had been motivated by "bad faith," "malice," and "animus." The court order blocked the proceedings, pending a hearing on the matter.

While courtroom maneuvers occupied the principals, Western Lead steadily sank on the LASE. By April 1, the stock had fallen to $1.07. Despite Julian's efforts to put a bright face on the decline ("That's the price you wanted to buy it at anyway, was it not?"), it signified a surrender in his war with the bears. In his April 2 advertisement, Julian revealed that between March 15 and March 26, he had bought almost one million shares of Western Lead stock and expended over $2 million to prop up stock prices before withdrawing from the fray.

On April 3, a federal judge rejected Julian's injunction request. In a lengthy decision on a matter he deemed "of great importance," Judge Paul McCormick reaffirmed the powers of the corporation commissioner. The federal government might intervene only if the state official acted arbitrarily or in bad faith. While Judge McCormick implicitly questioned some of Daugherty's actions, he dismissed Julian's conspiracy charges. He ruled that Daugherty had not demonstrated malice or ill will, "at least to a degree of certainty that would justify the Federal Court in tying the hands of a State officer." McCormick dissolved his temporary restraining order, removing the last remaining obstacle to a state hearing.

## IV

On Monday, April 5, 1926, the curtain rose on the Western Lead hearings. "Diamond wearing brokers, tired laborers whose mites are invested in Western Lead, [and] smart looking, weary looking women of all ages and characteristics" jammed into a cramped, smoke-filled chamber to catch the latest installment of the long-running Julian-Daugherty serial. C. C. Julian spoke the opening lines. Although theoretically addressing the bench, he stood and turned toward the audience. "I come as president and personal representative of Western Lead," he declared. "I want a voice. I want to present witnesses and cross question those put on the stand here to testify against Western Lead. Will the commission rule on whether or not I can make an appearance?"

Deputy Commissioner Earl Adams, the Corporations Department's prosecutor, interrupted Julian and called broker A. W. "Abe" Morris, the first of the brokers under investigation, to the stand. Julian persisted. "I ask for a general ruling. Am I to believe this honorable commission is going to conduct a one-sided hearing?" he asked sarcastically. Presiding commissioner Walther, described by Julian as "a young, inexperienced lad just out of college," ruled that Julian's right to question a witness or produce addit'onal evidence would be determined on a case-by-case basis. Smiling broadly, Julian withdrew to his table, where he sat loudly tapping a pencil, his trademark gesture throughout the hearings.

Commissioner Adams then questioned Abe Morris, who proved a revealing, if unwilling, participant. A broker in Los Angeles for ten years, Morris recently had won election to the LASE board of governors. When named in the probe, Morris had replied truculently, "I think the action of the Corporation Commissioner is misdirected. We act merely as brokers. The Corporation Commission might as well hold hearings over Los Angeles butchers to prevent them from selling pork to Jews." Morris testified that on January 27, three days before Western Lead had appeared on the exchange, he had purchased 5,000 shares from Julian at $1.35, fifteen cents below the opening price. A few days later he bought 2,000 shares on similar terms. In addition, his son, who worked at his brokerage, had received 1,000 shares from Julian in November for only fifty cents a share, which he sold for $1.50 on January 30. These transactions guaranteed Morris, who had voted to list Western Lead on the LASE, an immediate profit when the stock was called.

The practice of distributing stocks below market price to preferred customers before the listing had grown increasingly common on both the New York and Los Angeles stock exchanges. Recipients often included brokers, politicians, and other "insiders." Nonetheless, Julian's use of this ploy, and the selection of members of the LASE board of governors for special treatment, cast grave doubts on the propriety of the listing.

The second day of hearings brought surprising concessions from the major antagonists. Walther allowed that Julian would be able to cross-examine witnesses from "time to time." Julian, in turn, "inasmuch as he could not get on the stand any other way," waived his rights to immunity from criminal prosecution and took the witness chair. Julian's testimony recounted the history of Western Lead, the reorganization of the company under his leadership, and his acquisition of the lion's share of the securities. Julian noted that only his personal holdings had been put on sale. The company itself neither traded stock nor received proceeds from its purchase. He admitted selling stock to five stockbrokers for fifty cents a share prior to the listing of the company, and while rejecting the term "wash sales," confessed that he had manipulated stock prices by arranging "matched sales," selling shares through one broker and buying them back through another. The gallery of spectators laughed and applauded as Julian punctuated his testimony with facetious asides.

On April 7, the Corporations Department called state engineers Abel and Wyman to present their negative evaluations of the Leadfield claims. Both repeated the reports they had filed in February, but Julian subjected them to a rigorous cross-examination. Wyman admitted that he had had little experience with lead mines and that he had spent only eleven or twelve hours on the property. He had covered only a fraction of the territory without the assistance of maps or surveys. Abel also conceded that his visit had been short—only seven and a half hours—and that he had seen only one-third of the holdings. Both men argued that the showings were so minimal that additional examination had not proved necessary.

Those impressed by Julian's deflation of the state mining engineers could not, however, ignore the mounting evidence of irregularities in the listing of Western Lead. Frank Gardner, an LASE governor who had voted favorably on the matter, admitted buying stocks at fifty cents a share and selling them for $1.60 on opening day. Raymond Reese, secretary of the LASE listing committee, had received 1,500 shares of the stock at reduced prices.

A. C. Wagy, a former member of the board of governors, profited from similar transactions. President Pettingell testified that another broker had "invited him in" on 2,000 shares of Western Lead stock at $1.25 a share prior to the first day of call. Nonetheless, Pettingell insisted that the exchange had listed Western Lead in the regular manner.

For the spectators, the highlight of the hearings occurred on April 8, when Commissioner Walther asked Julian to call his first witness. Walther had expected Julian to produce his mining experts, but Julian turned to the crowd and announced, "I'll call C. C. Julian." The people roared with delight. "I'm going to cross-examine myself," added Julian as he took the stand. "I suggest," stated Commissioner Walther, "that you make a statement; don't go through the ridiculous procedure of asking and answering your own questions." But Julian insisted on his right to question himself. "I object to this commission referring to anything I may do or say as ridiculous." Walther conceded, but commented, "It's certainly not in keeping with any court procedure I've ever seen."

For the next fifteen minutes, interrupted by frequent rounds of laughter, Julian interrogated himself:

"What is your name?" he began.

"C. C. Julian."

"Where do you live?"

"In Los Angeles."

"How old are you?"

"I'm over twenty-one."

"I want the right answer."

This badgering proved fruitless, as Julian still refused to answer the question.

Julian brought a loud cheer from the crowd when he told of his first visit to the Western Lead Mines, descending the Titus Canyon astride a mule. Julian then described the richness of the claims. Commissioner Adams objected to this testimony. Julian, a layman, argued Adams, was not an mining expert. "I don't know why I'm not capable of testifying about what I saw," countered Julian. "I'm 21 years of age, reasonably intelligent and bright, and as president of this company should know what there is on my property."

"You cannot give testimony which an experienced mining engineer can give," ruled Commissioner Walther. "It is not the best evidence."

"Well," sneered Julian, amidst laughter and applause, "your alibi doesn't suit me." Julian then dismissed himself and departed the stand. Walther adjourned what one reporter labeled "The C. C. Julian Circus" until the following Monday.

As Julian entertained the hearing-room crowd, *The Truth* returned to southern California. A new edition of the newspaper that Julian had created during his 1924 battles with the Department of Corporations hit the streets on the morning of April 9. The sixteen-page tabloid, which Julian distributed by the hundreds of thousands, presented his version of the Western Lead controversy and conspiracy under incendiary headlines like:

"HARRY CHANDLER, POLITICAL BOSS";
"DAUGHERTY: MAN OR MOUSE";
"MIKE AND HARRY TRAPPED TOGETHER";
"TRY AND IMAGINE THIS IN RUSSIA."

Julian reprinted the text of his Superior Court suit against Daugherty and Chandler in its entirety and depositions from stockholders who alleged harrassment by Corporations Department agents. *The Truth* sketched out the details of the alleged conspiracy against Julian, adding to the cabal, along with Daugherty and Chandler, "a clique of L. A. Bankers." In Julian's prose, these financial institutions more closely resembled "a bunch of hock shops" whose principles "would put a pawnbroker to blush." They feared that a Western Lead that delivered profits to thousands would give Julian "a toe-hold with the voters" of southern California. Representatives of these banks had reportedly met to "devise ways and means to stop him." Unable to discover legal methods, they decided to "frame" Julian instead. For assistance they turned to Chandler, a director at several of the banks. Chandler, the "king of the conspirators," then concocted the scheme to trap Julian.

Julian reserved his most pungent attacks for Chandler and Daugherty. Chandler appeared as a "God-fearing gentleman, who goes to church on Sunday, always carries that "Holier Than Thou" expression," but possessed a "heart of stone" and lacked "one admirable trait," making "money his god." Reputed to be worth $100,000,000, Chandler had accumulated his wealth by "despicable methods." The Los Angeles public viewed him as "an avaricious, insincere, hypocritical and unprincipled miser."

Chandler at least possessed guile. Julian depicted Daugherty as "the laughing stock of every lawyer, every judge and every intelligent business man" in the state. "We find in this personality an object as totally unfit to occupy a position of authority as it would be possible to select were a search made throughout the world," wrote Julian. "If one has the slightest knowledge of physiognomy or phrenology, all one has to do is to take a look at the shape of Mike Daugherty's head or a peek into his countenance to fully appreciate the utter lack of ordinary intelligence." Julian reiterated this portrait in a free-verse poem:

If you have ever seen
Mike's picture

His head
Looks just like
A Woo-Hoo.
So we can't
Be disappointed
In his capacity . . .

Julian criticized the current hearings as a "Kangaroo Court" presided over by a deputy "with no more knowledge of the law than a burro." The object of the charade, he asserted, was not to elicit facts from the brokers, "but rather to get a lot of adverse statements against Western Lead on record." Before the hearings Julian had held the opinion that "if one were to crack open Mike's head, together with the heads of his various deputies, the result would be that we would find combined brain matter sufficient to fill one eggshell." Now he felt that "a thimble would carry their combined gray matter."

Amidst the name-calling and allegations of conspiracy, Julian also raised questions that sounded valid to his supporters. Why, in nearly five years as corporation commissioner, had Daugherty never before called any broker for a hearing to find out whether any of the thousands of stocks that they traded in were "fair, just or equitable" to the purchasers? Why had none of the other mining stocks listed under the same status as Western Lead been asked to defend their assets? Hadn't Daugherty himself, under oath, admitted that he knew little about the hundreds of stocks traded daily on the LASE? Thus, concluded Julian, "Is it not a fact that in your present investigation . . . you singled [Western Lead] out for no other reason in the world than that your heart is full of malice for this writer?"

The appearance of *The Truth* and the coincident break in the hearings gave southern Californians a weekend to absorb and debate Julian's charges. On Monday, April 13, the combatants returned for what the *Record* described as "the wildest session" yet, featuring "near stampedes, heckles, loud laughter and even cat-calls." Professor Bruce Clark, once listed as a proponent of Western Lead, appeared as a hostile witness, accusing Julian of "exaggerations and inaccuracies" in his advertising. Julian aggressively attacked and ridiculed the paleontologist. Clark admitted that he had visited Leadfield and initialled every page of the report submitted in his name. When Julian accused the corporation commission of dispatching agents to browbeat the witness, the audience burst into applause. Walther silenced the crowd and ordered one particularly boisterous spectator to "leave the room until he could behave." Julian resumed his vigorous attack on Clark, prompting Walther to warn the promoter that his "unappropriate and uncalled for remarks would not be tolerated."

Julian called additional witnesses to the stand, all of whom testified to the prospective wealth of the Leadfield claims, only to have their credentials sorely questioned. One "expert," who according to Julian's ads had been employed by two railroad companies to assess the properties, admitted that he had been engaged by neither and confessed to two mail fraud convictions in other cases involving mining ventures. Another "expert," whom Julian had called a "foremost engineer," revealed himself to be a former Julian Petroleum employee and a man of only "practical experience," who had based his estimates on "guesswork."

The next two days proved relatively uneventful, enlivened only when Julian unsuccessfully attempted to call Governor Friend Richardson and Commissioner Daugherty as his final witnesses. By this time, the hearings had degenerated into an only occasionally entertaining farce. Few people even remembered that broker Abe Morris, and not Julian, was the official target of the investigation. The *Record* reporter offered an unflattering rendering of the daily scene:

> Each day, like the proverbial dripping of water that wears away the proverbial rock, C. C., that master of wise cracks and promotion plans, sauntered in, parked his walking cane, joshingly referred to in his best humor as "my overcoat," on his favorite door-knob, nodded to the "audience," which really . . . was his own gang come up from the highly polished "Julian office" on Spring street, and then began casting and re-casting hypnotic glances at the press table.
>
>                                    . . .
>
> Department engineers testified. Then Julian engineers called them, politely enough, liars. Court reporters worked diligently at $15 a day, while we slept at the press table.

The *Record*, which consistently presented Julian's case in the most favorable light, also raised the issue of the mounting costs entailed in prosecuting Western Lead. "Thousands have been spent by the careful but over-zealous corporation department to prove that C. C. Julian is pulling a fast one on Mr. and Mrs. Gullible," complained the *Record*, usually a supporter of regulatory agencies. The investigation had become an "injustice to the public which pays and pays and pays."

If the price of Western Lead on the LASE represented a scorecard for the hearings, Julian had rallied after a dismal start. On April 6, Western Lead had dropped as low as sixty cents. During the remainder of that week it closed no higher than eighty-six cents. On Monday, April 13, however, heavy trading resumed, pushing the stock to $1.28. News from Leadfield gave the stock a further boost. "I received a telegram . . . stating that they had just opened up the most wonderful ore body," reported Julian in his April 16 ad. "It appears to be tremendous." With the hearings in recess, Julian hastened to Death Valley. The following day he wired home his advertisement:

> JUST TELL THE FOLKS THAT THE BOYS AT THE MINE HAVE CERTAINLY UNCOVERED DURING THE PAST FEW DAYS WORLDS OF THE FINEST LEAD SILVER ORE IMAGINABLE. . . . ANYONE THAT INSPECTED WESTERN LEAD IN THE PAST SAW NOTHING COMPARED TO WHAT IS IN SIGHT NOW. A DOZEN OF REAL MINING MEN WHO HAVE SEEN HER YESTERDAY AND TODAY INSIST THAT SHE WILL DEVELOP INTO THE BIGGEST LEAD MINE IN THE US.

Upon his return, Julian described the discovery of a " 'MAMMOTH CAVE' that extends upward of 300 feet . . . [and] is 60 feet across. Everywhere you look you can see evidence of silver lead. . . . " Julian's new mine superintendent boasted,

"C. C., you can tell the world for me that I believe that we are sitting on top of the greatest lead-silver deposit on earth."

When the hearings resumed on on April 27, Ira B. Joralemon, one of California's foremost mining engineers, made an unexpected appearance. Joralemon described the Western Lead mines as "a good prospect meriting careful development," but one that justified an expenditure of no more than $100,000. "If no ore is found with that money," advised Joralemon, "the property should be abandoned." State Engineer Lloyd Root took the stand the next day and virtually echoed Joralemon's testimony, urging caution, but calling surface showings at Leadfield "very satisfactory." Julian, noting that he had always stressed the speculative nature of Western Lead, hailed this testimony as substantiation "for every claim I have ever made for this project."

After three weeks of hearings, the spectacle had lost its luster. The *Record* correspondent reported, "I am broken mentally and physically because of the insane comedy you have forced me to cover. . . . . I have become dull. I am not myself. I have been numbed and dumbed." While Julian might have exaggerated in his ads, contended the writer, "Mr. and Mrs. Public like to read them and has shown they want no protection by buying and buying and buying." If Julian had bought and sold his own stock, "Who cares?" asked the newsman.

To the relief of the reporter and everyone involved, April 29 brought an end to the hearings. The Department of Corporations closed the case against Abe Morris, and the other brokers under subpoena agreed to abide by any decision reached regarding Morris's handling of Western Lead. After three weeks of testimony that put Los Angeles brokerage practices in the worst possible light, Deputy Commissioner Adams nonetheless guaranteed them that neither their licenses nor their "reputation, honesty or integrity" were at stake. The hearings, he admitted, as Julian had frequently charged, were the only means open to the Corporations Department to investigate Western Lead. Adams promised a decision in ten days, and the latest "C. C. Julian Circus" folded its tent.


# V

The ten days promised by the Corporations Department dragged into weeks, but few people doubted the inevitable verdict. Only Julian's most fervent supporters, of whom there remained many, could have followed the proceedings and retained any faith that Western Lead was anything more than a stock swindle. Even if one questioned Corporations Department tactics and accepted the conflicting testimony on the Leadfield claims as honest differences of professional opinion, the hearings had laid bare Julian's unscrupulous conduct to all but the most gullible. He had misrepresented laymen as experts, sold discounted stock to guarantee a listing on the LASE, and shamelessly manipulated stock prices to create an artificial market. While defenders cited Julian's multimillion dollar repurchases of Western Lead stock in his battle with the bears as proof of his honest intentions, Julian's net profits in stock trading ran as high as two million dollars for a few months' work.

Had the commission desired further evidence of Julian's duplicity, they might have examined John Salsberry, C. C.'s partner in Western Lead. Salsberry had grown increasingly disenchanted with Julian. According to a suit later filed by Salsberry, Julian never paid for hundreds of thousands of shares he gained control of, and he repeatedly failed to provide adequate development funds for the mines. With only 100,000 shares remaining in the corporate treasury, Julian requested that the board sell these shares to him. Salsberry refused to approve the transaction.

On May 4, several days after the conclusion of the hearings, Julian called Salsberry in San Francisco and asked to meet with him. Salsberry found an intoxicated Julian accompanied by Western Lead partners Staunton and Berger. Julian proposed that Salsberry join him in a $10 million project consolidating Western Lead and other mining properties. Salsberry declined to participate. Julian then offered to buy out Salsberry. Salsberry said that he responded, "I don't think that you are in a position to buy me out," and reminded Julian that he still owed money for earlier stock purchases. Julian grew abusive. "I will be eating chicken and turkey when others here will be eating liver," he vowed. Salsberry alleged that after further squabbling Julian struck him. Salsberry demanded that Julian turn over 155,000 shares of Western Lead stock, which Julian held in trust for him, and $59,250 he claimed Julian owed for unpaid stock purchases. Julian refused, and Salsberry resigned as a Western Lead director.

The final act in the Western Lead tragedy had yet to be played out. On May 8, a New York financial reporter approached LASE president Frank Pettingell with questions about the Western Lead listing. How, he asked, with its insufficient capitalization and dubious history, had Western Lead been allowed on the exchange? Pettingell declined to answer. The next day Pettingell left his office at 5:30 p.m. and hailed a taxi. He gave the driver no destination, asking simply to be driven about in the evening air. At one point he asked the driver to stop at pharmacy for some "port-olive," a remedy for indigestion. After almost three hours of traversing the city, the driver, sensing no movement in his cab, turned and found Pettingell slumped dead in the back seat.

Rumors spread that Pettingell had been poisoned, but the coroner attributed his death to heart disease. At his funeral Pettingell was hailed as a "man of honor, integrity and clean principles." On May 24 the LASE elected John Jardine, one of the few directors who had voted against the Western Lead listing, to replace Pettingell. Three days later, Deputy Corporation Commissioner Walther ordered all sales in Western Lead halted. The department also announced that it had turned over the transcript of the hearings to the district attorney for possible criminal prosecution.

By this time Western Lead was essentially a dead letter. On April 29, the day the hearings ended, a massive "bear onslaught" had dropped the value to fifty-eight cents. On May 13, in his first advertisement in two weeks, Julian made a halfhearted attempt to rally the troops. "If the mine looked like a hundred million dollar property two months ago," he proclaimed, "today it looks to me like a three hundred million dollar dish." Buying Western Lead at its present prices "is like picking up gold in the street" Julian reported. To prove his point he offered to buy every available share. The market responded mildly to Julian's efforts. On May 28, when Western Lead was called for the final time, the closing bid was seventy cents.

Julian made one last attempt to defend himself and restore his image. "For the life of me, I will never be able to figure out just how the Corporation Commissioner justifies his ruling," he commented in his June 1 ad. Julian expressed confidence that he could win an injunction in the courts: "But in the final analysis, what would all that get you or me?" he asked. "Clothed with authority and dignity of the State . . . the Corporation Commissioner comes out and tells the world that 'WESTERN LEAD' is worthless, robs you of the opportunity of even offering for sale . . . the security that you laid down your elegant money for, and . . . makes me look like an unscrupulous thief." Julian admitted that "this next move has been a most difficult one to figure out," but to "retain the steadfast confidence of the people," he would buy back all outstanding shares of Western Lead for two dollars a share. "I want you to believe in my integrity. I want you to believe in my honesty," he exclaimed.

On the surface, Julian's offer seemed a bold indication of defiant confidence. In reality, he had conceived yet another grandstand gesture. Julian had already accumulated the majority of Western Lead stock at low market prices. In addition he proposed buying the remaining stock, not with cash, but on his personal note, payable in twenty-four months at ten percent interest. During the next seven days, Julian accumulated 93,752 shares under this plan.

On June 12, Julian closed his offices until September 1, by which time, he promised, "I will be able to show you at least a million tons or ten million dollars worth of commercial ore blocked out and ready for the mill." In his final advertisement, Julian offered a telling salutation: "I will admit they have slowed me up a wee bit this time, but he who laughs last laughs longest." On June 29, a Superior Court judge dismissed Julian's damage suit against Chandler and Daugherty, ruling that a Corporations Department investigation could in no manner be construed as a conspiracy to discredit a company's stock.

In Leadfield, the boom began to peter out. The proposed forty-room hotel settled for only four rooms. Dave Poste's schoolhouse was never completed. Townspeople, however, retained their faith in Julian and his mine, and enough of them remained working for a post office to be opened in August 1926. The postmaster reported that 200 people received mail on opening day. Mining operations continued, but the "mammoth cave" yielded little more than crystal, and a second tunnel found scant lead ore. By January 1927, only one person received mail at the Leadfield post office. Ten years later the Leadfield ghost town ranked among Death Valley's outstanding points of interest.

Some Death Valley denizens, however, remained convinced of Leadfield's potential. "They just BROKE Julian," lamented Anna Poste. In the 1930s, desert historian Bourke Lee wrote, "Leadfield did not die because of lack of ore or transportation. Leadfield died of complications. Miners tell me Leadfield will be a big mine one day." In 1938, the California Division of Mines delivered a postmortem assessment. The Leadfield claims held lead, silver, and zinc, but all low-grade and in insufficient quantities to yield profitable development.

Scattered ruins mark the Leadfield site today, but Julian's road endures as the most lasting monument to his venture. The dusty, precarious slope, open only to four-wheel-drive vehicles, offers tourists the only access to the spectacular Titus Canyon, final resting place for the speculative dreams of C. C. Julian's Western Lead investors.

# 9 A THOROUGHGOING BUSINESSMAN

I

Adolph Ramish personified the California dream. Born to affluent immigrant German-Jewish parents in the Gold Rush town of Grass Valley in 1862, Ramish had successfully navigated the eddies of California's commercial, entrepreneurial, and mineral booms to a position of wealth and prominence. After an early education that included several years in private schools in Germany, Ramish found employment with San Francisco sugar baron Claus Spreckles. His experience in the sugar trade led to an appointment as assistant United States consul to the Sandwich Islands. Ramish spent his early twenties in the South Seas and then settled in Los Angeles to seek his fortune.

Ramish organized a successful contracting business in the 1880s and 1890s, but early in the twentieth century he embarked on a bold new venture. Anticipating the boom in motion pictures, Ramish built, owned, and managed the Belasco Theater. He subsequently built two "Adolphus" theaters. By 1912, Ramish ranked among the leading citizens of Los Angeles. A charter member of the Native Sons of the Golden West (a state fraternity), he served as the First Grand Officer of the prestigious organization. Investments in the burgeoning film industry complemented his theater business. Ramish had also become, according to one biography, "heavily interested in oil and mining stocks." Over the next decade, prosperity in the film and petroleum industries reinforced the wisdom of these ventures.

As 1925 drew to a close, Ramish could look with satisfaction at six decades of achievement. Yet the desire to expand his fortune remained strong. Among his holdings was an interest in Empire Drilling, a concern with several wells in the Torrance field. In December, his close friend Barnett Rosenberg, the principal owner of Empire Drilling, approached Ramish with a proposition. Julian Petroleum wanted to buy these wells but needed a loan to facilitate the transaction. Could Ramish provide the necessary funds?

Ramish met with S. C. Lewis, who made him an unusual offer. If Ramish would lend Lewis $200,000 for a short period (Ramish later claimed it was ninety days, but it was more likely thirty or forty-five days), Lewis would guarantee Ramish a $50,000 profit. To secure the note, Lewis offered shares of Julian stock. If, at the end of ninety days, the stock profits exceeded $50,000, Lewis and Ramish would split the excess return. If, on the other hand, stock prices failed to rise, Lewis would still pay Ramish the principal plus his $50,000 profit.

To Ramish, the offer seemed too good to refuse. No similar investment could offer anything approaching that rate of return on such a short-term basis, and the persuasive Lewis had millions of dollars in assets to support his proposal. Ramish's attorney drew up a contract specifying the terms, and Lewis delivered approximately 12,000 shares of Julian stock.

When Ramish's note came due in early 1926, Lewis had yet another surprise for the theater magnate. Rather than cash in his shares and depress the Julian market, Ramish could accept the $50,000 profit and hold on to his stock as security against another $200,000 loan with similar guarantees. Indeed, if Ramish would lend Julian Petroleum additional money, under the same terms, Lewis would secure that with more stock. Within a few months, Ramish had lent Lewis $550,000. Lewis had secured these loans with over 90,000 shares of Julian common stock, which Ramish believed gave him a controlling interest in Julian Petroleum.

At some point, Ramish must have suspected irregularities in these arrangements. Despite his "controlling interest" in Julian Pete stock, trading on the LASE remained heavy. But Lewis delivered his profits regularly, so Ramish did not complain. Instead, he began to subcontract his shares to family members and other acquaintances. "I got brothers, uncles and aunts and all in the stock," he later admitted. Among those he recruited was Carl Laemmle, founder and president of Universal Studios.

More significantly, Ramish also brought Edward H. Rosenberg, son of his Empire Drilling partner, into the Julian fold. Ed Rosenberg, who had briefly worked for C. C. Julian and Company, found an immediate kinship with Lewis's top associate, Jack Bennett. The two men, of similar age and ethnicity, shared a talent and energy for financial manipulation. More significantly, Rosenberg, raised in Los Angeles, had precisely what Bennett, the New Yorker, lacked: a knowledge of the local milieu. From his father's connections, Rosenberg knew men of wealth in Los Angeles who might be tempted to invest in Julian Pete. Rosenberg also could be a "financial buffer" or "pinch hitter" for Bennett with brokers and other investors who no longer trusted Bennett after the events of October 1925. Those who did not wish to trade directly with Bennett or have his name appear on their books dealt more readily with Rosenberg. In addition, Rosenberg understood the labyrinthine connections between politics and the underworld that dominated Los Angeles public life and knew how Lewis and Bennett might curry favor with local power brokers. Bennett offered Rosenberg a commission on all loans he might arrange. Rosenberg moved into the Pershing Square Building office as another unsalaried member of the Julian Petroleum team.

Rosenberg immediately displayed a talent for negotiating short-term loans with local banks to fuel the Lewis-Bennett operations. In March 1926, he negotiated the first of a series of thirty-day loans for $100,000 to $200,000 each through Arthur Adkisson, a junior official in the Merchants National Bank. Adkisson and bank vice-president Thomas Morrisey each received handsome commissions. Rosenberg ultimately made similar arrangements with other local financial institutions. These loans were usually secured by either Julian securities or the personal notes of either Rosenberg or Ramish.

Neither Ramish nor the Rosenberg loans could fully satisfy Lewis's and Bennett's desperate need for cash. Bennett raised additional funds by issuing new stock certificates from his ex-officio chambers at the Pershing Square Building. Between December 1925 and February 1926, approximately 160,000 shares rolled off the printing presses. By the end of February, 483,594 shares of Julian preferred, triple the authorized amount, were outstanding. In addition, according to transfer clerk C. T. Harris, Bennett began to raid the preferred stock escrowed by the stockholders.

Thus, by early March 1926, Lewis and Bennett had instituted a moneylending and stock-kiting scheme of ever-expanding proportions. To meet its expenses, Julian Petroleum borrowed money from local banks, Ramish, and other individuals, guaranteeing extravagant profits. To secure these loans, Lewis and Bennett offered counterfeit Julian stock. When the loans came due, they would make good on their guarantees by arranging additional loans or issuing more stock, which they sold on the open market. The system provided Lewis and Bennett with millions of dollars to operate Julian Petroleum and engage in stock manipulations, but, like a Ponzi scheme, it required a constant pyramiding of loans and stock issues to create the illusion of perpetual prosperity.

Lewis camouflaged these activities with the ultimate proof of corporate achievement. "Our company has been very successful during the past year," he reported in December 1925. "We have new wells, better service stations, better gasoline, and we have been expanding rapidly in the development of this, the largest Independent Oil Company in Southern California." In January 1926, Lewis, proclaiming that Julian Petroleum was "out of the woods," called a February 11 shareholders' meeting. "This particular meeting will go down in my memory as the greatest outstanding event in my life," stated Lewis, "because it is the 'MARKER' of an accomplishment that I will always be proud of, namely the stage where the 'JULIAN PETROLEUM CORPORATION' goes on a dividend paying basis."

On February 11, more than 13,000 people braved a rainstorm to attend the shareholders' gathering. A special detail of police handled the throngs outside Olympic Auditorium, while in the arena, green-jacketed ushers, with a large "J" emblazoned on their sleeves, guided people to their seats. The enormous crowd surrounded the stage, extending to the last rows of the balcony, requiring Lewis to turn and repeat his presentation to the opposite side of the arena.

With great fanfare, Lewis reported that the Julian Pete assets had been appraised at almost $17 million, excluding lucrative recent developments at Signal Hill, Costa Mesa, and in Louisiana. By Lewis's accounting, the corporation had turned a $166,000 profit in 1925. As a result, on March 31, the company would pay a cash and stock dividend on all preferred stock.

The crowd roared its approval, and Lewis then revived one of his perennial promises, a $5 million bond issue from his New York banking friends within fifteen days. Although none of Lewis's earlier promises of Eastern financing had been fulfilled, Julian Pete partisans hailed his prediction of additional funding. The loudest ovation erupted when, towards the end of the meeting, Lewis mentioned his predecessor, C. C. Julian. The audience responded with continuous cheering, until Julian himself, in the midst of his Western Lead campaign, stepped onto the stage and praised Lewis.

Lewis's performance won the accolades of less partial observers as well. "S. C. Lewis . . . has proven himself a thoroughgoing businessman," reported *Oil Age* in its February issue, "for he has succeeded in caulking up the leaks, mended all the sails and for several months past has been carrying cargo with a profit to such an extent that . . . the corporation is beginning to earn real money. . . . "

The promise of dividends also fueled the voyage of Julian Pete on the LASE. On January 17, Lewis had sounded the clarion call for a new bull market. Anticipating that the forthcoming appraisal would reveal assets in excess of $50 a share, Lewis predicted an immediate doubling of current prices. "I sincerely believe that at the present price Julian preferred will bring you greater and quicker profits than any other security procurable on any market in the world," advised Lewis. Overnight, the feverish trading resumed. Within a week, in part due to admitted buying by Lewis and his associates, over 30,000 shares of Julian preferred had changed hands, making it the most active local stock on the LASE.

Heavy trading in Julian Pete and modest price rises continued into February, despite the competition and broader excitement created by C. C. Julian's Western Lead offering. According to Lewis, Julian's re-entry into stock promotion violated their agreement terminating the latter's connection with Julian Pete. When Lewis protested, Julian told him to "tend to my own knitting [and] he would attend to his." Lewis claimed that thousands of Julian Petroleum shareholders had swapped their stock for Western Lead. Julian, in turn, dumped the oil securities on the market.

Indeed, throughout February and March the two Julian-inspired issues, Western Lead and Julian Petroleum, dominated LASE trading. Over a million shares of the two Julian stocks crossed the boards in these months. Lewis, writing more and more like Julian himself, cheered the onslaught. "There is a certain amount of Julian security procurable, most of which is in the hands of speculators, many of whom do not even believe there is a 'SUPREME BEING,' much less believe that our Preferred stock is actually worth four times what it can be picked up at," he advertised on January 30.

The February 11 stockholders' meeting dramatically elevated stock prices. Within two weeks Julian preferred rose to $26 per share. "In my judgment you will never be able to purchase Julian Stock at a price lower than it is selling on the Exchange today," Lewis warned on February 25. Lewis also resuscitated reports of imminent listings on the New York and San Francisco exchanges. Julian preferred reached 30 on March 2. Trading in Julian common also soared, doubling prices from 4 to 8. The market peaked on March 12 when the preferred stock hit an all-time high of 35. To reach this level almost 300,000 shares had been traded in six weeks. Put another way, each legally authorized share would have had to have been traded an average of almost two times during this period.

Both real and artificial forces supported the new bull market. To the unwary observer, the Julian Pete engineering appraisal and 1925 profit statement justified higher stock prices. In addition, the announced dividend stimulated greater activity. Lewis and Bennett supplemented this "natural" demand with increased buying and other manipulative ploys. Ed Rosenberg later explained that Bennett would guarantee large purchasers against loss. This arrangement proved highly effective in stimulating stock sales. DeKalb Spurlin, a wealthy real estate developer from Monrovia, California, for example, began a long-term relationship with Bennett on March 1, 1926. Bennett gave Spurlin assurances that if his stocks dropped below their purchase price, Bennett and Lewis would redeem his stock at its original cost on demand. Over the next year, Spurlin would purchase another $140,000 worth of stock under these conditions.

Lewis also attempted to raise funds through a second alliance with stockbrokers Wagy and Chessher. Lewis distributed 70,000 shares to A. C. Wagy and Company with instructions to sell the securities on margin to their house accounts, but not to allow the stock to reach the open market. Lewis promised to steer potential customers to Wagy and Company. In February, however, Wagy and Company began to trade heavily in Western Lead and simultaneously dumped its Julian shares, double-crossing Lewis once again.

The second betrayal by Wagy and Chessher marked a major turning point in the affairs of Julian Pete. "I recognized that with Wagy and Company, with 4000 customers . . . against us, in the future negotiations no market could ever be maintained on the Los Angeles Stock Exchange," Lewis later explained. Lewis determined to purchase Wagy and Company, and working surreptitiously, paid $350,000 to gain a controlling interest. Senator King's cousin, Culbert Olson, acted as attorney and trustee for the Wagy firm.

To retain Wagy and Company's position on the LASE, Lewis recruited Raymond Reese, a member of the exchange, to serve as president. Reese, however, was only a figurehead; all orders emanated from Lewis or Bennett. Although all stock certificates transferred by the brokerage featured Reese's signature (affixed with a rubber stamp and guaranteed by his father, Charles Reese, the firm's cashier and vice-president), Reese played no further role in the company. On March 18, 1926, advertisements in the local newspapers announced the retirement of A. C. Wagy "due to ill health" and revealed the creation of a new partnership. No mention of Lewis's clandestine involvement was made.

By this time, the latest Julian balloon had deflated. On March 13, Julian preferred slid from 35 to 28; Julian common fell below 5. Lewis attributed the slide to a "temporary reaction" triggered by "the natural anticipation" that "no stock could have a continuous rise . . ." and to profit-taking and short selling by speculators and bears. Despite his reassurances the decline snowballed. On March 18, Julian preferred closed at 20⅛. Lewis blamed false reports that dividends would not be declared and excoriated "those who desired a small profit, PLUS those who desired to sell cheap, PLUS those who listened to rumors AND those who always see short."

On March 23, Lewis hastily assembled the Julian Pete board of directors in New York to officially declare a cash dividend of $1 per share and an additional dividend of common stock. Lewis's efforts momentarily halted the downward plunge, but on March 27, Julian preferred dropped to 15¾, establishing a plateau in the mid-to-high teens, where the stock remained for more than a month.

Several factors contributed to the collapse of the second Julian bull market. The decline coincided with the burgeoning Western Lead scandal, which doubtless had negative effects on all Julian securities. The betrayal of Lewis by Wagy and Chessher and bear raids by other brokers also took a toll. The underlying cause of the March slide, however, was the largest overissue of Julian preferred to date. To meet the two major financial commitments assumed by Lewis, the cash dividend and the A. C. Wagy buyout, Bennett generated over 200,000 new shares of stock. By itself, the March overissue exceeded the authorized number of shares and increased the already inflated supply of stock by almost 50 percent.

Lewis's acquisition of Wagy and Company ranked as the most significant of these developments. He now controlled a major brokerage with thousands of regular customers. Wagy and Company could act as a conduit for their spurious stock issues, while legal buying and selling of massive quantities of stock through the Wagy brokerage could regulate price manipulations on the LASE. Lewis established several special accounts to hide questionable transactions.

Nine days after the buyout, Wagy and Company inaugurated a new "JULIAN NO MARGIN CALL PLAN." In consideration of the fact that Julian Pete was selling for "much less than its real intrinsic value," Wagy and Company offered lots of five to one hundred shares of Julian securities on a 33 percent margin, rather than the usual 50 percent. More significantly, the brokerage pledged that there would be no margin calls in the event of declining prices. Investors could thus enjoy the benefits of low-margin buying without the commensurate risks. From this moment on, A. C. Wagy and Company became known throughout Los Angeles as the "Julian Petroleum brokers."

## II

In April 1926, inspiration struck S. C. Lewis. Julian Petroleum securities had momentarily stabilized on the LASE, but they rested precariously on an ever-expanding pyramid of overissued stock. Lewis had delivered on his promise of dividends, but his pledges to the stockholders of an imminent listing on the New York

Curb Exchange and millions of dollars in Eastern financing remained unfulfilled. Julian Petroleum service stations flourished, but, despite continuous property acquisitions and drilling operations, petroleum supplies languished far behind demand. The short-term resolution to these problems, Lewis surmised, might lie in a merger, or at least the prospect of a merger, of Julian Petroleum with another, more solvent oil company.

Lewis had used the merger ploy on a small scale in the creation of Lewis Oil, uniting unprofitable companies into a broader money-losing endeavor. Julian Petroleum, with its greater assets and substantial marketing division, offered greater possibilities. On April 29, following weeks of rumors, Lewis returned from a New York trip and announced that Julian Petroleum would merge with an unnamed Eastern corporation, creating a new national concern.

By injecting a potential merger into the Julian equation, Lewis demonstrated a keen understanding of the economic temper of the decade. The 1920s were an age of consolidation. Thirty years earlier the nation's first great merger movement had triggered fears of corporate domination and provoked the passage of anti-trust legislation restricting combinations "in restraint of trade or commerce." By the 'twenties, however, social theorists praised the productive, rationalized, and liberating tendencies of the modern managerial corporation. Mergers uniting firms performing similar services in different communities theoretically eliminated the alleged inefficiency of local, rather than regional or national, management. Reflecting the new collective wisdom, Supreme Court decisions had eviscerated anti-trust restraints. Between 1920 and 1928, over 5,000 mergers occurred, eliminating more than 10,000 firms. Consolidations happened so frequently, remarked one businessman, "that he considered it as a loss of standing if he was not approached at least once a week with a merger proposition."

Mergers required new injections of capital and the issuance of new securities, stimulating fervent interest among stock and bond investors. The Julian Pete situation proved no exception. "Rumors of Julian Pending Merger Offer Remarkable Speculative Possibilities for Julian Common," advertised one stockbroker. "If current reports are well founded, price levels of two or three times present prices seem possible." Both common and preferred stock moved higher in the final week of April.

Like most of Lewis's projections, however, the impending combination proved far from imminent. On May 13, two weeks after the initial merger announcement, Lewis advised stockholders that, not one, but two Eastern oil companies had submitted propositions to Julian Petroleum. As a result he had suspended efforts to list Julian securities in the East and to secure additional funding pending the outcome of these negotiations. "One of the companies offered an extraordinarily attractive proposition," reported Lewis, and preliminary contracts had been drawn and agreed to.

The target of Lewis's takeover effort was the New England Oil and Refining Company, which operated a large refinery in Massachusetts but relied primarily on outside purchases for its oil supplies. According to Lewis's later account, three Boston financial institutions representing the New England firm had approached him about a merger. Lewis claimed to have negotiated an agreement that would provide Julian shareholders with both stock in the new venture and cash. A Chicago financial house agreed to underwrite a $10 million bond issue in conjunction with the Boston interests.

In late May, however, the proposed merger collapsed. Lewis later stated that the deal dissolved when the Boston investment companies suffered a multimillion dollar judgement against them in an unrelated stockholders' suit. While this account might contain elements of truth, the papers of Frank Arnold, one of California's foremost geologists, reveal another reason for the failed merger. In April 1926, New England Oil and Refining engaged Arnold to assess the "value and possible rate of recovery" of the oil in the Julian Petroleum holdings. Arnold examined approximately 200 lease agreements and a similar number of oil purchase contracts. His May 1 report offers an impartial appraisal of Julian Petroleum at that time.

Lewis's advertisments had repeatedly boasted of a "drilling campaign . . . which compares favorably with that of any other company in Southern California." Arnold noted, however, that drilling alone did not represent an asset. He considered "only those properties containing actual production" as having value. Other holdings, wrote Arnold, "should rightfully appear as liabilities until production is established."

Arnold found substantial production in the Signal Hill and Huntington Beach fields but was unimpressed by other leases. "The production from the Athens group of wells is so low and their apparent mechanical condition so bad, that little need be said regarding their future," stated Arnold. The wells at Torrance, he reported, "are so located structurally as to render them insignificant producers . . . and therefore of no importance in evaluating the assets of the company." Arnold also dismissed Lewis's heralded Costa Mesa discoveries. "While the wells at Costa Mesa are shallow and may be drilled cheaply, their oil is of low gravity and of no value to a refinery," he explained. "Their productivity is so extremely small and the cost of operating so high, that is doubtful if the total recoverable oil from one of these wells will pay the cost of drilling and operation."

Arnold's assessment projected an average 1926 daily production of only 1,824 barrels and valued Julian Petroleum's total oil reserves, the bulk of which he predicted would be dissipated within three years, at a paltry $1.7 million. This situation forced Julian Pete to purchase over 200,000 barrels of oil each month for its marketing operations. "Taking everything into consideration," concluded Arnold, "the present condition of the company seems to be that of having good marketing and refining facilities but very inadequate and uncertain source of supply in the shape of refining crude." In short, the weaknesses of Julian Petroleum paralleled those of New England Oil and Refining, making the two concerns an imperfect match and dooming the merger.

The Spring Street community had tracked Lewis's latest maneuvers with sharp anticipation. By May 1926, all but the dullest of LASE traders had to have recognized the probability of a Julian Pete overissue. The situation offered, not only added opportunities for commissions, but tantalizing openings for those who could skillfully "bear" the stock at strategic intervals. Although Lewis had removed a formidable opponent by acquiring control of Wagy and Company, other powerful brokerages wanted their share of the spoils.

Harold J. Barneson and C. C. Streeter led the Julian Pete bears. The thirty-year-old Barneson, son of a celebrated California oil pioneer and a graduate of Stanford University, had begun his career with General Petroleum, rising quickly through the ranks to become a director of the company. By 1924, Barneson had opened a brokerage house and gained a seat on the LASE. Streeter posted $5,000 to join Barneson and Company as a partner in early 1926. Together, Barneson and Streeter expanded the firm into a nationwide operation. By 1927, Barneson and Company would also have offices in San Francisco and New York, and Barneson would become the first broker to sit on all three exchanges.

One other characteristic distinguished Barneson. He reportedly hated S. C. Lewis "like poison." Nonetheless, Barneson and Streeter dealt actively with Jack Bennett in Julian Pete stock. The Spring merger negotiations, however, created a chance to strike at Lewis and profit at the same time. Lewis had to maintain the price of Julian securities to make his merger viable. The original merger announcement had pushed Julian preferred over 20, but during the first two weeks of May the stock slipped steadily. Lewis lashed out at "idle rumors emanating from a clique of speculators who are constantly hammering the market . . . to make their last big clean-up in Julian Pete before the merger takes place."

At this moment, according to Lewis, Streeter demanded that Lewis sell him 50,000 shares of stock at $4 below the selling price. "If you don't sell me stock under the market," Lewis quoted Streeter as threatening, "I can break the market anytime I want to." Lewis accused Streeter of "highway robbery" and refused to deliver the stock. Lewis charged that Barneson and Streeter immediately began to borrow stock from other brokers, dumping it on the market and forcing the price down. By May 16, Julian preferred had dropped to 11, its lowest level in eight months.

Streeter denied Lewis's charges of blackmail, but both he and Barneson readily admitted leading the ferocious bear raid. Operating through the Wagy brokerage, Lewis succeeded in mobilizing the bull forces. By the end of the month, both Julian preferred and common had rallied. These fluctuations occurred amidst spectacular trading. Over 253,000 shares of Julian preferred worth $3.7 million crossed the boards in May, accounting for more than half the total volume of oil transactions. An additional $302,000 worth of common was traded.

The late May recovery buoyed Julian backers, but early June brought another onslaught by the bears. On June 3, Julian preferred plummeted again. Lewis again took the offensive, attacking the "same crowd of stock manipulators . . . who would throw in all the stock they could borrow, buy or accumulate, if there appeared any weakness." Lewis promised to "buy all stock that was offered, at whatever price they offered it," thereby removing it from "the hands of the professional, who has habitually badgered it from one to another for the past several days for the purpose of creating the impression that a large volume of selling existed."

Lewis's efforts temporarily restored Julian preferred. But on June 7, Barneson and Streeter struck again. They drove the preferred stock back down to 12½. Common stock received an even sharper jolt, sinking to 2½. Lewis lashed out at "certain well-known brokers who in concerted plan habitually lend their customers' stock to the 'Bear' clique," which spread "various rumors and gossip calculated to create general selling." Lewis vowed to name the brokers who had lent Julian securities to the bears.

Whether Lewis's threat had its effect, or the bear market had run its course, remains unclear, but during the following week Julian Pete once again began to rise. "Satisfactory progress is being made in CLEANING the market of the element heretofore so injurious," reported Lewis on June 10. Three days later, he withdrew his threat to identify the errant brokers. "Life is entirely too short for one to engage in controversies which as a rule, are of a destructive nature rather than constructive," he explained in his June 13 advertisement. "The majority of the members of the LASE are high-class reputable brokers." He had received their "personal assurances" that they would no longer supply stock to those seeking to undermine Julian Petroleum.

The latest round of stock struggles had drawn to a close. Julian preferred settled into relatively narrow range in the mid-teens, where, despite a trading volume of hundreds of thousands of shares a month, it remained throughout the summer of 1926. Julian common, on the other hand, never recovered from its battering. With rare exceptions, it languished between 2 and 3 for the remainder of the company's history.


## III

After the collapse of the merger with New England Oil and Refining, S. C. Lewis wasted little time in identifying his next takeover target. In June 1926, even before the stock-selling residue from his first consolidation plan had cleared, Lewis set his sights on the Marine Oil Company.

Marine Oil shared a common heritage with Julian Petroleum, but their respective histories had diverged dramatically. Like Julian Pete, the enterprise had emerged from the "town lot" drilling boom. The original Marine Oil investors had secured a lease at Signal Hill and, like C. C. Julian, struck a massive gusher on their first well. Marine Oil No. 7, drilled on a fifty-foot lot, yielded one of the steadiest producers on the hill. By 1926, over $4,000,000 worth of oil and gas had flowed through its pipelines. As a result, Marine Oil had delivered the dividends Julian once promised—a gaudy 360 percent in its first year of operation. Marine Oil's drilling success continued. It expanded operations on Signal Hill and also prospered in the Athens field. At the close of 1925, Marine Oil possessed thirty-four producing wells, and annual profits totaled $794,000, yielding a healthy dividend to investors.

While Marine Oil drilling activities generated an ample supply of petroleum, it lacked the refining and marketing facilities necessary for further expansion. Its modest string of twelve filling stations could not dispense its total production. In January 1926, Marine Oil acquired Sierra Refining (a Julian Petroleum supplier), but Sierra's small plant only partially remedied the processing problems. This combination of surplus supply and inadequate distribution offered a mirror image of Julian Petroleum. With three refineries and thirty-five gasoline outlets, Julian Pete lacked the resources to meet customer demand. The merger of the two firms seemed an ideal solution to their respective limitations.

Lewis also liked several other features of the Marine Oil picture. The majority of the stock rested in the hands of a trust controlled by three shareholders. Associated Oil, an affiliate of the Southern Pacific Railroad, had a substantial minority interest. The combined assets of Julian Pete and Marine Oil would enable Lewis to more easily secure loans from California bankers. Furthermore, the merger would allow Lewis to achieve another increasingly important goal: eliminating the Julian name from his corporation. With the growing furor over Western Lead, the Julian connection had become far more of a liability than an asset.

On June 29, 1926, after a week of negotiations that had propelled the stock prices of both companies upward, Lewis announced that Julian Petroleum and Marine Oil would merge into a bold new enterprise, the California-Eastern Oil Corporation. Lewis indicated that this consolidation was only the first step in the construction of a new industry giant and floated rumors that Cities Service, one of the nation's largest and most widely respected oil companies, would join the California-Eastern alliance. Invoking his oft-mentioned but seldom-materialized New York connections, Lewis boasted that a leading Wall Street financial house had agreed to underwrite the merger.

Lewis also sought to open avenues of local financing long closed to Julian Petroleum. Associated Oil, which as a minority Marine Oil stockholder had a vested interest in the merger outcome, was a valued client of the Pacific Southwest Trust and Savings Bank, a branch of First National Bank, the fastest-growing financial institution in Los Angeles. Marine Oil had its headquarters in the Pacific Southwest Building. Lewis transferred his personal offices from the Julian Pete suite at the Pershing Square Building to the Marine Oil offices, where he could better court the bank's officers.

Pacific Southwest Trust and Savings was the brainchild of Henry Mauris Robinson, one of the forgotten giants of pre-Depression America. "At a time when the characters of many public men in America are undergoing the fierce assaults of destructive criticism," wrote John McGroarty in 1921, "Henry M. Robinson stands out in contrasting relief with a proved record of administrative skill and knowledge of domestic and international affairs." Over a half-century, Robinson amassed a record of private success and public service that rivaled that of more celebrated contemporaries.

Born in Youngstown, Ohio, the son of the city's leading attorney, Robinson attended Cornell University and entered his father's law firm in 1890. Uncomfortable as a trial attorney, young Robinson directed his energies into corporate law. "I didn't survey the whole world, choose one line of endeavor, and stick to it with grim perseverance," he later explained. "Business law . . . carried me into many fields."

As the century drew to a close, mergers in the steel and metal industries, and the considerable profits that accrued to the legal specialists who engineered them, attracted Robinson's attention. He played a crucial role in the creation of more than half a dozen major companies, including the American Tin Plate Company, the American Can Company, and Republican Iron and Steel. In 1900 he opened offices in New York City, where he participated in the culminating event of the merger era: the birth of United States Steel, the nation's first billion-dollar corporation. During the next five years, attorney Robinson rode the merger tide, amassing a considerable fortune.

By 1906, the wave of consolidations had subsided, and Robinson, just thirty-eight years of age, terminated his New York law practice and moved to Pasadena, ten miles east of Los Angeles. From this time on, Robinson's life replicated a pattern typical of many of the men who built southern California. They arrived as prosperous, accomplished businessmen determined to retire at an early age. But the sirens of economic opportunity lured them back to work and brought them wealth, prestige, and influence beyond their greatest expectations.

"I had no intention of jumping into active harness," Robinson later commented about his move to Pasadena. "Mrs. Robinson and I agreed that the time had come for us to take life leisurely." But within a short time, Robinson could be found pursuing an astounding variety of ventures. Publishing pioneer Bertie Forbes, who profiled Robinson in *Men Who Made the West*, attributed the transformation to California's fabled climate. "The climate of California took hold of him. It rejuvenated him. It gave him fresh pep. Ambition again began to course through his veins," explained Forbes.

More significantly, the opportunities and challenges offered by the American West proved too tempting for Robinson to resist. Even before arriving in California, Robinson had acquired mining properties (as well as ostrich ranches) in Arizona and engineered a merger of northern California lumber companies he retained an interest in. His earliest activities in southern California revolved around his new hometown of Pasadena. In 1907, Throop Polytechnic University, a local college, embarked on a program to become the "M.I.T. of the West." Throop recruited Robinson for its board of directors, and he spearheaded the transition of the tiny, unheralded university to the world-renowned California Institute of Technology. Robinson played a critical role in fund-raising and recruiting first-rate scientific talent for Cal Tech. "No man deserves more credit for the creation of the California Institute of Technology," commented Robert Milliken, the school's longtime president.

Robinson's expertise in mergers also made him a valued commodity in the Los Angeles business community. "We had no one here of quite his calibre and experience," stated one local citizen. "He had been accustomed to handling huge prop-

ositions in the East and elsewhere. So he wasn't afraid to launch big things here."
In 1910, the First National Bank, one of the city's oldest and largest, was seeking
to expand. First National "begged" Robinson to join its board, according to Forbes.
Within a short time, he had helped merge First National with Los Angeles Trust
and Savings to create the city's first complete banking service. In 1912, Robinson
became a director of the Home Telephone and Telegraph Company. Four years
later he united the firm with its major rival to form the Southern California Telephone Company. Southern California Edison, seeking to acquire Pacific Light and
Power, named Robinson to its board to facilitate the merger. Robinson also served
as a director for Union Oil and other companies. He continued his participation
in nonprofit enterprises as well. His generosity, expertise, and organizational abilities
proved instrumental in developing the Henry Huntington Art Gallery and Library
and the famed Palomar Observatory at Mount Wilson.

As American entry into World War I grew imminent, Robinson moved more
fully into public life. President Woodrow Wilson named him to the Council of
National Defense and put him in charge of coordinating defense efforts at the state
and community level. During the war, Robinson served on the United States Shipping Board. Following the 1918 armistice, Robinson, though a Republican, received
a dizzying succession of posts from Democrat Wilson: as one of six members (along
with Bernard Baruch and Herbert Hoover) of the Supreme Economic Council at
the Paris Peace talks; as delegate (with American Federation of Labor president
Samuel Gompers) to the 1919 International Labor Conference; as commissioner
of the United States Shipping Board; as member of the President's Second Industrial
Council; and as chairman of the Bituminous Coal Commission, which settled the
tempestuous 1919–1920 coal strike. For his work at the Peace Conference, Robinson was named *Chevalier de la Légion D'Honneur* of France and was decorated
by King Albert of Belgium.

Robinson's wartime experiences also cemented friendships with two men who
would greatly influence his subsequent life: Herbert Hoover and John E. Barber.
Robinson worked closely with Hoover on the Supreme Economic Council and in
Hoover's celebrated food and relief efforts in postwar Europe. Hoover's dynamic
performance won Robinson's ardent admiration. During the next decade, as his
political career unfolded, Hoover would rely on Robinson as friend, adviser, and
fund-raiser. John Barber formed an even closer bond. Although eighteen years his
junior, Barber shared a common Ohio heritage and Ivy League education with
Robinson. After seven years with a New York financial firm, Barber had taken wartime leave to work for the U.S. Shipping Board and, later, as attaché to the Paris
Peace Commission. In Paris he and Robinson discovered an immediate kinship.
For Robinson, who had no children, Barber became not just a friend and associate
but a surrogate son.

During the early months of 1920, with both war mobilization and peace talks
behind him, Robinson received a summons to return to southern California to face
yet another challenge. Banking in California was undergoing a major transformation. In the years before the war, San Francisco banker A. P. Giannini had pioneered

the development of branch banking, opening offices of his Bank of Italy in urban neighborhoods, suburban towns, and agricultural communities. In 1913, he had located his first branch in the Los Angeles area. Several Los Angeles banks, including Los Angeles Trust and Savings, also opened auxiliary offices. The war years slowed the spread of branch banking. But in 1919, Giannini's Bank of Italy announced plans for a full-scale invasion of southern California. To counter this onslaught, local banks would have to launch their own expansion and reorganization.

Giannini's declaration came at an inauspicious time for Los Angeles Trust and Savings and its parent, First National Bank. The longtime presidents of the two institutions had both fallen ill and needed to be replaced. Robinson, with his legal expertise, merger experience, and political influence, was the ideal candidate. The boards of directors of the two banks asked Robinson, who was still in Washington, to assume the presidencies of both institutions.

President Wilson, however, was not prepared to let Robinson leave. Wilson requested that Robinson replace the ailing Franklin K. Lane as Secretary of the Interior. Robinson declined. Several days later, Secretary of State Robert Lansing resigned; Wilson offered Robinson this position. But Robinson had already decided to accept the banking challenge. In March 1920, Robinson returned home to Pasadena and assumed the presidencies of the First National and Los Angeles Trust and Savings Banks. He persuaded John Barber, his protégé and trusted confidant, to accompany him. Barber purchased a home near Robinson's in Pasadena; each day the two men would drive to and from work together.

Robinson acted with characteristic vigor to thwart Giannini's ambitions. The economic recession of 1920–1921 weakened many local banks, and Robinson took advantage of their plight to absorb them into his growing empire. Within fifteen months Los Angeles Trust and Savings had thirteen branches. In September 1921, Robinson created the First Securities Company as an investment banking subsidiary, and named John Barber its president. Barber also became a vice-president of First National. In 1922, Robinson invited more than score of struggling banks in the communities surrounding Los Angeles into partnership with Los Angeles Trust and Savings. These banks formally merged on July 1, 1922. In recognition that the bank no longer simply represented the city of Los Angeles, the expanded institution soon changed its name to Pacific Southwest Trust and Savings.

Robinson also moved to block Giannini on the political front. Along with rival Los Angeles banker Joseph Sartori, Robinson advocated the concept of territorial banking, wherein branch expansion would be confined to specific regions. They received a sympathetic response from the State Superintendent of Banks, Charles F. Stern. At Robinson's urging, Stern, a former grocer and highway commissioner with no banking experience, drew an unofficial, and probably unconstitutional, dividing line for branch banking at the Tehachapi mountains, creating northern and southern regions. Shortly thereafter, Stern resigned his post to become vice-president of Robinson's banks. His intimate knowledge of the financial health of other banks would prove instrumental in identifying and acquiring further takeover targets.

Stern's successor, Jonathan S. Dodge, formalized the regional banking policy. In November 1921, Dodge promulgated the *de novo* rule, which stated that branch banks might only be created in the city or locality of the parent bank, "except by purchase of, or consolidation or merger with an existing bank in any such city or locality." In the battle over the lucrative southern California territory, this rule favored Robinson, whose expansion strategy stressed mergers, over Giannini, who had hoped to open branches of the Bank of Italy. By 1923, Robinson seemed to have won the war. His Pacific Southwest Trust and Savings had fifty-three branches in Southern California. Giannini's Bank of Italy had only three.

His banking task nearing completion, Robinson returned to the international arena. President Coolidge named Robinson, future Vice-President Charles Dawes, and General Electric president Owen D. Young as a three-man American delegation to the 1924 Inter-Allied Reparations Committee. The committee was assigned to determine the ability of Germany to pay war reparations. Robinson traveled to Europe, accompanied by Barber, and played an instrumental role in developing the "Dawes Plan," which sought to stabilize German currency through a combination of international loans and banking reorganization and established a flexible schedule for reparation payments. The popular success of the plan further enhanced Robinson's reputation.

Upon his return from Europe, Robinson continued his withdrawal from banking responsibilities. He had assembled a prestigious cadre of corporate vice-presidents whom he felt capable of directing the activities of his banks. In addition to Stern and Barber, Robinson also could rely on the Flint brothers—Motley, a longtime Los Angeles Trust and Savings vice-president, and Frank, the bank's legal counsel. William Rhodes Hervey, a former Superior Court Justice, headed the bank's trust division and was generally regarded as one of the city's foremost experts on trust law. "What 'Rhodes Hervey' might say about a trust matter was usually accepted in Los Angeles as 'the last word,'" wrote journalist Guy Finney.

In May 1924, Robinson reorganized Pacific Southwest Trust and Savings. He became chairman of the board and named Charles Stern, despite his relative lack of banking experience, as president. Robinson remained president of First National, but played a less active role in the affairs of both banks.

Thus, in June 1926, when Lewis took up occupancy in the Pacific Southwest Building, he found Stern, Barber, the Flint brothers, and Hervey at the bank's helm. Lewis approached them in an indirect way. After all, as he repeatedly reminded them, the Julian-Marine Oil merger had already acquired backing in New York and required no local support. At the same time, Lewis charmed the Los Angeles bankers with his self-confidence, phenomenal memory, and mastery of details. Unleashed in the bankers' den, according to journalist Walter Woehlke, Lewis displayed the supersalesman's "gift of plausible persuasion; he had the constructive vision and the ability to make others see the dream structures he built." "Talk to him an hour and your head is swimming," wrote one Lewis admirer. "He talks in millions, never in thousands." Charles Stern later recalled, "I took an interest in the man because of his very peculiar characteristics and his tremendous mentality. Mr. Lewis is a very able man."

The bankers at Pacific Southwest snapped at Lewis's bait. Why, they reasoned, should they let outside capital finance the merger when First Securities, their investment affiliate, could do the job just as easily? The combined assets of Julian Petroleum and Marine Oil should more than secure any loans they might make. According to Lewis, John Barber, president of First Securities, approached him and asked, "Why not let me do the financing?" Lewis cagily demurred. Time was of the essence, he objected. He had already taken one extension on his deadline to purchase Marine Oil, and his financing was in place. He could not risk further delays. Barber, however, was insistent. In addition to offering long-range financing, First Securities would arrange for short term-loans that would enable Julian Petroleum to consolidate its other obligations. "Reluctantly," Lewis accepted Barber's offer.

On August 7, with the support of Pacific Southwest, Julian Petroleum borrowed $2,325,000 from the Anglo-London-Paris Bank of San Francisco, another institution with close ties to Associated Oil. This loan was secured by a deed of trust on virtually all of Julian Petroleum's physical assets and 100,000 shares of California-Eastern stock. Motley Flint witnessed and signed the agreement.

The following day, Lewis announced the terms of the Marine Oil takeover. Marine Oil shareholders would receive an option of cash or stock in California-Eastern for their holdings. Those opting for cash would receive $1.65 a share; those who chose to receive stock would get the equivalent of $2.00 per share. Since the trust holding the majority of the stock had agreed to the buyout, Lewis was guaranteed control of the company. On behalf of California-Eastern, Lewis would make a similar offer to Julian Pete stockholders, as soon as its properties

could be appraised.

To facilitate these transactions, California-Eastern issued 750,000 shares of stock to Lewis in exchange for properties owned by Lewis Oil in Louisiana. One hundred thousand of these shares secured the Anglo-London-Paris loan. A similar block of shares would be transferred to Marine Oil stockholders. (Unlike most Marine Oil investors, Associated Oil wisely insisted on a cash settlement. Lewis borrowed $1.5 million from two New York banks to meet this obligation.) The remaining shares were placed in trust with Lewis. Lewis Oil, whose properties had provided the basis for the stock issue, received nothing.

Thus, while the plans to merge Julian Pete and Marine Oil remained incomplete, pending Pacific Southwest's investigation of the two companies, Lewis had secured the involvement of one of the city's largest banks and several of its most influential bankers. Any misgivings possessed by Barber, Stern, Flint and associates about the Julian company and its stock situation had been overcome by Lewis's reassuring manner.

One feature, however, continued to puzzle the men at Pacific Southwest. Their negotiations with Lewis were supposed to have been confidential to prevent stock manipulation. Yet reports of the meetings repeatedly reached Spring Street. Barber speculated that perhaps someone had bugged the meeting rooms. He asked young Rockwell Hereford, who had recently joined First Securities and had a degree in electrical engineering and experience as a ham radio operator, to comb the conference room for microphones. Recording devices, although relatively crude and bulky did exist in the 1920s, but, searching carefully, Hereford found nothing. The leaks continued, but the bankers from Pacific Southwest pursued the negotiations nonetheless. They apparently never suspected that S. C. Lewis, their charming confidant, might be the culprit.

# 10 THE MILLION DOLLAR POOL

I

The windows of the Julian Petroleum executive suite on the ninth floor of the Pershing Square Building overlooked Pershing Square, the central plaza adorning downtown Los Angeles. From this vantage point, one could watch investors and stockbrokers scurrying across the square to the Spring Street brokerage houses, just two blocks distant. Directly across from the Julian offices loomed a grayish building housing the prestigious California Club. Each day Julian Pete officials could watch Henry Robinson, Joseph Sartori, Harry Chandler, and other members of the southern California élite enter the club for luncheons and other social engagements.

For more than a year, S. C. Lewis and Jacob Berman had shared this view from adjoining desks in their luxuriously appointed office. In June 1926, however, when Lewis transplanted his base of operations to the Pacific Southwest Trust and Savings Building, he left Berman, alias Jack Bennett, behind. Ed Rosenberg moved in to occupy Lewis's former desk. Together, Bennett and Rosenberg coordinated the issuing and distribution of Julian Pete stock.

Journalist Guy Finney later dubbed this Julian inner sanctum the "bubble mill." During the summer of 1926, hundreds of freshly minted stock certificates, accounting for thousands of shares of Julian preferred, flowed from these offices each day. The volume required the efforts of two different printing companies to meet the demand. Certificates bore the rubber-stamped impression of vice-president H. J. Campbell or the forged handwritten signature of corporate secretary T. P. Conroy. Some certificates headed north to San Francisco, or east for sale on Wall Street. The vast majority, however, remained in Los Angeles for local distribution.

Operations at the "bubble mill" settled into a steady routine. Rosenberg dealt

with the brokerage houses, most of which refused to deal with Jack Bennett. Ed's brother Jack or Bennett's brother, Louis Berman ("He is rather a dumb boy," S. C. Lewis described the younger Berman, "He is not bright like the other one"), often acted as messengers, dashing off to Spring Street with crisp stock certificates to fill orders. When Lewis desired a show of market strength, Rosenberg would purchase stock in his own name from different brokers, elevating prices to a more desirable level.

Bennett handled transactions with individuals, haphazardly recording these deals in a notebook or on a random sheet of paper. When in need of immediate cash, Bennett would walk to Spring Street armed with a bundle of stock certificates, which he peddled on the spot, often at prices below the prevailing market rate. At the day's end he would appear at one of several banks with checks totaling tens of thousands of dollars. Usually Bennett would deposit the checks, but on some occasions he would simply cash them, leaving no paper trail of his income. Bennett also dealt directly with several preferred brokers. C. C. Streeter, allegedly Lewis's avowed enemy, frequently called Bennett to acquire counterfeit stock. According to Fred Packard, "Bennett would start cussing him. . . . That was for the effect of the people sitting in the room." Streeter would later secretly visit the office and receive the shares he had requested.

For his most prized group of customers, Bennett offered the lucrative three-point bonus agreement originally received by Adolph Ramish. If the stock did not rise within a fixed period of time, Bennett promised to buy it back for three dollars a share above the purchase price. When these arrangements came due, Bennett would either pay the bonus and extend the contract or, if necessary, redeem the stock in full. The most prominent beneficiary of these dealings was director Cecil B. DeMille, whose investment company purchased $62,000 worth of stock on June 30 and redeemed it for a $12,000 profit after forty-five days. A second arrangement involved "collateral agreements," in which Bennett would borrow money at high rates of interest for thirty to forty-five days, posting Julian stock as collateral. To fulfill these obligations, Bennett issued and sold new stock, perpetuating the cycle.

Lewis, Bennett, and Rosenberg also made sure that Julian stock, complete with guarantees, found its way into the hands of influential politicians, newspapermen, and underworld figures who might aid their cause. As Robert Shuler later observed, "Lewis was as big hearted as any exploiter who ever came to any city. He divided with any and all whose influence or power, political, financial or judicial, might be needed in the consummation of his enterprise." In Los Angeles during the 1920s, this meant dealing with the shadowy political machine headed by Kent Parrot.

Kent Parrot had arrived in Los Angeles from his native Maine in 1907 to attend law school and play football at the University of Southern California. Standing six feet two inches tall, Parrot possessed what the *Times* described as "a swaggering, more or less insolent and altogether colorful personality [of] imposing physique and magnetism." Amidst what one author called "the daffy and always heterogeneous political elements that made up Los Angeles," Parrot mastered the art of the unorthodox floating coalition, merging liberals with conservatives, church leaders with underworld figures, union officials with open-shop zealots, and prohibitionists with liquor interests.

Parrot first demonstrated his talents in the 1921 mayoral elections when he managed George Cryer's upstart challenge to incumbent mayor Meredith "Pinky" Snyder. Parrot skillfully assembled an odd menagerie of political supporters. Cryer received endorsements from the Central Labor Council and the archconservative Better America Federation; the Anti-Saloon League and liquor industry advocates. Several leading ministers, including the increasingly influential Reverend Shuler, backed Cryer, as did the *Times*, which published daily stories exposing corruption in the Snyder regime. On election day, Cryer narrowly outpolled Snyder.

Parrot quickly emerged as the power broker of the Cryer administration. He involved himself most heavily in the affairs of the Police Department and on the Harbor Commission, which awarded lucrative construction contracts. "Mr. Parrot's sinister shadow has fallen across the path of the Harbor Board at nearly every turn," complained one disgruntled commissioner. In 1924, the *Record* described him as the city's "De Facto Mayor."

Parrot financed his political machine with funds from the local underworld. Vice in Los Angeles operated under the aegis of Charlie Crawford, nicknamed the "Gray Wolf." Crawford, wrote one contemporary, "was a soft-voiced old gentleman, big of stature with a bland expression and a lock of silvery hair that curled over his forehead and gave him somehow a priestly look," but was nonetheless "a vicious old scoundrel, fully capable of framing his own mother." He had come to southern California via Seattle, where he had made a fortune operating dance halls during the Klondike gold rush. His lavish Northern Club became a meeting place for Seattle gamblers and politicians. After an investigation into his activities led to the conviction of one city official and the recall of Seattle's mayor, Crawford fled to Los Angeles. There Crawford opened the Maple Bar, outfitting the lower floor as a gambling casino and the upstairs rooming house as a bordello. Like its Seattle predecessor, the Maple Bar attracted local politicians, judges, and public officials, especially those with ties to Kent Parrot.

Crawford forged alliances with other local vice lords. The Gans brothers, Joseph and Robert, oversaw the distribution of slot machines. Zeke Caress, a former accountant, ran the bookmaking and betting operations. Former police officer Guy "Stringbean" McAfee supervised other gambling activities. Crawford imported a Seattle associate, Albert Marco, to become the "caliph of the local prostitution industry." Policemen in uniform often served as messengers and collectors for the syndicate. Prohibition offered additional opportunities for illicit profit, but Crawford and his associates had to share the hotly contested liquor trade with various contending interests.

In Seattle, Crawford had been the quintessential gambling lord—tough, hard-fisted, bedecked with bold clothes and flashy diamonds. In Los Angeles, he sought greater respectability. He toned down his wardrobe and opened a real estate office.

Speculation in downtown properties enhanced his fortune. As his ties with Parrot grew, Crawford often described himself as a "politician."

Parrot naturally preferred to keep his underworld associates at arm's length. He rarely appeared in public with Crawford or the others. When he met with them, he did so in the privacy of his apartment at the Biltmore Hotel, where he entertained a wide assortment of cronies, public officials, and favor seekers. "Every one in the State of California has possibly been there in the official line," he once boasted.

Remarkably, Parrot accumulated his power and influence while extensively overhauling the Cryer political coalition. In 1921, Cryer's opposition to municipally owned power had won him the support of the *Times* and other conservatives while alienating old-line Progressive leaders. In 1923, however, Parrot, sensing a shift in the political mood, persuaded Cryer to endorse municipal ownership of utilities in his re-election campaign. This transformation earned Cryer the support of the Progressives but alienated Chandler. The *Times* backed the mayor in 1923, but the following year turned against Cryer and Parrot with a vengeance. In 1924, the *Times* published a seventeen-article exposé of corruption in the Cryer administration, identifying Parrot as the malevolent source of evildoing. Chandler, Henry Robinson, and other leading business figures spearheaded the formation of a civilian Crime Commission, which called for a crackdown on local vice activities.

The decision to back municipal ownership and the blistering attacks by the *Times* solidified the coalition between Parrot, Cryer, and Progressive politicians. Progressive newspapers, W. R. Hearst's *Examiner* and *Herald*, and the *Los Angeles Record* defended Parrot and Cryer. At the state level, Parrot allied himself with Hiram Johnson, longtime archrival of the *Times*. By 1925, Parrot admitted to two obsessions: the breakup of the Crime Commission and "to place tacks in spots where he thinks Harry Chandler, publisher of the *Times*, is apt to sit down."

The 1925 mayoral race pitted Cryer against Chandler's handpicked candidate, federal judge Benjamin Bledsoe. Bledsoe, who also won the endorsement of former Parrot ally Robert Shuler, campaigned on the crime issue. Cryer stressed municipal ownership. The election marked the high point of Parrot's political power. Cryer overwhelmed Bledsoe, and most of Parrot's designees for the city council also triumphed.

Thus, in the summer of 1926, when Lewis and Berman sought to buy "protection" in local political circles, they lavished their attentions on the members of the Parrot machine and Cryer administration. The names of Parrot, Crawford, and Marco all appeared in the books of the A. C. Wagy brokerage. Bob Gans bought $100,000 worth of Julian stock on June 15 and an additional $27,500 on July 11. Although market prices remained relatively stable during the next two months, Gans redeemed his stock for $150,000 during the first two weeks of August, thus earning $22,500, or 20 percent, for his fifty-six-day investment. Judges and other public officials also reportedly received stock. As the Reverend Shuler observed, Lewis shrewdly "knew that thousands of men who will not accept a bribe outright will take a few shares of stock in which there is guaranteed large profits." Walter B.

Allen, president of the Los Angeles Harbor Board, and automobile dealer Perry Greer, an important Parrot supporter who later ran for mayor, pooled their resources for two stock purchases totaling $62,500 in late June. After holding this stock for less than three weeks, they received a $7,500 profit.

Lewis also paid local reporters, usually those allied with the Parrot machine, to plant favorable items about Julian stock. Rumors circulated that at least three local journalists became "almost rich enough to purchase a paper each for themselves." Morris Lavine, the ace newsman for Hearst's *Examiner*, received regular pay-offs from Lewis. According to Shuler, "one newspaper reporter [presumably Lavine] with a monthly salary of less than $250 owned a bank deposit book that reads like a romance." While no direct link between Julian Petroleum and the *Record* has ever been established, the generous distribution of stock probably explains that newspaper's dogged defense of the company.

In the summer of 1926, Parrot attempted to expand his influence into statewide politics. Incumbent governor Friend Richardson faced re-election, but his conservative fiscal policies and the erratic performance of state regulatory agencies had generated widespread opposition to him. The Progressive wing of the Republican Party, led by Senator Johnson, supported Lieutenant Governor C. C. Young against Richardson in the primary election. Young received financial support from banker A. P. Giannini, who hoped that a governor from the north and a sympathetic state superintendent of banks would reverse the policies that had blocked expansion of the Bank of Italy into southern California.

In Los Angeles, opposition to Richardson revolved around Parrot's Progressive allies and local businessmen seeking to replace Corporation Commissioner Daugherty. C. C. Julian, who had maintained a low profile since the collapse of Western Lead, joined the attack on Richardson in a series of advertisements in the *Record*. Governor Richardson, he charged, "jumps through a hoop every time [Harry] Chandler . . . cracks [his] whip" and "is as totally unqualified to hold the high office of Governor of California as a Jack Rabbit is unfit to be King of the forest." Parrot conducted an aggressive fund-raising campaign on Young's behalf, and rumors circulated that S. C. Lewis, who feared investigation by the corporation commissioner, ranked among the heaviest donors.

On August 31, Young polled 51 percent of the vote, narrowly defeating Richardson in the Republican primary. In California, where only one Democrat had sat in the governor's chair since 1887, this victory assured Young's election in November. In the months after Young's triumph, S. C. Lewis was heard to boast that the problem of the corporation commissioner had been resolved.

One other group of investors increasingly found themselves beneficiaries of Bennett's largesse during the summer of 1926. In June, Bennett and Rosenberg began recruiting individuals and borrowing money from a coterie of Jewish businessmen recommended by Ramish and Rosenberg from their acquaintances. These men purchased stock individually or in groups for thirty- to forty-five-day periods, after which they were guaranteed a 10 percent profit. At the end of each thirty-day period, they collected their interest but renewed the original agreement. Jacob Farbstein, owner of a pipe and supply concern, invested $27,500 in Julian stock as part of a pool on June 18. One month later he received a $2,500 payment and "reinvested" the original amount under the same terms. Over the next nine months, Farbstein participated in a total of twenty-one such arrangements and received $41,280 in profits. Clothing manufacturer E. F. Hackel engaged in at least nineteen deals and earned over $100,000.

These transactions provided Bennett with a steady stream of money, but required a constant flow of payments to those whose agreements had expired and the issuance of thousands of counterfeit shares. Between May and August 1926, 1.1 million new shares of Julian preferred came on the market. By the end of the summer, 1,773,502 shares were outstanding, more than eleven times the authorized issue.

In later years, S. C. Lewis would claim that leaving Jack Bennett unsupervised in the Julian Pete offices in June 1926 had proved his undoing. Without his guidance, argued Lewis, Bennett expanded the overissue beyond manageable limits. The events of the summer of 1926, about the time Lewis moved to the Pacific Southwest building, lend a modicum of credence to Lewis's version, but the systematic overissue had begun long before Lewis departed. Transfer clerk Pat Shipp later testified that he had informed Lewis of the overissue via telephone from New York in June 1926. According to Rosenberg, Lewis directed stock transactions to manipulate market prices. Lewis also accepted millions of dollars in cash payments from Bennett to finance his merger activities. The largest monthly overissue, almost 370,000 shares in August 1926, coincided with the completion of the Marine Oil pact. When asked if he knew where Bennett's money came from, Lewis answered evasively, "Well some of it I did and some of it I did not."

Substantial portions of the money that filtered through the "bubble mill" found its way into the personal coffers of Lewis, Bennett, and Rosenberg. All three men became renowned for carrying large sums of money and settling expensive transactions with cash. According to Bennett's later testimony, "Lewis was a big liver. . . . He would buy large amounts of jewelry, automobiles, and just use the money as if it were water, gave it to anybody if they were the female species. A man had to work hard to get money out of Lewis. A woman did not." Lewis bought a large home in Beverly Hills and remodeled it with marble bathtubs, gold faucets, and expensive furnishings. His wife also spent money freely on jewelry and travel. "Her demands on him for money were terrific," recalled Bennett. Bennett himself also ostentatiously displayed his newfound wealth. He provided generously for both his and his wife's extended families. At the same time he had evolved into a dapper bon vivant and escort of Hollywood starlets. One jaundiced observer described Bennett on the town as a "roly-poly, smiling and nodding 'duckling' . . . a bending-fashion plate, whom one would spot as a gigolo or 'breezy tout.'" Rosenberg accumulated a $40,000 home, two automobiles, and over $400,000 in investments.

By September 1926, Julian stock activities had distributed millions of dollars of profits into the hands of favored investors and brokers. With Bennett's birthday approaching, Ed Rosenberg asked him what type of a gift he desired. Bennett modestly requested a Rolls-Royce. According to Rosenberg, "A popular subscription was gotten up in which my various clients and brokers . . . each contributed. We raised in the neighborhood of $13,000. . . . The price of the car was $16,000. We took Jack's Packard car and gave that in as a partial payment and paid for his Rolls-Royce." Bennett received his car on the morning of one of the Jewish high holidays. Like most of the Los Angeles Jewish community, Bennett and Rosenberg were not observant Jews, but they nonetheless attended services on the holiest days of the year. After taking delivery on the car, they headed to the Congregation Sinai to celebrate Yom Kippur, the Day of Atonement.

## II

In September 1926, S. C. Lewis could survey his evolving Julian Pete labyrinth with satisfaction — and foreboding. Less than two years earlier, he had arrived in Los Angeles as president of the worthless Lewis Oil Corporation, devoid of standing or cash resources, another relatively insignificant hustler capitalizing on the insatiable 1920s' demand for instant wealth. Since that time, Lewis had sculpted two overlapping confidence games, each promising multi-million-dollar pay-offs. The Julian Petroleum–Marine Oil merger and his newfound friends at Pacific Southwest Trust and Savings had elevated him to the front ranks of Los Angeles businessmen. Simultaneously, the "bubble mill" operated by Jack Bennett kept Lewis supplied with an unending flow of cash for his personal and business expenses. How much longer he could navigate the maze, and what his ultimate escape route might be, remained uncertain. For the moment, however, Lewis stood unscathed and emboldened as new corridors opened before him.

Amidst the intrigue of high and low finance, Lewis strove to retain an image of constant corporate expansion. Even while negotiating the creation of the California–Eastern Oil Corporation, he had purchased control of several new properties for inclusion in the new concern. Julian Petroleum expanded operations in southern California fields and also acquired 600 acres of "proven area" in Montana for a reported $3.25 million. The transaction proved typical of Lewis's dealings: a down payment of $100,000 secured an option to buy; the balance was added to Julian Pete's growing debt. Lewis also secured agreements with two Texas companies with a reported daily production of 17,000 barrels a day. Lewis incurred $11.5 million in obligations for these firms.

At Pacific Southwest and its First Securities investment affiliate, preparations for completion of the Julian-Marine merger continued apace. Investigators and appraisers began assessing the holdings of the two companies, projecting a final approval in November. Barber hired William C. Kottemann, president of the local chapter of the Certified Public Accountants Association and vice-president of the state society, to audit the Julian Petroleum books. Kottemann simultaneously worked for Lewis auditing the records of Wagy and Company.

Bank officials, however, grew alarmed about the rumors arriving from Spring Street. Tales of an overissue and Jack Bennett's loans and investment pools generated concern. On September 14, John Barber and Motley Flint reportedly raised the issue with Lewis. According to Lewis, Flint labeled the pools "gyp loans" and argued that they reflected poorly on both Lewis and his company. If not brought under control, warned Flint, the pools would ruin Julian Petroleum and derail California-Eastern.

At this juncture either Lewis or Flint made the suggestion that changed the course of Julian Pete affairs: Why not stabilize stock transactions and consolidate Bennett's myriad small pools and loans by creating one overriding investment coalition? This syndicate would buy up outstanding stock and withhold it from the market, pending merger approval in November. Why not create a "Million Dollar Pool"?

The idea of a massive stock pool was by no means original or unique. Stock pools were time-tested strategies, increasingly common in both New York and Los Angeles. Groups of investors would "pool" their resources to buy a particular security, placing their holdings in trust under the control of a "pool manager." The purchase of a large block of stock would drive up prices, which would attract still other buyers. This would propel the stock still higher. The pool manager would then slowly liquidate its holdings at significant profits for its members. As Walter Woehlke explained, "The art of pool manipulation consists in knowing when and how to sell without causing a violent slump." When handled properly, both investors and pool operators made money. The latter, however, made far more.

The "Million Dollar Pool" negotiated by Lewis and Flint offered several twists on traditional arrangements. Lewis had offered Flint the same "three-point" agreement available to Ramish and other Bennett regulars, which guaranteed a $3 profit on each share. According to Lewis, Flint questioned the legality of this arrangement. He suggested that, if Lewis agreed to protect the syndicate against loss, the pool might reap the same benefit by purchasing the stock for $3 below the market rate. Julian preferred was currently selling at about $18. Flint promised to recruit a group of investors who would purchase $1,000,000 worth of stock at $15.50 a share. Lewis, in turn, agreed to repurchase the securities after forty-five days, at either the prevailing market rate or, if the price had dropped, at no less than the original purchase price. At the expiration of the pool agreement, Lewis and the Wagy brokerage would gradually return the stock to the market. Should the price after forty-five days exceed $18.50, the additional profits would be split: 80 percent to the pool and 20 percent to Wagy and Company. If the price reached $22.50 prior to expiration date, the pool manager and Wagy and Company could agree to begin immediate liquidation. In the event that Lewis and Wagy failed to repurchase the stock, the pool manager would sell the shares and distribute the assets proportionately. In addition to providing the 64,517 shares of Julian preferred required by the deal, Lewis promised that Wagy and Company would place an equal number of shares in escrow as security protecting the pool against loss.

In later years neither Lewis nor Flint would accept responsibility for suggesting the "Bankers' Pool No. 1." Each claimed that the other had lured him into a trap. In September 1926, however, the notion of a massive pool appealed to the immediate interests of both men. For Flint, the idea offered an opportunity for substantial profits for himself, his colleagues at Pacific Southwest, and other local friends and businessmen. Lewis would reap a quick million dollars and, more significantly, enlist a level of businessman previously outside his reach in his conspirational circle. As Robert Shuler later observed, Lewis saw the "Bankers' Pool" as a reinforcement of the wall of protection he had erected among local politicians. "Every millionaire, every banker, every movie producer, every judge, every leading lawyer, every professional man of every character who made an investment in these pools," wrote Shuler, "became by that act a party to the giant conspiracy and made more impossible the prosecution and conviction of anyone connected therewith."

Armed with a no-lose moneymaking scheme, Flint had little difficulty forming his syndicate. Flint readily recruited fellow Pacific Southwest vice-presidents William Rhodes Hervey, P. L. McMullen, and Herbert A. Bell. Flint, Hervey, and Bell each committed $100,000 to the pool. McMullen, a junior vice-president, added $20,000. Flint then turned to his cronies from the local businesses, social clubs, and fraternal orders. Harry M. Haldeman, president of the Pacific Pipe and Supply Company, later described the recruitment process:

> In September last friends of mine of high standing told me that the stock of the Julian Petroleum Corporation was intrinsically worth much more than the market price, and that refinancing of the company through a bond issue being handled by the First Securities Corporation and others was in progress, upon the completion of which the stock would probably materially advance.
>
> My friends also told me that a certain amount of Julian company stock was in the hands of professional traders, who were forcing the market up one day and down the next, and that a pool of $1,000,000 had been formed to buy up this stock and hold it for a limited period, and thus stabilize the market.

Haldeman invested $20,000. Stockbroker Alvin M. Frank and realty baron W. I. Hollingsworth each posted $100,000. Movie mogul Louis B. Mayer, president of Metro-Goldwyn-Mayer, and attorney Henry S. McKay, Jr., purchased $50,000 shares. Lewis brought in realtor-banker Joe Toplitsky, who paid $75,000, and Adolph Ramish, who became the largest single investor at $150,000.

The members of this "Bankers' Pool No. 1," as it later became known, represented an impressive cross section of the Los Angeles business élite. Mayer was the best known but not necessarily the wealthiest or most prestigious. Hollingsworth, a director of Pacific Southwest, had reigned as one of the city's leading realtors since the boom of the 'eighties. Toplitsky had also made his fortune in real estate, ranking, according to one source, "among the ten largest . . . operators in the country." Best known for his role in the development of the city's showcase Biltmore Hotel, Toplitsky also served as the director of banks, oil companies, theaters, and other ventures. "At one time," reported Guy Finney, "his name, in electrical signs and on dead walls, was displayed more numerously than any other Los Angeles broker." Frank and McKay were the sons of two of the city's pioneer figures. Alvin Frank's father, Herman, had founded the durable Harris and Frank chain of clothing stores in the 1870s en route to presidencies of the powerful Merchants and Manufacturers Association and Los Angeles Board of Education. Young McKay was the son and name-

sake of Henry Squarebriggs McKay, Frank Flint's partner in the city's most prestigious law firm. McKay, Jr., a partner in the practice, was also Frank Flint's son-in-law
(and Motley Flint's nephew). At his Uncle Motley's urging, McKay became the
trustee and manager of Bankers' Pool No. 1.

In assembling this syndicate, Lewis and Flint had enlisted men of wealth, prestige, and power. While Bennett's stock activities had brought in leaders of city
government and the Progressive wing of the Republican Party, the "Million Dollar
Pool" attracted several active supporters of Harry Chandler's brand of right-wing
Republicanism. Haldeman, the forefather of a line of Republican activists (including H. R. Haldeman of Watergate fame), was president of the arch-conservative
Better America Federation (BAF), dubbed by the *Nation* a "super-virtuous radical
baiting guardian of California's patriotism." The BAF denounced advocates of publicly owned utilities as Bolsheviks, purged liberal books and magazines from the
schools, and played a central role in the often violent repression of labor unions in
Los Angeles.

The members of the "Million Dollar Pool" all considered themselves pillars of
morality in the community. Several belonged to local Masonic lodges and the élite
California Club. Hollingsworth was often described as a "noted churchman." Flint,
the city's "Santa Claus," and Haldeman, once named "the most useful citizen of
Los Angeles," both participated in a vast array of charitable activities. Many of them,
as one observer commented, "could have lived a thousand years on their income."
Yet the lure of fast, easy profits on a sure thing proved too tempting. Most rationalized that the respected Flint would not lead them into corruption. The blue-ribbon
nature of the syndicate itself seemed to bequeath its own mantle of probity. If any
of them questioned the ethics of the pool, they did not dwell on it.

Yet certain elements of the arrangement were clearly questionable. The below-
market purchase price, the no-lose guarantee, and the large amount of stock
required to secure the pool should have, and in some cases did, raise eyebrows.
Several pool members consulted attorneys, who assured them that the proposal was
legal. As frequent participants in the local stock scene, most of these men had to
have heard rumors of an overissue. Nonetheless, they accepted Lewis's assurances
that the shares they had purchased were legitimate. In reality, most of the 129,000
shares provided by Lewis probably came from Bennett, who issued almost 120,000
new shares of Julian preferred in September 1926.

Subscriptions to the Bankers' Pool No. 1 closed on September 21, but the
demand for additional pools rapidly escalated. Businessmen who had missed the
opportunity and members of the Million Dollar Pool seeking a bigger piece of
the action clamored for more investment possibilities. On September 28, I. Linden
Rouse, a junior vice-president of First National Bank, became trustee of a $250,000
Bankers' Pool No. 2, also known as the "Rouse Pool." This coalition included
businessman A. W. Hackel; Benjamin Platt, the owner of the city's largest chain of
music stores; and Motley Flint, as a trustee for Louis B. Mayer.

The Rouse Pool revealed certain strains in the stock-issuing procedure. Lewis
certified in the agreement that the total number of Julian preferred shares outstand-

ing did not exceed 486,000. This marked both an admission of an overissue (though due to the confusion over the original issue, Rouse and the others might not have realized it) and an evasion about its magnitude. Rather than print more certificates to cover this pool, Bennett and Rosenberg asked two of their regular customers, Perry Greer and Walter B. Allen, to return the stock securing their investments. Rosenberg later explained, "I called up Mr. Allen and Mr. Greer . . . and I told them that Lewis needed the stock. . . . I asked them if they wouldn't return that collateral stock they had on their transactions. They agreed to it provided I would guarantee their accounts." Greer and Allen returned 16,241 shares. They nonetheless retained their investments, relying on Bennett to make good on the stock guarantee.

Within weeks several other pools sprang to life. Businessman Albert Lane arranged a personal $75,000 investment with Lewis. In addition, Ed Rosenberg assembled more than a dozen of his regular clients into a $570,000 pool, alternately known as "Pool No. 3," the "Rosenberg Pool," or because of it was composed entirely of Jewish businessmen, the "Jewish Pool." (The existence of this syndicate gave rise to the oft-repeated, but erroneous, charge that Jews were excluded from the Bankers' Pools due to anti-Semitism among the predominantly Protestant Los Angeles élite. In reality, Jews appeared alongside Protestants in both bankers' pools.) A fourth pool, organized by attorney Mendel Silberberg, a prominent conservative leader in state and local Republican politics, purchased over $585,000 worth of stock. This group, which allegedly included Charlie Crawford, Bob Gans, and nine other "representatives of a powerful political machine, wealthy Main Street pawnbrokers, and Tia Juana concessionaires," became known as the "Tia Juana Pool." Thus, by mid-October, syndicates totaling $2.5 million and secured by approximately 300,000 shares of Julian Pete had been formed. The Julian pool mania had begun.

### III

In late August 1926, A. C. Wagy & Co. celebrated the opening of its new headquarters. Occupying an entire sub-floor of the Stock Exchange Building, the Wagy offices featured gold-inlaid ceilings, ornate lighting fixtures, and lavish mahogany furniture. The complex offered "the last word in modern brokerage facilities," including ticker and tape to hasten market reports. The brokerage invited its "friends and clients" to inspect these "unusual facilities."

In September and October, Julian Pete stockholders and speculators received a sampling of how unusual these facilities truly were. On September 1 the stock escrow plan initiated two years earlier under C. C. Julian expired. Thousands of original shareholders who had answered Julian's call could now claim their stock. While bound to honor this pledge, S. C. Lewis lacked the cash to redeem these shares. In a series of advertisements, Lewis praised the stockholders for their "patience and splendid cooperation." He warned that, in light of current low stock prices and the impending merger, which "in a few short weeks" would fulfill the goals of the escrow plan "in a manner and extent which I know will be highly gratifying to you," it would not be "reasonable" to expect shareholders to sell at this time. Lewis urged those who wished their stock returned to visit the Julian offices and learn the details of the merger plan. If they remained unconvinced, he would personally purchase their shares at the current market price. Those who appeared at the Julian offices, however, rarely received their stock. More often, high pressure salesmen urged them to use it as collateral for additional purchases.

Meanwhile, Wagy & Co. began to push Julian securities more actively on the LASE. In early September, Julian preferred, which had languished in the low to mid-teens throughout the summer, had begun a slow rise. On September 18 the stock closed at 17. Three days later the Bankers' Pool took effect. On September 24, Wagy & Co. announced that although it had never previously endorsed any individual security, the opportunities offered by Julian stocks were too good to ignore. The impending merger would push Julian preferred as high as $50 a share, Wagy & Co. told its clients, and the climb would start that day. "It is our earnest advice that you purchase all the stocks of the Julian Petroleum Corporation possible," advertised the brokerage.

The Wagy position heralded a new bull market, stimulated, as in the past, by massive buying by Lewis and his cohorts. On September 24, Julian preferred jumped across the 20-point plateau for the first time since April. During the next two weeks the stock rose steadily. Despite the removal of shares from the market by the pools, Julian preferred remained the most heavily traded security on the LASE, with tens of thousands of shares transferred each day. On October 6, the stock closed at 22⅝.

The next day, the Lewis forces withdrew their buying power, and the market in Julian preferred opened down almost six points. As "speculators stood spellbound outside the rail running around the floor" of the LASE, brokers surrendered large blocks of stock, driving prices still lower. Rumors floated that many of the stock pools had disbanded or that shareholders had simultaneously opted for an orgy of profit-taking. "Unable to understand where the stock was coming from," reported the *Times*, "it did not take the spectators long . . . to join in the mêlée." At the day's end Julian preferred had lost 9⅜ points, closing at 13¼.

Lewis attempted to control the damage. "Stockholders needn't become panicky," he stated, blaming the drop, as usual, on stock exchange manipulators. Lewis urged investors "TO HOLD YOUR STOCK," promising that it would be purchased or absorbed by the syndicate representing California-Eastern within a few days, which would automatically steady the market.

Small investors seeking to unload their stock found they had no alternative but to follow Lewis's advice. Etta Demange, a charter stockholder in Julian Petroleum, had a typical experience. Demange had invested $4,000 in 1924 and escrowed the stock at Julian's behest. In September 1926, when the escrow period ended, a Julian Pete salesman persuaded Demange to use her original preferred shares as collateral on the purchase of an equal amount of new stock. The salesman pledged that

Demange could sell her securities through Wagy & Co. at any time. On October 6, before the market broke, Demange ordered the brokerage to liquidate her holdings. Two days later she attempted to collect her money. The firm refused to pay or to confirm whether or not it had sold the stock. Repeated inquiries by phone and in person yielded no further information. Neither Wagy nor Julian Pete representatives would speak with Demange.

Other minor stockholders found it equally difficult to dispose of their stock. One elderly couple, retirees from a New York tailor shop, invested their $3,000 savings in Julian stock during the September rise. "It was the advertisements in the papers. I never saw such things in New York," explained the wife. "I bought at $16 and it went up to $20. I went down to Wagy's and said I wanted to sell. Four salesmen crowded around me and told me it would go up to $25. They said I'd be a fool to sell." When the market broke, the tailor and his wife crowded into the Wagy offices with other dismayed investors. They found Lewis awaiting them, at his persuasive best. "Mr. Lewis made a speech blaming the brokers," lamented the tailor's wife. "He kept us from selling by saying it would go up to $45."

Jack Bennett's regular clientele and A. C. Wagy insiders fared far better. Although the drop in the market initially caused sharp losses among those who had participated in Bennett's summer pools and loans, the Wagy brokerage reportedly accommodated these "few favored friends" by recording fictitious sales dated October 6, when the market had peaked. Among those who allegedly received this largesse were Charlie Crawford, who pocketed $41,250, his henchman Albert Marco, and Parrott machine regulars Greer and Allen. Other alleged beneficiaries included newsman Morris Lavine; FBI investigator Lucien Wheeler; First National banker I. L. Rouse, and DeKalb Spurlin. Several Wagy employees participated in the payoff, as did Bertha Kottemann, wife of accountant William Kottemann, whom First Securities had recently employed to audit the Julian books.

Pacific Southwest president Charles F. Stern also unwittingly profited from these market manipulations. Neither Stern nor John Barber, the senior officials at Pacific Southwest and First Securities, participated in the Bankers' Pool, but both nonetheless dabbled in Julian stocks. Stern opened an account with Wagy & Co., which was surreptitiously carried in the name of his private secretary. "I had no agreement with Lewis guaranteeing a profit or against loss in the transaction," Stern claimed. After the October collapse he received a notice that Wagy & Co. had sold his 2,000 shares of Julian stock at the peak price. "I had the choice, of course, of saying that I didn't want my stock sold or take the proceeds, and I preferred to take the proceeds," he later explained. Stern netted over $10,000. He subsequently asked Lewis about the transaction. Lewis replied that he "had taken out some of his friends at the top of the market because he proposed to break it himself." Stern confessed, "I saw then that I was in a stock rigging operation and I got out and stayed out." This episode, however, apparently did nothing to sour Stern on Pacific Southwest's participation in the California-Eastern merger. Efforts to arrange financing continued unabated.

Behind the scenes at Wagy & Co., the events of October led several key figures to question the status of Julian securities. Manager John Ethridge discovered on October 6 that the Wagy brokerage alone held 420,000 shares of Julian preferred. He called Adolph Ramish and confirmed that Ramish also held tens of thousands of shares. Ethridge reported these figures to Lewis and expressed his suspicions of an overissue. Lewis glibly explained that 161,000 of Wagy's shares represented stock awaiting return to the Julian Pete transfer department for cancellation. The following day, 161,000 shares were removed from the company vaults. From that moment on, Ethridge was denied access to all Wagy records. "My job as manager practically ceased then," he later testified. On December 6, Ethridge severed his connections with the firm.

Accountant William Kottemann also worried about an overissue in the wake of the October collapse. On October 9, in the presence of Pacific Southwest vice-president Rouse, Kottemann dictated a report to be delivered to the bank. "The quantity of stock held by Wagy & Co. throughout June, July, August, and September has always been large enough to have raised in our minds the question as to where all the stock came from," stated Kottemann. As of September 20 (prior to the creation of the Bankers' Pools), his audit had revealed almost 300,000 shares on hand at Wagy & Co., an additional 284,000 shares occupied in pool agreements, and 60,000 additional shares held by Ramish or in escrow. The total exceeded the 600,000-share figure generally accepted as legal. In addition, noted Kottemann, "the above tabulation does not take into consideration any stock held by other brokers or held by individuals."

According to Kottemann and Rouse, Lewis demanded that the report be withheld from Pacific Southwest officials, and Motley Flint ordered Kottemann to filter all subsequent reports through Lewis. Kottemann attempted to clarify these instructions. "You two gentlemen are agreed that any further information we have to give must be given to Judge Lewis and that you will seek the information from Judge Lewis and not from me," stated Kottemann. Flint, reported Kottemann, responded, "That is right—heed it."

The decline in the price of Julian preferred also sent shock waves through the recently formed pools. Their creation had been predicated on the assumption that the impending merger and withdrawal of stock from trading would precipitate a sharp increase. Events before October 6 bolstered their optimism. When the stock reached 22⅝, Bankers' Pool manager McKay urged Lewis to exercise the escape option and begin liquidation. Lewis refused, predicting higher prices. Indeed, fears of a premature end to the Million Dollar Pool probably led Lewis to pull the plug on his latest stock-buying foray. As the weeks passed and the pool period drew closer to its November 9 expiration date, Julian preferred languished below the original purchase price. At best, syndicate members would receive Lewis's guarantee of $15.50 a share—or no profit for their consideration. At worst, should Lewis fail to repurchase the stock, they would have to sell both their shares and a portion of the shares securing the pool to recoup their losses.

In early November their direst fears materialized when Lewis met with Flint, Toplitsky, and Herbert Bell and informed them that, due to the drop in prices, neither he nor Wagy & Co. could afford to redeem the stock. Lewis urged them to extend the pool for an additional thirty days, by which time the merger would presumably be completed. In exchange, he volunteered a "three-point bonus"—$3 a share for each thirty-day period the pool continued. Thus, on December 9, the Million Dollar Pool would receive $190,000 for their forbearance and an option to liquidate or further continue the pool.

This bonus proposal bore a striking resemblance to the pool arrangment Flint had initially rejected due to its dubious legality. Yet, when faced with the prospect of selling at a loss or garnering additional profits, the pool members, at the urging of Flint and Toplitsky, unanimously agreed to grant an extension. During the next few weeks, Lewis reached similar accommodations with Rouse's Bankers' Pool No. 2, Rosenberg's "Jewish Pool," and Albert Lane for his personal account. Mendel Silberberg's "Tia Juana Pool," on the other hand, opted to divest its holdings. As Lewis later described, "It matured. No extension was granted, no bonus demanded, received or accepted, and Wagy & Co. bought back the stock."

While negotiating pool extensions, Lewis worked the more respectable side of his two-way street with John Barber. During September and October, the two men had assembled an impressive group to join them on the California-Eastern board of directors. From the petroleum industry they recruited independent oilman E. J. Miley, southern California's most successful wildcat driller. In October, attorney C. W. Durbrow, counsel and chief valuation attorney for the Southern Pacific Railroad, became a director. (Durbrow also arranged loans for Lewis with the California Bank of San Francisco. Lewis paid him $95,000 for his efforts.) Durbrow, in turn, enlisted L. J. King, vice-president and director of field operations for Associated Oil, a Southern Pacific affiliate. As a major Marine Oil shareholder, Associated had a vested interest in the merger and encouraged King's participation. Former senator Frank Flint, attorney for First National Bank, climbed aboard as the firm's legal counsel. On October 31, California-Eastern opened new offices in the Pacific Southwest Building and introduced its executive board at a ceremony hosted by Senator Flint.

During the following weeks, at least one of the new board members developed doubts about California-Eastern. When King joined the board, Barber informed him that the financing for the merger would be arranged by December 1. When that date came and went, King sought further assurances. Barber explained that First Securities had agreed to provide $7.5 million in financing through the sale of bonds, but that Lewis was responsible for raising an additional $5 million. Barber claimed that Lewis had failed to fulfill his commitment. As a result he had tentatively arranged for a local investment firm to assume this financing. Barber promised King a December 20 completion. At the year's end, however, the situation remained unresolved.

The delay stemmed in part from problems in the ongoing investigation of Julian Pete assets. First Securities had unleashed a small army of appraisers, engineers, auditors, attorneys, and title searchers on the California-Eastern project. For more than two months they had attempted to fix an appropriate valuation for the company's properties. According to Walter Woehlke, "These investigators struck one snag after another. [Julian Petroleum] was a one-man concern. Lewis had made dozens of deals, contracts and agreements which appeared on the books in a most sketchy manner. His marvelous memory contained all the facts . . . , but the owner of that memory was a busy, very busy man. He and the facts were hard to reach, and the work dragged on far beyond the expected time."

As the probe continued, Barber and Stern grew concerned about the mounting costs of the appraisal. In mid-December they requested that Lewis post an additional $100,000 to cover these expenses. The purpose of these funds would later be disputed. Lewis alleged that the $100,000 included, not only the usual trust fees, but a bonus for Barber and Stern. Jack Bennett, who raised the funds by selling over-issued stock, alleged more bluntly, "That $100,000 was to be a gift to them to use their influence on anything they could to put over the merger deal." The payment of such a bonus clearly violated California banking laws, and Barber and Stern denied any illegal intent. The bankers claimed that that the $100,000 represented a retainer needed to cover their expenditures in the event the deal fell through. "That is a very usual thing in underwriting," Stern explained.

Stern and Barber deserve the benefit of the doubt, but the peculiar history of the $100,000 lends credence to the bonus charges. Lewis allocated the money not with Julian Pete funds but with a cashier's check drawn from the Merchant's National Bank and paid for with a personal check from Adolph Ramish. Lewis delivered the Merchant National draft, which with his endorsement could be cashed by anyone, to Barber and Stern. They deposited it with First Securities. Lewis received no receipt for this transaction.

Had the funds remained at First Securities, no charges of impropriety would have arisen. But Barber ordered the check to be transferred to the Pacific Bond and Share Company, another First National subsidiary. Barber had recently purchased the worthless Pacific Bond and Share from businessman Harry Bauer in a transaction that Bauer described as "a standing joke" between the two. Pacific Bond and Share retained the check as the basis for a special account that monies could be drawn from. Over the next several months, approximately $63,000 in "miscellaneous expenses" flowed from this account. Most of this amount apparently paid the fees of engineers, attorneys, and auditors working on the merger. But Pacific Bond and Share also used the fund to purchase 10,000 shares of Julian preferred. The Merchants National draft remained uncashed. Its circuitous handling would earn it the title of the "orphan check."

The $100,000 payment to First Securities was only one of several extraordinary expenses incurred by Lewis in December 1926. On December 9, the thirty-day extension on the Bankers' Pool No. 1 expired. Once again Lewis pled poverty. Lewis could pay the three-dollar-a-share guarantee promised for prolonging the contract, but neither he nor Wagy and Company could redeem the entire block of shares supporting the pool. Lewis delivered $193,548 to pool manager McKay and requested another thirty-day extension. For pool members the payment represented a fabulous return on their money. The $100,000 shareholders each received $19,355, while retaining their original investment. Ramish, the largest individual speculator, received over $29,000. Smaller investors received prorated amounts. All of the original pool members, except Toplitsky, agreed to the extension.

Toplitsky's defection produced several curious twists in the Julian Pete saga. To replace himself, Toplitsky and Alvin Frank assembled four local businessmen to subdivide his share. For this service, Lewis paid Toplitsky and Frank a bonus of several thousand dollars. During the next several months Toplitsky and Frank recruited numerous investors into new pools and stock deals and received payments totaling tens of thousands of dollars for this service. Lewis related, "Frank would take an interest in a pool, then he would sell it out to his clients. . . . He did what was referred to as refinancing it." O. Rey Rule, general manager of the Pacific Finance Company, Republican state leader, and prominent member of the city's leading country clubs, participated heavily in these deals. For Frank, the combination of high interest rates and stock and cash bonuses provided lavish profits. It "was just like shooting fish," he stated.

Toplitsky, despite his withdrawal from the Banker's Pool, retained a high profile in Julian Pete affairs. By his own admission, he wanted "to see the thing through." In addition to lining up new investors, Toplitsky, a director of the Merchants National Trust & Savings, used his influence to arrange a series of dubious loans for Lewis that he and Frank received substantial commissions for. Lewis estimated that Toplitsky arranged over $300,000 in loans. By late 1926, Merchants National had become a clearinghouse for numerous Julian-related transactions.

Like Bankers' Pool No. 1, the other major pools also collected their three-point guarantees, extended their contracts, and grew more complex. In late December, Lewis again demonstrated his ability to lure the powerful and well-connected—in this instance, Senator Hiram Johnson's son, Archibald, a San Francisco attorney. Johnson and a colleague agreed to purchase 20,000 shares of Julian preferred for ninety days, after which Lewis would buy back the stock for a "three-point advance." "We did not know S.C. Lewis and had never seen him before," explained Johnson. "We thoroughly investigated him through certain prominent San Franciscans and certain banking institutions both here and in Los Angeles. . . . The reports received by us were such as to entitle him to our confidence." The "San Francisco Pool" invested over $250,000 in Julian preferred.

Underwriting these activities was Jack Bennett's "bubble mill." To finance the increasingly bewildering morass of payments for loans, pools, corporate activities, merger acquisitions, and other deals, Bennett kept printing stock certificates. During the last three months of 1926, Bennett generated over 450,000 new shares of Julian preferred. Pat Shipp later claimed that he called Lewis from New York in December 1926 to inform him of the extent of the overissue. Fearing that someone might be listening on the line, Shipp referred only to the color of the stock certificates in his report—shares "in the blue" meaning preferred stock and shares "in the yellow" for the common. "How much of a certain party?" asked Lewis, referring to Bennett. Shipp reported approximately 2,300,000 preferred shares outstanding and 1,700,000 of the common. Bennett had exceeded the total combined authorization by 1,100 percent; yet his most prolific period still lay ahead.

# 11 IT APPEARS THAT THERE IS AN OVERISSUE

**I**

On January 4, 1927, Clement C. Young confidently took the oath of office as the twenty-sixth governor of California. None of his predecessors could match the breadth of experience and achievement that Young brought to the statehouse. A former high school English teacher, whose 1904 poetry text remained a standard primer in California schools, Young had first won election to the state assembly in 1908. He allied himself with the insurgent Lincoln-Roosevelt League, which two years later led the Progressive political revolution that transformed California politics. A key leader of the Progressive forces in the legislature, Young served as speaker of the assembly from 1913 to 1918. Elected lieutenant-governor in 1918 and re-elected in 1922, Young presided over the state senate for an unprecedented eight years. Following his narrow defeat of incumbent governor Friend Richardson in the August 1926 Republican primary, Young romped to victory in November over his Democratic opponent, polling 71 percent of the vote, the largest margin in California history.

In winning the Republican primary, Young had incurred debts that his supporters now expected to be repaid. A. P. Giannini, whose endorsement and contributions had proved crucial in the election, sought a superintendent of banks who would be more sympathetic to the Bank of Italy's ambitions in southern California. On January 20, 1927, Young appointed Oakland mayor Will C. Wood to the post. Seven days later Wood approved the merger of three of Giannini's banking systems into the Liberty Bank of America. Shortly thereafter Wood abrogated the *de novo* rule, which had blocked Bank of Italy expansion, and on March 1 he approved the merger of the new Liberty Bank with the Bank of Italy. These rapid-fire events gave

Giannini a decisive victory in his struggle with the Sartori-Robinson forces in Los Angeles. Giannini suddenly controlled a statewide system with 276 branches, making the Bank of Italy the largest bank in the United States, outside of New York City. Two years later Giannini's brainchild would shed its ethnic origins when he reincorporated his holdings under a title more befitting his burgeoning national empire, the Bank of America.

Like Giannini, Kent Parrot, Mayor Cryer, and their Los Angeles political allies also laid claim to a key appointment in the Young administration. They set their sights on the corporation commissioner post vacated by Mike Daugherty. Parrot desperately wanted Los Angeles city prosecutor Jack Friedlander named to the job. Governor Young reacted coolly. He had never met the thirty-seven-year-old Friedlander nor heard of him prior to the election. He questioned Friedlander's youth and inexperience. Although seemingly an able public servant, Friedlander had never dealt with the private business affairs that fell within the purview of the Corporations Department.

Parrot unleashed a formidable campaign, mobilizing all of the resources at his command to persuade the new governor to appoint Friedlander. He importuned myriad local businessmen and public servants beholden to his political machine to write on Friedlander's behalf. Young found himself inundated with letters from Los Angeles bankers, merchants, judges, and office-holders praising the city prosecutor.

Parrot also attempted to enlist Senator Hiram Johnson into the fray through Johnson's southern California confidant, Frank Doherty. Doherty, a Los Angeles lawyer who regularly corresponded with Johnson to apprise him of the local scene, wrote to the senator at Parrot's request. Doherty warned that Parrot had threatened that "if the Governor does not appoint Friedlander . . . [he] is going to get off the reservation and have nothing to do with Young." Parrot noted that "he raised a very large sum of money to help fight Young's battle. . . . If he is going to be punched in the nose by the Governor, he wants to know it real soon." Parrot wanted Johnson to intercede to urge Friedlander's selection. "It is no secret that the Governor would not appoint Friedlander under any circumstances if it were not that the Mayor and Parrot had so strongly urged him," reported Doherty. "It is equally a fact that the ungracious way the governor has acted toward Parrot and Cryer in this matter has cooled their enthusiasm for Young almost as if he had rejected Friedlander and appointed someone else."

On February 8, with Young still refusing to commit to Friedlander, Parrot wired Johnson, beseeching his support. "It is my understanding that definite action on his appointment will probably be taken early next week and I suggest that whatever you do in this matter be done immediately," pleaded Parrot. On the same day, Doherty wrote to the senator, "I have just talked to Kent. He is still left out in the cold. He has been treated in my opinion, worse than a stepchild. I cannot figure out the Governor's attitude." Doherty complained that Young had failed to consult Parrot on any appointments made for southern California, adding, "Kent is still endeavoring to have his man named corporations commissioner, but up to the present time he has not put him over. The fact is the Governor does not want to appoint him."

Doherty attributed Parrot's vehement insistence about Friedlander to local political concerns. The *Times* and Edward Dickson's *Los Angeles Express* had published stories indicating that the appointment was a test of Parrot's influence. If Young failed to choose Friedlander, admonished Doherty, "it will be seized upon by the Express and the Times as a repudiation of Kent by the Governor." Parrot protested to Johnson that the "Chandler-Dickson combination is attempting to prevent his appointment with the idea of destroying the morale of our organization."

Rumors circulated, however, that considerations beyond politics accounted for Parrot's adamant stand. The corporation commissioner controlled investigations of errant business ventures. With the Julian pools gaining momentum and profiting many of Parrot's political allies, a sympathetic and politically indebted ear at the Corporations Department could prove useful. S. C. Lewis had reportedly contributed heavily to both the Young campaign and the Parrot machine. Had Parrot promised Lewis a compliant corporation commissioner in exchange?

The war of nerves between Parrot and Young continued for two more weeks. Young later reported receiving petitions, letters, and telegrams endorsing Friedlander from almost 500 people, including bankers, businessmen, judges, and attorneys. On February 16, Governor Young arrived in Los Angeles and checked into the Biltmore Hotel, where Parrot maintained his fabled suite. Parrot and others pressed their case. Young tried to maneuver around the issue by asking H. L. Carnahan, California's first corporation commissioner, to return to the job. Carnahan declined the governor's offer.

On February 25, Governor Young finally capitulated. He named Jack Friedlander corporation commissioner. Young made no effort to hide his misgivings. Noting the hundreds of people who had attested to Friedlander's qualifications, Young stated, "In view of these facts I feel that I have no right to hesitate any longer or oppose my own lack of personal acquaintance with Friedlander to the verdict of his own fellow townsmen who know him best." As Young doubtless feared during the long dispute, the Friedlander appointment would prove the most fateful of his administration.

Lewis, meanwhile, had begun to feel the pressure of his creditors. On January 10, Bankers' Pool No. 1 received its second $190,000 bonus, extending the arrangement for another thirty days. The other large pools formed in September also claimed the second installment of the three-point bonuses. In addition, Lewis owed hundreds of thousands of dollars in payments on the oil properties he had acquired in late 1926, and local banks had begun to enforce credit limits on Bennett and him.

Nor did completion of the merger seem imminent. On January 19, Lewis postponed a scheduled shareholders' meeting, called to reveal details of the consolidation, until February 26. Lewis, with consummate redundancy and tortuous syntax, blamed the delay on "the inability to complete or conclude the audit and engineering appraisals." In order to give Julian Pete stockholders "the benefit of the enhancement in the values of some very substantial properties made so by recent developments," appraisals would now be updated through December 31, 1926, rather than the original August 31 target.

On the LASE, Julian securities experienced an unusually uneventful period. Since the collapse of the October bull market, Julian preferred had stabilized between 13 and 14½. Julian common held steady in the 2 to 2¾ range. Although Julian stocks remained among those most heavily traded, the volume was relatively modest by past standards. The calm came to an abrupt end on February 9 when almost 55,000 shares of Julian preferred flooded the market floor. The stock closed down two points at 11½. The next day the sharp slide continued. With hundreds of spectators seeking explanations for the crash, Julian preferred dropped as low as 7½, its worst showing in seventeen months. The common stock closed below 2 for the first time in its history. Over 76,000 shares of preferred stock crossed the board. The *Los Angeles Times*, evincing the prevalent confusion over the amount of authorized stock, erroneously reported that the volume "approximat[ed] one-sixth of the total number of shares outstanding." In reality, about one-half the legal issue had been traded.

Lewis charged that "a prearranged drive" by speculators seeking to capitalize on the financing delay had broken the stocks. Wagy and Company cited "wild, unfounded rumors of every description." Heavy buying campaigns brought Julian preferred back as high as 10 on February 13, but on the next day the collapse resumed, dropping prices as low as 6½. The stock rallied late in the week, "enter[ing] a calmer sea," in the words of the Wagy copywriter, leveling off at about 8½. In the two weeks starting on February 9 almost 600,000 shares changed hands.

Trying desperately to present a silver lining, Wagy and Company actually hailed the collapse of the stock, predicting that if the "wholesale dumping of stock" continued, its clients would gain "complete control of the floating supply of Julian Preferred . . . at ridiculously low prices." By this time, however, controlling "the floating supply" of Julian preferred had become an impossibility. Jack Bennett was pumping new stock into the market almost as fast as the brokers and speculators could buy it. The most recent price collapse had resulted from the most spectacular binge of counterfeiting he had yet undertaken. Tens of thousands of shares flowed from the "bubble mill" each day. On February 14, Bennett staged his own Valentine's Day massacre, issuing hundreds of stock certificates and untold thousands of shares, most of them antedated to the eleventh day of each month from November 1926 to February 1927. By the month's end, over 500,000 new shares had accrued to the Julian overissue, raising the total outstanding to over 3,000,000.

The existence of an overissue now was undeniable. On Spring Street everyone from brokers to bootblacks openly discussed it. One brokerage house reportedly had handled 600,000 shares in one day. A stockbroker told Lewis that he believed the supply of Julian preferred ran into millions of shares. LASE officials, however, seemed oblivious.

Bankers' Pools No. 1 and 2 and other individuals and syndicates accepted their usual February extension payments. More cautious investors decided to forgo the lucrative three-point bonuses and cash in their increasingly suspect investments. In San Francisco the ninety-day agreement signed by Archie Johnson was due to expire. Lewis informed him that he could not afford to redeem his pledge and offered an extension. Johnson and his partner refused. "They said everyone was telling them there was an overissue," recalled Lewis. He paid $60,000 to cover the three-point bonus, and Johnson sold his stock, netting a substantial profit. According to Guy Finney, "Johnson later found occasion to chuckle over the incident when he related the details and his winnings to friends at the Bohemian Club."

Beneficiaries of the major Julian pools also grew increasingly restive. On February 26, the day Lewis was to address Julian stockholders, Alvin Frank hosted a meeting at which he, Toplitsky, Ramish, and O. Rey Rule, who would soon be added to the California-Eastern board, attempted to "back Lewis in a corner" and "come through with a statement" to ease their anxieties. With the amount of Julian preferred they held individually and as members of pools, these men calculated that they alone held over 400,000 shares.

Confronted by the overwhelming evidence, Lewis readily conceded an overissue. According to his account of the meeting, Lewis admitted that 825,000 shares had found their way onto the market. Those present assumed this represented a 300,000-share surplus. Lewis explained that the Julian stock transfer department had regularly overissued securities due to his habit of canceling excess stock from his own holdings. He assured the gathering that the current overissue could also be disposed of from his personal cache, but warned that the redemption of shares securing the major pools "will break me and the friends who are going to stay in the proposition." Lewis laid the primary blame for the situation on John Barber and First Securities for delaying the merger financing.

Ramish, who had as much as $2 million tied up in Julian stock, had become ever more nervous about his investment. Throughout the past months he had repeatedly sought reassurances about the stability of Julian Petroleum. Lewis, Rouse, Stern, and Barber had all sought to allay his fears. According to Lewis, Ramish now threatened to use his knowledge of the pools to "blow the ship up," if he did not receive his profits. He warned that he would withdraw his own stock and sell it. Ramish later admitted that at the February 26 meeting he did "a little hollering." Those present urged Ramish not to do anything drastic that might wreck the company. Lewis reminded Ramish that "he had made more money out of it than any other living man." The meeting ended, Lewis recalled, with an agreement that Barber would have to do the financing within ten days and that each person would use his influence to pressure the banker.

Ramish remained unplacated. He demanded his stock from the Bankers' Pool. "He said he didn't give a damn how much stock was overissued," reported Lewis, "he was going to get his money out of it." Since the stock certificates securing that pool remained in a vault, Lewis ordered Bennett to turn an equivalent number of shares over to Ramish. Bennett delivered 23,000 shares. During the next few weeks,

Bennett sold 1,000 shares a day for Ramish. Lewis estimated the value of these sales at $12,000 to $16,000 each day, a total of up to $800,000.

Other participants at that meeting, however, unabashedly resumed their Julian stock transactions. Frank, despite definitive knowledge of an overissue, continued his daily transactions with Bennett, many of which involved Rule's Pacific Finance Company. Julian Pete had survived the latest stock maneuvers, but the day of reckoning was drawing near.

## II

The Julian stock collapse of February 1927 also triggered a reassessment among California-Eastern board members and Pacific Southwest bankers. The inexplicable downward trend of Julian securities, the persistent rumors of an overissue, and the inability of Lewis to deliver his share of the financing cast the future of the merger in grave doubt.

Stern and Barber still seemed blind to the overissue. When the merger talks had begun in August, Lewis had promised to return the corporate ledgers from the New York offices for verification. Accountant Kottemann, already admonished for his efforts to audit the Wagy & Co. books, patiently awaited the return of the stock records. The ledgers finally arrived on January 24, 1927, stored in eleven separate trunks weighing 3,600 pounds. Lewis instructed transfer clerk Pat Shipp, who had returned from New York with the records, to secrete them at his home. The books remained in Shipp's garage while Lewis procrastinated and repeatedly reassured the bankers. Kottemann made no effort to accelerate the process. When questioned about an overissue by an Oakland brokerage firm, he replied that he had discovered no "serious irregularities."

By February, First Securities investigators had delivered favorable reports on all other aspects of the merger. The only remaining uncertainty involved the share count. In and of itself this was not unusual. According to Rockwell Hereford, who worked for First Securities, the stock situation "is the last thing normally looked into in an underwriting and then only to determine whether the total is within the . . . authorization and which stockholders are authorized to vote." Following the price skid on the LASE, Barber pressed Lewis to release the records. In mid-February, Lewis finally complied, delivering the books to Kottemann piecemeal.

The bankers simultaneously moved to solve two other problems. Barber and Stern recognized that a firm hand was needed at the helm of California-Eastern to neutralize Lewis. They also conceded that Lewis would be unable to arrange acceptable secondary financing. Barber volunteered to organize a syndicate to handle these debentures if L. J. King, who had expressed severe misgivings about the progress of the merger, could recruit a prominent and qualified individual to serve as California-Eastern chairman of the board. King recommended Harry J. Bauer.

Bauer seemed an inspired choice. A successful corporate attorney, Bauer already was board chairman at Southern California Edison. Like other local business figures he served as director and officer in numerous firms. Among his business achievements, the most relevant to the Julian situation was the rehabilitation of the Pacific Gasoline Company. In 1921, Bauer had taken charge of the struggling firm and within a few years engineered a $20,000,000 sale to Standard Oil. King and Barber hoped that Bauer might perform the same magic for California-Eastern.

At King's urging, Bauer also had become a major speculator in Julian Pete securities. According to Bauer, King had approached him in December 1926 and asked him to "buy and carry some Julian preferred for me." Bauer expressed reluctance. "I said I didn't think the stuff was any good," he recalled. King reassured him that as a director he could vouch for Lewis and the company. Bauer bought 20,000 shares. Since that time, King, Bauer, and businessman Joseph Dabney, whom King also sought to put on the board, had accumulated a considerable amount of additional stock. King also had been prodding Bauer to take a more active role in California-Eastern.

According to his later account, Bauer first met S.C. Lewis under peculiar circumstances. When the price of Julian securities had plummeted on February 10, Bauer asked to be introduced to Lewis. "I took a look at him and I said, 'I understand you are a crook,'" Bauer reported. "Well, some people say so," responded Lewis. Bauer asked Lewis how much he owed stockbrokers as a result of margin calls due. Lewis estimated that $90,000 would cover his immediate needs. "I suppose I may lose this money," Bauer told Lewis, "but I am going to loan you $90,000." Lewis secured the loan with the coin of his realm, Julian stock certificates, and promised repayment later that day. In addition to lending money to Lewis, Bauer ordered his broker to purchase thousands of shares of Julian preferred to help prop up the market.

Bauer later offered several explanations for this unusual spur-of-the-moment transaction. Despite Lewis's shortcomings, Bauer believed that the Julian Pete president retained the stockholders' loyalty, and his presence remained essential to a successful merger. Bauer also feared that he would not be able to unload his own stock unless the market for Julian issues could be restored. "I was sure that if Lewis could not meet his brokers' bill on that day, the whole thing would be done for," stated Bauer. Another non-financial motivation prompted Bauer's loan. "I wanted to find out just how he handled himself. I knew that I could not get in . . . as the head of that organization and accomplish anything if I was going to have the sand cut out from under me by Lewis . . . and the first thing I was going to do was to find out what sort of fellow Lewis was and determine whether or not I could rely on what he said."

Lewis evidently repaid the loan, and Bauer moved closer to assuming the chairmanship of California-Eastern. The more carefully Bauer examined the merger proposal, he claimed, the more feasible it looked and the more the challenge appealed to him. In addition, Bauer's stock purchases had given him a substantial stake in the outcome. At Lewis's behest, Bauer had continued to boost the market in Julian securities. "Mr. Lewis would call me up from time to time and he would say 'The market is going to the devil this morning. . . . Will you jump in and help?'" In exchange, Lewis would refund one dollar per share to Bauer. Bauer ultimately purchased 60,000 shares of Julian preferred.

King and Barber encouraged Bauer to chair the board. First National president Henry Robinson also intervened. Robinson had close ties to Bauer. They served on several boards together, including Southern California Edison, and had several country club memberships in common. Thus far Robinson had played no visible role in the Julian project, but the venture clearly had at least his tacit approval. On most mornings Robinson would still ride to work with Barber. It seems inconceivable that they would not have discussed California-Eastern. In February, Barber brought Robinson directly into the picture to recruit Bauer.

According to Bauer, Robinson invoked the needs of the 40,000 stockholders and Bauer's commitment to public service. "This is the biggest job that has ever been in Los Angeles," Robinson told his colleague. In late February, Bauer agreed to head the company. In addition he promised to personally subscribe for $1 million of debentures and raise an additional $300,000. Bauer put two conditions on his acceptance, however. He demanded that the bankers relieve him of his Julian stock at its original purchase price (an estimated $500,000) and that they pay him $250,000 in cash.

Bauer's stipulations caused dissension among California-Eastern insiders. Frank Flint viewed them as extortionate and vowed he "wouldn't stand for it," according to Lewis. King objected that he "had done as much as Bauer had in the reorganization" and argued that no one individual should have a financial advantage over the other participants. Henry Robinson also balked but ultimately capitulated. Bauer said that Robinson told him, "That is a lot of money. As far as your stock is concerned, we will take it off your hands. As far as the $250,000 is concerned that is a lot of money. I don't want to see you get any less, and I don't want to see you get any more."

Bauer defended his position as a matter of principle. "I was not interested in making that $250,000, because my present income is more than I can spend," he explained innocently. The down payment represented the sincerity of his commitment. Bauer's protests notwithstanding, even for a man of his substantial wealth, the $750,000 worth of considerations were no mean concession.

Throughout these negotiations Frank Flint figured ever more prominently in the California-Eastern picture. Several ties linked the former United States senator to the new corporation. As legal counsel to First National Bank, Flint had taken an interest in the project. Lewis had subsequently engaged him as his personal attorney. Both Flint's brother Motley and son-in-law and partner Henry S. MacKay, Jr., were heavily involved in the Million Dollar Pool. Perhaps most important, Flint had succumbed to the charms of S. C. Lewis. Despite the rumors swirling around the Julian Pete president, Flint remained a stalwart Lewis supporter. He viewed the financing delays and recruitment of Harry Bauer as an effort by Barber to "sell out Lewis" and steal the corporation. Flint accepted Lewis's assurances that any overissue was minor and would be easily absorbed in the merger.

To ease persistent anxieties about an overissue, Flint invited Lewis, board members King and Durbrow, and MacKay, Jr., to his luxurious Flintridge home. Flint railed against Barber and Bauer. Lewis denied the existence of a significant overissue. To demonstrate his sincerity, Lewis agreed to place both his Julian Petroleum and California-Eastern holdings, including any overissued stock, in trust with Flint. The former senator, whose integrity was unquestioned, would hold Lewis's shares until completion of the merger and use them to offset any overissue.

These developments cleared the way for the long-awaited February 26 Julian Pete stockholders' meeting. Lewis, Flint, and the Pacific Southwest bankers hammered out the final elements of the pact shortly before the meeting commenced. At 8:00 p.m. Lewis took the podium at the Olympic Auditorium, filled to its near-12,000-seat capacity. For two hours Lewis delivered what the *Times* described as a "fiery and impassioned" address extolling the successes of Julian Petroleum and the propects for California-Eastern. An audit performed for First Securities by one of the nation's leading accounting firms had estimated assets at almost $30 million and 1926 net earnings at over $3.2 million. Annual oil production had reached 6.5 million barrels.

Lewis then revealed the details of the California-Eastern offer. Julian Petroleum stockholders could exchange each preferred share for a minimum of three shares of California-Eastern stock, at a par value of $10 per share. Thus investors would receive at least $30 worth of stock for each share. Julian common could be redeemed for one-half share of California-Eastern. Julian Petroleum would be completely absorbed by the new company, and the Julian name would disappear from all gas stations and properties. California-Eastern had fifteen days to finalize the deal. But this, stated Lewis, was a formality. Stockholder approval would complete the merger.

Lewis called for a vote on the proposal. A scattering of "noes" rumbled from the crowd, only to be drowned out by "a thunder of 'ayes.'" Almost four years after its tempestuous birth, reported its longtime nemesis, the *Los Angeles Times*, the Julian Petroleum Corporation would fade into oblivion.

Throughout this period the volume of trading on the LASE remained heavy as reports of an overissue circulated incessantly. In early March, former Wagy & Co. bookkeeper Wray Berthold desperately tried to expose the fraud. Berthold engaged attorney Ben Beery and the pair informed L. J. King of the overissue. King and Bauer immediately challenged Lewis. The Julian president revealed that Wagy & Co. had recently suspended Berthold for conducting fraudulent transactions. Berthold, said Lewis, simply sought revenge. Berthold admitted that he had been dismissed for buying stock in his wife's maiden name without putting up any margin but defended the veracity of his tale. King and Bauer dismissed the allegations as inconclusive.

Beery and Berthold next wrote to the Department of Corporations. Their letter charged that the Wagy & Co. ledgers revealed that that brokerage alone had 667,000 shares on its books. "It will be impossible for the stockholders . . . to ever realize on their money or procure their stock upon payment of balance due on their accounts," warned Beery and Berthold. On March 11, Beery and Berthold met with the head of the department's auditing division, who expressed skepticism about their charges. "I figured that statements by a man who had recently been discharged should not be given serious consideration," the auditor reported to Commissioner Friedlander. After reviewing the evidence, he concluded that, while "on the face of it his reasoning seemed fair," Berthold had not considered that a "short condition . . . might be disclosed by a complete examination of all the ledgers," or that "repeated resales of Julian Petroleum stock might show a great many shares more than the entire capital stock of the company." Berthold's allegations, he decided, did not prove an overissue, but "they might indicate an oversale by the broker." Friedlander refused to take any action prior to the completion of the ongoing Kottemann audit. U.S. attorney Samuel McNabb also ignored their entreaties, informing Berthold and Beery that federal authorities could only act upon the complaint of an actual fraud victim.

On March 25, 1927, the new California-Eastern board of Directors convened to ratify the agreement accepted by Julian Petroleum stockholders. Harry Bauer presided as chairman of the board. Other directors included Lewis, who would also serve as company president; Barber, King, Durbrow, Miley, Dabney, and Rule. The directors voted to assume all assets and liabilities of Julian Petroleum. Immediately after the meeting, Barber formally announced that First Securites would underwrite the new company. First Securities would issue $7.5 million in bonds secured by a first mortgage on all properties and $5 milllion in five-year, 7 percent debenture bonds.

"The provision of $12.5 million places the California-Eastern Oil Company in an enviable financial position," stated Barber. "The entire proceeds of the financing will be used exactly for the corporate purposes of California-Eastern. There will be no promotion stock issued, no bonuses, and no stock commissions will be paid to anyone." Thus, reported the *Times*, after seven long months of uncertainty the Julian Petroleum-California-Eastern consolidation was "virtually" complete.

### III

The March 25 announcements by California-Eastern and First Investors gave the Julian Pete faithful cause for celebration. "If there was any doubt existing as to the future success of . . . California-Eastern," advertised Wagy & Company, "the personnel of its Board of Directors is in our opinion a complete answer to such 'doubt.'" Preferred stock rose above the ten-point mark. On March 30, S. C. Lewis announced that, "in line with [his] . . . policy of taking no step without full authority of the shareholders," he would submit the final merger proposal to a special meeting of Julian investors on April 14.

At this time, while the corporate venture that had spawned it seemed closer to reality, the Bankers' Pool No. 1 disbanded. The Million Dollar Pool had espoused two original purposes: to maintain the price of Julian preferred and to enrich its participants. On the first count it had failed miserably; on the second it surpassed all expectations. Since December, pool members had divided four separate $190,000 payments from Lewis, bringing the total return on their investment to $760,000. In the aftermath of the February meeting when Lewis conceded an over-issue and Ramish angrily defected, Alvin Frank, Motley Flint, Haldeman, and others had withdrawn and cashed in the shares backing their investments. Flint, upon collecting his portion of the sell-off, departed for an indeterminate stay in Europe.

By March 30, only pool manager McKay; Pacific Southwest vice-presidents Bell, Hervey, and McMullen; realtor W. I. Hollingsworth; and film magnate Louis B. Mayer remained in the syndicate. A total of 107,317 shares of Julian preferred secured their $420,000 interest. If the price of Julian securities kept dropping, this collateral would no longer cover the original investment. McKay decided to slowly liquidate these remaining shares and close out the surviving accounts.

Before McKay could initiate this action, Louis B. Mayer erupted with rage. Why hadn't his close friends Frank and Toplitsky told him earlier of their departure from the pool? he asked angrily. He demanded the immediate return of his original investment. If the funds were not forthcoming, Mayer warned, he would withdraw all M-G-M accounts from the Pacific-Southwest Bank. Mayer's threat set in motion a bizarre version of monetary musical chairs. A panic-stricken Lewis asked Toplitsky to lend him the money to pay Mayer. Toplitsky called in Frank. Frank turned to Rule, who informed him that his Pacific Finance Company "had all the Julian paper they wanted." Toplitsky then remembered that he had recently procured some large insurance policies for the Pacific Indemnity Company. Perhaps Pacific Indemnity would oblige them. The company president agreed to provide the funds if Frank and Toplitsky would sign the note securing the loan. Within a day Mayer received his money.

Mayer had invested $65,000 for six months and had received a $50,000 profit. He nonetheless remained dissatisfied. In the final accounting, he claimed, the pool had shorted him. Louis B. Mayer, president of M-G-M and one of the richest men in the world, demanded an additional $39.50. The request so stunned S. C. Lewis that he made a rare gesture. Lewis paid Mayer with his personal check.

Between March 30 and April 14, pool manager McKay sold the collateral shares back to Lewis and Bennett for a total of $401,000, which he distributed to the remaining pool members. An additional $20,238 was owed to McKay, but that, according to his final report to Lewis, "has been taken care of by an agreement between you, myself, and Jack Bennett." On April 15, McKay terminated Bankers' Pool No. 1. "I wish to state," he wrote to Lewis, "that while this transaction has been a tremendous and difficult siutation upon my part to handle and has consumed and occupied a great deal of my time and thought, nevertheless it has been a pleasure to have been able to serve you and the other gentlemen interested in the Pool." In less than seven months the Million Dollar Pool had returned to these "gentlemen" their original subscription and a staggering 76 percent profit.

During this period, even as Bennett continued to pyramid additional loans and pool arrangements, the other major syndicates also liquidated their holdings. Both the Rouse and Rosenberg pools cashed in and paid off their participants in full.

Julian preferred felt the effects of these transactions. By April 8, stock prices had dropped to 7½.

On April 12, two days before the proposed final meeting of Julian stockholders, California-Eastern announced that a "nationwide banking syndicate," including three prominent New York financial houses, would participate in underwriting the $7.5 million mortgage bonds in the East. As a result, there would be no public offering of California-Eastern stock. The only way to acquire shares in the corporation would be to convert Julian Pete securities into California-Eastern stock. Shares in Julian Petroleum would continue to be traded on the LASE until gradually removed from the market at the soonest possible date following completion of the merger.

On April 14, on the eve of Good Friday, an unusually small crowd of 5,000 people attended the stockholders' meeting. A combination of the upcoming holiday and growing frustration over the repeated delays limited the turnout. Nonetheless, those present enthusiastically approved the latest developments.

While most southern Californians enjoyed the Easter weekend, the architects of California-Eastern confronted their final obstacle: dealing with the Julian stock overissue. Despite the languid pace of the Kottemann audit, the existence of an overissue could no longer be questioned. Only the magnitude of the problem remained unresolved. Lewis acknowledged the stock surplus and, as always, offered a plausible explanation. Over the years, he told the bankers, he had advanced the company millions of dollars of his personal funds. He had issued himself stock in exchange for these loans. In order to provide additional resources for Julian Petroleum, he had borrowed money through the various pools, secured by this personal stock. Lewis admitted the irregularity of this arrangement but argued that his unorthodox actions had benefitted the corporation. Once final financing had taken place, he promised, he would easily be able to cancel the excess shares and make good on the remainder with his California-Eastern stock, thereby eliminating the overissue.

Lewis and the bankers, however, had never resolved the issue of reimbursement for the money he had advanced the corporation. Lewis insisted that California-Eastern assume the indebtedness to him along with other Julian Petroleum liabilities and redeem it with stock. This stock he promised would be assigned to Frank Flint and held in trust for the benefit of stockholders with overissued stock. The bankers had predicated their financing on the assumption that Lewis would take a loss on these funds in exchange for the corporate presidency. Without the California-Eastern stock, objected Lewis, he would be unable to cover the overissue. (In reality, as Lewis well knew, even with this concession, he would barely be able to scratch the surface of the overissue.) Negotiations were stalemated on this matter.

At some point during these discussions, First Securities president John Barber finally had his moment of revelation. Barber always denied that he knew of the overissue before this Easter weekend. Yet, as Robert Shuler later wrote, if Barber did not know what was going on, he should have been a blacksmith rather than a banker. In late February, Lewis had admitted the surplus to the Bankers' Pool

members, several of whom were Pacific Southwest vice-presidents regularly in contact with Barber. King and Bauer had heard the tales of Wagy and Company clerk
Wray Berthold one month earlier. The volume of trading on the LASE should have
erased all doubts.

First Securities employee Rockwell Hereford offers some insight into Barber's
state of mind during April 1927. President Coolidge had recently named Henry
Robinson to head the United States delegation to the International Economic Conference in Geneva in May. Robinson departed in mid-April, removing his services
and expertise at a critical time. In Robinson's absence, Hereford, Barber's Pasadena
neighbor and family friend, would drive Barber to work. "John rarely talked business," recalled Hereford, "but one morning he could not resist saying that all the
(Julian) stock activity made it look like the rumor of an overissue of stock was true."
However, remarked Barber, bank auditors had thus far discovered evidence of only
one minor error, which would be easily corrected. Hereford was so impressed that,
after refraining from the Julian market to that date, he purchased some stock for
himself and advised several friends to do the same.

Lewis's insistence on the reimbursement for his loans to cover the stock surplus
ended Barber's California-Eastern reverie. Lewis had confirmed the overissue and
plunged First Securities underwriting into doubt. Barber now felt compelled to
inform LASE President John Earl Jardine of the situation. "It was about [Monday]
April 18 . . . that I first got the information that this overissue might be of a serious
nature," Jardine recalled. "Mr. Barber called me by phone and hinted at an overissue. He told me that the offering of bonds was postponed for 30 days and that the
overissue might be enough of a serious matter so that the bond issue might not
come off at all." Barber also warned that he might ask for the suspension of all
trading in Julian Petroleum.

Less than a year had passed since Jardine had replaced the late Frank Pettingell.
A twenty-three-year member of the exchange and president of the William R. Staats
Company, which, as much as any other brokerage in Los Angeles, represented the
Spring Street establishment, Jardine owed his selection to the need to restore
respectability to the LASE following the Western Lead debacle. Thus far, however,
Jardine's ascension seemed little more than window-dressing. The laissez-faire phi-
losophy of his predecessor still prevailed. In the face of the tumultuous trading in
Julian securities and rumors of an overissue, Jardine had neither interfered with nor
investigated the unusual circumstances transpiring on his watch.

After receiving Barber's call, Jardine could no longer completely ignore the
obvious. He assembled Barber, Stern, and Bauer to discuss the Julian Pete problem.
The trio revealed that they had no information on the extent of the overissue or
whether or not Lewis's own holdings would cover it. They made no request for
Jardine to withdraw the stock from the LASE. Despite the ominous impending
tempest, Jardine maintained his steady course of concerted inaction.

During the days after April 18, the California-Eastern merger unraveled. First
Securities and Pacific Southwest withdrew from the fray. Barber informed Lewis
that under the present circumstances they would not underwrite the consolidation.

Harry Bauer resigned as chairman of the board. On April 22, Lewis and Frank Flint announced a "postponement" on the sale of California-Eastern bonds by First Securities, promising to present a revised plan of financing "as advantageous to the stockholders as the former one" the following week. Julian preferred closed the week at 5, a record low; Julian common dropped to 1.

Jardine responded to the latest events by calling a meeting of the LASE on Saturday, April 23. Lewis composed a letter to be read at the gathering denying reports of an overissue and predicting that the Kottemann audit would show the company was "in good condition." After sending the statement Lewis had second thoughts and frantically tried to retrieve the letter before it could be entered into the record. His efforts failed, and the LASE secretary read Lewis's reassurances to the assembled stockbrokers along with a message from Kottemann that his employees, working day and night on the Julian audit, had yet to discover any irregularities. The LASE members, most of whom knew of the overissue from their own records, accepted these statements at face value and allowed trading in Julian securities to continue.

Lewis now realized that his fragile fortress might collapse at any moment. After months of procrastination, Kottemann had finally received all of the Julian Petroleum ledgers from Pat Shipp's garage. The completion of his report could not be delayed much longer. On Sunday, April 24, Lewis called Shipp into his office and advised him, "Pat, if anything comes up, you remember you were just a clerk in the office . . . so far as you know there is no more stock outstanding than 600,000 shares of preferred and 600,900 shares of common. Now that is your story . . . stick to it."

The next morning, when Lewis arrived at the Pacific Southwest Building, hoping to hammer out a new financing scheme, he found stockbroker A. W. Morris, a longtime speculator in Julian securities, awaiting him with a wad of checks that had bounced because of insufficient funds. Lewis later claimed that Morris was the only man he knew who could dominate Bennett. Morris had been squeezing Bennett with short-term loans, while simultaneously collecting mounting commissions on stock sales. Lewis phoned Ed Rosenberg and asked him to raise $400,000 as soon as possible. Rosenberg, displaying the talents that had made him invaluable to Lewis and Bennett, frantically scoured the city for funds. He borrowed $125,000 from the Union Trust and Savings Bank, $100,000 from slot-machine king Robert Gans, and an additional $130,000 from his brother and friends. Rosenberg posted the remainder of the money from his personal funds and delivered the $400,000 that afternoon.

While dodging creditors Lewis also faced down a palace revolt among the California-Eastern directorate. After disposing of Morris, Lewis discovered Flint, King, Durbrow, and Miley lying in wait for him at the company offices. The group had spent the preceding days crafting a new reorganization plan. According to Lewis's self-serving account, the directors now abandoned this plan and replaced it with a statement announcing the completion of the California-Eastern deal, which, in effect, repudiated the overissue. Lewis claimed that he refused to endorse this scheme, asserting that "as long as I had any money or any resources or was not locked up in jail they would never repudiate any of this overissued stock that got into the hands of innocent persons." The directors demanded Lewis's resignation on the grounds that he had failed to comply with contractual obligations to purchase California-Eastern stock, falling $50,000 short of the required payment. Lewis dramatically produced $50,000 in cash, forcing the board to withdraw its resignation demand. Flint accused King and Durbrow of double-crossing him by not supporting the ouster, stated Lewis. After the meeting, Lewis, convinced that Flint could no longer serve as both his personal attorney and counsel for First National Bank, replaced him with the firm of Loeb, Walker, and Loeb.

Unable to remove Lewis, Senator Flint publicly unveiled the previously agreed-upon reorganization of California-Eastern. Flint officially announced the withdrawal of First Securities as underwriter and Barber and Bauer as directors. Flint himself replaced Bauer as chairman of the board. Senator Flint also presented a new plan to finance the corporation, eliminating both the first mortgage bonds and debentures. California-Eastern would now issue almost 3,000,000 shares of stock, most of which would be exchanged for Julian securities. Lewis would receive an additional 661,222 shares to compensate him for the more than $6 million he claimed to have advanced Julian Petroleum. Lewis assigned these shares, along with those he had acquired in August 1926, to the trust controlled by Flint to protect against the Julian Pete overissue. Applications would be made to list the stock on the New York, San Francisco, and Los Angeles exchanges. The new plan earned the immediate approval of Corporation Commissioner Friedlander, who issued a permit for the sale of securities.

Senator Flint, who fifteen years earlier had won acclaim for reconstructing the affairs of the Los Angeles Investment Company, doubtless saw the new plan as a last-gasp effort to resurrect the company and a chance to duplicate his earlier triumph. Lewis knew better. The 3,000,000 California-Eastern shares authorized would not absorb the overissue even if exchanged on a one-for-one basis with Julian Petroleum stock. The financing plan approved by Julian Pete stockholders called for the distribution of 3.4 shares of California-Eastern for each share of Julian preferred and an additional half-share for the common stock. For Lewis the latest announcement simply bought additional time.

With Julian Petroleum hurtling toward destruction, Jack Bennett launched one last foray on Spring Street. Since the massive February overissue Bennett had continued to print new stock at a healthy rate. During March an additional 285,000 shares of Julian preferred had entered the market. Throughout April tens of thousands of shares continued to flow from the "bubble mill." During the last week of April, according to Bennett, whose account was adamantly denied by Lewis, "Lewis came to me and begged me to flee the country. I protested but he insisted in his pleading." Lewis offered Bennett a "vivid picture of the financial catastrophe which was near at hand and insisted that it would be futile to attempt to salvage any of the wreckage with me available to the authorities."

Bennett claimed that he resisted Lewis's entreaties, but evidence exists that he had long since prepared his departure. A Julian Petroleum stock clerk later testified that ever since Bennett had dispatched his pregnant wife to New York two months earlier, transfer office employees "were just waiting for him to leave daily. . . . He had a lot of guarantees out among us boys. We were all waiting to collect and we thought he was going to try and get away and not pay us."

Lewis later testified that the first week of May found Bennett sick in bed, troubled by his indebtedness to stockbroker Abe Morris, and tormented by heart and stomach pains. Bennett's activities belie this description. Bennett spent these days frantically peddling stock certificates and borrowing money secured by counterfeit Julian securities, offering additional "guarantees" to all available takers. A. W. Hackel, for example, had provided Bennett with short-term loans and received lucrative returns for several months. During the last days of April, Hackel advanced Bennett $92,000. Bennett also reportedly sold large blocks of stock through the Morris and Barneson brokerages.

Throughout the week, according to Bennett, he and Lewis "were in one conference after another." Both men continued to reassure those inquiring about overissue rumors. On Friday, April 29, C. C. Julian, who had continued to speculate in the stock of his old company, called Bennett. "I . . . asked him if everything was all right,'" related Julian. "He said, 'A thousand percent, C. C., step on it.'" That same day Julian Petroleum secretary Conroy told King that he believed that the overissue ran into millions of shares. King challenged Lewis with this information. Lewis, reported King, "went into a rage [and] declared there was nothing to all this business."

On Saturday, April 30, Bennett continued his whirlwind fundraising. Long Beach real estate man Daniel Barnes was among the victims. Barnes appeared at Wagy and Company seeking to buy 2,000 shares of Julian preferred. Bennett sold him the shares, offering a pink receipt that could be redeemed with a stock certificate within fifteen days. Barnes paid with a $10,000 check, which Bennett promised not to cash until Monday. Upon returning home Barnes received a call from his banker informing him that Bennett had already presented the check for payment. After receiving assurances from Bennett that stock in his name was already available, Barnes naïvely allowed the check to be cashed.

Last-minute transactions like the Barnes affair closed out Bennett's April activities. He had pumped almost 300,000 more shares of Julian preferred into the market, bringing the total number of outstanding shares to 3,614,283. On Sunday night, May 1, according to Bennett, Lewis told him "absolutely and emphatically that he was in a position to fix the whole mess and prevent it from ever becoming a public scandal." Bennett reported to his office the following morning to close out his affairs. When Alvin Frank, a mainstay of the Bennett operations, attempted to collect on two loans totaling $75,000, Bennett instructed his messenger to return at four in the afternoon. At 3:00 p.m., Bennett departed, leaving Frank and the others in the lurch. Three hours later he reportedly boarded the Sante Fe Chief at Union Station and headed east. Around his waist was a money belt allegedly holding an estimated $625,000 in currency.

As Jack Bennett fled Los Angeles, S. C. Lewis confronted an angry California-Eastern directorate. King, Durbrow, Miley, and Rule had assembled at Senator

Flint's office to demand an accurate accounting from Lewis. Lewis remained adamant in his denials of an overissue. King also questioned the stability of the Wagy
brokerage. Lewis called in accountant Kottemann. Kottemann, as always, testifed
on Lewis's behalf. Wagy and Company, he declared, was financially sound. Even
if Julian Pete securities depreciated to no value, he assured them, Wagy & Co.
could simply close out the margin accounts and continue business as usual. A full
accounting would take five or six more days, maintained Kottemann, the master
procrastinator.

By this time the grand sellout in Julian stocks had begun, as Julian insiders,
privy to the impending collapse of the company, began to dump their stock. On
Monday, May 2, tens of thousands of shares crossed the boards on the LASE as
Julian preferred fell to 4¼. Tuesday trading shaved another point off the stock. LASE
officials continued to ignore the situation.

The plunging prices sent Lewis scurrying for additional funds. On Tuesday
morning he phoned Ed Rosenberg and told him that Bennett had gone to San
Francisco "to get some money" and would return to Los Angeles that evening.
Lewis asked Rosenberg to take care of Bennett's obligations. Rosenberg arranged
loans to cover these needs but the following day he called Lewis and informed him
that with Bennett gone he could no longer raise money. "I . . . told him it was
utterly impossible for me to sit in and take care of those things," Rosenberg later
testified. Lewis suggested that Rosenberg sell off whatever shares he held to "reimburse yourself to some extent until I can get things straightened around."

On the morning of May 4, Lewis placed large advertisements in the local newspapers to counteract "rumors running rife" that Julian shareholders would not
receive their California-Eastern stock. "I want to give you the my assurance that in
the period of reorganization . . . *every interest* of yours was safeguarded to the *fullest
extent*. Every resource at my command was pledged to guarantee that you receive
*every share of stock* to which you are entitled." Lewis denied reports of financial
difficulties and pledged that "this deal will work out *exactly* in accordance with the
contract." He implored stockholders to "*sit steady*, in the boat. . . . My faith in the
future is *more firm* today than at any time in the past. With the continuation of your
loyalty for another 60 days . . . *I shall have delivered the goods*." Lewis's entreaties
had little effect. The price of Julian preferred dropped as low as 2⅛.

About this time John Barber also took a fateful step. Rather than leave Lewis'
"orphan check" in the accounts of the Pacific Bond and Share Company, he transferred the draft, still uncashed, back to Pacific Southwest Savings where it could be
applied to the Julian Petroleum accounts. This seemingly innocuous transaction
would have dire consequences.

For eight months bankers, brokers, state officials, investors, and speculators had
anxiously awaited the stock audit by William Kottemann, one of the state's most
respected accountants. On May 5, with the Julian Pete stock situation rapidly dete-
riorating, he could no longer conceal his knowledge. That evening, in the presence
of Lewis and Lewis's new attorney, Edwin Loeb, Kottemann disclosed the full extent
of his findings. Over 3,000,000 shares of Julian preferred were now outstanding,

confirmed Kottemann. An additional 1,275,000 shares of the common stock had been issued. On Loeb's advice Lewis immediately informed Frank Flint and other Pacific Southwest officials. According to Lewis, Bauer offered to buy up the entire overissue at current low market prices on behalf of the bank. Lewis and Loeb protested that under this arrangement tens of thousands of people who had purchased the stock at higher levels would suffer losses.

That evening Flint hastily assembled the California-Eastern board. He announced that, due to the discovery of an overissue, he had demanded Lewis's resignation as California-Eastern president. The board elected L. J. King to replace Lewis at the helm of the shattered company.

On Friday, May 6, as heavy selling continued to drive Julian securities ever lower on the market, Lewis wrote to LASE president Jardine. Due to "indications of an overissuance," advised Lewis, all trading in Julian Petroleum stocks should be halted. Jardine could no longer evade his responsibilities. He called the LASE board of governors to meet that evening. In the meantime, trading continued. When the closing bell sounded on the market floor, Julian preferred, which had once sold for $36 a share, stood at 2⅞. Julian common had sunk to 75 cents.

The LASE board of governors convened on Friday night and quickly passed the following resolution:

> *Whereas*, from a communication made to the Los Angeles Stock Exchange by S. C. Lewis, president of the Julian Petroleum Corporation, it appears that there is an overissue of the stock of the Julian Petroleum Corporation, and
> *Whereas*, the exact amount of such overissue cannot be immediately ascertained, Now, therefore, it is hereby resolved that trading on the the Los Angeles Stock Exchange in stock of said Julian Petroleum Corporation is suspended until further order of the board of governors of said Los Angeles Stock Exchange.

The following morning Corporation Commissioner Friedlander suspended the permits of both Julian Petroleum and California-Eastern. "The Great Los Angeles Bubble," as Guy Finney would dub the Julian Pete swindle, had finally burst.

# PART III  SCANDAL

# 12 THE GREATEST SWINDLE EVER PERPETRATED IN AMERICA

## I

During the early months of 1927, C. C. Julian had watched the unraveling of his namesake corporation with growing consternation. In the days immediately following the May 6 suspension of Julian Pete securities, Julian observed the growing despair and confusion spreading through Los Angeles. Crowds gathered along Spring Street, seemingly more stunned than angry. Police officers kept the masses moving in a promenade of uncertainty. Rumors flew through the city, offering a half-dozen explanations for the collapse. Reports of an impending receivership and a dozen unfolding investigations by federal, state, and local officials fueled the sense of dismay.

Like others, C. C. Julian had regularly speculated in the Julian Petroleum market. On one occasion, Julian claimed, he had given a desperate S. C. Lewis a $100,000 interest-free loan to meet the Julian Pete payroll. But his efforts to reverse his own business fortunes, rather than the affairs of Julian Pete, had preoccupied Julian in the months leading up to the crash. Since the collapse of Western Lead in May 1926, Julian had devoted himself to innovations in mining: pioneering techniques to tunnel around the Corporations Department and unearthing new methods to excavate money from his faithful following. Most significantly, C. C. Julian had discovered the potential of radio broadcasting.

Julian had spent the summer of 1926 consolidating his worthless Western Lead claims with other mining properties into a grand new company. "I recalled reading," he later wrote, "of all the great mergers of different big industries in the United States . . . but never a 'MERGER OF BIG MINES' . . . so I says to myself, says I, 'CC, you're the bird to pull a Mine Merger.'"

Julian acquired the rights to mines in the Southwest and British Columbia. For his centerpiece, Julian set his sights on the Monte Cristo Mines of Arizona, owned by Phoenix hardware merchant Ezra Thayer. The Monte Cristo mines included 600 acres of claims producing gold, silver, and copper. Perhaps most attractive to Julian, Monte Cristo had been incorporated in Washington, D.C., under a code established by Congress. As a federally chartered enterprise, he reasoned, it fell outside the jurisdiction of the corporation commissioner.

To acquire the Monte Cristo Mines would require an extraordinary effort. Thayer had reportedly turned down numerous offers for his properties. Julian employed Jacques Van Der Berg to help weaken the hardware merchant's resolve. Van Der Berg, who traveled under the aliases Jac Van and Baron Jac Ferdinand Van Viletzy Brugges De Frelinghuysen, had a long history of arrests for embezzlement and confidence games. Julian equipped Van Der Berg with a luxury automobile, chauffeur, valet, and secretary, and $12,333 in expense money, and dispatched him to Phoenix to pose as a member of the Dutch nobility seeking to acquire the Monte Cristo Mines. Van Der Berg failed in his mission, but in early September, Julian nonetheless persuaded Thayer to sell him a controlling interest in the company.

Julian linked Monte Cristo with Western Lead (where most work had ceased the preceding month) and other mining properties to form the Julian Merger Mines. Incorporated in Washington, D.C., Julian Merger Mines surfaced in September 1926 as the master promoter's latest effort to provide his constituents with "THE POT OF GOLD AT THE RAINBOW'S END." To re-enlist Western Lead shareholders in his campaign, Julian pulled a classic trick from the promoter's carpetbag, the "reloading" scheme. Reloading involves the exchange of worthless stock, in this case Western Lead, for the right to invest in a second, seemingly more promising, issue. Julian ads in the *Los Angeles Record*, the only local newspaper that would carry his messages, offered Western Lead stockholders the opportunity to exchange their securities for shares in Julian Merger Mines and to buy additional shares for twenty-five cents below the proposed listing price.

In October, Julian extended the sale of Merger Mine stock to the general public. "Folks," he proclaimed, recycling an old motto, "today opens the offering of what I actually believe will prove to be the greatest money-maker of the age." As a reminder of his previous success, Julian released another $71,000 in dividends from the Santa Fe Springs wells, promising "all the dividends I have ever paid in the past will look sick compared to our dividend rates on 'JULIAN MERGER' beginning inside of seven months."

In a particularly revealing advertisement Julian claimed that he had contemplated financing the venture on his own. "Why let the public in on such a plum?" he mused. "Then I started to think of the people of Southern California and the loyal way you have stood behind me.... I thought of the times you had won on me, and and I thought of the times you had lost on me...." At this point Julian injected an element of hyperbole, which subconsciously revealed more than he had planned. "I tried to make it balance," he wrote, "but ... I couldn't satisfy my conscience to pass you by, after all your kindness to me, *on what I know better than anyone, is the only real legitimate offer to make something worthwhile that I have to offer you in these six years" (italics added).*

On October 15, interim corporation commissioner Clifford J. MacMillan brought the Julian Merger Mines promotion to a halt. Relying on the United States Attorney General's opinion that incorporation under an act of Congress did not remove a company from state regulation, MacMillan ordered all sales of Julian Merger stock immediately discontinued. Julian filed suit to block MacMillan's action. Whatever the verdict, however, Julian had fared well. In less than a month he had sold 1,286,000 shares at prices ranging from one dollar to $1.25.

The corporation commissioner had stopped sales before Julian could enact a bold new strategy for marketing his securities—radio broadcasting. During the past half-decade, radio had revolutionized American leisure patterns. From an experimental industry that boasted just four licensed stations in early 1922, radio had blossomed into a nationwide phenomenon, with thousands of stations competing for millions of listeners. Entrepreneurs and educational institutions, churches and newspapers, and businesses large and small had leapt into the fledgling industry. In Los Angeles the audience for radio broadcasts had expanded and diversified rapidly. *The Los Angeles Times* had a station; so did the *Express*. A local laundry had founded KUS. Evangelist Aimee Semple McPherson owned, operated, and regularly preached on KSFG. If these interests and others expounded their messages via the airwaves, why not a stock promoter? Julian had long ago recognized the marketing potential of expanded newspaper advertising; now he leapt aboard the radio bandwagon.

On October 16, one day after the termination of Julian Merger Mines stock sales, Julian purchased radio station KMTR. "There is no plan for using the station for Mr. Julian's private interests," the station manager reassured reporters. Julian, however, clearly had other ideas.

By spring 1927, Julian had incorporated a third mining enterprise, the New Monte Cristo Mining Company, which included all Julian Merger Mines and Western Lead properties, as well as several other mines thrown in for good measure. On April 29, he launched his new effort with radio and *Record* advertising. "FOLKS, WATCH THIS BABY MAKE WINNING HISTORY," he beckoned. "I honestly believe that even $500 invested in New Monte Cristo today . . . will mean a meal ticket to you for life and . . . will be paying substantial monthly dividends long after you and I start to push up poppies." As journalist Walter Woehlke would later write, the same people "with their gums still bleeding" from Western Lead and Julian Merger Mines flocked to hand their money over "to the man with the long vulpine nose and the close-set eyes who did all the extracting, but never put in the promised gold bridge-work."

Julian's timing proved unfortunate. One week after New Monte Cristo reached the market, Julian Petroleum collapsed, casting fresh doubts on all speculative ventures bearing the Julian imprimatur.

For several days after the crash Julian remained silent as the principal figures in the Julian Petroleum and California-Eastern corporations rushed to pacify investors. S. C. Lewis donned the garb of benefactor to the victims of the fraud. "My purpose is to protect each and every person holding Julian stock," he declared, promising to work with "the ablest financiers and best legal ability in Los Angeles" to reconstruct the company. Wagy & Co. advertisements pronounced, "We are absolutely capable of handling the present situation satisfactorily. . . . Marked progress is being made and every precaution has been taken to protect the interests of the shareholders." The brokerage called for "calmness and deliberation . . . until dependable facts can be submitted." Frank Flint improbably reassured investors that the Julian Petroleum situation had no relationship to California-Eastern.

On Sunday, May 8, Julian broke his silence. The preceding day he had promised to reveal "a few interesting facts" on his next radio broadcast. Thousands tuned in to hear his version of events. Julian offered minimal enlightenment, but unveiled his own probe into the scandal. "My investigators have been on the job for the past forty-eight hours and within the next twenty-four hours I will know the exact situation in the Julian corporation," he pledged. "If the matter looks crooked to me I will land every man connected with it behind bars." Julian promised to speak again on May 10, when he would offer "every detail of the present status of [Julian Petroleum] and let the pieces fall where they may."

While southern Californians awaited C. C.'s next broadcast, confusion characterized Julian Pete affairs. The directors of California-Eastern expressed "a great uncertainty . . . as to the legality of the board . . . to authorize stock transfers" and "some doubt as to the legality of ownership of stock" between California-Eastern and Julian Petroleum. To resolve these issues, Lewis announced that he had retained the law firms of Flint & McKay and Loeb, Walker & Loeb to assist in piecing together the financial puzzle. While emphasizing that California-Eastern was "altogether solvent," the attorneys requested a receivership for both Julian Petroleum and California-Eastern. The following morning Lewis publicly announced his resignation as director, president, and executive committee member of California-Eastern, promising "to devote my exclusive time and attention to the unraveling of the situation which exists . . . in the best interests of the shareholders."

Amidst the posturing, anticipation escalated as the moment approached for Julian's Tuesday broadcast. " 'DYNAMITE,' " he promised in a *Record* advertisement. "It will be worth your while to tune in and get all the dirt." During the past two days Julian had amassed information and prepared his performance. He had also made contact with several businessmen involved in Julian Pete affairs. Each side presented different versions of these discussions. Walter Woehlke later asserted that emissaries from Julian had approached bankers and pool members with an offer to tone down his attacks if they would use their influence to open local newspapers to his advertising and intervene on his behalf with the corporation commissioner. Julian claimed that the business leaders had offered to buy his silence. "They offered me a million dollars in one thousand dollar bills if I would keep quiet," volunteered Julian. "But they couldn't have shut me up for forty millions. I had 'em where I wanted 'em. For six years I'd been waiting for this chance. They didn't have enough money to buy me off."

For those without radios Julian had erected voice-magnifying horns outside the KMTR station. As the nine o'clock broadcast time neared, crowds began to gather, blocking traffic for a half-mile in every direction. Hundreds of thousands of others reportedly tuned in at home to hear Julian's exposé.

Julian surpassed their wildest expectations. He unleashed a vicious attack against the Pacific Southwest "banking crowd," charging them with a conspiracy to loot Julian Pete. He revealed their participation in investment pools collecting usurious rates of interest. He accused the bankers of extorting illegal bonuses, commissions, and gifts, most notably in the instance of what he dubbed "the orphan check." To underscore his accusations, Julian dramatically identified the culprits: Charles Stern, Motley Flint, and a half-dozen other Pacific Southwest vice-presidents, spelling out each name, and pointedly adding adjectives like "thief," "crook," or "scoundrel." He reserved particular venom for John Barber and Henry Robinson, whom Julian blamed for quietly recruiting federal officials to investigate his earlier promotions. Julian exposed the pool participation of Harry M. Haldeman and took several shots at Harry Chandler. He demanded a grand jury probe and an indictment of the "banker crowd," pledging to assist in imprisoning the lot.

To many, Julian's scattershot performance drifted over the line between free speech and slander. Julian's radio station, protested Guy Finney, "enables him to strike from ambush those he regards as enemies." But California law in the 1920s had not yet recognized the dangers posed by the radio revolution. While written and printed charges might constitute criminal libel, no restrictions existed on the spoken word. "Any judgment-proof individual having access to a radio station may with perfect impunity tell an audience of 100,000 (anything)," complained Woehlke, who charged that Julian had invented, twisted, and distorted facts. Even more stringent libel laws, however, would probably have proved of little avail. Julian may have exaggerated, but he had based his exaggerations on kernels of truth.

Julian's charges, particularly those implicating Robinson, sent shock waves throughout southern California. They successfully diverted public attention away from the overissue, which had wrecked Julian Pete, and directed it at the subsidiary irregularities engaged in by the bankers. A public uproar arose to punish all of those responsible, no matter how rich or how powerful.

Julian's attacks received national and even international attention. Upon hearing of Julian's broadcast in Geneva, Switzerland, Henry Robinson, heading the American delegation to the International Economic Conference, dispatched a letter to Stern marked PERSONAL AND CONFIDENTIAL:

I have had three open and one code telegram from you and Barber; also two letters from Barber, which give me on the whole a fairly adequate picture of the difficult situation that has developed in the California-Eastern. Assuming that the values are sound, I am not inclined to believe that the elusive attack of Julian would have any serious repercussion though I judge from Barber's telegram that you are alive to such possibilities. At any rate, I intend to decline to lose any sleep as I could accomplish nothing beneficial by worrying and in addition have full confidence that you and

Barber and the other officers, together with the extremely loyal and effective directors, will succeed in preventing any serious ruction.

Robinson expressed regrets about "being absent at this juncture" and revealed his longstanding suspicions about Lewis, whom he described as Stern's "prize devil." "I should have been more determined in my attitude that he was what he is, and with that thought have stayed home to help in the war," wrote Robinson, "but on the other hand, it is pretty hard to conceive that so awkward and difficult a situation would arise." Robinson closed his missive with the cryptic statement: "It is pretty late in life now."

In the days following Julian's radio charges, federal authorities quietly moved to secure evidence. Lucien Wheeler, an FBI agent stationed in Los Angeles, had long known of the impending disaster. According to Wheeler, a confidential source (probably Ed Rosenberg) had kept him apprised that Julian Petroleum affairs "were in a bad way" and that there had been a considerable overissue. This had not prevented Wheeler from speculating in the stock through the Wagy brokerage. Rosenberg, along with Julian transfer clerks Pat Shipp, C. T. Harris, and Fred Packard, had holed up in a Long Beach hotel since the Julian Pete crash. On May 11, Wheeler's informant told him that the four men "were anxious to confide to the proper authorities all they knew concerning illegal operations" and that they "had in their possession certain confidential records which would reveal the fraudulent transactions" of Julian Petroleum officers. The following day Wheeler and two other federal officials seized the records stored in Pat Shipp's garage.

On May 13, United States District Court judge Paul J. McCormick placed Julian Petroleum in receivership. To handle this delicate assignment, Judge McCormick asked longtime friend, attorney Joseph Scott, to serve as receiver. Scott was a legendary figure in Los Angeles. Renowned for his dynamic oratorical skills and defense of underdog clients, Scott had won acclaim for representing minority stockholders in the Los Angeles Investment Company collapse in 1917. In addition, Scott had been out of the country for several months and had no entangling alliances with the Julian Pete situation. Scott reluctantly agreed to accept the position if McCormick could persuade another person to share the responsibility. McCormick and Scott agreed on H. L. Carnahan, the former state corporation commissioner, as the second receiver.

The selection of Scott and Carnahan won widespread approval. Both had impeccable reputations. They also had close ties to the Republican political establishment. In later years, however, some would question the rationale behind McCormick's decisions. Receiverships, according to one Los Angeles attorney, "had been practically unknown in this community" prior to the Julian Pete collapse. In addition, neither Scott nor Carnahan had much experience with either bankruptcy law or running an oil corporation. Indeed, the individual who stood to gain the most was S. C. Lewis. Lewis had originally suggested a receivership. As the only person with an understanding of the complicated Julian Petroleum affairs, his assistance would be essential to the receivers. McCormick's choice of lawyers, rather than oil men, would render Lewis's counsel even more critical. Thus, by promoting the creation of a receivership, Lewis guaranteed that he would have a strong say in the ultimate dispensation of all disputed assets.

Attention also focused on the A. C. Wagy brokerage. President Raymond Reese resigned his seat on the LASE, and on May 15, Lewis revealed that he owned A. C. Wagy & Co. That night Julian took to the airwaves again, repeating his charges of pool manipulations and excoriating the brokerage for its role in the overissue. Julian demanded that Corporation Commissioner Friedlander close down Wagy & Co.

Julian's broadcast produced immediate effects. Although Lewis denounced these "false attacks," at noon on May 16, the Wagy brokerage ceased operations. Clients found the following notice posted on the glass doors: "The ownership of A. C. Wagy & Co., Inc., has heretofore been assigned to the Julian Petroleum Corporation for the benefit of its shareholders. . . . The statement of C. C. Julian for the purpose of wrecking this institution has been averted by our action in placing this company office in the hands of the receivers for the protection of all interests." Scores of investors milled about the padlocked company or drifted over to Julian's nearby offices for advice.

On May 16, the Los Angeles County grand jury launched its investigation of the Julian Pete fiasco, but another sensation stole the headlines. That evening a taxi driver noticed two bodies slumped in the front seat of a new Cadillac sedan. Looking inside he found a middle-aged man and young woman shot to death. An automatic pistol dangled from the woman's hand. Police identified the bodies as George Powell, Jr., vice-president of a local insurance firm, and Margie Pike, a twenty-four-year-old stenographer with that company.

Rumors instantly began to connect the deaths to the Julian scandal. Pike had left messages indicating that she had received death threats. Stories also began to circulate along Spring Street that Powell had participated in the Julian pools and that he had possessed sensational information about them. Reports reached the district attorney that Powell, whose offices were three floors beneath those of Lewis's attorneys, had used a dictaphone to tap their conversations. Some suggested that Powell was the primary source of Julian's revelations. Speculation ran rampant that Julian Pete figures had murdered Powell and Pike to silence them.

Upon investigation, none of the rumors held up. All indications pointed to a murder-suicide unrelated to the Julian affair. The district attorney's office decided that the evidence to this effect was "so plain . . . it would only confuse the [coroner's] jury to give them [the] additional evidence" involving the Julian Pete rumors. But in the heated atmosphere of the ten-day-old Julian scandal, many people believed that the district attorney's office had withheld evidence and that those seeking to suppress the full Julian story had silenced Powell and Pike. Yet, as succeeding weeks would demonstrate, the flow of rumor and revelation had barely begun.

## II

On May 17, FBI agent Lucien Wheeler filed his first report to his superiors in Washington about the blossoming Julian Pete scandal. Wheeler's terse prose captured the growing chaos pervading Los Angeles:

> . . . this case is of extreme importance in this community. There are several of the largest banks in Los Angeles involved, and through the operation of the stock pools a number of wealthy bankers are now being charged, through the press and over the radio with having charged usurious rates of interest. The scandal is far-reaching and has become so serious that both Federal Judges McCormick and Janes have called this agent into many conferences in order to determine the policy to be pursued with reference to Federal prosecution.
>
> Charges and countercharges are being made to the general effect that the district attorney of Los Angeles County, as well as the State Corporation Commissioner are involved and it is whispered that an honest prosecution by the State authorities is unlikely because of the prominence of the parties involved in banking and financial circles.

Within days of Wheeler's dispatch, the worst of these rumors burst full-bore into the public press. On May 17, the day of Wheeler's report, attorney Ben Beery revealed that he and Wagy & Co. clerk Wray Berthold had attempted to expose the overissue in March 1927, implicating Corporation Commissioner Friedlander in a cover-up. Friedlander denied any wrongdoing and announced his own investigation.

On May 18, receivers Scott and Carnahan released the official estimate of the Julian Pete overissue. The totals of over 3,000,000 shares of preferred stock and 1,265,000 shares of common far exceeded previous reports, staggering the public imagination. When expressed in terms of the original $50 par value of Julian preferred stock (and many commentators used that computation), the fraud exceeded $150,000,000. On May 19, Judge McCormick increased Scott and Carnahan's responsibilities, adding California-Eastern to their receivership.

At this point the Julian Pete pools took center stage. Los Angeles newspapers confirmed Julian's reports of pool and loan operations involving scores of prominent business leaders. Since these men had profited at the expense of Julian Petroleum, Scott and Carnahan immediately took aim at the accused looters. The hopes of the Julian shareholders, they announced, now depended on the ability of California-Eastern to carry on its business as a producing concern. Immediate funds were necessary to complete drilling on wells in the newly discovered Alamitos field. These funds should be contributed by the pool members.

Local officials injected an element of coercion into these suggestions. On May 23, Chief Deputy District Attorney Harold Davis announced that the grand jury had issued subpoenas to bankers and brokers "to tell what they know about this financial tangle." Davis advised that "some of the biggest money handlers in the city" would be called. Simultaneously, City Prosecutor E. J. "Doc" Lickley entered the fray. Lickley, a Kent Parrot crony widely rumored to have higher political ambitions, revealed that his agency had investigated the pool and loan schemes and charged that the various extensions, guarantees, and bonus payments had transformed all pool investments into usurious loans.

A 1918 state law had defined "usury" as the collection of interest payments in excess of 12 percent per annum. The returns on the Julian transactions had far exceeded this figure. While usury remained a misdemeanor, violators were subject to fines and six-month jail sentences. More significantly, noted Lickley, "a conviction provides the foundation for suits against those convicted for the return of three times the amount of usury they charged." Lickley placed the illegal interest payments as high as $11 million.

The city prosecutor made it clear that he was hunting "big game." The targets of his probe, he stated, came from "the high financial circles of the city." "I consider these gentlemen who loaned money to the promoters at usurious rates of interest more culpable, if possible than the wreckers of the concern," charged Lickley. "They allowed greed to override their judgment, their conscience, and their business sense . . . . These men are what Roosevelt would have called 'malefactors of great wealth.'" Lickley predicted that arrests could begin within forty-eight hours.

With Lickley's threats of prosecution providing the proper inducement, Scott and Carnahan instituted a "restitution" campaign. "Large sums of money on pooling arrangements and otherwise have been wrongfully and illegally obtained," announced the receivers. "Those who have been parties to such financial transactions . . . and who have profited thereby, whether their intentions were innocent or otherwise, have the opportunity to save the receivers considerable time and money if they will immediately . . . get in touch with us for an adjustment of such matters." The receivers intimated that civil suits for triple damages would be filed against those who failed to cooperate. The program got off to a promising start. Within three days the receivers reportedly collected $250,000.

Many Julian Pete stockholders, meanwhile, awaited further guidance from C. C. Julian. Julian had left Los Angeles to examine mining properties in British Columbia. He almost did not return. On May 23, his chartered plane dropped its landing gear upon takeoff. Unequipped with a radio, the pilot could not be warned of the danger. On landing in Redding, California, the aircraft "toppled head first to the ground . . . [and] nosed into the earth," smashing the propeller and front of the ship. Julian and the pilot emerged unscathed. "We were just a little shaken up, but it wasn't half as bad as what's going on in Los Angeles," quipped Julian. "There should be fifteen or more in jail by this time and yet nobody has been arrested. I intend to stir things up a little bit."

Julian chartered another plane and flew to Los Angeles on May 26. Two days later he met with Prosecutor Lickley to substantiate his radio charges. Lickley enthusiastically endorsed Julian's allegations. "I believe everything Julian has said is true," pronounced Lickley. "I believe his statements can be substantiated by documentary evidence." Lickley noted the enormity of the task before him. "If I prosecuted every person that sapped money from the organization as Julian charges, I would be busy for more than two years," stated the city prosecutor. Julian predicted that "these thieves" would pay back millions of dollars to evade "the prosecution they deserve."

Throughout the early weeks of June, revelations about the pool operations poured forth daily from the city prosecutor's office. Lickley announced that he had uncovered 400 money lenders operating in 100 stock pools with transactions totaling as much as $100 million. He estimated that pool profits amounted to $18 million. (To reach these figures, Lickley had lumped both investor pools and individual loans together as pools. This perception would become fixed in the public mind.) Lickley set a June 4 deadline for moneylenders to make restitution to the receivers, before he began to levy usury charges.

Confronted by Lickley's ultimatum, several pool members hastened to square their accounts. Most held back, however. On June 6, Lickley played his ace. He exposed the operations of the Million Dollar Bankers' Pool No. 1 and drew up usury complaints against Harry Haldeman and two minor pool investors. He charged these men with collecting interest rates prorating at 228 percent annually. Formal charges against other Bankers' Pool participants were imminent. "The evidence is concrete and conviction would only be a matter of form," pronounced Lickley.

Haldeman quickly visited the receivers and refunded his profits. Adolph Ramish, although not yet publicly named, returned $57,000 to Scott and Carnahan. Scott hailed these developments. "This unfortunate experience of Haldeman," he hoped, "might arrest the attention of other people who have honorable intentions in the matter, but have neglected to realize the tense situation which confronts us as receivers." On June 7, Lickley threatened actions against the members of Bankers' Pool No. 2.

The restitution payments were just one of several promising developments for Julian/California–Eastern receivers Scott and Carnahan. Julian Pete gasoline sales seemed unaffected by the stock swindle, they discovered, but the operational side of the company had been badly neglected, and Julian Petroleum was losing thousands of dollars each day. The receivers quickly laid off 118 employees and cancelled unprofitable leases and contracts. Since demand for gasoline still exceeded the company's available supply, they focused their attention on increasing oil supplies. Their efforts bore immediate fruit. On June 1, drilling crews completed a 1,400-barrel well at Alamitos Hills. Nine days later a second well came in at 3,000 barrels. Carnahan hailed these achievements in an effusive statement reminiscent of the early C. C. Julian. "On a couple of hills between the municipal golf course and the sail-dotted lagoon, which is Alamitos Bay, there is a forest of oil derricks, dirty and oil-splattered," wrote Carnahan. "It is a little forestry in which the real Julian battle is being fought." Upon the "simple announcements" of completed wells, stated Carnahan, "lies the only real hope of the 40,000 stockholders . . . for in the end it is only through the production of crude oil and the sale of its refined products that actual values can be put back in the stock. . . . A thousand barrels a day means approximately $365,000 a year."

These successes and the promising restitution campaign braced the receivers for their next challenge, an impending stockholders' revolt. Since the Julian Petroleum collapse, victims of the crash had haunted downtown Los Angeles, reciting their tales of woe to all who would listen. "All morning long women, many of them weeping, stood before the grand jury room telling each other of the thousands of dollars they claim to have lost," reported the *Times*. When District Attorney Asa Keyes rashly promised that his office would respond to inquiries to distinguish genuine from forged stock, hundreds of certificate-waving shareholders, including widows and other "elderly women . . . attired in somber garb," and those representing family members "too ill to make the trip" invaded the Hall of Justice on June 2. "Every upward trip of the battery of elevators . . . augmented the crowd," reported the *Times*, until Keyes announced he could not provide the desired information.

Amidst the escalating furor C. C. Julian appeared ready to seize command. "I am sitting on the sidelines, waiting for results," he advertised on June 3, "and the moment those results do not satisfy me, I will call you together and advise you as to future moves. Some of the newspapers say the stockholders have no leader, but some of the newspapers will find they are plenty wrong this trip."

The notion that Julian himself might emerge as a protector of stockholder interests struck many, in journalist Finney's words, as an "effrontery" and "an ironic jest; with the irony unfortunately hitting an army of unwary." The *Times*, in its first editorial comment on the Julian Pete affair, dismissed Julian's proposal as "colossal impudence." Reviewing his career as one that had left "his pockets bulging with money which he has taken from his own shareholders in enterprises the wreckage of which strews the investment field," the *Times* concluded that Julian "is the last man who should have charge of any part of such proceedings."

Before Julian could act, however, DeKalb Spurlin stepped forward. Spurlin, a wealthy realtor, had been an early recipient of Jack Bennett's loan schemes. For over a year he had invested and reinvested his profits into Julian Petroleum stock until the crash wiped out his interests. On May 19, he filed suit against Lewis and Bennett claiming an aggregate loss of $135,000, which he later described as "virtually his entire fortune." On June 2, Spurlin opened the offices of a Julian Stockholders' Association, "for the benefit of every stockholder regardless of the size of his holdings." Spurlin promised that his organization would "be operated entirely without profit" and would remain "positively free from exploitation by any individual."

During the ensuing days, Spurlin's office became a gathering place for the victims of the Julian Pete crash, who sorrowfully repeated their tales of hardship: one man had attempted to jump from the eighth-story window of Wagy & Co.'s San Francisco office; an aged woman with an invalid husband had tried to poison herself; another man had surfaced, "partly demented," in San Bernadino. When penniless or hungry shareholders appeared, the association treated them to meals.

Spurlin attempted to formalize his association at a mass stockholders' meeting on June 10. The 3,000-seat arena overflowed as a jazz band "gorgeous in red silk and high black Turkish hats" entertained the entering throngs. "Twenty Patrick Henrys flashed across the horizon," as Spurlin attempted to maintain order amidst the "wriggling, buzzing, cheering, hissing" crowd. "Victims described themselves as "shorn lambs" and "Julian goats." "I am not a member of one the pools, I am one of the fools," boomed a voice from the audience. "The pep of politics has gone into Julian Pete," reported the *Record*.

Spurlin read progress reports from the receivers and District Attorney Keyes and called for the election of a five-man committee to head the association. Before his action could be approved, C. C. Julian materialized, walking "dramatically down the center aisle, cane in hand, attired as out of a bandbox." People had speculated that Julian would use the shareholders' association to regain control of his corporation. Most of the audience seemed to welcome this prospect. They burst into applause as Julian climbed onto the stage. When Spurlin tried to prevent him from speaking, they chanted, "Julian, Julian. We want C. C. Julian." "Mr. Chairman, isn't it possible to have Mr. Julian represented on the committee?" shouted a woman.

Overruled by his constituents, Spurlin permitted Julian to speak. Julian launched into his now-standard attack on the "banker crowd," praising City Prosecutor Lickley, and promising to swear out warrants personally against the usurers. Suddenly the auditorium plunged into blackness. "The bankers got their agents here all right," shouted a voice in the dark. Several policeman and others began to wave their flashlights above the crowd and the band pitched in with "Cuddle Up a Little Closer." When the lights came back on, the band switched to "Hail, Hail, The Gang's All Here" and the crowd enthusiastically joined in the chorus.

Julian announced that he would not serve on the steering committee but the audience would have none of it, voting overwhelmingly to expand the comittee to include him. Shortly thereafter the meeting dispersed. In the eyes of the *Record* the "good natured . . . laughing" throng represented "a marvel of tolerance and sportsmanship. A sort of vindication of genus homo." While they had established no clear line of command, the stockholders had assembled and organized themselves. Thus, on June 10, 1927, with the receivers issuing rosy reports, the restitution campaign and Lickley prosecutions making impressive progress, and a shareholders' association in place, some of the gloom had lifted for the victims of the Julian Pete swindle. The break in the clouds would prove short-lived.

## III

On June 14, 1927, Harry Haldeman threw a curve at the Julian Pete receivers. Haldeman, who had earlier refunded his Bankers' Pool profits, demanded the return of his $15,200 check. Haldeman's attorneys had advised him that his Julian Petroleum stock transactions could not be construed as usury "in any shape, manner, or form." Furthermore, restitution might be seen as an admission of guilt. "The amount of money involved is insignificant," stated Haldeman's lawyer; "to pay it to the receivers would be easier and cheaper than the course Harry Haldeman is pursuing. There is involved, however, that which transcends in importance any

C. C. Julian strikes a defiant
pose during his confrontation
with Corporation Commissioner
Mike Daugherty in October, 1924.
(USC Regional History Center)

Two views of the spectacular fire at Bell No. 2 in February, 1922.
(USC Regional History Center, Hathaway Ranch Museum)

People crowded into a "sucker tent" to buy oil units at Santa Fe Springs.
(Hathaway Ranch Museum Collection)

C. C. Julian wells No. 2 (in foreground) and No. 1. Many of the original orange groves remain near the wells.
(Hathaway Ranch Museum)

The densely packed wells at the Four Corners section of Santa Fe Springs in 1923.
(UCLA Special Collections)

Mary Olive Julian.
(USC Regional History Center)

C. C. Julian in 1926.
(USC Regional History Center)

S. C. Lewis.
(USC Regional History Center)

Julian Petroleum stock transfer clerk Pat Shipp poses with a thick volume of company stock sales records. (UCLA Special Collections)

Jacob Berman, Jack Rosenberg, and Ed Rosenberg at a 1928 bail hearing. District Attorney Asa Keyes is seated at the extreme left. (USC Regional History Center)

Henry M. Robinson (seated) at a bankers' luncheon in November, 1927. In the background, behind the speaker, are Frank Flint (far right) and Charles Stern.
(USC Regional History Center)

A panorama of Pershing Square in the 1920s. The Biltmore Hotel is on the far left and the Pershing Square Building, which housed the Julian Petroleum offices, is the tall building on the right behind the Hotel Portsmouth. The prestigious California Club is directly across the street from the Pershing Square Building.
(USC Regional History Center)

Members of the Bankers' Pool.

Motley Flint.
(USC Regional History Center)

Harry M. Haldeman.
(USC Regional History Center)

Louis B. Mayer.
*(Southwest Jewry)*

Adolph Ramish.
*(Southwest Jewry)*

A Julian Petroleum service station featuring Lightning Gasoline in 1927.
(USC Regional History Center)

Receivers Joseph Scott and H. L. Carnahan (seated, left to right). Standing behind them are Jacob Berman, attorney John Murphy, and Julian Petroleum vice-president T. P. Conroy.
(USC Regional History Center)

(*Left*) Arthur Loeb in 1927 after losing his left eye in a fight at the offices of the Julian Petroleum Stockholders' Committee. *(Right)* Arthur Loeb at the 1930 stockholders' suit hearings. (USC Regional History Center)

The crowd outside the courtroom at the 1928 Julian Pete trial. (USC Regional History Center)

Prosecution and defense attorneys at the 1928 trial of the Julian Petroleum defendants. Harold L. Davis, Julian Richardson, and A. H. Van Cott of the District Attorney's office are seated at the left. S. C. Lewis, who defended himself, is at far right. Also depicted are pool member Mendel Silberberg (sixth from left), and in the rear (left to right), Julian Petroleum and A. C. Wagy & Co. officials R. M. Reese, H. F. Campbell, Charles E. Reese, William C. Kottemann, and Louis Berman. (USC Regional History Center)

S. C. Lewis, acting as his own attorney, demonstrating Julian Pete stock transactions at the 1928 trial. (USC Regional History Center)

*(Left)* Former district attorney Asa Keyes addresses the jury in his own defense at his 1929 bribery trial. *(Right)* Ben Getzoff, whose tailor shop became a brokerage house for bribery, outside the grand jury room. (USC Regional History Center)

Los Angeles District Attorney Buron Fitts (l.) and former FBI agent Lucien Wheeler, head of the District Attorney's Bureau of Investigation. (USC Regional History Center)

"Perpetual surprise witness" Leontine Johnson, former secretary to S. C. Lewis. (USC Regional History Center)

The Reverend Robert Shuler poses by his jail cell while serving his sentence for contempt of court in 1930. (UCLA Special Collections)

Artist's rendering of the murder of Motley Flint. (USC Regional History Center)

Frank Keaton hearing himself being sentenced to death for the murder of Motley Flint.
(USC Regional History Center)

Accused murderer "Handsome" Dave Clark looking relaxed and confident during his 1931 trial. (USC Regional History Center)

amount of money—Harry Haldeman's good name. . . . Under such circumstances his honor and self-respect required that he should refuse to pay in order to avoid prosecution."

From its inauguration the restitution program had aroused concern. Critics complained that the receivers required repayment of only the profits exceeding legal rates of interest. Furthermore, Lickley's actions seemed to absolve the usurers from criminal liability and collection of triple damage awards. Like Haldeman, former pool members and moneylenders also began to have second thoughts. Several local attorneys, including Walter Tuller, chairman of a statewide crime commission that had called for harsher penalties to combat crime, now advised their white-collar clients against cooperating with the receivers or Lickley. The day after Haldeman rescinded his refund, receiver Joseph Scott acknowledged that he had also returned a $57,000 check to Adolph Ramish.

Lickley issued a warrant for Haldeman's arrest. "We are engaged in a little civic experiment here in Los Angeles to determine whether or not it is only the obscure . . . that the statute of usury in this state affects, or whether too, it applies to Mr. Haldeman," challenged Lickley. Scott and Carnahan threatened to file civil suits demanding triple damages against those who failed to voluntarily "put back" their profits. But the momentum behind the restitution campaign had dissipated.

The Julian Petroleum Stockholders' Committee also staggered after its promising start. Within days of the mass meeting, conflict between Julian and Spurlin rent the organization. Spurlin, who had always opposed Julian's participation, explained, "I do not think anyone who has had official control or connection with the corporation at any time, ought to be on a shareholders' committee." Prior to the June 10 conclave, Spurlin had attempted to embarrass Julian by distributing forty questions about his activities. Julian repeatedly refused to answer them. When the elected executive committee convened on June 15, Julian reportedly insisted upon being named chairman. According to Spurlin, Julian wanted "to be the whole show or nothing." After a heated debate, C. C. stormed out, loudly announcing, "I'm through, I'll mail you my resignation." Spurlin placed ads in local newspapers urging all stockholders to sign proxies authorizing him to "vote my stock and represent me in carrying out any plan" the committee decided on.

Meanwhile, C. C. Julian prepared yet another radio address, promising more revelations. "Folks, don't miss tuning in because I'm about fed up on this 'SOFT PEDAL' stuff on the Julian Petroleum scandal," he advertised. On Monday, June 20, listeners reported hearing strange noises on all stations for a half hour before the scheduled 8:00 p.m. start of Julian's talk. A humming sound briefly struck each station and then moved on, as if somebody were trying to zero in on a specific frequency. As Julian began to speak, a cacophony of "howls and shrieks . . . sound[ing] like a dozen fire sirens and as many boiler factories" erupted on KMTR, drowning out his words. Listeners reported hearing whistling noises "like canaries" over the airwaves. The noise continued until nine o'clock when Julian went off the air.

Someone with a powerful transmitter had aimed a carrier wave directly across

KMTR's frequency, causing a heterodyne that blocked Julian's message. Julian lost no time in naming the saboteur. "I see before me the initials KHJ," he said, giving the call letters of the *Times* radio station. "But I never knew before tonight that they stand for 'Kill Honorable Justice.'" Julian noted that KHJ, which used singing canaries as a trademark, traditionally refrained from broadcasting on Monday evenings, leaving its transmitter free to interfere. He said that he had traced the source of the disturbance to the area of the KHJ station. A Julian shareholder claimed to have telephoned KHJ during the broadcast. "The noise and whistling almost knocked me down," he alleged.

The *Times* adamantly dismissed Julian's charges. "The utterly unscrupulous falsity of Julian's charges . . . suggests that he may have selected Monday night for his address knowing the regular absence of a program at KHJ would give him an opportunity to charge that the *Times* station was improperly used on that occasion," suggested a statement broadcast over KHJ. "This would not only allow Julian to pose as a victim of unfair tactics, but would make possible a well-advertised second delivery of his speech." Both the *Times* and Julian demanded a thorough investigation by the Federal Radio Commission. Julian called for criminal action against the "radio pirates."

It is measure of the *Times'* lack of credibility that so many people accepted Julian's charges. Thousands sent telegrams denouncing the newspaper. The *Los Angeles Ledger*, a local weekly, reported, "from a careful canvass of men and women along the way . . . the *Ledger* is forced to the conclusion that in the vast majority, in fact, without exception, the word of Julian, at least in this instance, is held as superior to that of the *Los Angeles Times*." After all, noted the *Ledger*, the *Times* and the city establishment had "securely muzzled the daily press of this city against reflecting conditions as they actually exist," while thus far developments had "sustained [Julian's] every utterance."

The morning after the abortive broadcast violence erupted at the Stockholders' Committee offices. Committee secretary Louis Horchitz accused Spurlin of working in league with Harry Chandler and demanded that Spurlin resign. Spurlin countered that Julian had planted Horchitz and committee member Arthur Loeb to create friction and immediately fired Horchitz. According to Horchitz, Spurlin shoved him out of the office, overturning furniture and shattering the glass door in the process. Horchitz placed committee records in his possession in a safe deposit box, and Spurlin filed a complaint calling for the arrest of Horchitz.

Later that day Loeb confronted Spurlin, now accompanied by his friend H. J. Kimmerle. Spurlin told Loeb that he had also removed him from his position because of his ties to Julian. When Loeb refused to leave, Kimmerle shoved him violently. Loeb's head smashed against the doorjam, shattering his eyeglasses into his left eye. Kimmerle allegedly threw the bleeding Loeb over a railing, chased him down the hall, and warned him "to get the hell out of the building and stay out."

Police rushed Loeb to the hospital and arrested Kimmerle for felonious assault. Three days later, doctors removed Loeb's eye. Spurlin nonetheless defended Kim-

merle and himself. "Those people [who] . . . have admitted that they were placed
. . . by Julian to further his interests have been harassing us for a couple of weeks,"
stated Spurlin. "We just cleaned house, that's all there is to it." Spurlin was now as
"cock of the walk," but the fracas had shattered the unity essential for a successful
stockholders' committee. "The greatest swindle ever perpetrated in America," commented *Express* columnist A. S. Tully about the stockholders' brawl and the Julian
radio dispute, "apparently is fast slipping into the vaudeville stage."

With passions running high, Julian brought his soft shoe back to the airwaves
on June 22. Once again thousands of people, anxious to hear the revelations oblit-
erated by Monday night's "malicious heterodyning," lined the streets outside KMTR
to listen to his broadcast over conveniently placed loudspeakers. But while Julian
ardently renewed what the *Times* called his "campaign of vilification," he offered
little new information. Julian again denounced the Pacific Southwest bankers and
devoted a half hour of vitriol to their "buddy" Harry Chandler and the *Times*. But
given the promise of an exposé and the controversy over the Monday broadcast,
Julian's performance proved anticlimactic.

Chandler and the bankers bore the brunt of Julian's assault, but the promoter
also targeted S. C. Lewis. "Lewis today says he's broke," challenged Julian. "Maybe
he is. But if so, I want to know who got the dough." Others wondered the same
thing. During the weeks since his fall from grace, Lewis had struggled to restore his
reputation and credibility. He had placed his "assets" at the disposal of the receivers
and pledged full cooperation in all efforts to assist shareholders in recovering their
losses. He reportedly turned down offers from well-known attorneys to defend him,
promising to dust off his lawyer's skills and represent himself. "While they can indict
me, it will be a long hard fight before I am convicted of any crime," he promised.
"The people who are convicted should be the ones who got the money. If they can
show where I got any money out of it I am willing to go to prison, but on the
contrary, if I didn't, then those who did get the money should go to jail." Employing
a curious bit of legal logic, he argued, "No crime was committed . . . in the actual
overissue so long as the public was not defrauded thereby, and so long as some
individual or officer of the corporation did not profit or appropriate money received
to his personal use or benefit."

Throughout May and June 1927, Lewis evolved a public defense for his conduct
of Julian Pete affairs, the most thorough version of which appeared in a biography
that ran in the *Record* from June 9 through 22. Absolving himself from all sins save
bad judgement, Lewis assessed blame in three directions: the Los Angeles brokerage
community, which incessantly manipulated Julian Petroleum stock; the "banking
burglars," who repeatedly postponed financing for California-Eastern, allowing the
moneylenders and pool sharks to capitalize on the delay; and his subordinates, Jacob
Berman and Ed Rosenberg, who had failed to advise him of the extent of their
activities. Lewis denied knowing of the overissue prior to February 1927. In the
end, explained Lewis, "It was a consistent, concentrated, determined attack made
by outside forces . . . which produced a financial crisis in the affairs of the corpo-

ration." In addition, he alleged "I was double-crossed by those who were associated with me." These parties, he charged, were "moving heaven and earth" to transfer the blame to him.

Lewis had presented th's version of the Julian Pete saga to the Los Angeles grand jury in May. For six weeks the nineteen-member grand jury listened to over fifty participants, attempting to decipher the Julian mosaic. Walter Woehlke would later complain that the order in which the jury called the witnesses, with Julian first and Lewis second, had prejudiced their deliberations. Julian's accusations "produced the desired psychological effect," and Lewis, while seemingly defending local financiers, "gave the impression that he had been betrayed, that the greed of the financiers and moneylenders had been the primary cause of Julian Pete's undoing." By the time the bankers came to testify, wrote Woehlke, "they faced nineteen pairs of hostile eyes."

The grand jury foreman offered a more complex description of the dilemma facing the panel. "About the time one group of witnesses had succeeded in painting a well-rounded picture and the jury was preparing to vote an indictment or two," he explained, "along came another witness with an entirely different 'set-up' of the whole situation and as a result the jurors' previous concept of the responsibility of the overissue would be smeared up. Then would come another picture."

At 5:00 p.m. on June 23, the Los Angeles County grand jury handed down the largest indictment in the history of southern California. Fifty-five men, including many leading businessmen, faced criminal charges. Those directly affiliated with Julian Petroleum and A. C. Wagy & Co.—including Lewis, Berman, Rosenberg, Reese, and Kotteman—were indicted on three counts of conspiracy: to violate the corporate securities act, to violate state usury acts, and to obtain money under false pretenses. Members of Bankers' Pools Nos. 1 and 2—including Motley Flint, Rouse, Haldeman, Ramish, Toplitsky, Mayer, and Hollingsworth—and a variety of other moneylenders, large and small, faced trials for conspiracy to violate the usury act. The grand jury charged bankers Charles Stern and John Barber with embezzlement and violation of the banker's bonus law for their handling of the "orphan check."

City detectives armed with arrest warrants fanned out over Los Angeles County, visiting Spring Street offices and palatial homes. The following morning the "Big Parade" began, as defendants represented by the city's legal élite nervously filed into court to post bail. The sudden competition for bail bond services drove down rates from the usual 10 percent to a low as 3.5 percent.

Nor did the indictments salve the fears of those who had gone unnamed. The grand jury promised to continue its probe, extending its investigations north to San Francisco; Prosecutor Lickley advised that the indictments would not deter his civil suits against the usurers; and C. C. Julian, whose original accusations had been validated by the grand jury, announced that his staff of investigators had now accumulated evidence against at least twenty of the most prominent Spring Street stockbrokers, who, claimed Julian, had sold stock to innocent purchasers for eight months after they had learned of the overissue. The aura of uncertainty and unreality that had prevailed since the Julian Petroleum crash continued to hover over Los Angeles.

# 13 THE INSTITUTIONS THEMSELVES MUST NOT BE TRIED

## I

On June 27, 1927, stockbroker Donald O'Melveny addressed a "SPECIAL MEM-ORANDUM" to his sales personnel. When the Julian Pete scandal had begun to unfold, he had warned his employees that until better information emerged, "the sporting and gentlemanly thing for each of us to do would be to refrain from gossip and to steady the people we talk to rather than inflame them." To spread the rumors running rampant in Los Angeles, argued O'Melveny, would destroy confidence in the financial institutions and have serious consequences for business. He urged his salesmen "to be constructive rather than destructive." In the wake of the June grand jury indictments, O'Melveny reiterated his fears and admonitions. "These rumors and accusations have had a tendency to destroy the confidence of the community in its institutions and its leading citizens and in the things that make for sound business and prosperity," wrote O'Melveny. "Everyone of us [who] . . . comes in contact with other people is a power for good or evil. You must not try these people in your own minds . . . and what is more serious, the institutions themselves must not be tried."

In these memoranda, O'Melveny expressed the foremost fears haunting the Los Angeles business community in the aftermath of the Julian Pete collapse. The spectacle of leading citizens first accused by C. C. Julian and now hauled before the bar of justice jolted the city's booster optimism. "The actual damage done by the collapse of the oil company itself is trifling in comparison to the injury which is being attempted by willful wreckers whose activities are aimed at the financial foundations of the city," worried the *Times*. The sums lost in Julian Pete investments, warned Chandler's paper, "are inconsiderable in contrast with those placed in jeopardy" by the blows against "the cornerstone of the business structure of Los Angeles."

Some observers questioned the essential moral fiber of the Los Angeles community. "The really important thing to my mind," wrote Charles Stern to a friend, "is the fact that a splendid, prosperous community such as this can be first defrauded and then misled by the propaganda of unprincipled and irresponsible mountebanks. . . . A community so susceptible to this type of ignorant hysteria can hardly be said to have reached the age of competence." *Saturday Night*, a weekly magazine, warned "Don't get stampeded. Don't go into a panic, and imagine everything is ready to bust . . . . Los Angeles is sound . . . most men are honest. . . . If a hundred men have done wrong in the Julian affair, remember that several hundred others refused to become involved. . . . Think of them and keep your faith in humanity."

Both before and after the grand jury indictments, the *Times* and other establishment voices hastened to defend accused civic leaders. The *Express* derided Julian's "attempt to drag and smear with the mud . . . persons whose character is the highest, whose integrity no one questions, whose names are synonym for honor and uprightness" as "vicious beyond ghoulishness." The "dastardly and cowardly attack" on "world citizen" Henry Robinson "when he was absent in Europe, representing the United States," particularly riled the *Express*. Charles Stern (and presumably others whose correspondence has not survived) received dozens of telegrams and letters of support from business leaders throughout the state.

Stern, the *Times*, and others repeatedly argued that preoccupation with the stock pools was a ruse created to obscure the true culprits. "We are covered with soot from the smokescreen under which the real crooks are attempting to escape," protested Stern. Julian and Lewis, stated the *Times*, were the prime villains of the piece, and the overissue "the major parent crime from which proceeded all the rest of the horde of big and little difficulties which have involved the company." In a delightfully creative editorial, the *Times* compared Lewis and Julian to the protagonists in Lewis Carroll's "The Walrus and the Carpenter," who invited the oysters to "come and walk with us" for "A pleasant walk, a pleasant talk upon the briny beach." The eloquent, spellbinding Carpenter thrilled the oysters with his high-pressure appeal. The Walrus employed subtler methods. Although the wise ones refused to follow, the unsophisticated "hurried up all eager for the treat." When feeding time arrived, the oysters grew concerned:

> "But not on us," the oysters cried,
> Turning a little blue.
> "After such kindness that would be
> A dismal thing to do."

These protests notwithstanding, the Carpenter devoured the small oysters, while the Walrus dined on the large.

> "Oh Oysters," cried the Carpenter,
> "We had a pleasant run;
> Shall we be trotting back again?"

But answer came there none—
And this was scarcely odd because
They'd eaten every one.

The *Times* proved less inspired in its attempts to find silver linings in the ever-thickening clouds. While "frauds of considerable proportions have been perpetrated," reasoned one particularly audacious *Times* editorial, "very little of the money paid for this stock actually left the community. The great bulk of it is still here, in active and useful circulation. Added to it are the proceeds from the sale of stock in other cities. . . ." Even the grand jury indictments, predicted the *Times*, would have a "salutary effect" by disproving "the most vicious misstatement of all— that the 'worst thieves' would be permitted to get away with their depredations by virtue of their wealth or prominence in the community."

The fears that O'Melveny, the *Times*, and other community leaders hoped so fervently to dispel, however, appeared with great regularity in the city's opposition press—the less widely circulated daily and weekly newspapers and journals. These periodicals rejected any exoneration of the pool profiteers. "Then the cry that the only wrong done was in the overissue of Julian stock. Yah. Usury and embezzlement and such like practices are only innocent sport. Huh! What guff," exclaimed *Saturday Night*. The *Ledger* dismissed the *Times*' attempt to blame Julian as "a herring across the trail of thieves."

The Reverend Bob Shuler, whose radio show and monthly *Bob Shuler's Magazine* had a broad audience, ranked among the loudest voices. Shuler called for a "score or more of gentlemen . . . [to] be retired from business and given free transportation to a place where stripes would adorn their portly forms." Other ministers joined the chorus, leading Charles Stern to complain about the "various Elmer Gantrys of the pulpit of our local churches" who, along with Julian, had "fanned public hysteria into flame."

Many of these critics from press and pulpit questioned the ability of those indicted to remain community leaders. "This nasty affair . . . has taken a shine off many a tin god," opined *Saturday Night*. "The Babbitts and flag-waving patrioteers come to light a pretty ragged bunch." These men, noted Guy Finney in the *California Graphic*, had led the city's " 'better America' movements, 'better business' and 'better advertising' campaigns, 'better service' ideals . . . and various other betterments that make effective business camouflage." Now, commented Finney, "for the future they are marked men . . . . We no longer think of them as competent to handle the reins of business leadership, neither in banking matters nor civic affairs." More pointedly Finney reflected:

Who . . . would think of Motley H. Flint as a wise and safe banking counsellor?
    Whose patience with truth is so sublime that he is not disposed to snicker at the incongruity and draw the curtain of doubt over the sincerity of Mr. Cecil B. DeMille's "King of Kings," with its heroic figure of Christ driving the money changers (they may have been usurers) from the Temple. . . .

Who but a hungry thespian seeking a movie contract could now take seriously the pretensions of Mr. Louis B. Mayer and his co-worker, Mr. Adolph Ramish, as leaders in the movement with Mr. Will Hays to establish higher ethics and better standards in the work of making films. . . .

Liberal voices took delight in berating Harry Haldeman for his hypocrisy. "Whose mind is so childlike in its faith as to accept without disbelief the picture of Harry M. Haldeman in the role of leader of the Better America Federation, which . . . offers itself as a staunch defender of Constitutional law, when they have before them his later picture as a contrite Julian pool member?" asked Finney. "It is barely possible that this is the sort of 'leadership' that makes men see 'red' in another way." The *San Francisco Daily News* lampooned the BAF and its call for "100 per cent Americanism." Although Haldeman, they gibed, "is the best of all Better Americans," his paltry $20,000 investment indicated "that maybe, after all, there were better Americans around than he." Noting the 228 percent interest collected by Haldeman and the others, the *Daily News* surmised, "While we were plodding along as 100 percenters, the Better Americans were making us look like pikers. . . . So they were just 216 percent better than the law allowed, and just 128 per cent Better Americans than the rest of us."

Nor did these periodicals share the *Times'* confidence that the grand jury indictments would result in justice. The *Los Angeles Daily News* contrasted the pool members to a sixteen-year-old farm worker arrested for stealing a watermelon on a hot summer day. "This was wrong; very, very wrong," chided the *Daily News*. Although he later paid for the watermelon, the boy received a six-month jail sentence. Another kind of "melon," the paper added, had been stolen in Los Angeles— "Full grown men did it and they passed pieces of the melon all around town to goodness knows how many folks." These men also hoped to make restitution; and none had gone to jail. Bob Shuler predicted, "There are men in Los Angeles in places of power and fortified behind the barred gates of office and position, who ought to go to San Quentin. . . . But they will not go. . . . The penitentiary will have to be satisfied to entertain men who have no money and no political backing."

By injecting the element of political influence, Shuler broached a subject the mainstream press largely ignored. "It is whispered about that men high in the the Los Angeles city administration have had their pockets well stuffed," reported Shuler. "What interest did Charlie Crawford have in this affair? Where was Kent Parrot sitting while this big steal was going on?" The *Express* also guardedly raised this issue. It repeatedly questioned the failure to indict or even name the members of the reputed "Tia Juana pool," which allegedly included prominent political figures. "Gosh! What do those reporters want us to do—expose our friends?" commented *Express* columnist A. S. Tully facetiously. Shuler also questioned the inaction of Corporation Commissioner Friedlander. Friedlander, he charged, had been appointed "for the express purpose [of] covering the guilty and preventing the actual thieves from being brought to justice."

The debate also revealed the ugly specter of anti-Semitism in Los Angeles. The recruitment of Jewish businessmen by Jacob Berman and Ed Rosenberg gave Jews a prominent position in the scandal. Several accounts made note of Berman's Jewish background. Rosenberg's Pool No. 3 was dubbed "The Jewish Pool." Shuler referred to Friedlander as "a certain Jew" and predicted none of "our big Jew money lenders" would go to jail. Guy Finney described music store owner Benjamin Platt of Pool No. 2 as an "honest confiding child of Israel" and in mock biblical terms encanted,

> Oh Julian, what monster of punishment art thou, that thou hast brought thy lash on the yielding flesh of this lowly purveyor of song, whose only crime has been that he loved money once and then too well.

Occurring during a decade in which anti-Semitism and social ostracism of Jews in Los Angeles had greatly increased, the Julian Pete scandal contributed in no small measure to southern California's growing intolerance.

Among the institutions most directly damaged by the Julian Pete stock collapse, Pacific Southwest Trust and Savings stood most desperately in need of public rehabilitation. The grand jury had indicted eight bank officials, including Stern, the president; Hollingsworth, a director; and six vice-presidents. Efforts to assure the public of the bank's stability quickly followed the indictments. State Superintendent of Banks Will Wood announced that all loans the bank had made to Julian Petroleum were secured by pledges of tangible property. Pacific Southwest, he stated, held no loans secured by stock and had not been involved in the pools. Several Pacific Southwest officals issued a joint statement labeling their indictments a "terrible wrong" and asking people to withhold any judgement until they had the opportunity to prove their innocence.

The major task of reconstructing the institution's reputation fell to Henry Robinson, chairman of the parent First National Bank. Robinson returned to the United States in mid-June, and John Barber met him in Chicago to update him on the debacle. Upon his June 16 arrival in Los Angeles, according to his biographer Rockwell Hereford, Robinson found numerous promises of financial support from banking allies throughout the nation.

On June 26, three days after the indictments, Robinson issued a statement presenting the bank's official position. Despite Julian's repeated accusations, Robinson had played a minimal role in the creation of California-Eastern, and the grand jury had not charged him. Nonetheless, Robinson felt compelled to reiterate his innocence, denying any personal involvement in pools and loans. He also dismissed allegations that the bank had any ulterior motives in the original merger proposals. He compared the Julian Petroleum rehabilitation project to that of the Los Angeles Investment Company a decade earlier, where his associates had taken a fraud-ridden enterprise and "built out of it a wholesome and prosperous institution." This had been the intent of the California-Eastern reorganization. If the project had been carried out, argued Robinson, it would have been of great benefit to both Julian Petroleum shareholders and the community. Indeed, Robinson maintained that a refinancing along the lines of that originally proposed by First Securities remained "the only way" for investors to recoup their losses.

Addressing the issue of pools and usurious loans, Robinson fully absolved the bank. "If the participation of any of the officers of the Pacific Southwest Bank in so-called pools is in question, it is their private affair involving their own money and has nothing to do with the bank's resources," he stated. Robinson expressed astonishment that this issue had diverted attention from the real reason for the downfall of Julian Petroleum. "I do not know how the public for a moment can fail to see that the cause of the trouble, which is perfectly obvious, is the overissue of the stock and nothing else," he concluded. Behind the scenes, Robinson accelerated existing plans to merge the First National and Pacific Southwest banks into one entity. This would remove the name of Pacific Southwest, the chief offender in most people's minds, from public attention. Robinson formally announced the consolidation on July 11.

Both the LASE and the Corporations Department also made characteristically bureaucratic face-saving gestures. The LASE passed a resolution requiring that all listed stock issues be registered with the exchange secretary and countersigned by a bank or trust company. The Julian Petroleum situation would never have arisen had these modest regulations been in effect, LASE officials assured investors. President John Jardine, while expressing confidence that LASE members "transact their business with the public in a highly efficient and honorable manner in every way," announced the creation of a special committee to draft a new LASE constitution and by-laws.

Similarly, the Corporations Department unveiled several new "safeguards." Corporations selling stock to the public would henceforth appoint a transfer agent to maintain records of all sales and transfers and would be required to file semi-annual reports. In July, Corporation Commissioner Friedlander instituted two new divisions: a complaint and investigating department and a public relations branch. Friedlander also called for the adoption of stricter regulations to give "more teeth" to the blue-sky laws.

Following the constant stream of revelation and activity in May and June, the Julian Petroleum affair unfolded more languidly in July. Attorneys for the fifty-five defendants unsuccessfully challenged the validity of the grand jury indictments. The Julian Petroleum Stockholders Association, shaken by the violent Spurlin-Loeb confrontation, sat dormant. Two rival groups, including an association representing Japanese shareholders, who reportedly lost $500,000 in the crash, were created. City Prosecutor Lickley's usury prosecutions ebbed after issuing several more complaints, the most notable lodged against film director Cecil B. DeMille, as Lickley entered the hospital on July 9 for throat surgery.

The ongoing grand jury investigation also produced few new sensations. Several grand jurors charged that, since the indictments, they had been shadowed by private detectives attempting to intimidate them from voting further indictments. Detectives had visited their neighbors and pried into their banking connections, wealth, and church and social affiliations. The *Record* pegged the detectives as "in the employ of powerful Los Angeles banking and financial interests." The grand jury foreman reported at least three instances of attempted bribery. An assistant district attorney stated that an investigation had confirmed the allegations, but that he had "obtained the promise of those responsible for it that it will cease immediately." Remarkably, given the gravity of the accusations, no names were mentioned and no charges preferred.

Those who had hired the detectives apparently had little to fear. Despite predictions of scores of additional indictments, the grand jury wound down its deliberations uneventfully. On July 13, they announced that both Archibald Johnson's "San Francisco Pool" and the mysterious "Tia Juana Pool" had accepted no usurious extensions and thereby stood as legitimate stock purchases. Three days later the grand jury exonerated the Los Angeles brokerage community of criminal wrongdoing. With the exception of a few minor investigations, the grand jury ended its probe on July 16.

Not even C. C. Julian could renew the earlier enthusiasms. He took to the airwaves on July 6 with a promise to embroil two more banking firms in the scandal, but failed to do so. The usually sympathetic *Record* branded his performance a flop. Julian postponed the following week's broadcast and on July 18 canceled his scheduled performance. Shortly thereafter he left town without comment.

## II

By early July 1927, receivers Joseph Scott and H. L. Carnahan had emerged as the only untarnished figures in the Julian Petroleum affair. In the public eye the pair appeared to have performed Herculean tasks. "In less than two months they had cleaned up the Augean stable, trimmed expenses and increased revenues sufficiently to turn the daily loss into a small profit," praised Walter Woehlke. In reality, Scott and Carnhan had just begun to probe the depths of the Julian Pete debacle. "It was a source of grief to me almost from the time I started," recalled Scott several years later. "If I had known as much as I do now, I would not have touched it with a ten-foot pole."

For Scott, and particularly for Carnahan, the Julian Pete receivership became an all-consuming affair. They divided responsibilities so that Scott handled the usury cases and other legal matters while Carnahan supervised the stock analysis and corporate operations. "I worked an average of 15 to 16 hours a day, not only on weekdays, but about three Sundays out of four," Carnahan later testified. Both men effectively closed down their existing law practices to administer the rescue operation. In early July, Scott and Carnahan assumed an added burden when the federal court added A. C. Wagy & Co. to their receivership.

Scott and Carnahan had already realized that the stock overissue represented only the most visible manifestation of Lewis's handiwork. Henry Robinson and others involved in the financing of California-Eastern had always argued that the physical assets of the company more than justified the merger; the stock manipulations had undermined a sound venture. This was not the case, however. Hidden from sight was a corporate iceberg.

The stockholders who had approved the creation of California-Eastern had acted on the basis of a February 1927 audit establishing the company's assets at almost $30 million. This figure was based on Lewis's creative bookkeeping techniques, which consistently overvalued virtually all of the company's properties. A lease in the Goleta, California, field valued at $2.4 million, for example, proved to be located on the wrong side of a fault, cutting it off from production and rendering it worthless. The vaunted Alamitos Hills properties, carried on the books at $1,275,000 and praised by Carnahan as the key to California-Eastern's future, seemed unlikely to repay even the cost of drilling. Sixty other leases had to be abandoned due to failed test wells. Two refineries acquired by Lewis were non-functional; a third was worth less than half its purchase price.

Virtually all of the out-of-state properties posed problems. The Montana wells, which Lewis had bought for $3 million and listed on the books at $4.7 million, would never produce enough oil to justify either of these figures. The Canyon Oil Company of Texas was worth less than a fifth of its $829,000 listing.

Even more perplexing was the status of the Louisiana oil leases allegedly acquired from Lewis Oil. The receivers found a contract dated July 3, 1925, transferring these properties to Julian Petroleum in exchange for 420,000 shares of Julian Pete stock. The directors who had taken over Lewis Oil, however, claimed that they knew nothing of this transaction. Lewis Oil had never received any Julian Pete stock and furthermore had continued to operate these properties as their own. On closer examination, Scott and Carnahan discovered that the contract lacked a corporate seal and bore a forged signature. They concluded that Lewis had prepared the document in November 1926 to include these leases in the audit. More remarkably, Lewis had already transferred these properties to California-Eastern in August 1926 as the basis for the initial 750,000-share stock issue of that company. Lewis had "purchased" the same leases twice, yet they still legally belonged to Lewis Oil!

The Louisiana lease chicanery had two implications. Since Julian Petroleum did not own them, this further deflated the physical assets of the company. More significantly, since they had never been transferred to California-Eastern, the 750,000 shares issued to Lewis to start the company were invalid and, in effect, like most Julian Pete stock, overissued. In addition, the February audit had not included over $6 million claimed by Lewis for monies he allegedly advanced Julian Petroleum. In April 1927, Lewis received 661,000 shares of California-Eastern stock to forgive this unrecorded debt. Thus, all of Lewis's stock held in trust for the benefit of Julian Pete shareholders was overissued and invalid. The receivers never publicized this fact.

Even a generous reappraisal of the California-Eastern properties placed their value at half the $30 million figure used for the merger. The corporation's liabilities dwarfed this sum. The company owed $9 million to the Pacific Southwest, Anglo-London-Paris, and other banks. These banks had wisely secured these loans with the company's physical assets. Lewis had also accumulated over $5 million in unsecured claims in the form of mechanics liens' and unpaid bills. Over and above this was $38 million worth of unsecured stock claims.

The massive stock overissue posed further complications. Auditing and assessing the authenticity of all outstanding Julian Pete stock would prove a daunting assignment. According to William Kottemann, "The preferred stock alone . . . consists of two volumes, typewritten, single space, on very thin paper, and each of these volumes is bigger than the Los Angeles telephone directory. There are two similar volumes covering outstanding common stock." To determine what price the stock was sold at and who benefited from these sales, advised an FBI accountant, "it will be necessary to classify each and every share of stock and then conduct an investigation of the books and records of Wagy and Company." Scott and Carnahan assigned three accounting firms and forty accountants to this chore.

A unique feature of California law raised yet another stock-related problem for the receivers. A 1917 state constitutional amendment (authored by Carnahan, ironically) held stockholders liable for the proportion of the debts and liabilities of a corporation represented by their stock. Furthermore, people who had purchased spurious stock were considered general creditors of the corporation, which entitled them to sue the bona fide shareholders to recoup their losses. Thus, those who had speculated in the blatantly overissued Julian Pete securities in the latter stages of the stock campaign could attempt to collect from those who, in good faith, had invested with C. C. Julian in his original venture. Indeed, the receivers, legally the representatives of Julian Petroleum and California-Eastern creditors, should theoretically have pressed these claims against the legitimate stockholders.

The overlapping difficulties of severe overvaluation, massive liabilities, and extreme overissue should have forced the Julian Pete receivers into the logical solution: a foreclosure on the liens of the secured creditors and a liquidation of the corporation. Politically, this course seemed unviable. The banks would recoup their investments, while thousands of legitimate shareholders would not only lose their initial payments but also become the target of lawsuits by holders of spurious stock and other unsecured creditors.

Carnahan, in particular, opposed this solution. "I thought the bona fide stockholders deserved more protection than the fellow who had gambled it in the end," he later explained. Carnahan and Scott were determined to find a remedy that would put all stockholders on a common footing and shield the original purchasers from any liability. Even after the stock audit allowed them to distinguish between the legitimate and the spurious shareholders, Carnahan and Scott refused to provide these lists to those wishing to bring suit. Judge McCormick supported them in these decisions. "I knew very well it was not the law," admitted Carnahan in 1933. "We made our decision as to what was the honest and decent thing to do . . . and we stood by it."

One other factor weighed heavily in the decisions reached by the receivers: the growing influence of S. C. Lewis on their deliberations. That Scott and Carnahan, confronted daily with the evidence of Lewis's fraudulent malevolence, should have fallen under his spell remains one of the more astounding features of their receivership. Yet, facing the complexity of the Julian Pete maze and the absence of coherent books and records, these "unsophisticated receivers," as Scott later described himself and Carnahan, came to rely on Lewis's counsel. "This man Lewis had a remarkable memory, one of the most ingenious minds I have ever met," recalled Scott. Lewis, operating out of offices adjoining those of Scott and Carnahan, further impressed the receivers with his characteristic energy and diligence. Regularly working sixteen- to eighteen-hour days, including Sundays and holidays, Lewis projected the image of a man determined to atone for past sins.

Lewis steered the receivers away from bankruptcy and foreclosure and advocated reorganizing California-Eastern into a new company. The details of Lewis's proposal bore startling resemblance to his original merger plan, which had generated the current crisis. Lewis suggested that Pacific Southwest and Anglo-London-Paris, the two banks that had underwritten his earlier efforts, lend $8 million to the new firm. These funds, when added to at least $5 million collected from those charged with usury, he argued, would be sufficient to rehabilitate the company. Most appealing to Scott and Carnahan, all stockholders, whether owners of real or spurious stock, would receive shares in the new enterprise on an equal basis in exchange for surrendering the right to sue.

Lewis's program thus offered something for everyone. The bankers received the opportunity to bail out the troubled concern and undo the negative publicity generated by the Julian Pete collapse. Those accused of pool operations could "put back" their profits and escape criminal and civil liability. Stockholders, both bona fide and spurious, might be able to recoup their losses. Conveniently overlooked was the fact that the physical assets and earning power of the reorganized company would be insufficient to cover its reconstructed obligations. By early July 1927, Lewis had persuaded not only Scott and Carnahan, but Henry Robinson and the shell-shocked bankers at Pacific Southwest, of the efficacy of his plan.

On July 7, two days after the U.S. District Court declared him legally bankrupt, Lewis unveiled the reorganization scheme at a meeting of 5,000 Julian Pete shareholders. A radio hookup carried his speech to thousands of others. Lewis had promised to identify those who had profited from the debacle, but once at the microphone he reneged. "I will not engage in personalities," he declared. Lewis instead called for leniency for pool members who offered restitution. "Cold money" rather than "a pound of flesh" would restore financial health to California-Eastern. Lewis promised, "If I fail within sixty or ninety days to rescue this corporation from its present condition, I will go before the court and ask that the judge accept my plea of guilty and I will take the maximum penalty for the crimes with which I stand charged." Despite sporadic heckling, the crowd demonstrated surprising affection for Lewis and warmly accepted his proposal.

The inclusion of funds returned by pool members and other moneylenders as part of the reorganization plan refocused attention on this aspect of the scandal. Joseph Scott had assumed responsibility for dealing with these issues. But, as with so many elements of Julian Petroleum, the deeper he probed into these matters, the more complex they appeared.

The receivers had obtained copies of the grand jury audit, which listed all known pool arrangements and collateral note transactions. Scott wrote letters demanding restitution to the over 100 people named in this report. The responses of the accused usurers varied widely. A handful of individuals, citing either an unwillingness to profit at the expense of the Julian shareholders or a desire to extricate themselves as painlessly as possible from the Julian mess, simply refunded their gains.

The five members of Bankers' Pool No. 1 affiliated with Pacific Southwest Bank—Flint, Hervey, Bell, McMullen, and McKay—had already agreed to return $235,000. Two other members of Bankers' Pool No. 1, W. I. Hollingsworth and Harry Haldeman, refunded their profits on July 8. Hollingsworth's attorney denied any attempt on his client's part to evade prosecution, noting that Hollingsworth had already been indicted. Nonetheless a general sentiment prevailed that those who "put back" would fare better in the courts than those who resisted.

Most other lenders rejected the demands of the receivers. Several argued that they had entered into transactions with Berman or Rosenberg as individuals, not with Julian Petroleum. Thus the receivers had no right to require repayment. Others pointed out that while they had profited in certain transactions with Berman, he had failed to pay off on others before he fled Los Angeles. The grand jury audit showed that Berman had paid out over $3 million to his investors, but these same men held notes for an additional $2.1 million Berman had failed to redeem. In some instances obligations exceeded the original profits. If the receivers were entitled to claim the payments, were not they also obligated to recognize the debts accrued by Berman on similar transactions?

Throughout the summer of 1927, the receivers filed civil suits against dozens of moneylenders, claiming triple damages. Scott and Carnahan recognized their fragile position, however. A lengthy and expensive court battle would be necessary to establish their right to these funds. The verdicts could go against them. Meanwhile, the receivership needed money to fund its operations. They decided that the wisest course would be to compromise on these claims as quickly as possible.

Scott held conferences with the moneylenders, usually attended by S. C. Lewis, to attempt to reach settlements. Scott settled most cases by accepting small sums to settle large claims. In the case of Jacob Farbstein, for example, the receivers sued for $162,140 but settled for $16,670. A. W. Hackel paid $24,255 on claims approaching $400,000. Others paid as little as $100 to evade suits running into thousands of dollars. Despite these generous compromises, most of the defendants still refused to pay. By the summer's end, the usury suits had produced only a little over $100,000 in restitution. Nonetheless, the receivers continued to pursue the ever-vanishing ghost of restitution while they traveled the reorganization course laid out for them by the ubiquitous S. C. Lewis.

### III

On July 21, 1927, thanks to the enterprising efforts of reporter *cum* attorney Robert Kenney, the Julian Pete scandal flamed anew. During the long weeks of investigation, the grand jury testimony had remained secret. Kenney, a *Los Angeles Herald* courthouse reporter recently admitted to the bar, realized that by law all defendants would receive copies of the grand jury transcripts. Kenney arranged with one of the accused to get his copy immediately upon its release. The transcript ran to 2,500 pages; copies for all fifty-five defendants and their attorneys reportedly weighed two tons. Kenney and another reporter pored over this massive tome, and the *Herald* brazenly headlined their summary in exclusive stories on July 21 and 22. Other dailies quickly followed suit, each attempting to outdo the other with its exposés. The *Examiner* ran lengthy verbatim excerpts from the testimony for over a week. The *Record* protracted its account into mid-August.

The daily doses of grand jury transcripts, with their detailed testimony on pools, loans, the overissue, and the California-Eastern negotiations, generated a renewed paroxysm of interest and outrage. The impact reverberated, not so much from the revelations—most of the information had already leaked out during the preceding months—as from the spectacle of the civic leaders' self-serving defense of their questionable activities. Harry Bauer's testimony brimmed with callous bravado as he boasted of his losses: "I have been bunkoed out of $593,413.25 in cold cash. You are looking at the biggest boob of all." When asked if he still had his shares, Bauer replied, "Yes, I have got it. I will bring it up here. IT WILL FILL THE WHOLE ROOM." In contrast, Adolph Ramish cried openly before his interrogators. "Pardon me, gentlemen, but I never was in a fix like this in my life before," he sobbed.

The transcripts also revealed the incredulity of the grand jurors as they listened to the varying accounts. When stockbroker Harold Barneson denied knowledge that an overissue existed, one exasperated juror muttered, "Innocents Abroad!" Alvin Frank contended that he did not know that loans to Berman and Rosenberg were destined for Julian Petroleum. A juror told him he "had a lot of gall." When Lewis attempted to defend his banking allies, he explained, "I want to be fair to these men, even if it is at my expense." "Be fair to all," admonished a juror.

Among the most damaging images was that of First Securities president John Barber attempting to explain his unorthodox handling of the "orphan check." District Attorney Buddy Davis grilled Barber about why he had returned the check to the Julian Petroleum account in early May. "When the thing blew up we preferred to be in a position of having collected no money from Lewis," confessed Barber. Davis persisted: "What is the real reason that you did it?" he asked. "Just the sentiment," answered Barber, as Davis audibly gasped.

In the final analysis, Lewis's testimony dominated the newspaper accounts of the grand jury hearings. Lewis had not only held the stand for the longest period of time, but his seeming lack of evasiveness and uncanny memory for facts, amounts, and transactions made him more believable than most other participants. Lewis succeeded in simultaneously defending the Pacific Southwest bankers while casting subtle doubts about their sincerity and intentions. His detailed descriptions of the pools, loans, and merger negotiations tended to heighten the significance of these activities and diminish the importance of the overissue in the public perception.

The grand jury transcripts, usury prosecutions, and earlier revelations allowed southern Californians to absorb the magnitude of the fraud and assess who had profited and who had suffered. While most public attention had focused on the losses of tens of thousands of small investors, many wealthy individuals, including Bauer, King, Dabney, and Spurlin, had each lost hundreds of thousands of dollars. Los Angeles stockbrokers clearly ranked among the big winners. Brokerage houses had earned millions of dollars in commissions on Julian Pete stock transactions. The more unscrupulous houses had also profited handsomely by "bulling and bearing" Julian securities. H. J. Barneson and Company showed $2 million in stock purchases and $750,000 in loans on its books.

Some pool members and moneylenders had also fared astonishingly well. The grand jury had employed accountant Fred Hahn to audit Jacob Berman's financial dealings. With cooperation from Ed Rosenberg and the evidence from millions of dollars' worth of canceled checks, Hahn skillfully re-created Berman's complex pyramid of stock manipulation. While only five actual stock pools had existed (just three of which had accepted extensions and thereby committed usury), scores of individuals had lent money to Berman under the "three-point 45 day plan" or for thirty-to-forty-five day promissory notes secured by Julian Pete stock. Berman had received approximately $26 million in loans, which he later repaid at exorbitant interest rates. The largest lenders included the Rosenberg brothers, who posted over $4 million (much of which they assigned to others); Adolph Ramish, whose transactions aggregated between $2 million and $3 million; and brokers Alvin Frank and A. W. Hackel, both of whom lent over $1 million to Berman. Many others, including Parrot cronies Bob Gans and Perry Greer, registered sums in the neighborhood of a half-million dollars.

The amounts handled by Berman stunned both grand jurors and onlookers. Hahn testifed that between January 1926 and April 1927, Berman had collected over $100 million in loans and stock sales. "One hundred million dollars?" questioned a disbelieving juror. Hahn confirmed the amount and revealed that Berman had deposited over $66 million in local banks. "He drew checks for the same amounts because in none of the accounts has he any balance," reported Hahn. An additional $34 million remained entirely unaccounted for. Hahn said he was "very dubious" that this money would ever be traced.

Hahn's accounting and the grand jury testimony answered many questions about the Julian Pete fiasco but left others unresolved. How had Lewis, who the transcripts revealed was broke when he arrived in Los Angeles, managed to fool the Pacific Southwest bankers? How had so many businessmen, who, as Finney noted, "were so wisely considered in the ways of finance that they should have known better," been lured into the trap of collecting usurious interest? As Finney later wrote in *The Great Los Angeles Bubble*:

> It still remains a puzzle to many why a group of presumably farseeing, thoroughly trained, admittedly intelligent, sophisticated businessmen would default in their responsibility. . . . It amazes one to think that men who had attained a conspicuous rank in a great community's financial affairs . . . would fall so completely under the influence of two itinerant sharpers. . . .
>
> It almost passes the bounds of belief that these "representative men" . . . would

knowingly accept vast quantities of the fraudulently issued stock of a company they must have known at the time was headed for a crash, and then passed the same stock to the unsuspecting public's hands.

It seems incredible, yet it is true, that the management of the Los Angeles Stock Exchange would have permitted enormous blocks of this spurious stock to pass daily from broker to broker and thence to the public without raising a protesting voice. . . .

Most intriguing was the question of how Lewis hoped to maneuver his way free of economic and legal disaster. Lewis always claimed that, while he was aware of an overissue, he had underestimated the diligence of Jacob Berman and misjudged the amount of stock circulating. Lewis might have believed, until a relatively late date, that he could absorb and retire the overissue. Given the relationship between the two men, the testimony of others, and the evolution of the overissue, this seems improbable. Perhaps Lewis and Bennett became victims of their own ever-escalating scam, plunging them ever more deeply into arrears until escape became impossible.

Reverend Shuler offered another scenario, which he claimed he heard from "a very knowing man, high in the world of finance in Southern California." Lewis or the Pacific Southwest bankers planned to "gut" Julian Petroleum, driving the price down to two dollars a share or less. At this time they would buy up the entire overissue for under $10 million, cancel the excess stock, and with a minimum investment, gain control of California-Eastern. Premature exposure of the overissue, however, had short-circuited this ploy.

At least one other alternative existed. Lewis and Berman had never planned to profit from either Julian Petroleum or California-Eastern but in the course of their multi-million-dollar transactions had salted away a substantial portion of funds into hidden accounts, which could be tapped once the merger had collapsed and the resulting storms had blown over. This possibility raised the most perplexing question of all. According to the final accounting, $34 million was missing. Where had this money gone? Most people believed that Jacob Berman held the solution to this mystery. But, three months after the Julian Pete crash, Berman had yet to be found.

# 14 YOU CANNOT CONVICT A MILLION DOLLARS

I

In the course of his voluminous grand jury testimony, S. C. Lewis had casually attempted to illuminate how his accomplice, Jacob Berman, alias Jack Bennett, had kept track of his complex web of pools, loans, and stock sales. Berman, explained Lewis, was a "very bright young man." The phrase captured the public fancy. In the months and years following the Julian Pete exposés, few newspapers referred to Berman without appending some variation on the appellation of "bright youngster" or "bright boy." By late summer of 1927, the elusiveness of the wily fugitive had done nothing to dispel this image. Widely regarded as the key to the missing millions, Berman bobbed in and out of sight, baffling his pursuers and frustrating Julian Pete investigators.

Berman had fled Los Angeles on May 2, four days before the Julian Pete crash. Lewis reported that Berman had gone first to San Francisco and later to New York in an attempt to raise money to bail out Julian Petroleum. On May 13, Berman surfaced in New York, where reporters discovered he had taken up residence at a friend's East Side apartment. By the time reporters appeared on the scene, however, Berman had fled in a chauffeur-driven automobile. The following day Berman telephoned a Los Angeles newspaper and reported that he was in New York for a vacation. "I am not in hiding and at any time the authorities wish to reach me they can do so through my attorneys," advised Berman. The "bright boy" promptly disappeared again.

On May 21, his attorney revealed that Berman had embarked on a "pleasure trip to regain his health and prepare for the battle that will free his name from any irregularities." Rumors spread that Berman had sailed for Europe on the liner *Berengaria*. Los Angeles district attorney Asa Keyes alerted Scotland Yard, which posted detectives to meet the *Berengaria* at Southhampton. Berman failed to materialize. French authorities repeated the scene when the ship arrived at Cherbourg but failed to find Berman. Berman's attorney confirmed that Berman had indeed sailed on the *Berengaria*, but had evaded capture and remained at large in Europe.

District Attorney Keyes employed Burns Agency detectives to track down Berman on the Continent. The investigators picked up Berman's trail in France and followed it to Austria and Italy before locating him back in Paris on June 13 where he had rented an "elaborately furnished apartment" in the luxurious Hotel Goya. Burns detectives reportedly had Berman under "constant surveillance." California officials launched extradition proceedings.

Berman's attorney claimed that his client ardently wished to return. "He has the interests of the stockholders at heart," promised his lawyer. "When he does come back he will reveal the names of several persons who made large profits and will be compelled to make restitution." Nonetheless, by June 17 Berman had vanished once again. French police reported that despite round-the-clock observation Berman had chartered an airplane and clandestinely departed France for an undisclosed destination. Burns agents found Berman in England, where he boarded the steamship *Paris* en route to the United States. On June 22, a cordon of New York City police officers and Burns detectives awaited the arrival of the *Paris* in hopes of apprehending Berman. The "bright boy" employed a ruse and once again evaded capture.

The following day Berman dispatched telegrams to Keyes, the receivers, and several Los Angeles newspapers explaining his flight. Berman contended that he could not return to California "because I have been threatened with assassination and because I do not intend to be made a scapegoat." Berman wired the receivers that he was willing to recover $8 million from usurers and others. For this service, he hoped, "proper consideration should be shown me." With these communications Berman dropped from view.

The prolonged hunt for the former Spring Street wizard triggered numerous rumors of his whereabouts and activities. Lewis alleged that several Los Angeles stockbrokers had wired Berman to travel to Russia and remain there until "this thing blows over." Newspaper articles emphasized the extravagant nature of Berman's travels and lodgings and circulated reports that the amount of money carried in the "chamois belt around his waist" totaled, not $625,000 as originally believed, but $10 million. Another report had Berman shipping trunks full of gold or currency to Europe and the Orient and stashing his booty in various international caches.

In late July, New York City police again thought they had Berman cornered. Detectives staked out an apartment house where they believed he might be hiding. Newspapers predicted an imminent arrest. Further investigation revealed that, while Berman's pregnant wife had established residence there, the fugitive had long since fled. Authorities nonetheless believed that Berman remained in the country, as he had arranged for his Cadillac and Rolls-Royce to be shipped east from Los Angeles. On August 5, Keyes dispatched Chief Investigator Ben Cohen to New York to "take personal charge" of the hunt for Berman. Cohen spent three weeks on the East

Coast but returned on August 29 "ruefully admitt[ing] that he had failed to get sight of the fugitive."

In Los Angeles the prime anticipation centered upon the impending Julian Pete criminal prosecutions. The trial of Lewis, Rosenberg, Kottemann, and other Julian Petroleum and A. C. Wagy officials was scheduled for September 8. Although observers feared that the absence of Jacob Berman would weaken the case against his former associates, the district attorney's office expressed confidence. On the trial date, defense attorneys won a four-day continuance, but the judge indicated that he would refuse further delays without "very good reasons." The best of all possible reasons materialized the following day. On September 9, Jacob Berman surrendered in San Francisco.

Berman's return caught most of Los Angeles by surprise, but the district attorney's office had apparently negotiated his reappearance during Ben Cohen's New York excursion. Chief Deputy District Attorney Harold Davis awaited Berman in San Francisco, and Berman dictated a 150-page statement to him before meeting with the local press. Berman regaled the reporters with tales of his life on the run. "Boys, I'm glad this jumping around is done," he grinned. "At first it gave me a thrill and was exciting. That was the game in it of course." He recounted attending a banquet in honor of Charles Lindbergh at the American embassy in Paris after the aviator's transatlantic flight and boasted that he had borrowed a match from a detective watching for him outside his wife's New York apartment. Berman said that he narrowly escaped death when his automobile overturned on a New England road. His surrender was delayed by both his recovery from injuries sustained in the accident and the birth of his son several days earlier.

Berman vehemently denied possession of the missing millions. "I am flat broke; penniless in fact," he told reporters. "There will be no recovery from me for none of that bunch of millions stayed in my pocket. The bankers and brokers got it all. . . . I expected to make a fortune; instead I am heavily in debt." The cash stashed in his money belt when he left Los Angeles totaled $67,000, "only enough for traveling expenses." Only the support of good friends had enabled him "to dodge so long, engage an attorney, or live in good hotels."

Deputy District Attorney Davis rejected speculation that his office had offered Berman immunity in exchange for his cooperation. Nonetheless, Davis was enthusiastic about Berman's testimony: "I was startled by the revelations in Berman's statement . . . and now realize that the Julian case is really in its infancy. The real story is yet to be unfolded." Berman's account, promised Davis, "clears up a lot of mysteries and uncertainties."

Berman's return moved him to center stage. The *Times* described the "boy 'Ponzi' of the Pacific Coast" at one court session as "jaunty and dapper . . . nattily dressed in a brown checked suit, [with] a tie of pinkish hue, and a soft brown felt hat. He was as debonair as pictured during the time he was fleeing over two continents." Out of the shadows and relieved of his odyssey, the vain, egocentric "bright youngster" stood poised to seize the spotlight.

In Berman's absence, Lewis had increasingly begun to shift the blame for the swindle to his confederate. Angered by this betrayal, Berman now turned on Lewis. In his San Francisco press conference, Berman unequivocally laid responsibility for the stock overissue at Lewis's feet and alleged that Lewis had urged him to flee Los Angeles. Lewis responded that Berman was a "Judas" attempting "to clear his own skirts" and stated that in the months prior to the Julian Pete crash, Berman had wired millions of dollars to his relatives in New York. Berman, in turn, declared that Lewis had known of the overissue for over a year, and "furthermore he over-issued the most of it. . . . He created a new block of stock every time he needed money to meet a payroll or for any other purpose."

Berman's reappearance nonetheless offered Lewis a reprieve of sorts as the Julian Pete defendants received a further continuance of their trial. Berman's appearance also renewed anticipation that illicit profits might be wheedled from the Julian moneylenders. The much-heralded "put back" campaign had once more become mired in obstruction as Haldeman and Hollingsworth had again rescinded their restitution and others refused to compromise their accounts.

City Prosecutor Lickley, recovered from his surgery, attempted to assist Scott and Carnahan with additional misdemeanor usury suits against moneylenders and pool participants. Lickley filed charges against what the *Times* described as the smaller fish of the "more or less 'pike' variety," many of whom had evaded grand jury indictments. By October 6, Lickley had filed 133 suits. Due to a relatively brief one-year statute of limitations on usury cases, Lickley issued many of these charges as "John Doe" complaints, to which names would be appended at a later date. Since most documents and affidavits pertaining to these charges were tied up in the criminal prosecutions, judges delayed all usury cases pending completion of the Lewis-Berman case.

Meanwhile, District Attorney Keyes unexpectedly announced on September 19 that he would recommend dismissal of usury conspiracy indictments against defen-dants who "put back" their profits. "This office recognizes the extreme need of the receivers for assistance in their efforts to protect the life savings of 40,000 or 50,000 shareholders from actually being wiped out," stated Keyes. Within days, several of the most prominent Julian Pete pool participants accepted Keyes's offer. Louis B. Mayer, Joe Toplitzky, and eight other investors in the Bankers' Pools refunded their profits in exchange for a dismissal of charges. "I never had any doubt as to the outcome," stated a less-than-repentant Mayer. "I am told the dismissal was based upon a thorough investigation . . . in which my innocence was established."

These efforts notwithstanding, on September 30 the receivers confessed that California-Eastern hovered on the brink of bankruptcy. Refunds totaled less than $500,000, and fewer than fifty of the alleged four hundred moneylenders had vol-untarily refunded their profits. Of these, only a dozen had deposited sums in excess of $10,000. Despite the fact that field workers had delivered ten new wells since May, declining oil prices had reduced revenues. The receivers needed an additional $150,000 to meet the October payroll.

Upon his return Berman had pledged to recover $1 million within ten days, but his promises dissipated amidst a bitter feud with the receivers. Scott and Carnahan wanted Berman to assist them in preparing usury suits on behalf of the receivership. Berman's attorneys insisted that since their client had acted as an individual in moneylending ventures, rather than as an agent of Julian Petroleum, all suits should be filed on his behalf, with proceeds turned over to California-Eastern.

The conflict raised other issues as well. Berman's attorney charged that the receivers had settled with certain lenders for "only a drop in the bucket" in relation to the usurious profits they had made. Scott retaliated that Berman's attorneys were simply attempting to protect their fees, which would be higher if Berman recovered the funds as an individual. Berman, furthermore, had demanded that the receivers release him, his father-in-law, and his brother-in-law from all claims against them, as a condition for his cooperation. "We replied that we could not pay Jacob Berman for telling the truth," said Scott.

In mid-October the warring parties declared a truce. The receivers agreed to allow Berman to file personal usury suits in certain instances, while Berman promised to extend his full efforts to assist California-Eastern. On October 13, Carnahan expressed optimism that the agreement would expedite a significant collection of funds. Two weeks later, with the criminal trial of Lewis, Berman, and others slated to begin, Carnahan supported defense requests for yet another delay, arguing that the cooperation of Berman and Lewis was essential to the receivership. "We hope to recover some $12 million to $13 million with their aid," testified Carnahan. Judge William Doran, who would preside over the trial, acquiesced. Doran set the new trial date for January 5, 1928.

## II

During the years in which the community's bankers, brokers, businessmen, and swindlers had juggled Julian Petroleum stock with the aplomb of circus performers, Los Angeles County had erected an $8 million white marble tower on the downtown corner of Temple and Broadway streets to house its Superior Court. On the morning of January 5, 1928, almost seven months to the day since the collapse of Julian Petroleum, hundreds of people flocked to the new courthouse to witness the opening of the trial of those deemed most responsible for the Julian Pete debacle.

The ten defendants included the prime stock manipulators and senior officials of Julian Petroleum and Wagy & Company: S. C. Lewis, Jacob Berman, and Ed Rosenberg; Julian Petroleum vice-president H. F. Campbell and secretary T. P. Conroy; A. C. Wagy president Raymond Reese and his father, Wagy vice-president Charles Reese; and accountant William Kottemann. Berman's brother, Louis, and Pacific Southwest banker I. L. Rouse rounded out the group. All ten were charged with conspiracy to violate the Corporate Securities Act and obtaining money under false pretenses. Most observers agreed, as Guy Finney later wrote, that "on the outcome of the trial . . . hinged the fate of the larger group awaiting trial . . . it being generally conceded that failure to convict the . . . principals would greatly lessen the probability of hanging a guilty verdict on the others."

The defendants had engaged several of the city's foremost attorneys on their behalf. William Neblett, partner of perennial presidential hopeful William Gibbs McAdoo, would represent Conroy and Campbell. Rosenberg had employed Republican influence-peddler (and former Julian pool member) Mendel Silberberg and onetime United States attorney Dudley Robinson. Two venerable and loquacious defense attorneys, LeCompte Davis and Joe Ford, were on hand to defend Rouse. Jacob Berman imported John Murphy from New York. S. C. Lewis would defend himself, as promised. Congestion at the defense table necessitated a one-hour delay at the trial's opening as attorneys and court officials tested alternative seating arrangements to accommodate the crush of lawyers and clients.

Arrayed against the defense armada stood the prosecution team assembled by controversial district attorney Asa Keyes. "Ace" Keyes (pronounced *Kies*) had served in the district attorney's office for a quarter of a century and had headed the agency since 1923. Few local political figures so divided public opinion. A 1927 biographer praised him as a native son (the first district attorney born in Los Angeles County) whose "popular confidence and esteem" offered a "pretty sure measure of the caliber of a man and his work in the community." The *Los Angeles Examiner* and *Record* regularly lauded his regime. The *Times*, on the other hand, had vehemently opposed Keyes's election and repeatedly attacked his administration for incompetence and dishonesty, charging that he made deals with the criminal underworld. Crusading evangelist Bob Shuler joined in these assaults. Shuler described Keyes as "putrid and filthy" and his office as "contemptible and nauseating . . . so full of perjury that it smells to heaven."

Keyes had a convincing rejoinder for his detractors. The *Times*, he said, had "marked [him] for vilification" because he would not do its bidding. Most southern Californians accepted this defense, and, at the dawn of the Julian Pete prosecution, Keyes "stood high in public favor," according to Finney.

Keyes had always displayed a keen eye for publicity. As a deputy district attorney he had won acclaim for his handling of the 1921–1922 murder trial of Madalynne Obenchain, a classic femme fatale accused of murdering a prominent socialite. Keyes impressed one observer with his "logic, good manners, and occasional flashes of wit." When District Attorney Thomas Woolwine resigned due to ill health in June 1923, Keyes, his most celebrated deputy, was appointed to succeed him. The following year Keyes won election in his own right.

In 1926, Keyes received nationwide attention for his indictment of evangelist Aimee Semple McPherson. McPherson had disappeared while swimming at Santa Monica beach and resurfaced in the desert two weeks later, claiming to have been kidnapped. When evidence mounted that the popular preacher had spent a romantic fortnight in a "love cottage" in Carmel with a married church radio engineer, Keyes charged "Sister Aimee" with filing a false police report. The preliminary hearing on the case dragged on for two well-publicized months. Several weeks later Keyes abruptly dropped the charges, to the dismay of some, but to the relief of most others. Keyes also reopened the most puzzling unsolved murder case in Los Angeles annals, the 1922 shooting of film director William Desmond Taylor. Claiming the discovery of fresh evidence, Keyes chartered special train cars to the East in pursuit of new leads. The "tax-paid junket," lavishly covered by the press, allegedly bore fruit, but the new evidence mysteriously disappeared from a police locker.

While preparing the Julian prosecution, Keyes immersed himself in the spectacular "Hickman Horror" case. In 1927, young William Hickman had kidnapped 12-year-old Marion Parker, the daughter of a Los Angeles bank manager. He demanded $1,500 for the girl's return and signed his notes "The Fox." The father paid the ransom, only to find the mutilated body of his daughter, with her arms and legs sliced off, wrapped in a blanket nearby. The discovery terrorized southern California. Hickman evaded capture for several days before being apprehended in Oregon. Keyes took personal responsibility for his prosecution. Although Hickman's trial would run simultaneously with the Julian Pete case, Keyes promised to also supervise the overissue prosecution.

Keyes assigned his top lieutenant, Harold "Buddy" Davis, to try the Julian Petroleum case. The choice raised several eyebrows. Four years earlier Buddy Davis had been a newspaper reporter attending law school in his spare time. In June 1924, he had signed on as Keyes's secretary. Upon passing the bar several months later, Davis became a deputy district attorney. Shortly thereafter, Keyes appointed the inexperienced Davis chief deputy. Many saw this as a blatant example of cronyism, a view reinforced in late 1927 when a resigning deputy cited Davis's "gross inefficiency" as the cause of his departure. Now the inexperienced Davis would handle the important and sensitive Julian Petroleum prosecution. To assist him Keyes assigned Julian Richardson and E. P. Van Cott.

Many questioned whether the public legal team could match the high-priced talent assembled by the defendants. Others, like Shuler, questioned the prosecution's sincerity. A public uproar after the dismissal of criminal usury charges against Mayer, Toplitsky, and others who had "put back" their profits had already embarrassed Keyes, and the grand jury had reindicted those whom Keyes had forgiven. "The District Attorney will stall and cavort about in an effort to make a great horseplay performance and thus seek to impress the public with the idea that he is making an honest prosecution," predicted Shuler. Lewis and his lieutenants might be sacrificed, said the minister, but not "a single financier" would be "'stuck' for a fine."

The early days of the trial presaged a lengthy siege. Officials transported over two tons of documentary evidence to the courthouse. Over 100 witnesses received subpoenas. Buddy Davis promised that the trial would reveal "startling" new disclosures of stock manipulations. For two weeks the battery of lawyers struggled to seat an acceptable jury. The large number of defendants and the difficulty in identifying jurors who had no Julian Pete stockholders among their friends and relatives or who had not formed opinions about the case lengthened the process.

On January 18, attorneys finally agreed on a jury. "The box is filled with aging men and women. Several of them are nearing the 'seer [sic] and yellow' period when inactive folk like to sit by the fire and browse," Guy Finney reported. "For the most part, they have been on juries before, and rather like the work. It gives them something to do . . . and besides there is the two dollars a day. . . . It is not much, but it will help."

The following day the prosecution called C. C. Julian as its first witness. "Did you have anything to do with the organization of the company?" Davis asked Julian. "Why certainly," replied C.C., "I had everything to do with it." The long-awaited trial was under way.

The first surprise came early. On January 20, District Attorney Keyes strode into the court for only the third time since jury selection had begun and addressed Judge Doran. Keyes moved to dismiss the charges against Louis Berman, Raymond Reese, and, most unexpectedly, Ed Rosenberg, one of the prime architects of the overissue. Speaking on behalf of Rosenberg, Keyes stated, "There is insufficient evidence in the indictment to show a crime was committed. I believe this man to be innocent." Keyes indicated that at least two of the men might appear as prosecution witnesses.

Keyes' motion stunned not only the audience but his own assistants. Deputies Van Cott and Richardson, according to the *Times*, "stiffened in their chairs at the counsel table as Keyes spoke, looked surprisedly at each other, and then glanced at their chief." Judge Doran, unpersuaded by Keyes's argument, took the motion under submission. Keyes, having made his plea, immediately left the courtroom, leaving the prosecution to his demoralized staff.

During the next several weeks, the Julian trial "advanced with all the delayed action of a slow-motion camera," according to Guy Finney. From January 24 to February 16, the accountant who had supervised the audit of the Julian Petroleum books waded through a morass of financial minutiae tracing the origins and evolution of the stock overissue, then withstood an unusual, and ponderous, eight-day cross-examination by Lewis. Exhibiting no visible sense of irony, Lewis attempted to discredit the testimony on the grounds that the Julian Petroleum books and records were so erratic and unreliable as to render any audit based on them invalid. Day after day Lewis repetitiously challenged the accuracy of each stock certificate.

Throughout February, March, and early April, a procession of former Julian Pete employees testified about irregularities, forged signatures, and overissued stock certificates. These accounts largely repeated previously published grand jury testimony. Lewis, acting as his own attorney, doggedly cross-examined each prosecution witness. "Your questions are entirely out of order and your emotion and enthusiasm is a little overdone," admonished Judge Doran on one occasion. Other defense attorneys took a far less active role.

Rumors persisted that Rosenberg would turn state's evidence once Judge Doran had ruled on Keyes's earlier motion to dismiss charges. Doran, however, remained silent on the matter until the prosecution completed the presentation of its case on April 11. The following morning Doran denied the motion for dismissal, ruling that sufficent evidence existed to allow the jury to decide the case. The trial now shifted to the defense.

S. C. Lewis was the first of the accused to present his case. On April 16, he made a long opening statement, placing the blame for the overissue stock, "if it was overissued," on Jacob Berman and the employees of the stock transfer department. Lewis subpoenaed twenty-eight witnesses on his behalf, most notably brokers C. C. Streeter, H. J. Barneson. A. C. Wagy, and H. B. Chessher. But the brokers, whom Lewis hoped to lay much of the blame upon, had all conveniently left the state. Only Streeter would ever testify.

Lewis called LASE secretary Norman Courtney as his first witness. Courtney, who had assumed his post on May 18, 1927, two weeks after the crash, provided one of the few trial revelations. He reported that the LASE had incinerated all records of Julian Pete transactions one month after the exchange had halted trading in the stock. Courtney admitted that "there was grave concern over the liability of those brokers who had been dealing in supposedly overissued stock" but denied that this had led to the destruction of the records. "There was an accumulation of possibly five years of records lying about and being kicked around the exchange and I just had them dumped into a draywagon and ordered them taken to the city incinerator to be burned," he explained.

While Lewis attempted to established his innocence, attorneys for most other defendants had decided to rest their cases with no or little testimony. From the start they had emphasized, in the words of attorney Joe Ford, that the defendants were "not being tried on a charge of responsibility for wrecking the Julian Petroleum Corporation, but are accused of conspiring to overissue the company's stock." Ford advised the jurors "to keep that clear in mind and remember the difference." Ford also hammered away at the distinction between a conspiracy to commit fraud and innocent misrepresentation with no criminal intent. Both jurors and courtroom observers found this subtlety hard to grasp. "Legal definitions may make a conspiracy seem a vaguely unreal charge and difficult of proof," warned journalist Finney. The defense attorneys hoped that this confusion would benefit their clients.

After Lewis had completed his defense, his former confederates quickly rested their cases. Jacob Berman's attorney took less than five minutes, calling but one witness and presenting a single document. Rosenberg and Louis Berman offered no defense. Kottemann's lawyer called just three witnesses to testify favorably on his audit of the A. C. Wagy books. Before any further defendants could be heard, Judge Doran surprised the court by abruptly dismissing charges against former Julian vice-president H. F. Campbell, A. C. Wagy president Charles Reese, and Louis Berman. Following this action the remaining defendants rested their cases.

On May 2, after almost four months of trial, the prosecutors began their final appeal to the jury. For four days District Attorney Richardson laid out the county's case and vilified the remaining seven defendants, reserving particular ire for S. C. Lewis. The former Julian Petroleum president countered Richardson with an impassioned three-day plea on his own behalf. Lewis described himself as a "martyr to a lost cause," a man so devoted to "developing and pushing the company forward" that he had failed to see the illicit activities of Berman and his cronies in the stock transfer department. Lewis asked the jurors to exonerate Rouse, Conroy, and Rosenberg, all of whom he stated were guilty of no wrongdoing. As he reached the conclusion of his address Lewis "broke into a near hysterical crying spell" in asking for a verdict of "innocent."

Most of the remaining defense attorneys, in keeping with their strategy, made arguments lasting less than an hour. Only LeCompte Davis, representing Rouse, made an extended plea. Davis's oration, which occupied an entire day, defended not only Rouse, who he asserted had acted under the orders of his superiors at Pacific Southwest, but the other accused men as well. Davis was particularly moving in expressing his feelings for Lewis. "I say now that while I am not representing [Lewis] he can count on me as his friend, for I believe that the evidence shows he is an innocent man," stated Davis.

The conclusion of the defense arguments set the stage for the grand finale, the closing address by District Attorney Asa Keyes. Several days earlier Keyes had announced that he would not stand for re-election but would retire from public service and enter private practice. The Julian Pete argument, he told the jury, might well be his final courtroom address as a prosecutor.

Keyes had barely attended the trial, and upon his courtroom visits had seemed distracted and uninterested. On one occasion, Judge Doran had halted the proceedings to admonish Keyes. "I want the record to show," stated Doran, "that District Attorney Keyes was present in this court while [Deputy] Davis was arguing with a difficult legal problem without once offering to give him the benefit of his knowledge or assistance." But even Keyes's detractors recognized the eloquence and abilities he had demonstrated in scores of criminal trials during his twenty-five years as a public prosecutor.

Keyes amply displayed his talents in his excoriation of S. C. Lewis. "He issued stock like sausage coming out of a mill," charged Keyes. "Day after day, hour after hour, minute after minute, this stock came out and Lewis, the president, in full charge of the operation and management of that concern, was responsible for that overissue." Even as he flayed Lewis, Keyes astounded the courtroom with his comments on the other defendants. Keyes declared Rosenberg and Reese totally innocent and expressed doubts about the culpability of Conroy and Kottemann. He mentioned Rouse only once during his argument and placed minimal emphasis on Berman. At times Keyes directly contradicted the evidence advanced by Richardson, his subordinate. After a scant two hours, Keyes concluded his baffling address.

The bewildered jurors listened to Judge Doran's instructions and then, at 2:00 p.m. on May 22, 1928, retreated to the jury room for their deliberations. The Julian Petroleum trial had lasted four and a half months, the longest in the history of the county. Ninety-nine volumes of trial transcripts had been recorded. The investigation and prosecution had cost the county in excess of $250,000. Southern Californians anxiously awaited the outcome.

The following day at 3:00 p.m. the jurors notified the court that they had made their judgements. Defendants and their attorneys filed into the courtroom. The court clerk rose and read the verdicts. The jurors, he announced to the hushed crowd, had found all seven remaining defendants, including S. C. Lewis and Jacob

Berman, not guilty. Amidst tearful, emotional celebrations, onlookers absorbed the impact of the verdict: One year and eighteen days after the collapse of Julian Petroleum, the perpetrators of the great Julian Pete swindle had walked off scot-free.

## III

One month before the dénouement of the Julian Pete trial, three thousand miles away in Washington, D.C., another jury deliberating on a more celebrated oil industry case had reached a similar verdict. In April 1922, shortly before C. C. Julian had secured his first lease at Santa Fe Springs, United States Secretary of the Interior Albert B. Fall had leased federal oil reserves at Teapot Dome in Wyoming to Harry F. Sinclair's Mammoth Oil Company. Shortly thereafter, Fall offered a similar arrangement to Los Angeles oilman Edward Doheny at the California Elk Hills Reserve. As the nation would soon learn, for his largesse Fall had received $360,000 in loans and liberty bonds from Sinclair and Doheny. On April 10, 1928, Sinclair stood trial for conspiracy to defraud the government on the Teapot Dome deal. Twelve days later, Sinclair was acquitted. United States Senator Gerald P. Nye commented, "This is emphatic evidence that you cannot convict a million dollars in the United States."

The Julian Pete verdict and its aftermath drove home Nye's point to southern Californians. *California Graphic* described the outcome as "preposterous" and a "travesty." "Is this monstrous crime forever to go unpunished?" asked the *Los Angeles Express*, which called the "fantastically ludicrous" trial "a remarkable indictment of the American system of dealing with criminals." On his radio show, Bob Shuler declared, "I consider the same rule works in Los Angeles as in Washington, New York, Chicago or elsewhere in our unfortunate country. It is virtually impossible to convict men of large financial holdings, it matters not as to their crimes."

Even Flora Pyle, the jury foreman, seemed embarrassed. "We felt you would be angry at us for rendering this verdict," she told Judge Doran. Pyle and other jurors argued that their decision was dictated by the conspiracy indictments. "We felt there was a great wrong done the public by some of the men mixed up in the Julian matter," stated Pyle, "but we all agreed that we could not convict any of these men on conspiracy under the indictment. . . . Only one who sat through the trial can realize what a hodgepodge was made of the prosecution." Another juror asserted, "The way the indictments were drawn made it that we had to find the defendants themselves conspired to sell the overissue. Well, the district attorney's office just didn't present convincing evidence of that kind."

Judge Doran, in an unorthodox trial post mortem, joined the chorus criticizing the district attorney. "After careful consideration and because of my conception of the functions and obligations of the office I hold, I have concluded that I cannot discharge my duties to the people of this county by remaining officially silent on the results of . . . the so-called Julian cases," explained Doran on May 24. The Justice laid the blame squarely on the shoulders of Asa Keyes:

The jury was deprived of a careful, effective analysis of the people's case, which could have been supplied by an adequate closing argument by the District Attorney. . . .

During the course of the trial the District Attorney was admonished in open court because of his apparent unwillingness to assist his deputies in the conduct of the people's case. His infrequent attendance at the trial seriously handicapped him in making a proper closing argument.

I felt that due diligence would have brought about a different verdict at least in the cases of some of the defendants who were primarily responsible for one of the most deplorable, unfortunate, and reprehensible episodes in the history of this county.

Despite the acquittal of the first group of Julian Pete defendants, forty-one men, including Lewis, Berman, Pacific Southwest officers John Barber and Charles Stern, and the Bankers' Pool élite, still stood charged with additional crimes. On May 29, six days after the trial had ended, Judge Doran dashed any lingering hopes of punishment for these men. Doran dismissed all criminal indictments save those charging Jacob Berman with embezzlement and forgery. "We have just concluded a long and costly trial," he declared. "The evidence in this trial was conclusive in that it showed that a gigantic fraud had been perpetrated. . . . I believe that the guilt of at least two of these defendants, S. C. Lewis and Jacob Berman, was clearly established. . . ." Doran termed the loans negotiated by these two, which most of the remaining defendants stood accused for, "merely incidental" to the ultimate outcome. "Most of the defendants now before the court are charged with conspiring to do an act which is in itself a misdemeanor," argued Doran. With Lewis and Berman free, "to proceed further with their prosecution would not to my mind be justice."

The dropping of criminal charges against the Julian Pete pool members complemented an earlier decision that had deferred the danger of civil liability penalties for usury. In August 1927, after W. I. Hollingsworth and Harry Haldeman had again reneged on their promises of restitution, receivers Scott and Carnahan had filed civil suits against them asking triple damages for their Bankers' Pool profits. The Hollingsworth case became a test case. On March 5, 1928, even as the trial of Lewis, Berman, *et al.* dragged on, Judge Leon Yankovitch dealt the receivers a severe blow. On one hand, he concluded that by accepting payments to extend their agreements pool members had clearly violated the state usury law. At the same time, however, Yankovitch ruled that the Julian Petroleum receivers lacked the authority to collect monies owed the company under a penal statute. Thus, while those accused of usury stood guilty as charged, the receivers had no way to secure restitution. If Scott and Carnahan wished to pursue the matter further, they would have to either appeal Yankovich's decision or petition the federal court for expanded receivership powers.

The combined effect of the jury verdict, Doran's dismissal of criminal charges, and the Yankovich decision seemed to doom all hope of retribution against those who had plundered Julian Petroleum. Misdemeanor usury charges filed by City

Prosecutor Lickley remained in the courts, but Lickley now seemed disinclined to prosecute. "There is now no hope of any of the money ever being recovered," reported the *New York Times* in June. "Financiers and others involved in the transaction have returned from vacations and retaken their places in the business and social world." The New York newspaper invoked Senator Nye in reporting the public reaction. The people, it concluded, "[accept] as axiomatic the proposition that it is impossible to convict a million dollars."

While the overissue trial dominated public interest, C. C. Julian wove in and out of the headlines. In late 1927, he had revived New Monte Cristo Mining, his latest promotion. To woo California investors he dispatched an eighteen-page single-spaced letter extolling its virtues. Julian's latest offering drew the satirical barbs of one Eugene Brown, who, in an "Open Letter" published in the *Times*, complained:

> Sometimes I think you do not love me for myself alone. You never seem to write me except when you need more money. . . .
>
> You are the most persistent worker I have met. You will take a man's money the ninth and tenth time in order to give him a chance to get even. You are as good-hearted as a roulette wheel in that respect. . . .
>
> You seem to be making a merger of all the liabilities you have behind you and listing them as a $10,000,000 asset. That is what we call high finance.

Julian's venture into "high finance" also attracted attention from the Corporations Department, which called upon District Attorney Keyes to investigate. Keyes' office acknowledged that Julian's activities were "undoubtedly ill-advised and constituted a technical violation of the Corporate Securities Act" but expressed doubt "whether a prosecution of Mr. Julian . . . would be successful or effective at this time," and declined to take action. Julian, meanwhile, seemed to have lost both his golden touch with investors and his good humor. A second mailing protested: "if the response I have received on the offering of these bonds up to date is a criterion of the support I can expect from you, you have lifted from my shoulders a great moral responsibility, for you are not worthy of the extreme, untiring effort, and the unselfish policy I have pursued in carrying on the operations of your company to its present stage."

New Monte Cristo faced an even more serious problem in Arizona. On December 29, 1927, minority shareholders in the original Monte Cristo Mining filed a complaint demanding a receivership for the corporation "to avert impending insolvency and disaster to said company and great and irreparable loss to its shareholders." The dissidents charged that Julian had purchased Monte Cristo "to further his fraudulent machinations and schemes in the sale of a huge amount of stock to the general public," rather than to "operate its property for the benefit of its shareholders or to develop said mine in a systematic or scientific manner." The suit cited evidence of "gross and fraudulent management" and "ruthless and reckless waste and squandering of money and property of the company and wanton malfeasance." Development work at the mines was said to have halted, and water had flooded the shafts.

Other court cases and hearings plagued Julian as well. A breach-of-contract suit by Jacques Van Der Berg, the "Dutch baron" who had helped Julian procure the Monte Cristo Mines, exposed Julian's unethical tactics. Several disgruntled investors also sued Julian. In addition, the public received a glimpse of the tormented affairs of the Julian household. On March 23, Julian committed his wife, Mary, to the county psychopathic ward and removed his teenaged daughters from their Los Feliz Boulevard home. Julian's formal insanity complaint alleged that Mary "throws anything she can lay her hands on and is quite destructive. She is a menace to herself, her family, and society in her present condition." Friends of Mary Julian's hired an attorney and arranged her release pending a hearing before the "Lunacy Commission." After five doctors adjudged Mary Julian sane, Julian countered with an affidavit of intemperance, stating: "She has been on a continual debauch for months, using whiskey to such an extent that she is beyond the point of reason. . . . She has been using stimulants to such an extent that she is a moral wreck and totally lost to all sense of decency or of her duties to her children and family."

At public hearings Julian stated that the proceeding marked the first time he had seen his wife sober in four years. He described Mary attacking him with fire irons and driving his friends from their house in "fits of wild drunken rage." Doctors who had examined Mrs. Julian declared her sane, but several cited indications that she suffered from either drink or drugs. When put on the stand Mary could not explain her husband's actions. "I can't believe that he is in his right mind to try to do this terrible thing to me," she testified. She nonetheless showed no ill will toward Julian. When her attorney attempted to prevent Julian from visiting her at the hospital, Mary replied, "I do not wish him restrained. I do not mind if he calls on me."

Julian's visits apparently resolved the situation, as the *Times* remarked, without any further "airing of the family skeletons." On March 29, Mary Julian dismissed her attorney and announced that she had placed her affairs in her husband's hands. Julian told the Lunacy Commission that Mary would take a "liquor cure" and had "taken the pledge never to drink again." When asked by the judge, "Have you joined her in this pledge?" Julian confessed, "No your honor, I have not." Mary Julian said little and the pair departed the courtroom together. The Reverend Shuler, a Prohibition supporter, summed up the sad affair cruelly but accurately: "Liquor is the fountain of almost any kind of woe. Julian accused his wife of being a drunkard and insane. If she is the former she is very likely the latter. And living about him, I would think she might be both."

Yet another element in the Julian Petroleum universe took a peculiar turn during these months. In September 1927, a jury had convicted H. J. Kimmerle of assault to do great bodily harm in the attack that had cost Arthur Loeb his eye. The court sentenced Kimmerle to two years in the county jail. Following unsuccessful appeals, Kimmerle began serving his sentence on May 18, 1928. Incredibly, the next day, District Attorney Keyes and Chief of Police Jim Davis, two members of the three-man Parole Board, surreptitiously approved a parole for Kimmerle, who was freed the following day, after serving only five days. Loeb later claimed that

Kimmerle had paid $5,000 to Keyes and Davis and threatened to "expose names if he didn't get out quick." After the loss of his eye, Loeb had anointed himself the avenging angel of the Julian Pete victims, vowing to expose and punish the perpetrators. In the aftermath of the Kimmerle parole, he added Asa Keyes to his long list of those awaiting retribution.

Meanwhile, S. C. Lewis and Jacob Berman discovered that their recent court victory afforded them only a brief respite from their legal travails. With the collapse of the Julian Pete prosecution, federal authorities moved quickly to indict the pair for mail fraud for their earlier activities with Lewis Oil. After a summer of delay, and despite protests from the receivers who requested Lewis's continued assistance, the trial began on September 19, 1928. During the next month and a half, witnesses imported from various parts of the United States and Canada re-created the high-pressure sales tactics and fraudulent promises made by Lewis and Berman in their 1925 to 1927 Lewis Oil Gold Note campaign. On October 30, the jury took only seven hours to convict Lewis and Berman of fifteen counts of using the mails to defraud investors. One week later a federal judge sentenced the former Julian Pete president and the "bright youngster" who had assisted him to seven years each in federal prison.

Although the Lewis Oil prosecution took place in Los Angeles, the trial commanded minimal attention. The conviction of Lewis and Berman for other crimes offered scant satisfaction from the disillusioned southern California populace. The bitter memory of how the Julian defendants had eluded punishment for the over-issue lingered like a disease in remission. On the day after the Lewis and Berman verdicts, the community would learn that the Julian Pete cancer had also invaded the body politic.

# 15 WELL, WHAT OF IT?

I

In the years before 1928, Milton Pike had led an anonymous existence, typical of millions of American immigrants. Born Marlo Amundsen in Sweden, he had moved to the United States, adopted a more American name, learned the tailor's trade, and traveled from city to city eking out a marginal living. In 1927, he settled in Los Angeles and secured employment as a salesman in a downtown tailor shop owned by Ben Getzoff. Each day Pike would work in the front of the store with tailors Joseph Sherman and John Rittinger, while Getzoff ran the business from a room in the rear. Although partitions obstructed Pike's direct view into Getzoff's office, reflections from a three-flanged mirror used for alterations and fittings allowed Pike, Sherman, and Rittinger to observe the events therein.

For Pike's boss, Ben Getzoff, the shop at 609 Spring Street, centrally located near the financial district and the courts, proved an ideal spot for both his vocation as a tailor and his avocation: currying favor with local politicians and public officials. Guy Finney described Getzoff as "an illiterate, weasel-like type of man in his middle fifties." To Detective Leslie White, he resembled a "shrunken old gnome." Frail and in ill health following three stomach operations, he always kept a healthy supply of "medicinal" alcohol on hand to share with his visitors. Among the most frequent patrons of this inner sanctum was District Attorney Asa Keyes. Keyes, a well-known "secret drinker" in Prohibition Los Angeles, often dropped in for a heavy dose of Getzoff's "medicine."

In the early months of 1927, Getzoff had become friendly with Ed Rosenberg and invested heavily in Julian Pete securities. When the stock crashed in May, Getzoff lost $25,000. He paid visits to Keyes and other politician friends and angrily demanded retribution, but found that amidst the prevailing confusion they could offer no help. Nonetheless, as the 1928 trial of the Julian Petroleum defendants had drawn nearer, Rosenberg again began to frequent the Getzoff tailor shop. Shortly after the trial commenced, Jacob Berman also paid call. Getzoff seemed to choreograph the visits so that Berman and Rosenberg never collided with Keyes. Pike grew intrigued by the procession. After the Julian acquittals he began to keep a written record on scraps of paper or the backs of business cards of the comings and goings, the conversations he heard, and the scenes he viewed through the mirror. Later he transcribed these notes into a single bound book.

Pike worked for Getzoff until July 18, 1928, when Getzoff sold the tailor shop. Two weeks later Pike bumped into Joseph Sherman. The two unemployed men expressed a shared distaste for their former boss. Pike suggested how they could gain revenge and make a profit on the side. He told Sherman of his diary and said that with Sherman's corroboration someone might be willing to pay handsomely for the contents.

Pike first attempted to sell his diary to the local Hearst newspapers, but these journals, supporters of Asa Keyes, declined to publish the story. Pike and Sherman next approached the Reverend Shuler, who they thought might pay to use their story over his radio station. The pair may also have attempted to blackmail Getzoff and Rosenberg. In the end, the *Times*, a long-time nemesis of Asa Keyes, offered Pike $2,000, contingent on publication of the diary.

Meanwhile, Arthur Loeb heard of Pike's attempts and brought Pike and Sherman to the attention of the county grand jury. Evidence exists that the Reverend Shuler also played a role in getting the tailors to testify. The *Times* apparently agreed to withhold its exposé, and the grand jury, which usually worked hand-in-glove with the district attorney's office, launched a probe cloaked in "utmost secrecy."

At 8:00 p.m., on the evening of October 31, 1928, just one day after S. C. Lewis and Jacob Berman had been found guilty of federal mail fraud violations, the grand jury filed from its deliberations room. Remarkably, no word of its sensitive investigation had leaked out. Before a stunned assemblage of reporters and courthouse workers, the grand jurors unveiled the earthshaking contents of the Pike diary. The jury alleged that the Spring Street tailor shop had been a brokerage house for bribery. Asa Keyes, with Ben Getzoff acting as middleman, had accepted sundry gifts and tens of thousands of dollars to guarantee acquittals for Ed Rosenberg and Jacob Berman.

"The ghost of the Julian Petroleum swindle would not down!" lamented Guy Finney. "Corrupt money . . . had revived the nauseous thing." The Pike diary, now excerpted in the *Times* and other newspapers, revealed a sordid saga of palavers, pay-offs, and drunken celebrations in the Getzoff tailor shop. Pike, whose laconic commentary added elements of irony to his chronicle, described angry arguments among the conspirators as Getzoff attempted to squeeze Berman for additional funds, and the "bright boy" tried to pay off his debts in bonds rather than cash. Commenting on Getzoff's burgeoning career as a bribery broker, Pike wrote, "Ben seems to be the busiest man in town. He seems so filled with big business, he wouldn't give the Prince of Wales a look."

In the early phases of the conspiracy, according to Pike, Rosenberg and Berman "would run like rats downstairs" to hide when Keyes entered the store. Following the trial, the conspirators became emboldened, meeting openly and addressing each other by first names. Pike old of gifts of an expensive wristwatch and a beautiful smoking jacket. "It looks like Keyes (who had announced his retirement) is going after all the easy money he can get before his term expires," concluded Pike.

Pike also revealed that county's chief law enforcer had blatantly violated the Prohibition statutes. One entry referred to "twelve quarts of booze" placed in Keyes's car for a two-day trip. Another spoke of a "bagful of booze." Pike described Keyes and Getzoff emerging from the shop one morning at 8:00 a.m. "both half-lit." After the Julian Pete acquittal, wrote Pike, "all the gang were present," and Rosenberg announced, " 'I am giving a big party tonight.' Congratulations were numerous and boisterous followed by plenty of drinks."

Based on the evidence in Pike's diary and corroboration from Sherman and Rittinger, the grand jury charged Asa Keyes with "willful and corrupt" misconduct in office for accepting bribes from Berman, Rosenberg, and A. I. Lasker, a local insurance man whose grand theft charges Keyes had dismissed. The jurors also filed additional counts for transporting liquor and issued criminal indictments against Jacob Berman; Ed Rosenberg and his brother, Jack; Ben Getzoff and his son, Dave; David Reimer, a detective in the district attorney's office; and Lasker.

The revelations, according to the *Record*, "broke like a bombshell." For the people of Los Angeles, wrote Finney, "a public idol of long-standing had been shattered." The atmosphere was flooded with ominous rumors and accusations as suspicion fell on other patrons of the Getzoff tailor shop. "The confidence of the public in its elected servants, the foundation stone of the community, has been undermined to the danger point," warned the *Times*, which feared that the failure to convict the Julian defendants had "bred in the public mind a dangerous contempt for the agencies supposed to protect the public against despoilers."

Keyes adamantly denied the charges. "I know my skirts are clean and that the confidence which thousands of my friends have put in me . . . will show itself to be fully justified," he declared. "Just think, I have arraigned thousands of men during my twenty-five years in service of this county . . . and yet I have been accused on testimony so flimsy as to repulse the common sense of any individual!" Keyes welcomed "the opportunity to appear before an American jury to disprove any false charges that have been concocted against me." He refused to resign and announced he would serve out the remainder of his term.

Despite these denials, the web of accumulating evidence ensnared Keyes ever more tightly. On November 17, the grand jury returned new indictments charging additional instances of bribery. Former deputy district attorney and Keyes protégé Buddy Davis joined the list of those testifying against the embattled prosecutor. Davis told the grand jury that Keyes's performance at the Julian trials and his insistence on Rosenberg's innocence had caused the deputy to offer his resignation. "I do not care to be a party to the insincere prosecution of the Julian cases," Davis had written to Keyes on January 10, 1928. Keyes refused to accept Davis's withdrawal. Ten days later, as they walked to the court, Keyes shocked Davis with the news that he was about to ask for dismissals of Rosenberg and two others. Only Judge Doran's refusal to grant the motion prevented Davis from quitting.

To add to the drama, Keyes would be prosecuted by his colorful and controversial successor as Los Angeles County District Attorney, Buron Fitts. Born in Texas and raised in Oklahoma, Fitts moved to Los Angeles as a teenager in 1906. Ten years later he graduated from the University of Southern California Law School. World War I interrupted Fitts's private law career and transformed his life. Fitts enlisted on the day the United States declared war and received a commission as a second lieutenant. At the battle of the Argonne in 1918, Fitts's right knee was blown away, and he suffered mustard-gas burns. Confined to a military hospital for over a year, Fitts ultimately underwent twelve operations to save his leg.

Fitts emerged from the war a decorated and celebrated hero. He immediately became active in the growing veterans' movement. A charter member of the American Legion, he actively lobbied for state and federal relief measures for war veterans, widows, and orphans. In 1920 he became commander of the California branch of the American Legion and two years later headed the California Veterans' Bond campaign to enhance benefits for returned soldiers. In addition to his American Legion activities, Fitts joined the district attorney's staff and rose rapidly through the ranks. In 1924, when Keyes became district attorney, he named Fitts his chief deputy.

Renowned as California's "most prominent veteran" and ardently supported by the increasingly influential American Legion, Fitts launched a political career in 1926. Fitts rode his record as a war hero to victory in the lieutenant governor's race, amassing the largest vote yet given any candidate for that office. His new job, however, provided neither the action nor the exposure that Fitts craved. When Keyes announced his retirement, Fitts tossed his American Legion hat into the ring. His popular presence intimidated potential challengers, and he ran virtually unopposed in the 1928 election. In November, when charges against Keyes became public, Fitts resigned as lieutenant governor to act as special prosecutor in the bribery case. On December 3 he formally assumed the office of district attorney.

The scandals surrounding the Keyes administration combined with Fitts's personal prestige and widespread press support to generate a mandate to "clean house." Within a week Fitts began to reorganize operations. He introduced a new Bureau of Investigation, modeled after the FBI. This staff of sixty to seventy men gave Fitts his own force of detectives, independent of either the Los Angeles police or sheriff's departments, and a fruitful source of patronage for his political allies and American Legion confederates. To head up the new detective bureau, Fitts recruited former FBI agent and one-time Julian Pete investor Lucien Wheeler. Immediately before accepting this assignment, Wheeler had worked as a private investigator for his good friend Jacob Berman.

Fitts proclaimed a crackdown on the vice that had flourished under Keyes. An avid Prohibitionist, Fitts warned the city's hotel and café owners that his office would "maintain order and see that prohibition law is observed to the letter." Shortly before

Christmas of 1928, his office staged several well-publicized raids on nightclubs serving liquor. In addition, Fitts reversed an earlier Keyes mandate legalizing slot machines. Leslie White, a police officer and aspiring mystery writer who enlisted with Fitts's Bureau of Investigation, captured the moral fervor of Fitts's ascension in his book *Me, Detective.* "I was for the 'cause' and I loved a fight," wrote White. His selection to work with Fitts seemed a "reincarnation," which he approached with "unbounded enthusiasm."

Fitts, whose prohibitionist, anti-labor, and anti-radical sentiments typified American Legion politics, became the darling of the *Los Angeles Times* and the Republican right wing. The rival Progressive faction viewed him differently. The *Record* repeatedly belittled Fitts. Frank Doherty, Senator Hiram Johnson's confidant, spiced his letters with caustic comments about the World War I veteran. He "permits it to be said and by inference suggests it himself, that he, and not General Foch, was the main factor in running the war," wrote Doherty. Doherty most feared that Fitts might challenge Governor Young in the 1930 elections. "[Fitts] is very ambitious and it behooves the Governor to be on his mettle. If Fitts has a successful administration, I am firmly of the belief that he will seek the Governorship two years hence," warned Doherty in January 1929. The first, and most important, stepping-stone would be the prosecution of his predecessor and former superior, Asa Keyes.

## II

The trial of Asa Keyes began on January 7, 1929, in the old courthouse where he had established his reputation. Jacob Berman; the Rosenberg brothers; the Getzoffs, father and son; and former detective Charles Reimer stood alongside Keyes as defendants. As had become common in all of the cases related to the Julian Pete scandal, masses of spectators hoped to observe the proceedings. Celebrities and movie stars attended the trial, and fights erupted along the courthouse hallway among others jostling to gain entry. It was, wrote Leslie White, "a spectacle I'll never forget."

Fitts sprang his first surprise as soon as the trial opened. As the proceedings began, Deputy Robert Stewart asked for the dismissal of all charges against Jacob Berman. The "bright boy," announced Stewart, had agreed to turn state's evidence and become a prosecution witness. Berman's defection marked a crushing blow to Keyes's defense. Previously the case against the former district attorney had rested heavily on the testimony of the tailors—Pike, Sherman, and Rittinger—whose accounts could be construed as motivated by animus against Getzoff. When Berman took the stand, he could offer not only corroboration but direct and expanded evidence of the machinations of the bribery plot.

Fitts and William Clark, whom Fitts had appointed as special prosecutor, quickly unfolded their case against the defendants. Deputy District Attorney Julian Richardson, who had borne the brunt of the Julian Petroleum prosecution, testifed how he had put in twelve- to eighteen-hour days, "bent the oar, you might say" to secure convictions in that case. Richardson revealed that at the start of the trial Keyes had told him, "I would like to do something for Ed Rosenberg, if I can." Shortly thereafter, Keyes requested that charges against Rosenberg be dropped.

The prosecution then put Milton Pike on the stand. Pike launched into a day-by-day retelling of the events in the tailor shop. Although the highlights of Pike's testimony had been previously published, his account had tremendous impact. Pike described checks' changing hands and arguments about delayed payments. Pike told of one occasion when Rosenberg threatened that, if Keyes did not dismiss his case, he was going to double-cross him. "Why you _____!" responded Getzoff, "If you do, I'll kill you." Berman also engaged in a heated argument with Getzoff over payment of the bribe money. "Do you think I was born yesterday? I want to know where I stand before I pay any money out," shouted Berman. On cross-examination Pike denied attempting to blackmail the Getzoffs. He closed his testimony by recounting the final entry in his diary: on July 28, 1928, the day he stopped working for Getzoff, Pike saw Keyes drive up to the store in a brand new Lincoln coupe.

Pike's testimony set the stage for Jacob Berman. For several weeks Berman had lived under the protection of detectives Blayney Matthews and Leslie White. "I nursed him during the daytime," recalled White. "Matthews relieved me at nights." White described his ward and his distasteful task in detail:

> Berman was a bumptious young Jew. . . . He was sleek and flashily groomed; he reminded me of a wharf rat that has just climbed out of the water. He had brazen, bulging eyes and the most obnoxious personality it has ever been my misfortune to contact. . . . It was always a relief to turn over this odious responsibility to Blayney Matthews. . . .
>
> What amazed me most about Berman was his colossal gall and attitude towards our office. I was even more amazed, however, at the attitude of our office toward Jake Berman. He was handled with "kid gloves." Instead of being thrown in a cell as would any impoverished material witness, he was allowed complete liberty, humored, and supplied with an armed guard to prevent what most of us would have considered as a justifiable homicide. . . .
>
> Although my task was to guard his life and, if necessary, sacrifice my own in so doing, I had no authority over Berman and his attitude toward me was a mix of condescension and contempt. I hated the man and would have been delighted had he been assassinated under the guardianship of *someone else*. . . .

However obnoxious and odious White found his charge, in three days on the witness stand Berman dazzled the spectators with his debonaire wardrobe and spellbound them with his sordid tale. Berman dated his participation in the bribery scheme to February 9, 1928, more than one month after the Julian Pete trial had begun. He testified that on that day, Ed Rosenberg, his onetime confederate and now estranged co-defendant, approached him in the corridor of the Hall of Justice. Rosenberg told him "that although I had double-crossed him and had not treated him right, that he was going to do something for me." Rosenberg informed Berman,

"You know I have made arrangements to have my case dismissed," and offered to introduce him to "the old man." Berman asked who "the old man" was. "The tailor . . . Getzoff," responded Rosenberg.

Berman and Rosenberg took a taxi to the Spring Street tailor shop. En route, according to Berman, Rosenberg boasted that he had paid $125,000 to have his case dismissed and that "Ace got it all." Berman protested that he could not afford that kind of payoff, but Rosenberg reassured him, "It will not cost you anywhere near that because you can ride through on what I pay."

At the tailor shop Berman met Getzoff. The wizened tailor warned him that, "if he could not do anything or would not do anything," Berman would be "convicted sure as hell." He demanded $10,000 in "preliminary expenses." When Berman asked what this would buy, Rosenberg responded, "It covers nothing." Berman would have to gamble the initial payment. If they decided to help him, the money would be applied to future arrangements. Berman testified that, when he pressed Getzoff on what this money would be used for, Getzoff replied, "It is being used to pay off Asa Keyes' gambling debts in town. He has been gambling pretty heavy, and he wants to get it cleaned up."

Rosenberg and Getzoff made additional demands. Rosenberg insisted that Berman relieve him of the obligation of paying $100 to $200 a week for Getzoff's entertainment expenses. Getzoff instructed Berman to dismiss his attorney and hire Dudley Robinson, Rosenberg's counsel. Robinson and Keyes were "buddies," stated Getzoff, and Keyes would do Robinson's bidding. When Berman protested that representing both him and Rosenberg would pose a conflict of interest for Robinson, Getzoff advised, "Ed Rosenberg will be dismissed in two or three days . . . and then [Robinson] can step in as your attorney." Berman testified that he paid Robinson a $5,000 retainer, but because Judge Doran had not released Rosenberg, Robinson never acted on his behalf.

The February 9 meeting proved only the opening gambit in Getzoff's efforts to squeeze money from Berman. Shortly thereafter Getzoff informed the Julian Pete defendant, "I want you to buy a gift for Keyes' home. . . . I want you to buy a chaise longue." Together the tailor and the swindler went to Brooks Brothers where they selected an appropriate style and fabric for the Keyes household. Berman paid $440 for the furniture. In late February, Getzoff complained that Keyes was receiving all of the money, while he was going broke. He demanded $500 to keep his tailor shop open. Berman gave him the money.

On yet another occasion, Getzoff anounced that he wanted to buy a new car for Keyes's daughter. "I think you ought to buy it for her as a gift, as it will please Ace and help the situation along," suggested Getzoff. Berman authorized an expenditure of not over $1,000, and Getzoff ordered a new Chevrolet. The tailor apparently found this idea so appealing that he demanded a new automobile for himself. When Berman again attempted to impose a $1,000 limit, Getzoff laughed and protested, "I wouldn't ride around in a cheap automobile." Berman increased his allowance to $1,500.

According to Berman, Getzoff extracted one more major payment during the course of the trial. One night he called Berman at home and asked him to visit the shop the following day. Getzoff demanded an immediate $10,000 so that Keyes could get the home he had recently bought in Beverly Hills out of escrow. When Berman objected that he did not have the money, Getzoff replied, "I don't care where you get it. You have got to beg, borrow or steal it." Berman sold 1,000 shares of United Artists stock and delivered the cash to Getzoff. The tailor grabbed the phone and called Keyes. "Hello, Ace, I have got the money," he exclaimed excitedly.

Despite these expenditures, Berman testified that he had received few tangible benefits. Unable to have the charges against Rosenberg dismissed, Keyes dared not make a similar motion on Berman's behalf. Nonetheless, Keyes had helped him with his rambling closing address to the jury. "I thought if he wanted to, he could have crucified me," stated Berman. Instead, Keyes had "gone easy."

Berman corroborated Pike's testimony that after the Julian Pete verdicts the conspirators grew less cautious. On May 24, 1928, Berman and Rosenberg met Keyes at the tailor shop. Rosenberg presented a wristwatch to Keyes and hailed him as "one of the best fellows that he ever met." Berman thanked Keyes for "not bearing down on me." He asked the district attorney when the last remaining charge against him, a forgery and embezzlement indictment, would be dismissed. Keyes promised to act "in due time."

With this additional favor outstanding, Getzoff continued to pressure Berman. On June 20, he suggested that Berman buy Keyes a gift of golf clubs. Getzoff called the prosecutor for the appropriate weights and sizes, and Berman made the purchase. On another occasion, Keyes, Getzoff, and their wives had dinner at the home of Jacob Berman's brother, Louis, while Jacob was in New York. The trio called Berman long distance to pressure him for $3,000 outstanding on his promised payments. Berman estimated that overall the bribing of Keyes had cost him $40,000. Ed Rosenberg, he said, had paid far more.

Berman's tale left the courtroom stunned. While his credibility and motives for testifying remained suspect, his detailed account proved totally convincing. For two days defense attorneys, including Keyes himself, attempted to break down Berman's story. Berman repeatedly denied that Fitts had promised him immunity in exchange for his testimony. At one point, Berman, almost shouting, charged that he had given written statements to Keyes's deputies revealing "where the Julian money went," but that Keyes had not presented this evidence at the trial. "For every question that was shot at him," reported the *Times*, "'the bright young man' of the Julian cases had a ready come back and proved as good a witness for the State on cross-examination as he was on direct." Trial judge Ben Butler later said of Berman, "I never saw a witness bear up better under such a bitter crossexamination than he did. Not once did he contradict himself on a major point and no man could have concocted the story he told and tell it with the precision he did, if it wasn't the truth."

Those who followed Berman to the stand corroborated much of the preceding testimony. Louis Berman, who had acted as his brother's go-between and banker in these transactions, offered especially damaging evidence. Since his brother had no checking account in Los Angeles, Louis had written numerous checks to Getzoff. He identified these documents for the prosecution. Louis admitted paying for a lamp selected by Keyes and his wife at a downtown store and purchasing a set of wooden golf clubs inlaid with the initials "A. K." and delivering them to Keyes's home. Louis Berman also described the audacious dinner party he had hosted for the Keyes and Getzoff families. Carl Vianelli, Louis Berman's houseman, substantiated the events of the dinner party.

Buddy Davis testified about his misgivings on the handling of the Julian Pete prosecution. Representatives of two auto dealers confirmed the purchase of Elizabeth Keyes's Chevrolet and Asa Keyes's Lincoln, paid for in cash by Ben Getzoff. Another local salesman confirmed the purchase of a $1,150 radio by Ed Rosenberg for delivery to Keyes. Bank officials and realtors revealed details of the purchase of Keyes's home that gave credence to Berman's accusations.

On January 22, the prosecution rested its case. Fitts and his deputies had presented a damaging array of evidence against Keyes, Rosenberg, and Getzoff, but had produced little to implicate the three lesser defendants, Dave Getzoff, Jack Rosenberg, and Charles Reimer. Judge Butler dismissed charges against these men, and the defense announced that it would call Asa Keyes as its first witness.

The spectacle of a former district attorney testifying on his own behalf attracted the largest throngs yet to the courthouse. Those who gained admittance saw a cool and deliberate Asa Keyes, tempered by a quarter of a century of courtroom experience, systematically attempt to debunk the charges against him. Keyes readily admitted his friendship with Getzoff, whom he viewed as a wealthy patron, and his frequent visits to the Spring Street tailor shop. "Ben Getzoff was always an affable, sociable, jolly fellow and he was always a gentleman around me and my family," explained Keyes. The one-time prosecutor denied all charges based upon unsubstantiated testimony. He had never accepted cash from Getzoff or the others and never had accumulated gambling debts. Getzoff had not asked him to "lay down" in the Julian cases.

But Keyes had great difficulty explaining his receipt of the gifts numerous witnesses had described. He had assumed that the chaise longue and radio were simply gifts from Getzoff. He denied knowing of their connection to Berman and Rosenberg. The Lincoln coupe was purchased with a loan from Getzoff, which Keyes intended to pay back, although the tailor had repeatedly deferred repayment. Keyes admitted meeting with Berman and Rosenberg at the tailor shop after the trial and dining at Louis Berman's home, but claimed that he saw no impropriety in this.

Keyes also defended his handling of the Julian Petroleum case. He had avoided the courtroom because he had been immersed in the Hickman trial. He attempted to assist Rosenberg because Rosenberg had provided invaluable information and had been willing to assist the prosecution "to the limit" in demonstrating the connection between Lewis and Berman. Keyes attributed his brief closing arguments to the length of the trial and the verbosity of the other attorneys. "When it came my turn I did not want to tire the jury any further and so decided to make a short, snappy, general résumé," he explained.

Keyes's defense proved so flimsy that the prosecutors wasted little time in cross-examining him. Deputy William Simpson, however, could not resist ridiculing Keyes's visits to the tailor shop for refreshment. Simpson asked, "What kind of drink do you mean?" Keyes heatedly replied, "A drink of whisky, of course." When Keyes stated that Getzoff had obtained the liquor for medicinal purposes, Simpson snapped sarcastically, "So you were helping Getzoff to drink up his medicine, were you? What kind of medicine was it—scotch, rye or bourbon?"

Other defense witnesses proved ineffective, at best, and terribly damaging at their worst. Attorney Dudley Robinson admitted accepting $5,000 from Berman, even though he had never acted on his behalf. "I took the fee only for preparing to enter the case in an emergency," contended Robinson. Ed Rosenberg denied all allegations against him but admitted presenting a $630 watch to Keyes. The inscription, he revealed, read, "To A. K., a real he-man—E.H.R.—5/23/28." The date referred to the day of the Julian Pete acquittals.

The prosecutors began their final arguments on February 1, 1929. Deputy Simpson opened with a caustic account of the Rosenberg-Keyes connection. Keyes, he said, had made the motion to dismiss charges against Rosenberg, "not because he needed him as a witness in the Julian trial, but because in the dimly lighted back-office of a Spring Street tailor shop your public officer had sold himself." Chief Deputy Robert Stewart followed Simpson with an even more scathing attack. When faced by Getzoff's machinations, Keyes should have shouted, "Get thee behind me, Satan." Instead, the two had become as close as the figures in the Al Jolson song "Me and My Shadow." Stewart charged that "the gifts, the associations at unusual places, the liquor, show the breaking down of the moral fabric of that once trusted servant."

The defense focused its attack on the credibility of Keyes's accusers. Jed Rush, Keyes's chief counsel, described Pike and Sherman as greedy, grasping men who had fabricated their stories to enrich themselves. Berman was a "master criminal" whose story was not to be believed. "The leopard doesn't change its spots," shouted LeCompte Davis, Rosenberg's attorney. "You can't expect sweet and wholesome water from a sewer."

Defense attorneys also appealed to the jurors' anti-Semitic prejudice. They made frequent allusions to the fact that most of Keyes's accusers, and "mastermind" Ben Getzoff, were Jewish. William Beirne ironically depicted Berman concocting his story "to make peace with his God," and depicted Fitts's deputies listening to his confession, nodding "Ay, ay Jakie, we believe you." Attorney Al McDonald, ostensibly representing Getzoff, described his client as "of a nationality not yours or mine" and said of Asa Keyes, "It should not be held against him because he had a Jewish friend."

In his summation, McDonald made the greatest gaffe in the trial. Attempting to dismiss the significance of the gifts that Keyes had received from Getzoff, McDonald admitted the transactions and then asked boldly, "Well, what of it? He had the money to pay for them." The phrase "Well, what of it?"—erroneously attributed to Keyes in subsequent trial accounts—came to symbolize the shameless arrogance of the Keyes-Getzoff conspiracy.

On February 6, after one month of trial and five days of concluding arguments,

Asa Keyes rose to speak on his own behalf. Keyes told the jurors that if he were found guilty he would "go to the penitentiary and rot with a full knowledge in my own heart that an innocent man had been convicted." Repeatedly referring to himself in the third person, Keyes attempted to belittle the charges against him. "If Keyes had fallen so low, lost all idea of honor and manhood, would he have gone to Getzoff's tailor shop with the bright sunlight streaming in, the employees and customers about, and been as big a fool as to take bribe money? Why the very statement makes it out a lie."

Keyes assaulted Pike and Sherman as "the damnedest liars God ever made," and asked, "If you don't believe them what is left? Nothing but the testimony of Jacob Berman," whom he branded "a liar, double-crosser and thief." Keyes challenged the jurors to "remember this one thought" when they made their deliberations: "Would you like to be sent to the penitentiary, or would you like to have your son or your daughter or your brother or your sister or any friend of yours condemned to prison on the testimony of such men as Pike, Sherman, Rittinger and Jack Bennett?" Keyes limited his remarks to two hours, the same length as his maligned summation against the Julian Pete defendants.

With the defense attorneys having spoken, the prosecution resumed its offensive. Special Prosecutor William Clark reminded the jurors that, even without the testimony of Berman and the tailors, the circumstantial evidence justified a guilty verdict. Clark drew derisive laughter as he depicted Keyes "on his way to heaven—wearing a smoking jacket, listening to an $1150 radio with a set of golf clubs in one corner, a Lincoln roadster on behind as a trailer, and in his hands a sheaf of Ben Getzoff's liquor prescriptions."

Clark's speech primed the courtroom for the performance of Buron Fitts, making his first address to a jury as district attorney. Through the early part of the trial, Fitts had remained relatively inconspicuous. "He does his work on the radio by addressing luncheon clubs, etc., upon crime conditions," wrote Frank Doherty, whose distaste for Fitts seemed to grow with every letter to Senator Johnson. Later in the trial, Doherty reported, "Buron is in the court every day and at times . . . rises to his feet in discussing a legal problem and states, 'In all my experience in the trial of cases, etc., etc.' When he proceeds this far, [defense attorney] Paul Schenk usually asks him 'when and where.' Buron probably never tried more than three or four cases in his life."

Doherty may have questioned Fitts's legal talents, but when given the opportunity for a public oration, the "soldier-prosecutor," as the *Times* described him, rose to the occasion. Repeatedly invoking wartime analogies, Fitts labeled Keyes "the greatest deserter in the army of peace—a deserter who prostituted the law . . . as great a traitor to the people . . . as Benedict Arnold." Fitts acknowledged that at the start of the trial he had hoped that his former superior would be able explain the charges leveled against him. Now Fitts said, "He has utterly failed . . . even in the smallest detail." The defense had portrayed prosecution witnesses as perjurers, argued Fitts, but Keyes "is a greater perjurer than any one of them. I'd rather go to face my God in a pauper's sack than from the Beverly Hills home of Asa Keyes. . . . He has debased and rendered contemptible this county in the eyes of the world. He has made the Temple of Justice a den of thieves. By your verdict you will say whether this sordid, rotten, putrid scheme of things as carried on by Asa Keyes will continue in Los Angeles County."

On the evening of February 8, Judge Butler passed the case to the jury. After reviewing the evidence for ninety minutes, they cast their first and only ballot. All twelve jurors voted the three defendants guilty as charged. At 9:45 p.m., the jurors returned to the courtroom. The short deliberation period and their stern faces foretold their verdict. The formal reading nonetheless crushed the defendants. Keyes sat stoically, but all color drained from his face. Ed Rosenberg began to sway in his seat, his pallor turned a "greenish white," and then he fainted at the defense table. Ben Getzoff exclaimed, "That settles it!" then slumped back in his seat "visibly crushed." On the way to the jail, he collapsed completely.

The conviction of Asa Keyes, reported the *Examiner*, "rocked" Los Angeles. The events of the next seventy-two hours jolted the city anew. Less than an hour after the verdict, Detective Blayney Matthews prepared to move Jacob and Louis Berman from the Hollywood apartment house where he and Leslie White had protected them during the trial. Normally, when they departed, Jacob Berman exited first. On this evening, according to Matthews, "some hunch caused me to step in front of Berman and open the door. As I did so, I noticed a man level an automatic pistol at us." Matthews jumped back into the apartment and slammed the door, pushing the Bermans to the floor. The gunman disappeared into the night.

Meanwhile, at the county jail, while Keyes calmly smoked cigars in his private cell and continued to maintain his innocence, Ben Getzoff writhed in pain. The strain of the trial had reawakened his stomach ulcers. Getzoff moaned that after two weeks in jail he would be dead. On February 9, the day after the verdict, prison doctor Ben Blank told reporters that Getzoff had collapsed for a second time, and he feared the old man would suffer a complete mental and physical breakdown.

Two days later Getzoff could no longer endure the agony. "I want to see Fitts," he told the guard. The district attorney rushed to Getzoff's cell accompanied by a stenographer. For two hours, during which Getzoff repeatedly collapsed and had to be revived, Fitts listened while Getzoff gave a full confession of his activities as a bribery broker. At one point Getzoff fainted and could not be roused. Fitts summoned Dr. Blank to restore Getzoff to consciousness so he could resume his revelations.

"Getzoff told a story that will shake this county to its very foundations," proclaimed Fitts. The tailor had named fourteen men, half of whom had never been mentioned before, as members of a "fixing ring" emanating from his tailor shop. Fitts reported that these men included prominent politicians, former deputy district attorneys, a well-known private attorney, and a figure who held a higher position in the community than any of the others. Getzoff claimed that Keyes was the true instigator of the plot and that the former district attorney had received other gifts undisclosed at the trial. At Keyes's urging, said Getzoff, he had proposed bribery schemes to many wealthy defendants for sums totaling hundreds of thousands of dollars.

The Keyes trial and the Getzoff accusations reopened the wounds inflicted by the Julian Pete scandal. The *Times*, in an editorial entitled, " 'Well, What of It?' " noted, "It is unfortunate that the source of the bribe money—the millions of dollars stolen from the public by the Julian swindlers—cannot be recovered and restored to the victims." To the *Record*, Keyes had become the "sacrifical goat" for the Julian affair: "He prosecuted the friendless, disgusting Hickman with relentless fury, but he fawned before the mighty. . . . There are few men in this community—or in any other . . . who would have fearlessly and vigorously prosecuted every defendant in the Julian cases. They were too powerful, with too many alliances reaching out into every angle of our business and civic life."

Buron Fitts emerged as the clear beneficiary of these events. The *Express*, in its editorial, "Well Done Fitts," wrote, "The victory was won by the courageous and brilliant district attorney," who had left state office "as a public duty, in the face of one of the most desperate criminal-political situations this section of the state has ever known." In his closing address, Fitts had "delivered an argument of such force and masterfulness it is bound to go into the annals as one of the greatest and most effective ever heard in a local court."

On February 20, Keyes, Rosenberg, and Getzoff appeared in court for sentencing. Getzoff appeared in a wheelchair. When called to rise, "his face worked with nervous tension, his legs shook and he clung to the counsel table with both hands." Judge Butler ordered each of the men to serve one to fourteen years in the penitentiary. Keyes and Rosenberg returned to the county jail pending appeal; Getzoff was released due to fears that he might not survive incarceration.

In early April it became apparent that for his cooperation Getzoff would receive considerations similar to those Fitts had granted Berman. On April 5, the sickly tailor appeared before Judge Butler again. Butler announced that he was suspending Getzoff's sentence at the district attorney's request. "It is quite manifest that he is sick . . . but the main reason that it affects my mind is that I believe the man is penitent," reasoned Butler. "I think that he has seen the evil of his ways."

Others were not so certain. Many questioned Fitts's generous use of immunity policy, suspecting that Getzoff had fabricated tales in exchange for his freedom, and that Fitts, ever ready to generate publicity, had leaped to accept them. The paltry results produced by Getzoff's confession made this view plausible. Former Keyes aide Buddy Davis became the primary victim of the Getzoff-Fitts alliance. On May 1, 1929, the grand jury charged Davis with accepting $7,500 from Getzoff to "control and direct" the prosecution in favor of Ed Rosenberg in the Julian Pete trial. Davis charged that "the indictment was returned on the bought-and-paid-for perjury of the miserable and unspeakable Ben Getzoff, who has sold his testimony for his liberty."

Davis stood trial in July 1929. Getzoff described bribing Davis with money, whisky, reading lamps and cloth for a dress for Davis's wife. Under cross-examination the tailor conveniently collapsed on the stand. Joseph Sherman and Dave Getzoff offered supporting evidence. (Milton Pike was withdrawn at the last minute after Davis and his attorneys trapped him into accepting a bribe to alter his testimony.) Despite numerous witnesses who praised Davis's performance in the stock fraud prosecution, the jurors found him guilty. Davis blamed his predicament on Fitts's "personal animosity." The conviction was later thrown out on appeal. Only one other minor indictment emerged from the Getzoff confession.


### III

The Keyes trial and its aftermath had produced sensation after sensation. Other elements of the Julian Pete tangle played out with far less fanfare. In December 1927, Governor Young had appointed Julian receiver H. L. Carnahan to replace Buron Fitts as lieutenant-governor; Carnahan nonetheless remained as receiver. Throughout 1928, he and Joseph Scott labored to bring the reorganization to fruition. In January 1928 stockholders had approved S. C. Lewis's plan to refinance the corporation with funds borrowed from Henry Robinson's newly created First National Trust and Savings Bank (the result of the merger between First National and Pacific Southwest) and the Associated Oil Company, which assumed the liabilities formerly held by the Anglo-London-Paris bank. As in the earlier financing plans, these funds would be secured by the company's physical assets.

The reorganization plan also resolved the problems created by the stock overissue. The new company was authorized to issue 3.5 million shares of stock with a par value of ten dollars. A portion of these shares would go to unsecured creditors of the two original companies; the remainder would be divided among all Julian Petroleum shareholders, whether they held bona fide or counterfeit stock. Shares would be allocated in proportion to the amount of their original investments.

In February, Scott and Carnahan reincorporated Julian Petroleum and California-Eastern into the Sunset Pacific Oil Company. (Reverend Shuler proposed the name "Lollypop Oil Company" and the slogan "Original Suckers' Delight.") Judge McCormick approved the plan on May 24, 1928. During the summer months Scott and Carnahan dispatched contracts to all unsecured creditors and shareholders seeking formal approval of the reorganization. In signing these agreements these individuals forfeited all rights to file suits arising from their original transactions. This proviso notwithstanding, 90 percent of all stock claimants and those holding 93 percent of other unsecured debts approved the plan.

On November 19, 1928, the receivers put the physical properties of the Julian Petroleum and California-Eastern corporations on the auction block. The auction was the largest ever court-ordered sale in Los Angeles County. By prearranged agreement the Sunset Pacific Oil Company was the only bidder. With this maneuver, the Julian Petroleum and California-Eastern companies passed from existence.

A handful of major obstacles still confronted the Julian Pete receivers. Several substantial claims, including some arising from the usury suits, remained outstanding. The largest such claim involved Adolph Ramish. In August 1926, the receivers had sued Ramish for $456,000. Ramish, however, had always contended that his losses in unredeemed Julian Petroleum stock far exceeded the profits he had made. For eighteen months Ramish and the receivers had failed to reach a compromise.

In late December 1928, rather than delay the reorganization, Scott and Carnahan conceded Ramish's position, dropping their claims and recognizing a $770,000 debt to the theater owner. They agreed to issue Ramish 62,500 shares of Sunset Pacific stock to cover this liability. Ramish, one of the largest individual stockholders in Julian Petroleum, would now hold a similar position in the new company.

Most other usury cases remained unresolved. For all of Scott's efforts, the "put back" campaign had netted only $540,000, more than half of which came from the members of Bankers' Pool No. 1. Few of the hundreds of people who had participated in Berman's other transactions had cooperated. Eighteen suits remained pending, among them the Hollingsworth test case. In March 1928, Scott and Carnahan had petitioned for and been granted expanded authority to pursue these cases as part of their receivership. But, shortly thereafter, a judge ruled that California state law only allowed personal representatives of a company to file usury suits and that court-appointed receivers did not fall into this category. The receivers appealed this decision to the State Supreme Court, where its fate, and the fate of all related cases, remained unresolved.

One other large and controversial claim awaited settlement. In December 1927, Lewis Oil had filed a petition in intervention, seeking to establish title to its properties in Texas and Louisiana and claiming unpaid indebtedness for hundreds of thousands of shares of Julian stock acquired in its name but personally used by S. C. Lewis. Exactly who controlled Lewis Oil at this time remains unclear. Petitions filed in federal court stated that a new board of directors, "not under the control of S. C. Lewis," had initiated these actions. But communications from the company were signed by Axel Swanson, a long-time Lewis confederate, and Lewis had apparently recruited Guy Crump, the Los Angeles attorney who represented Lewis Oil. On December 17, 1928, one month after the auction of Julian Petroleum properties, Crump filed a second, amended, complaint that threatened to indefinitely postpone the reorganization.

As in the Ramish case, the receivers ultimately capitulated to Lewis Oil's demands. Lewis Oil received undisputed title to its properties in Louisiana and Texas and $900,000 worth of Sunset Pacific stock. Stockholder representative Arthur Loeb later charged that Lewis himself had manipulated this agreement and become its prime beneficiary. Loeb alleged that Lewis quickly dumped the stock for thirty cents a share, and once out of prison would reclaim the oil leases in Texas and Lousiana, which had proved to be far more valuable than the receivers had realized.

Indeed, while most people lavished praise on Scott and Carnahan, Loeb, who had become a full-time monitor of all aspects of the scandal, expressed nothing but disdain for their work. Loeb protested that while Carnahan was a hard worker who put in fourteen hour days, he was a "flop" as an administrator. Scott, on the other hand, particularly after the early flurry of usury cases, "did nothing." According to Loeb, Scott and Carnahan employed "rosy propaganda," sending out press releases that "painted pictures that were so beautiful" that the stockholders readily endorsed the reorganization. But Scott and Carnahan, like Lewis before them, had vastly overstated the assets of their corporation. The physical properties of Sunset Pacific were not worth the reported $19 million, charged Loeb, but a paltry $8 million. Its liabilities included an $8 million bonded indebtedness and a $35 million stock distribution. Thus the new company was no healthier than its predecessors.

Loeb's accounting proved to be fatefully accurate. Carnahan later admitted that Sunset Pacific lacked the assets to support its secured liabilities, let alone its stockholders. The purpose of the reorganization, allowed Carnahan, was to "give us time to see if we couldn't sell the company to save money for the stockholders out of it." Loeb protested the various inaccuracies in the receivers' reports and attempted to expose them in the press, to no avail. Despite his objections, Judge McCormick approved the reorganization.

In the final analysis, no creditor or stockholder of Julian Petroleum or California-Eastern "got a nickel . . . in cash," reported Carnahan. Stock issues covered all liabilities. The receivers and the attorneys who had worked for them, however, fared far better. Carnahan and Scott each received $105,000. Fees to other attorneys totalled over $275,000. All lawyers were paid in cash.

With the filing of a final report in the spring of 1929, Scott severed all connections with Sunset Pacific. "I wanted to get back to my own profession and business," he explained. "I am not a receiver. I am a lawyer." Carnahan, on the other hand, accepted the presidency of Sunset Pacific. He also remained as receiver for Julian Petroleum in order to pursue the outstanding usury cases. In addition to these posts, Carnahan continued to serve as lieutenant-governor.

Henry M. Robinson doubtless hoped that the reorganization of Julian Petroleum, financed by his new bank and often (inaccurately) attributed to his creativity, would erase the embarrassment of the massive scandal, but echoes of the Julian Pete debacle haunted him at every turn. In 1928, he published a brief book entitled *Relativity in Business Morals*, in which he applied Einstein's theory of relativity to "all of life's movements and trends." Notions of right and wrong, he argued, varied from era to era and place to place. Businessmen in the 1890s "engaged in practices that would today be considered reprehensible," argued Robinson. Morals in rural areas differed from those in the great metropolitan centers, where businessmen "are under relentless pressure from opportunism and expediency." Nonetheless, concluded Robinson, the "general tendency [of business morals] is up not down." Many southern Californians saw Robinson's arguments as an apologia for the Julian affair; others derided the very notion that the president of Pacific Southwest would dare lecture the nation on business morals.

In addition, what should have been the fulfillment of one of Robinson's fondest dreams instead brought cruel embarrassment. For a decade Robinson had served in a select group who had counseled and guided Herbert Hoover's political career. Following his election as President in 1928, Hoover hoped to reward Robinson with a Cabinet post. But after the Julian Petroleum scandal, Robinson feared he had become a liability. Rumors circulated that the missing Julian Pete money had found its way into Hoover's campaign chest. The President-elect received letters warning him, in the words of one correspondent, "Mr. Robinson's selection will be sincerely regretted by thousands of your supporters in California." The new President nonetheless offered his old friend the key post of Secretary of the Treasury. Robinson judiciously turned him down. In later years, Hoover would write that his inability to persuade Robinson to assume this position ranked among the greatest disappointments of his presidency.

Personal tragedy with Julian Pete overtones also struck Robinson. In early February 1929, Robinson's longtime friend, business associate, and legal adviser Frank Flint died of a heart attack at age sixty-six. According to Joseph Scott, "Senator Flint was aged ten years by his experience" with Julian Petroleum. Flint's health began to fail, forcing him to withdraw from the myriad affairs which had characterized his career. In January 1929, he had embarked on a round-the-world cruise for "rest, recreation and recuperation." One month into the tour, Flint died at sea. According to Guy Finney, many of Flint's friends regarded him as "the victim of the crushing weight he carried into the Julian Petroleum fiasco."

The hardest blow to Robinson, however, involved his greatest business achievement, the First National banking empire. The Julian Petroleum scandal had shattered public confidence in the institution. According to Rockwell Hereford, Robinson would not fire those implicated in the scandal, fearing that this might be seen as an acknowledgement of illegal activities. He had hoped that the total immersion of the offending Pacific Southwest Trust and Savings into the parent First National Bank in 1927 would eradicate the bank's negative image. But throughout 1928, First National continued to falter.

Fearing that the bank might collapse, Robinson considered two options. The first involved the sale of First National to A. P. Giannini's rapidly expanding Bank of Italy. Giannini had already benefited from the Julian Petroleum controversy by absorbing Merchants National Bank, which had been weakened by its questionable lending activities. The acquisition of the immense First National branch system would advance Giannini's conquest of southern California. But animosity between Robinson and Giannini after years of contention prevented a deal.

In desperation, Robinson turned to his leading southern California rival, Joseph Sartori, president of Security Trust and Savings. Even as they competed for customers, Robinson and Sartori had always cooperated in the fight to maintain regional banking and on issues related to the advancement of Los Angeles. Robinson and Sartori met quietly to discuss a merger of the city's two largest banks. The pair quickly hammered out an agreement on the financial aspects, but Sartori demanded one difficult concession. Sartori insisted that those implicated in the Julian swindle, whether guilty of wrongdoing or not, would have to resign. Robinson reluctantly complied, but sought an exemption for his best friend and protégé, John Barber. Sartori remained adamant; all taint of the Julian scandal had to be removed.

On January 25, 1929, Robinson and Sartori announced the impending merger. The new Security–First National Bank, with $600 million in assets, would rank second in the state, surpassed only by the Bank of Italy. It would be the eighth-largest bank in the nation. Robinson assumed the position of chairman of the board, and Sartori became president. The new bank, with Sartori's people holding all key positions, became a reality on April 1. The Julian Petroleum Corporation had dragged down with it the biggest bank in Los Angeles. Indirectly, it contributed to the creation of one of southern California's more durable financial institutions. The enhanced capitalization of the giant new bank enabled it to weather the Great Depression. In 1968, it became the Security Pacific National Bank. In a final irony, in 1992, Security Pacific merged with the Bank of Italy's successor, the Bank of America.

The consolidation of Security and First National also revealed the limited lessons that southern California investors had learned from the Julian affair. Rumors of the Robinson-Sartori entente and other mergers had stimulated widespread speculation in bank stocks in 1928. LASE patrons avoided oil securities in the aftermath of the Julian Petroleum collapse and funneled their funds into bank issues. This latest binge fueled the West Coast equivalent of the Wall Street bull market. Like its New York counterpart, the LASE grew exponentially in 1928. Sales volume exceeded the record 1927 figures (which had been inflated by the fraudulent Julian stock issue) by 400 percent. Sharp breaks in the market for bank stocks in June and December caused hundreds of millions of dollars in paper losses but failed to deter the Spring Street bulls. They continued to favor bank stocks and pour money into local financial markets. Only the national stock crash in October 1929 would douse the speculative fires that had raged throughout the 1920s.

# 16 WHEN JUSTICE FAILS

## I

With the conviction of Asa Keyes, the creation of Sunset Pacific Oil, and the demise of the First National Bank, the Julian Petroleum swindle had claimed more than its fair share of collateral victims. Beneficiaries, other than those who had profited from the original stock manipulations, proved far rarer. Among them, none had fared as well as the Reverend Robert Pierce Shuler. By mid-1929, "Fighting Bob," as his devoted patrons called him, had ridden the Julian Pete exposés to a level of local political influence unmatched by any American religious leader of the era. And with the help of District Attorney Buron Fitts, Shuler hoped to expand that power to the state government and perhaps beyond.

In a decade noted for its religious fervor, Robert Shuler personified the stereotype of the 1920s preacher. Born in a log cabin in Virginia's Blue Ridge Mountains and raised, in his own words, among "the poorest of the poor," Shuler prided himself on his strict adherence to the "old-time religion" he had absorbed as a boy. These beliefs stressed the divine inspiration and unquestioned accuracy of the scriptures. Shuler had preached these doctrines unswervingly since age seventeen, first on the circuit in Tennessee, Virginia, and Kentucky, and later from his own pulpit in Temple, Texas.

He arrived in southern California in 1920, bringing with him the full panopoly of beliefs and biases characteristic of 1920s fundamentalism. Shuler ardently supported Prohibition and regularly railed against evolution. He condemned the sinful manifestations of the Jazz Age — music, dancing, movies, and sexual promiscuity. Shuler made Manichean distinctions between good and evil and saw conspiracies lurking everywhere. He included blacks, Jews, and immigrants among his targets, but reserved his most vitriolic attacks for Catholics and the Roman Catholic Church.

Yet Shuler differed from most of his contemporary ministers in at least one significant way. He professed no strict adherence to the millennial and "born again" teachings so central to 1920s fundamentalism. He clung instead to the evangelical revivalism and social reform common to the Methodist Church of his youth as personified by William Jennings Bryan. In this configuration, writes Shuler's biographer, Mark Sumner Still, "ideals of Christian piety were combined with the ideals of 'the progressing and democratic American nation' to form a commitment to preserve traditional Christianity, broadly interpreted, with an emphasis on the practical and social."

Shuler saw himself as a champion of average, old-stock Americans in their battle against the wealthy and corrupt. About millionaires, he commented: "I've found very few . . . who didn't get their money in a manner that I doubted if God could own or bless." Shuler's commitment was to the poor and working-class white Protestants. "I have been an underdog all my life," he explained. "My sympathies and efforts will always be on the side of the common people." In Texas he had alienated political and church leaders alike with his outspoken challenges to authority.

Shuler believed that his 1920 assignment to Los Angeles was punishment for these Texas activities. His new pulpit, the Trinity Methodist Church, one of the city's oldest and most illustrious congregations, had fallen on hard times. In 1912, Trinity Methodist had mortgaged its future on a lavish nine-story building, the largest religious facility on the Pacific coast. Three years later it had to sell its ambitious headquarters and move to a modest church at Twelfth and Flower streets in downtown Los Angeles. When Shuler arrived, he found a discouraged congregation numbering only 600 active members and burdened by a $70,000 debt.

If Shuler's superiors had hoped to deflate the troublesome minister, they had misjudged both the man and the setting. Shuler proved an ideal match for Los Angeles on the eve of the great boom. Like his contemporary rival Aimee Semple McPherson, who preached a less contentious road to salvation, Shuler understood the insecurities and anxieties of the Midwestern Protestant refugees who had flocked to Los Angeles since the 1880s. As Edmund Wilson observed, "Bob Shuler managed a perfect appeal to these retired farmers and their families. . . . They were delighted to have their imaginations stimulated with ideas of sexual and other vices at the same time they were emphatically assured of their own superior righteousness and even allowed, as it were, to have a hand in bringing the wicked to judgment."

To Shuler, Los Angeles was "the only Anglo-Saxon city of a million population left in America. It is the only city that is not dominated by foreigners. It remains in a class by itself as the one city in the nation in which the white, Christian idealism still predominates." Shuler sought to become, in his own words, "one of God's watchmen on the walls of this city," an avenging angel poised to expose and punish the works of the sinful.

On December 27, 1920, exactly three months after his arrival, Shuler began his crusade with a Sunday sermon entitled, "Vernon Country Club vs. Decency—Will Los Angeles Stand for It?" Shuler offered a graphic description of a suburban Christmas party where a thousand people, "engaged in a drunken carousel . . . hugging, kissing, in drunken fashion; [with] women displaying their nakedness brazenly, openly, flagrantly and viciously . . . [and] the most suggestive dancing engaged in." The local press gave the Vernon Country Club sermon widespread coverage. Other Shuler sermons exposed high school girls who posed for nude photographs and public school teachers caught in extramarital affairs. In 1921, Shuler began to devote the first hour of his evening service to a discussion of moral issues. Crowds overflowed Trinity Methodist in anticipation of his exposés.

Not surprisingly, Shuler's constituency overlapped that of the rapidly growing southern California Ku Klux Klan. Shuler, while never a member, became a leading advocate for the "Invisible Empire." He argued that the Klan "is the result of conditions. . . . They have found the Jew gradually taking over the nation financially and the Roman Catholic Church as surely taking over the nation politically. So they are here." In 1922, Shuler launched *Bob Shuler's Magazine*, a monthly journal. The magazine became one of the primary outlets for Klan publicity in California. Klan members reciprocated by advertising in its pages. Shuler's support proved indispensable to Klan recruiting efforts.

Their common sympathies notwithstanding, by 1924, the Klan ran afoul of Shuler's stringent standards. Shuler had targeted the police department and public officials in his campaign to purify the city. The first indication of Shuler's growing influence occurred in 1923, when he succeeded in persuading Mayor Cryer to remove Police Chief Louis Oakes, a Klan member, whom Shuler charged with corruption. The following year, when Klan leaders endorsed District Attorney Asa Keyes, the preacher denounced this position as "despotic and unAmerican." Shuler and the Klan split bitterly on the issue. Without Shuler's support and racked by national scandals, Klan influence in southern California declined dramatically. Shuler's star, however, continued to ascend.

To supplement his journal, Shuler began to publish a series of pamphlets. "McPhersonism" attacked his rival Aimee Semple McPherson for her immorality, financial practices, and the healings and speaking in tongues that highlighted her services. "Beastism" assailed the Roman Catholic Church. His 1928 pamphlet "Al Smith", attacking the Catholic Democratic presidential candidate, "reveals Shuler at his worst, sometimes crossing the boundary into virtual paranoia," writes Still. These and other pamphlets, priced at 25 cents each, sold tens of thousands of copies.

In September 1926, an unsolicited donation laid the most significant cornerstone of Shuler's rise. Lizzie H. Glide, the wife of a Sacramento oil millionaire, gave Shuler $25,000 to build a radio station. Mrs. Glide (who also endowed the Glide Memorial Church in San Francisco) specified that the station be used for "the spread of the Gospel and the proclamation of righteousness, justice and truth" and forbade any teaching of modernism or principles contrary to fundamentalism. Aimee Semple McPherson had preached her version of the gospel over the airwaves since 1923, greatly expanding her following. Now Shuler would join her among the ranks of the pioneer radio evangelists. Shuler constructed two broadcasting towers above Trinity Methodist and on December 26, 1926, began broadcasting over station KGEF.

Unlike McPherson, whose radio broadcasts reflected the spectacle and glamor of her religious services, Shuler used his broadcast pulpit to excoriate an endless array of moral and political evildoers. Armed with information provided by an entourage of spies, both paid and unpaid, Shuler took on the city's power élite, public officials, rival religious leaders, and myriad others who violated his rigid precepts. Impetuous by nature, Shuler often failed to distinguish rumor from fact, printing or broadcasting all tales that crossed his path. To the *Los Angeles Times*, a frequent target, Shuler would become the "Champion 'aginner' of the universe" with a "penchant for attracting attention by denouncing everything." Others described him as a "moral racketeer" randomly wreaking havoc with his malicious reports. "No politician even dares to run for office if he has ever lunched publicly with a woman not his wife," reported Duncan Aikman.

Ironically, the individual whom Bob Shuler most resembled in his ascent to power was C. C. Julian. They arrived in Los Angeles at the same time and exploited the desires and vulnerabilities of a changing population removed from its roots and seeking economic or spiritual security. Each took advantage of the growing mass market for periodicals, Julian through his advertisements and Shuler with his magazine and pamphlets. By 1927, both men had discovered the power of radio to dispense information and sway popular opinion. They portrayed themselves as outsiders battling conspiratorial cabals controlling the city and directly challenged the alleged agents of these plots. Like Julian, Shuler loved a good fight and tended to personalize his disputes, creating a ready cast of villains and demons. And, like Julian, Shuler possessed a certain undeniable charm.

Unlike many other celebrated evangelists, Shuler adhered to the moral code he preached. He acquired no great wealth and remained untouched by personal scandal. "They can't find a scratch on my name if they hunt back to the cradle," he boasted. "I have never fooled with liquor, gambling, a woman or a dishonest deal." Even his opponents often gave him grudging respect. A reporter from the *Record*, which for years warred against the preacher, found Shuler "considerably more likable" than he expected and concluded, "If you admire complete sincerity you can heartily admire Shuler." Frank Doherty, who as a Catholic and political liberal disapproved of Shuler, wrote to Senator Hiram Johnson, "I have always been of the opinion that Bob too often accepted as truth the thinnest gossip and did a great deal of harm in his attacks over the radio. At the same time . . . I never questioned Shuler's sincerity or desire to accomplish good and remedy what he deemed to be wrong."

By May 1927, when Julian Petroleum collapsed, Shuler, not yet at the peak of his power, stood poised to enter the fray via pulpit, periodical, pamphlet, and radio. The Julian Pete scandal, with its conspiracy theories and lengthy cast of wealthy predators, Jewish moneylenders, high-living exploiters, and compromised public officials, offered ample grist for Shuler's radio and periodical mills. He repeatedly attacked the unholy alliance of businessmen and politicians who had assisted S. C. Lewis and Jacob Berman, predicting that none would receive the punishment they deserved. The collapse of the Julian Pete prosecution and dismissal of usury charges came as no surprise to Shuler's followers. The exposure of Asa Keyes vindicated Shuler's long campaign against him and enhanced the preacher's credibility.

The Julian Pete revelations also helped doom the "Cryer-Parrot political machine." The involvement of Charles Crawford, Bob Gans, Walter B. Allen, and other political insiders in the Julian pools lent support to Shuler's allegations of corruption. On February 28, 1928, Cryer announced that he would not run for reelection. His withdrawal created a vacuum, and fourteen aspiring politicos scrambled to fill the void in the 1929 election. The *Times* and its conservative allies backed John Quinn, a former national commander of the American Legion. Parrot sponsored automobile dealer Perry Greer, who, as a former Julian pool member, ran under a heavy burden. City Council President William Bonelli won the endorsements of liberals and labor unions.

Most observers assumed that the next mayor would emerge from this trio of aspirants. Few paid any attention to dark-horse candidate John R. Porter. Porter had few qualifications to run the city. A junk dealer who specialized in secondhand auto parts, Porter had migrated from Iowa, joined and then quit the Ku Klux Klan, and become active in the Federated Church Brotherhood. His only previous political post was as a member of the grand jury that had indicted Asa Keyes. Dull and colorless, Porter possessed but one major asset, the unswerving support of the Reverend Shuler. As the May 1929 election approached, Shuler alone predicted a Porter victory. Each Julian victim, he noted, cast as many votes as Harry Chandler, Kent Parrot, or Charlie Crawford. Porter polled 76,198 votes, more than 30,000 votes better than Bonelli, his nearest competitor. He then crushed Bonelli (who targeted his campaign against "Shulerism") in the run-off. In addition, eight of nine city council candidates endorsed by Shuler also triumphed. With the 1929 election, Shuler had emerged as the dominant force in Los Angeles politics.

Porter's success inspired imitation. In January 1929, when Governor Young appointed City Prosecutor Lickley to the Municipal Court, Mayor Cryer had named Lloyd Nix, yet another "professional Legionaire," to succeed Lickley. Nix discovered among the unfinished business in the prosecutor's office the 143 John Doe usury complaints that Lickley had filed but never prosecuted in 1927. When Jacob Berman promised Nix that he would take the stand against the accused usurers, the ambitious official tried to advance his political career by reopening these cases. At his request, the city council allocated $25,000 to employ a team of special prosecutors.

On April 12, Los Angeles sheriffs arrested brokers C. C. Streeter and Max Slossberg on the twenty-month-old charges. During the next several weeks, dozens of others who had loaned money or purchased stock from Jacob Berman were arraigned. While the indictments included several prominent individuals previously mentioned in the scandals, like Harry Haldeman and Cecil B. DeMille, the vast majority of those named were relatively anonymous investors and moneylenders.

Nix's crusade against the accused Julian usurers advanced at a promising pace.

Nix found a staunch ally in Judge Charles L. Bogue, who heard the initial cases. "The Julian case is a national scandal in all its ramifications," proclaimed Bogue. "Anyone connected with it who is accused by any theory of law ought to be put on trial and, if he is guilty, found guilty and punished." Bogue ruled that Lickley's promise of clemency for those who "put back" money was illegal and not binding on the court. Those found guilty would be jailed rather than fined, added Bogue. "A fine in these Julian cases would be more than ridiculous," he declared. "It would amount to the State's condoning and sharing in the illegal profits of those who added to the Julian debacle."

On May 27, 1929, apartment-house owner Raburn Eberts pleaded guilty to twenty-two counts of usury, giving Nix his first conviction. Eberts claimed to have acted on behalf of an unnamed banker and to still hold $46,000 in promissory notes from Berman. Judge Bogue showed no leniency. He sentenced Eberts to ninety days in jail. When Eberts protested that he was just another victim of the Julian swindle, Bogue replied, "We hope to get some of those 'big bugs' in court soon, and when we do they'll go to jail just as you are going." Eberts became the first person to serve time for activities leading to the Julian Pete crash. The following week Bogue issued a 180-day sentence to garment wholesaler Arthur Perry and accepted a guilty plea from Samuel Benson.

The prosecutions quickly unraveled, however, when they shifted from Judge Bogue's courtroom. On June 7, Judge Pope dismissed all charges against business-man Meyer Brown. "This case must be decided according to the law and the evidence," stated Pope. "It is not for the court to stretch a point merely because these cases have been attended with considerable publicity. . . . On its face this is a contract for sale and repurchase of stock and this is all the evidence sets forth. I cannot assume that this is usury."

Four days later Nix suffered another setback in Pope's courtroom when he called Jacob Berman to testify at the trial of Louis Rothman. Berman violated his earlier promise and took the Fifth Amendment. Lewis and Ed Rosenberg followed suit. On June 13, Judge Pope declared that the prosecution had failed to prove usury and acquitted Rothman. The case against C. C. Streeter met an identical fate. Nix put Berman on the stand, but he again refused to testify. Pope dismissed the charges.

The series of adverse decisions broke the back of the Nix initiative. On June 19, Nix announced that, while his office would continue with the remaining trials, he had discharged the special prosecutors. The unwillingness of Berman to testify had rendered their efforts "an unjustifiable waste of public money." Two days later Nix himself led the prosecution against Harry Haldeman in what was viewed as a test case for the Bankers' Pools. After listening to the evidence and reviewing the contracts, Judge Pope, in language that effectively doomed future cases, instructed the jury to acquit Haldeman. "It is my duty to advise you that under the law, the documents produced here were merely contracts of sales and repurchase," stated Pope. "It is incumbent on the prosecution to prove the illegality of the instrument and that they are in the character of a loan. This has not been shown. . . . Under the law the only verdict that can be brought is one of not guilty."

The usury prosecutions, which Nix had hoped would provide impetus to his political career, now stood in shambles. On June 24, Nix moved to dismiss the remaining 137 cases. Judge Pope shocked the courtroom by denying the motion and reprimanding the city prosecutor. Pope declared:

> To grant this motion would mean that all the actions which are before the courts growing out of the affairs of the Julian Petroleum Corporation would end. There would then remain on the public records as testimonial to the efforts of the law enforcement officers of this community . . . nothing other than that the District Attorney of this county was in prison. . . . In other words it would appear to indicate that a financial catastrophe had overtaken this community and that the administration of justice had broken down. When justice fails very little else remains.

Pope ordered Nix to try each of the remaining cases separately. When Nix protested that the judge had ruled "that there is no usury involved in these transactions," Pope responded, "I said there has been no proof of usury introduced into this court." Pope criticized Nix for his inability to get convictions, arguing that, given the magnitude and duration of Berman's transactions, evidence of wrongdoing must exist.

A chastened Nix continued to prosecute the usury cases and even won one conviction. The charade ended on July 5 when the California Supreme Court agreed to hear a petition on behalf of several defendants charging that the original issuance of John Doe complaints and the almost two-year delay in their service had violated their rights to a speedy trial. Nix suspended all prosecutions pending a decision. On February 25, 1930, the Supreme Court ruled in favor of the defendants, and Judge Pope quickly dismissed all charges.

The 1929 usury cases, in addition to embarrassing Nix, became yet another failed attempt to punish those who had traded with Lewis and Berman. While four relatively minor moneylenders were convicted or pleaded guilty and three actually spent time in jail, the remainder of the 143 defendants had again escaped retribution. The February 1930 Supreme Court decision, however, landed with no jarring impact. Due to the persistence of Arthur Loeb, the seemingly dormant Julian Petroleum volcano had erupted anew.


## II


Just as World War I had transformed Buron Fitts, the Julian Petroleum debacle remolded Arthur Loeb. The scandal produced an epiphany for Loeb—mobilizing him, maiming him, and giving him a cause that irreversibly altered his life. A former engineer who had worked on the construction of the Panama Canal, Loeb had invested tens of thousands of dollars in Julian Petroleum stock. The crash depleted his life savings and plunged him into the affairs of the Stockholders' Committee. After his violent clash with DeKalb Spurlin and H. L. Kimmerle cost him his left eye, Loeb became obsessed with the Julian affair. By his own account, after June 1, 1927, he "had no other occupation" other than monitoring the affairs of Julian Petroleum, its receivership, and, later, Sunset Pacific.

Acting at times as an official of the Stockholders' Committee and at other times on his own initiative and expense, Loeb had spent several years and thousands of dollars compiling lists of stockholders, coordinating their activities, and investigating leads on the Julian cases. He assembled a half-dozen scrapbooks of newspaper clippings and rented offices adjacent to those of the receivers to oversee their activities. In 1928, Loeb played a major role in exposing Asa Keyes. Later that year, he unsuccessfully attempted to block the creation of Sunset Pacific. Loeb vowed to continue his crusade until the money plundered from Julian Pete was reclaimed from those who had wrecked the company and restored to its rightful owners, the company's legitimate stockholders.

Like so many before him, however, Arthur Loeb could not resist the persuasive charms of S. C. Lewis. Loeb believed that Lewis held the key to unraveling the Julian mess. In his expense accounts he made repeated entries on Lewis's behalf, arguing, "it was necessary to keep him friendly as he had valuable information and was the principal witness." On various occasions Loeb paid Lewis's entertainment expenses, rent, and fees for Lewis's "corps of private investigators." But as he had done with the receivers, Lewis steered Loeb along a course which more closely adhered to his own hidden agenda than to that of the defrauded stockholders.

Since the crash, Lewis had repeatedly offered an alternative explanation for the Julian fiasco: local stockbrokers had repeatedly whiplashed Lewis and Berman, forcing them to overissue stock while pocketing millions of dollars from stock manipulation, commissions, and margin loans. These individuals, Lewis argued, "were the real 'profiteers' in the sale of this stock." Since the receivers' efforts to recover money from pool members and moneylenders had proven inadequate, Lewis urged Loeb to sue these brokers. Municipal court judges had repeatedly ruled against stockholders seeking to recover money paid to brokers in the Julian affair, reasoning that if brokers came by the shares honestly and acted in good faith they had no liability for the overissue. Lewis nonetheless maintained that, since these men knew the stock was overissued and yet continued to trade in it, a suit against the LASE and individual brokers could succeed.

In August 1928, Loeb engaged an attorney to investigate such a suit, but receivers Scott and Carnahan repeatedly blocked his efforts to examine corporate records. Lewis recommended that that Loeb consult Guy Crump, a former judge and current head of the Los Angeles Bar Association. Unbeknownst to Loeb, Crump had previously represented Lewis Oil. In June 1929, Lewis composed a four-page single-spaced memorandum for Crump addressing the legal issues involved and the liability of several individual brokers and the LASE as an institution. In a separate letter, Lewis pledged "to render any and all cooperation . . . gratis" in such a suit. After reviewing Lewis's arguments and supporting documentation, Crump agreed that either Sunset Pacific, as successor to Julian Petroleum, or individual stockholders might sue the brokers and the LASE. Crump warned that Lewis would also have to be a defendant. Loeb later alleged that Crump refused to file a similar complaint against the First National Bank because "it was too much dynamite."

Armed with Crump's favorable opinion, Loeb attempted to persuade Carnahan, now president of Sunset Pacific, to file suit on behalf of the company, but Carnahan declined to participate. Instead, Loeb and Lewis persuaded several of the larger Julian shareholders, including Adolph Ramish and former California-Eastern directors Joseph Dabney and L. J. King to post a $20,000 retainer to Crump. On October 8, 1929, Crump filed suit on behalf of Loeb and seven other Julian Petroleum shareholders against the LASE, forty-six individual brokers, and nineteen partnerships. Lewis, Berman, and several employees of the Julian Petroleum transfer department were also named as defendants. Demanding $12 million, the suit alleged that the brokerages had continued to sell Julian Pete securities long after they knew of the overissue. "The stockholders who are bringing this suit take the position that all who knowingly made money from the fraud should return it to the corporation," announced Crump.

The pretrial proceedings began three months later on January 13, 1930, amidst a tense atmosphere. That morning a federal grand jury indicted Lewis, Berman, and Ed Rosenberg on twelve counts of using the mails to defraud during their Julian Petroleum stock-issuing binge. Three days later the federal grand jury filed income tax evasion charges against C. C. Streeter for the period covered by the Julian Pete stock manipulations. The lawsuit against the brokerages, however, quickly seized center stage, imploding with reverberations unforeseen by any of the participants. Crump opened with an attempt to obtain a deposition from Jacob Berman. When Berman refused to testify on the grounds that he might incriminate himself, Crump charged that the defendants had silenced Berman with bribes. To substantiate his allegations, Crump introduced a new player in the drama, S. C. Lewis's former private secretary, Leontine Johnson.

Over the next week the attractive, auburn-haired Johnson became what the *Examiner* dubbed a "perpetual surprise witness," unveiling a "flood of sensations." Johnson testified that stockbroker H. B. Chessher had approached her in early October, even before the stockholders' suit had been filed, and offered her $30,000 to provide information about Lewis, Crump, and their legal preparations. Several other brokers later supported Chessher in this arrangement. Johnson testified that she accepted a $5,000 down-payment and a weekly salary but simultaneously informed Lewis of the plot. Lewis encouraged her to continue working with the brokers so that he might feed them false information. Lewis and Johnson put together two suitcases of documents from his files. Most of these materials were legitimate, but Lewis also dictated new correspondence indicating that Berman was helping the stockholders and double-crossing the brokers.

Johnson liked to portray herself as an innocent. "I have been kind, benevolent and generous," she described herself in one letter. "I came from a little Georgia town . . . and lived a clean, upright and honorable life." As a secretary she had worked for film industry censor Will Hays and the president of the DeSoto Motor Company, and boasted "some of the finest letters of recommendation in my possession any girl ever received." But her dealings with Lewis and the stockbrokers belied this image. She testified that in December 1929, Chessher asked her to sign an affidavit naming Lewis as the instigator and financier behind the stockholders' suit and the controlling figure in a new California Stock Exchange that would soon begin operating. Johnson demanded $100,000 before signing; Chessher told her that the original $30,000 offer "was plenty of money for any little girl." At this point Johnson developed second thoughts about her relationship with Lewis. "When the deal became so alluring and mention of $100,000 was made . . . I stopped telling Lewis everything," she revealed. "I went into business myself."

Remarkably, defense attorneys corroborated Johnson's astounding story but offered their own interpretation. They alleged that Johnson and Lewis had hatched a blackmail plot, and they had cooperated to smoke them out. "We knew it was the rankest kind of shakedown, but we played along in the hope that something of value might turn up and to determine just to what lengths the thing would be carried," explained one attorney.

Johnson's spellbinding tale of "intrigue and doublecross" opened the lid on a Pandora's box of revelations. Johnson reported rumors that Berman and Lewis had paid $10,000 to bribe a woman juror in the original Julian trial. She also recalled hiring an instructor in Slavic languages to coach Lewis so that he could flee to Russia. After two days on the stand, Johnson took a day off due to "nervous collapse" but returned on January 17, armed with a "little gray account book" with the names of those contributing to the stockholders' "war fund" and two suitcases of documents belonging to Lewis, Julian Petroleum, and Crump's law firm. After defense attorneys demanded that they be allowed to examine the contents of the suitcases, the judge placed the satchels under the jurisdiction of Johnson's attorney pending a decision.

After a weekend recess, Johnson resumed her exposés on January 20. According to Johnson, she, Lewis, and her former husband, H. R. Rule, had organized a brokerage house, H. R. Rule and Company, to trade on the new California Stock Exchange. Lewis owned 50 percent of the company and also controlled fifty-two seats on the new exchange. Despite her repeated denials that Lewis, rather than Loeb, had instigated the stockholders' suit, Johnson's testimony gave substance to defense charges that the real purpose of the suit was to discredit the LASE and advance the interests of the California Stock Exchange.

On January 21, Lewis replaced Johnson on the witness stand. Lewis had promised to "waive all my constitutional rights and immunities in court . . . and tell everything I know inside and outside." But Lewis, while acknowledging a major role in the California Stock Exchange, embarked on a rambling account of his dealings with various stockbrokers from 1925 to 1927 and their culpability for the overissue. Lewis continued his testimony for over a month.

As Lewis droned on, Leontine Johnson plotted to capitalize on her intimate knowledge and the two suitcases of documents. On March 10, the *Examiner* unveiled the first of a series of articles written by Johnson as part of a book entitled *My Three Years with S. C. Lewis; or, The Truth about Julian Pete*. The *Examiner* promised "amazing inside details" of the Julian Petroleum collapse, but Johnson's first installment dealt solely with Lewis's post-Julian activities. According to Johnson, Lewis had spent money so rapidly that, when arrested, he needed her to post his bond and pay his expenses. During 1929, Lewis had sold his seats on the California

Stock Exchange for more than $120,000 and made another $55,000 on stock manipulations on an independent oil company. Lewis "squandered the money as fast as he got it," renting three suites a day in expensive downtown hotels and "entertaining himself with 'wine women and song,'" while simultaneously sending thousands of dollars a month to his wife, touring Europe.

On the afternoon of March 10, before a second installment could appear, *Examiner* reporter Morris Lavine, who had ghostwritten Johnson's account, appeared at underworld boss Charlie Crawford's Sunset Boulevard offices. Lavine was no less a local legend than Crawford. Leslie White wrote that Lavine, "with his peculiar mincing gait and arrogant manner, was an institution around the courts . . . and a power in political circles." At the the old courthouse press room, recalled Robert Kenney, a large streamer signed "Morris Lavine" ran across the ceiling proclaiming, "Thou shalt have no other God before me." Lavine came to Crawford's office not as a reporter, however, but as a blackmailer.

According to testimony later released by the grand jury, Lavine and Johnson had hatched a scheme to extort large sums of money from Julian participants in exchange for keeping their names out of the *Examiner* series. The preceding week they had approached Jacob Berman and asked him to act as a middleman. "I know a lot of things about Lewis and Charlie Crawford and Abe Morris and Kent Parrot and Jack Friedlander that wouldn't be so good if they were put on paper," Johnson allegedly told Berman. Berman contacted Crawford and Morris, who held a midnight meeting with Parrot and Friedlander, who had recently resigned as corporation commissioner. The quartet decided that, rather than submit to extortion, they would attempt to entrap the blackmailers.

Parrot and Friedlander informed Buron Fitts of the plot. When Lavine and Johnson called to arrange the pay-off, Fitts's deputies listened in on the line. They claimed to hear Lavine tell Crawford to "have $75,000 by noon Tuesday as the first payment and I will take care of you and your group." He later instructed Crawford "to line up the rest of the Julian crowd, including the brokers and bankers." Berman withdrew at this point, explaining that Crawford did not trust him with the money and had insisted that Lavine come to get it. On Monday morning, March 10, Johnson and Lavine ran their first article, to "prove they meant business." That afternoon Fitts's detectives, Leslie White and Frank Meyers, hid in an office adjoining Crawford's to overhear Lavine's conversations; Blayney Matthews was posted outside.

Crawford "talked shop and politics" to Lavine before the reporter asked for money. Crawford counted out seventy-five crisp $1,000 banknotes, but, according to the anti-Semitic White, "Lavine's racial avidity cropped out . . . and he counted the bills himself," dividing them into two piles, "fifty thousand to divide with the girl, and the other twenty five was his own 'hold-out.'" "Now, Morrie, I don't want any more chiseling, see?" commented Crawford. Lavine replied, "You don't have to worry, I understand." Lavine then walked out of the office into the waiting arms of Detective Matthews. "Let that be a lesson to you, Morrie," drawled Crawford ironically. "It pays to be honest." One hour later, detectives arrested Johnson.

Both Johnson and Lavine denied any attempted extortion, each blaming the other for their predicament. "I didn't know what Mr. Lavine was doing. He was acting entirely on his own in any money transactions," stated Johnson. Lavine said he went to see Crawford at the request of Johnson, who had threatened to halt their articles if he did not comply. Both also claimed to be victims of a frame-up. Lavine argued that their intent was, not to blackmail the men involved, but to sell them the contents of the two suitcases produced in the stockholders' suit. Crawford, who had been indicted the previous year for attempting to frame a city councilman on sex charges, had set it up to look like an extortion attempt to prevent the *Examiner* series from continuing.

Johnson offered an alternative explanation. The purpose of the arrests was to destroy her credibility and provide attorneys defending the stockbrokers and LASE access to the suitcases and other papers in her possession. "The charge of extortion," she later wrote, "gave them the chance to . . . go through my apartment and trunks and bring to the district attorney's office every paper, letter, and everything else they saw fit to bring in. . . . All of my documents, letters; in fact, everything I was saving for years was brought in, dumped on the table in the District Attorney's office, and all of the lawyers invited to a feast for the eyes."

The Lavine-Johnson arrests staggered a populace already reeling from an overload of Julian Pete sensations. Not only was one of its most prominent reporters charged with blackmailing political figures, but the suddenly omnipresent suitcases now threatened to unveil further scandals. "WHAT WERE THE FACTS THE PROMISED SUPPRESSION OF WHICH JUSTIFIED THE POSSIBLE PAYMENT OF A $75,000 BRIBE?" wondered the *Examiner*, after suspending Lavine and canceling the Johnson articles. The *Express* described the suitcases as "loaded with dynamite . . . the 'spark which will set off the entire Julian case' anew." The satchels, reported the *Express*, held "evidence which has never been revealed . . . reported to be the most sensational yet unearthed in the Julian case," implicating "men high in civic and business life." With these materials in the hands of Buron Fitts, "for the first time . . . a sincere effort will be made to unearth the full story of the Julian crash—the who, when, and where of the entire affair." "This phase of the case," warned the *Express* in a forecast that would prove prophetic, "is but a 'drop in the bucket' to the revelations expected to be made."

## III

"What is the matter with justice in California? Why has its machinery slipped so many cogs in prosecuting the Julian case?" asked the *Examiner* in the aftermath of the Lavine-Johnson arrests. Buron Fitts hastened to reopen the Julian Pete case, but not in time to escape the scathing denunciations of his opponents. "Mr. Fitts has been in office now . . . fifteen months," attacked the *Record*. "Nobody is in prison except Asa Keyes." The tabloid noted that the three-year statute of limitations on crimes related to Julian Petroleum stock manipulations would expire on May 6,

1930, putting "beyond the reach of the prosecution all the higher-ups who participated in this looting of the Julian company." In subsequent front-page editorials,
the *Record* began counting down the number of days left to secure indictments; a
series of scathing cartoons showed "Jack" Berman taking his "immunity bath."

For Fitts, who planned to challenge Governor C. C. Young in the 1930 gubernatorial race, the clamor for action assumed added significance. The new Julian
Pete prosecutions offered unparalleled opportunities for implicating Parrot, who
had coordinated Young's 1926 campaign, and Friedlander, Young's appointee as
corporation commissioner.

Before Fitts could capitalize on this windfall, however, the *Examiner*, assisted
by Arthur Loeb, exploded another bombshell. On March 19, the *Examiner* published an affidavit from Carl Vianelli, an employee of Jacob and Louis Berman,
alleging that the brothers not only paid off Asa Keyes, but also conspired with private
detective Louis Krause to bribe two jurors in the 1928 Julian Petroleum trial.
Vianelli, who had worked as handyman, bartender, and watchman at the Bermans'
Hollywood home, reported that during the trial months Krause had approached
Berman and offered to "help him with the jury." Berman instructed the detective
to recruit "a couple of jurors on his side" and promised "to stand the expense,
whatever it was." Krause received $100 a night to entertain jurors with dinners and
theater tickets, and, related Vianelli, "Under instructions from Bennett I gave Krause
two to three bottles of wine . . . also two or three bottles of whiskey, either Scotch
or Bourbon." Ben Getzoff also visited the Berman home to tap the liquor supply.
"I had to keep Keyes drunk, as that was the only way I could handle him," he told
Vianelli.

According to Vianelli, Krause identified juror John Groves as a likely accomplice and plied him with drinks (and, it later developed, $5,000 in cash) throughout
the trial. Meanwhile, Berman, in the constant company of an attractive movie
actress, "gave house parties almost continuously," drawing from a $4,000 stash of
illegal liquor. On the night of his acquittal, Berman threw a lavish celebration.
Krause and Groves sat in the breakfast room, both "very drunk." Groves returned
to Berman's home several days later as a dinner guest.

Berman also had planted Krause in an apartment in a four-unit house owned
and occupied by juror Frank Grider. "Jack Bennett put me in here and even bought
this furniture because he wants me to be near this juror Grider, so that I can work
on him," Krause told Vianelli. Krause arranged for Berman, using his dentist as a
front, to purchase Grider's house at an inflated price. The dentist later deeded the
property to Louis Berman, who transferred it to his brother's bail bondsman.

The *Examiner* exposé upstaged Fitts's own no-less-spectacular revelations of jury
tampering. That afternoon Fitts issued arrest warrants for those named in the
Vianelli affidavit, and also for Lewis; a third juror, Caroline Love; and Jack Weaver,
the alleged go-between in an attempt to bribe Love. Weaver, who like Vianelli had
been unearthed by Loeb and the *Examiner*, was a sixty-year-old San Pedro rooming-
house resident who had previously dealt with Lewis and Berman in Julian Pete

securities. According to the new charges, he had approached Lewis at the trial and informed him that he could negotiate a deal with Love.

Caroline Love seemed an improbable candidate for corruption. A plump, middle-aged mother of four who spoke with a soft Southern accent, Love boasted a long record of community service in her San Pedro hometown. She was a trustee of the First Presbyterian Church; a former president of its ladies' aid society, the Cheerful Workers; and a former official of the Los Angeles Federated Women's Clubs. But Love's husband, an inveterate drinker, had become obsessed with the possibility of profiting from the trial. He and Weaver convinced her she could collaborate.

According to Weaver, Love initially asked for $20,000, but, when Lewis balked, agreed to $10,000. Lewis paid the money to Weaver, who appropriated it for his own use. After three days Lewis confronted an inebriated Weaver in a hotel and demanded the return of the money. Weaver, who had already spent $2,900, turned over the remainder. At this point, Love took matters into her own hands. She wrote to Lewis, advising him that her husband and Weaver were unreliable and proposing direct negotiations. If Lewis was interested, advised Love, he should wear a white flower in his buttonhole at the trial. Lewis did so, and the two later met and agreed on a subsequent payment. Love, however, got "cold feet" and declined the money, though she nonetheless voted for acquittal. "I was glad it turned out like that, for I had a feeling Mr. Lewis was innocent," she told Fitts.

While Lewis shouted that he had been framed, Fitts obtained confessions from Weaver and the Loves, as well as corroboration of the Vianelli affidavit from Jacob Berman and Krause. Declaring that this "was but the start," Fitts promised to "tear the lid off the whole affair." But Fitts had apparently known of the bribery plots for over a year, and had failed to act. Vianelli and Krause had both come to Fitts with their stories, and the *Times* had forwarded Viannelli's story to the district attorney six months earlier. Whether due to a lack of evidence, or, as some charged, to protect his star witness in the Keyes case, Jacob Berman, Fitts had ignored the reports of jury bribery.

Speculation ran rampant that the district attorney would once again grant Berman immunity. The *Record* called Berman "Immunity Jack, and reported that "Jack Bennett is a 'big shot' at the Hall of Justice these days. The bright boy . . . has a very assured manner as he prances about the marble halls of the official building." Even the *Times*, a Fitts supporter, noted the reappearance of Berman at the district attorney's office, armed with recently rediscovered documents, and attributed "Berman's new desire to talk" to his expected "bid for immunity." The "Julian mess," added the *Times*, had become an apparently "bottomless pit."

Fitts's allegiance to the "bright youngster" became apparent on March 24 when Judge Walton Wood assigned Berman's bribery trial to his colleague Marshall McComb. Several weeks earlier Judge McComb had attempted to force Berman and Ben Getzoff to stand trial on their original indictment for bribing Asa Keyes. The case had been continued ten times but never dismissed. When Fitts asked for dismissal of the charges based on his promises of immunity, McComb refused.

Attorneys for Berman and Getzoff entered a plea of double jeopardy, since their clients had been co-defendants in the earlier trial, and filed affidavits of prejudice against McComb. On March 1, the Appellate Court accepted the double jeopardy argument, forcing McComb to dismiss these charges, but exonerated McComb of any prejudice.

When Judge Wood again placed Berman's fate in McComb's hands, Fitts vehemently objected. "I have had difficulties with the judge in that court," argued Fitts. "I want to ask this court that the case be assigned to any of the other forty-four divisions of the Superior Court." Wood defended McComb and denied Fitts's request. "Before I'll go to trial before Judge McComb, I'll go to [the federal penitentiary at] McNeil Island and serve my seven years rap for the Federal conviction," announced Berman. "Neither will I talk to the grand jury until my case is set in another court." Fitts pronounced the Julian investigation "at a standstill."

Fitts's dispute with Wood and McComb reflected the broader battle prompted by his gubernatorial aspirations. This clash pitted Fitts and Shuler on one hand and Fitts's opponents, most vocally the *Record*, on the other. Both sides vociferously demanded prosecution of the "big boys" in the Julian Pete scandal. "We don't want little minnows. We want the big fish. We want the whales, and some of them are big enough to swallow Jonah," declared Shuler in a radio address. The *Record*, however, countered that, under Fitts, "not a single indictment of a big Julian crook will be returned and successfully prosecuted to the point where the big crook enters San Quentin prison." But the rival factions could never agree on the identity of these "big crooks." Shuler and Fitts implicated the bankers; the *Record*, the stockbrokers. Both sides agreed on the culpability of Lewis and Berman, but each sought to protect one or the other of these two to procure his critical testimony.

On March 27, Judge McComb withdrew from the bribery cases. With this issue resolved, Fitts brought Berman before the grand jury. Within twenty-four hours the panel issued the long-awaited Julian Pete indictments. Those hoping for new revelations were sorely disappointed. The grand jury simply reindicted the thirty-five members of the Bankers' Pools on the same conspiracy charge that Judge Doran had dismissed in 1928. "Again the district attorney's office has blundered in the prosecution of the Julian 'overissue' cases," protested the *Examiner*, fearing another long parade of unsuccessful prosecutions.

Fitts promised more indictments on stronger charges, but other developments again put him on the defensive. When Fitts moved to prosecute Lewis for attempted bribery of Caroline Love, Lewis played the hole card that Berman had previously bluffed with. On March 28, Lewis, "preferring Uncle Sam's frying pan to the District Attorney's fire," surrendered to federal officials and asked to begin serving his sentence for the Lewis Oil conviction.

"I'm tired of it all," stated Lewis. "For three years I've fought a losing fight. I've tried to do right by the 40,000 persons who lost money when Julian Pete went under. But I'm through. I can't go on." Lewis attacked his bribery indictment as an attempt by the stockbrokers seeking to "protect their own ill-gotten gains." Lewis also said that the recent usury indictments had precipitated his flight. "The District

Attorney is attempting to make a goat out of me and through me to strike at men who are innocent of any wrongdoing in the Julian matter," stated Lewis. By entering prison he could thus "attempt to help men who should never have been charged."

Fitts vowed to bring Lewis to trial. Meanwhile, he secured a second set of indictments from the grand jury. On April 1, Lewis, Berman, Rosenberg, Motley Flint, A. C. Wagy officials Charles and Richard Reese, and surprisingly, H. S. McKay, Jr., who had headed Bankers' Pool No. 1 but had no other connection to Julian Petroleum, were charged with violations of the Corporate Securities Act.

Throughout these chaotic events "Fighting Bob" Shuler had continued his radio harangues denouncing the Julian wreckers, attacking the *Record* and other remnants of the "Cryer-Parrot gang," and defending Fitts. On April 3, Shuler announced the imminent publication of a new pamphlet, "Julian Thieves". "The most sensational booklet I've ever turned out," promised Shuler on his evening broadcast, "written by me as a messenger of righteousness under God."

"Julian Thieves" featured Shuler at his best and worst. The booklet alternated perceptive analysis with broad accusations based on rumor and innuendo. Shuler warned that people should not be misled by the *Record* and *Examiner* into thinking that Lewis and Berman were the only thieves. "There are banker thieves and broker thieves and lawyer thieves and thieves that belong to many other lines of endeavor and enterprise," he wrote. "If the present Grand Jury does not bring in 200 indictments, it will be because they grew weary before their work was done."

Shuler recounted how Berman, a "human scavenger," "took in" the banking crowd, while Lewis allowed them to gut and bleed the corpse of the company. Receivers Scott and Carnahan had "finished the job." He described the stock holdings of Kent Parrot and his allies, questioned the activities of local judges, some of whom, he charged, had invested in Julian Pete stock, and attacked city prosecutors Lickley and Nix for their inability to secure favorable verdicts in the usury cases. Shuler rehashed the familiar story of the overissue and the pools and speculated on where the money had gone. "In my opinion," he wrote, "two of the richest men in America, when the seven years at McNeil Island are up, will be S. C. Lewis and Jack Bennett." But, charged Shuler, "it is certain . . . that neither Lewis nor Bennett salted the whole steal away. There are a good many millions somewhere else."

Amidst this tirade Shuler defended Buron Fitts, who was "too brave, too splendidly fine, too promising in the paths that lead the way to the future," to be made a fool of by even a "cunning human rat" like Jacob Berman. "If he must grant him immunity that the corruption of the community be purged, well and good," allowed Shuler. "Get all you can out of him, Buron, that will help to produce justice and vindicate right."

The most striking sections of "Julian Thieves", however (which by Shuler's reckoning sold 5,000 copies in two weeks and almost double that number during the first month), contained sensational new charges directed at Governor C. C. Young. These allegations opened a new phase of Julian Pete controversies that ultimately would decide the outcome of the 1930 gubernatorial race and the political future of California.

# 17 NEW VISTAS

# OF ROTTENNESS

I

As the new decade dawned, California governor C. C. Young could look back at a term of substantial achievement and forward to renomination and re-election in 1930. Young had rejuvenated the spirit of Progressivism in California. His administration had completed the consolidation of government agencies, secured passage of conservation measures and humanitarian reforms (including the nation's pioneer old-age pension law), and restructured California's tax and budget practices. Young faced stiff opposition in the Republican primary from San Francisco mayor James Rolph and Los Angeles district attorney Buron Fitts, but he hoped to attract northern California Progressives with his record and southern Californians with his support for Prohibition. Two elements, however, complicated his campaign: general discontent stirred by the deepening economic depression and the gathering storm over the Julian Petroleum scandal.

"Unless all signs fail, the Julian debacle will elect the next Governor of California," predicted the weekly *Los Angeles Review* on March 20, 1930. The *Review* resuscitated the three-year-old rumors of irregularities in the appointment of Jack Friedlander as corporation commissioner, reporting that Governor Young would have to answer charges that Julian Pete interests had contributed thousands of dollars to his 1926 campaign in exchange for a pledge that Mike Daugherty would not be retained. Three days later, "Fighting Bob" Shuler broadcast the *Review* charges to his radio audience. Young's appointment of Friedlander, at the behest of Kent Parrot and Charles Crawford, had made the entire Julian affair possible, proclaimed Shuler.

Armed with illegally acquired information from unpublished grand jury transcripts, Shuler gave substance to these charges in "Julian Thieves". He reported that Jacob Berman had testified that Lewis had donated $100,000 to the Young campaign in exchange for promises that Friedlander would be appointed and that, thereafter, state investigations of Julian Petroleum would cease.

On April 6, one day after "Julian Thieves" went on sale, Buron Fitts released Berman's testimony, which went beyond even Shuler's account. Berman reported that he had given Lewis between $200,000 and $250,000 for Young's campaign fund. Lewis told him that, in exchange for these donations, Jack Friedlander would become the corporation commissioner and that Friedlander "would make no trouble in arranging any stock structure that [Lewis] wanted in any of the merger negotiations to take care of the over-issue, and that I wasn't to worry about anything." According to Berman, when the appointment was in doubt, Friedlander would call Lewis daily to find out if Young had named him.

Berman further implied that Young had personally accepted funds from Lewis. When Young had visited Los Angeles in February 1927, shortly before appointing Friedlander, Lewis asked Berman for $10,000. Lewis then met with Young at the Biltmore Hotel, after which Lewis told Berman that "he had the direct word" of Governor Young that he would appoint Friedlander.

Berman's account triggered an outcry of denial and protest. Governor Young, who had previously remained silent about the Julian Pete fiasco, replied with vehement indignation. He called Berman's testimony "an absolute and malicious falsehood, politically inspired and without the slightest foundation in fact." Young reported hearing rumors of a conspiracy against him, but, in a barb at Fitts, he refused to believe "that any candidate for high office could lend himself to so shameful a plot." Young also pointed out that Fitts had supported Friedlander in 1927. William J. Carr, treasurer of Young's 1926 campaign, dismissed the charges as frivolous. "The amount which the Julian interests are said to have contributed to the campaign is about as much as the total amount expended," said Carr. Lieutenant-Governor Carnahan reported that Berman had given thorough accounts of his money management to the receivers two years earlier and never mentioned any political contributions. Lewis, Parrot, and Friedlander also denounced Berman's story.

Even Fitts's supporters seemed embarrassed by the allegations. The *Express* called the idea that Young had promised the appointment of Friedlander in exchange for a Lewis contribution "too grotesque to need refutation." The governor, noted the *Express*, rather than promising to appoint Friedlander, had only done so with great reluctance and after intense pressure from southern California businessmen. The *Times*, noting that in exchange for immunity Berman "apparently tried to give authorities their money's worth," also made light of the attack on Young.

Yet, if no one accepted Berman's tale at face value, the controversy brought into question both Young's appointment of Friedlander and the corporation commissioner's handling of the Julian affair. Why had Parrot and his allies insisted so doggedly on Friedlander? "I am not charging that Governor Young is corrupt in this or any other act of his official life," argued Shuler plausibly. In exchange for Parrot's support in the election, Young had simply granted him influence over certain appointments. In response to Carr's defense that, as treasurer of Young's campaign, he would have known of such a large contribution, Shuler pointed out, "It is possibly true that not one cent of Julian money passed through the hands of the state political organization. . . . Kent Parrot and his associates were given a free hand in Southern California. . . . Possibly a hundred thousand dollars was collected and disbursed in that campaign, which amount may never have officially entered upon the books of the Governor's State Campaign."

Nor did anyone charge that Friedlander had committed any overt acts to assist Lewis. "This publication does not mean to impugn [sic] that Jack Friedlander was corrupt," maintained the *Los Angeles Review* in its original article. "But the new corporation commissioner did not have the background of knowledge that Daugherty had. . . . When Friedlander was named . . . he did not cooperate with the Julian promoters. But he did not suspect them; and they got away with murder." Shuler agreed. "If there had been a thoroughly conscientious, sane and courageous Commissioner of Corporations," he argued, events would have evolved differently.

To Shuler, the Friedlander selection illustrated a pattern. Young had also named receiver H. L. Carnahan, "who said he would like to see the Julian usurers go if they gave the money back," to succeed Fitts as lieutenant governor, and E. J. Lickley, "whose relationship to the Julian cases was a running sore," to the Municipal Court. Shuler charged that Lickley was "rewarded" for quashing the usury prosecutions and blocking exposure of the Tia Juana pool.

If Fitts had hoped to use the Berman revelations to weaken Young's campaign, he had succeeded. But he simultaneously damaged his own credibility. Fitts's opponents alleged that he had granted Berman immunity in exchange for fabricated testimony. The *Times* called these charges "too silly for serious consideration," but they are substantiated by other evidence. Leontine Johnson later claimed that upon her arrest she was "dragged in, virtually by the hair of my head and asked how much money was spent for campaign expenses on behalf of C. C. Young . . . and offered a chance to 'turn State's evidence.'" Ray L. Donley, who served on the 1930 grand jury, wrote in 1935 that, although he had been "an enthusiastic admirer" of the district attorney, "It was not long before it became apparent to many of us that the Grand Jury was being used as a political club to advance the candidacy of Mr. Fitts." The Julian Pete case, said Donley, was "the particular vehicle on which Mr. Fitts was supposed to travel . . . . We were asked to bring an indictment on Governor C. C. Young on evidence that was not only unconvincing, but was on the other hand truly rediculous [sic], and apparently manufactured for the express purpose of destroying the character and reputation of the governor and ruining his candidacy."

Both Johnson and Donley, whom Fitts later unsuccessfully indicted (and possibly framed) on corruption charges, had ample reason to smear the district attorney. But their accounts cannot be lightly dismissed. Whatever the truth, the clear victor was the third candidate, James Rolph. "One may imagine the glee of Mayor Rolph . . . at a spectacle involving the partisans of Young and Fitts in a controversy not related to the Governorship race in any respect," observed one political commentator.

"It is to be hoped that the end of the Julian scandal is at least distantly in sight," opined the *Times* amidst the latest controversy. "So far, every turn in the crooked and slimy trail has opened new vistas of rottenness." But developments continued to unfold at a dizzying pace. On April 9, S. C. Lewis departed Los Angeles for the federal penitentiary at McNeil Island. "I don't . . . relish the idea of being made a political sacrifice to Fitts' ambition to become Governor," stated Lewis. "I do not believe that I should be forced to stand trial on these various charges which have been brought against me by Jacob Berman in an effort to save his own skin."

Two days later, Judge McComb thrust Berman's skin into renewed peril. McComb discovered that an indictment against Berman dating from May 27, 1927, just weeks after the Julian Pete crash, was not covered by Fitts's grants of immunity. The indictment charged Berman with fraud and embezzlement for his pre-flight sale of stock to investor Daniel Barnes in April 1927. McComb placed it on the court calendar and ordered Berman held in lieu of $250,000 bail. Fitts angrily denounced McComb's "incomprehensible action," which "places in serious jeopardy, if it does not actually destroy, all hope of convicting any guilty person in any way connected with the gigantic Julian Petroleum Corporation crash." The embezzlement charge, claimed Fitts, had merely been a legal subterfuge to allow Berman's extradition. No one in the district attorney's office believed that a conviction could be obtained on this "unimportant" case. Berman spent a brief time in jail before securing his release on a writ of habeas corpus. "I'm being made a political football of," he protested.

On April 14, Governor Young sent Fitts an eleven-page report prepared by Fred Athearn, Friedlander's successor as corporation commissioner, naming more than a dozen stockbrokers who had illegally profited from Julian Pete stock manipulations. "It appears that the real conspiracy was a conspiracy to overissue stock and that the other crimes for which indictments have thus far been returned were only by-products of this main conspiracy," wrote Young in a direct slap at Fitts. The governor called for "exceedingly prompt measures" due to the impending May 6 statute of limitations.

Fitts countered by flying in C. C. Julian from Oklahoma to testify before the grand jury about bribes paid to an unnamed state official. Before Julian's testimony could be assessed, however, "Death, sudden and tragic stalked into the Julian case snarl to give it the final element of possible sinister mystery," as the *Examiner* melodramatically reported. On the morning of April 15, police found the body of Robert Bursian pinned in the tangled wreckage of a Buick coupe standing upright at the bottom of a 100-foot embankment. Fitts immediately announced that Bursian was an undercover agent working on "a particularly vital phase of the Julian investigations." According to Fitts, Bursian had recently uncovered "extremely important information" which he was to have presented to the grand jury that day. "If it is not foul play, it is a remarkable coincidence," asserted the district attorney.

The initial investigation appeared to bear Fitts out. Police discovered a glass with traces of cyanide in Bursian's car, leading to speculation that Bursian might have died prior to the crash. Rumors spread that Bursian had a special assignment to meet a woman and "under the guise of lovemaking" elicit information linking a prominent politician to the Julian scandal. The appearance of a "young and beautiful brunette" at the district attorney's office and reports of liaisons with a blonde former stockbroker's secretary enlivened the scenario. Within days, however, Bursian's link to the Julian Pete case quickly dissolved. The brunette turned out to be a hostess at the Metropole Cafe whom Bursian (despite a marriage only six months earlier and a second wife in the East) had been dating for several months. She reported that Bursian had been despondent and spoke of suicide. The coroner's report found traces of chloral hydrate "knockout drops" in Bursian's blood but no sign of cyanide. Within a week, investigators had ruled out foul play and were inclined toward theories of suicide or accidental death.

Fitts, while adhering to his claim that Bursian worked for his office, never produced any evidence connecting Bursian to the Julian investigation. The *Record*, always quick to criticize the district attorney, questioned the whole affair. "Fitts produces sensation after sensation which may or may not divert the public mind," charged the *Record*. "He reads in the papers that an obscure bookkeeper . . . has died under tragic circumstances. Immediately this modest dead man, who had no connection with the Julian Pete crimes or criminals, becomes a 'key witness.'"

Meanwhile, Fitts took aim at two of Governor Young's political liabilities, Jack Friedlander and Charlie Crawford. On April 16, the grand jury indicted Friedlander, Crawford, and Lewis on charges of bribery and influence peddling. Based on testimony from Jack Roth, former C. C. Julian business associate and current partner of Morris Lavine, the grand jury alleged that Lewis and Crawford had acted as middlemen for Friedlander in a scheme to shake down applicants for permits from the Department of Corporations. Roth reported paying bribes to either Lewis or Crawford on fourteen separate occasions, with Lavine serving as pay-off man.

Fitts, however, suffered a severe setback on another front. On April 17, the Appellate Court ruled that Judge McComb had acted properly in reinstating the forgery and embezzlement indictment against Berman. McComb ordered Berman to stand trial four days later. Fitts announced that his office would cooperate fully, but the *Record* warned, "Unless you, the public, watch closely, Professor Houdini Fitts may put over another Applesauce trick." The *Record* prediction proved right on target. On the morning of the trial, Berman staged a disappearing act. The "bright boy," although supposedly under continuous guard by Fitts's detectives, fled to Seattle in hopes of joining S. C. Lewis at McNeil Island. Thus, explained Frank Doherty, "We have two instances where two arch crooks broke all speed records attempting to get into the federal penitentiary not because they liked the place, but really because it furnished them a safe haven at least temporarily against all pending state charges here."

Determined to regain custody of Lewis, Fitts also surfaced in Seattle, seeking to spring Lewis from the penitentiary at the same time that Berman was desperately trying to break in. Berman surrendered to federal marshals in Seattle, only to discover that he lacked the appropriate court orders. To avoid state arrest, Berman begged local officials to allow him to remain in the sanctuary of the federal building. When the building closed for the evening, however, waiting deputy sheriffs arrested him. He languished in jail while two conflicting sets of documents, one ordering commitment to the penitentiary, the other extradition to California, raced through the courts. Whichever arrived first would determine his fate. Lewis, meanwhile, claiming to have found in prison his "first peace of mind" in years, desperately fought to stay there. On April 23, however, Fitts prevailed when the United States Attorney General ordered Lewis returned to Los Angeles to stand trial for jury bribery. The following evening Berman "crashed" into prison when a federal judge accepted his commitment papers and directed him to McNeil Island.

In Los Angeles all parties involved charged the others with protecting the "big fish." The *Record* wondered why all of Fitts's sensations involved crimes post-dating the Julian Pete crash and not subject to the impending statute of limitations, which would allow "the BIG Crooks to go scot free." Lewis contended that lawyers for the LASE had arranged his indictment to silence him in the stockholders' suit. Fitts and Berman viewed McComb's actions as designed to block his crucial testimony in the usury and jury-bribing trials. Reverend Shuler denounced a "conspiracy as deep and black as perdition" designed to shield "a score of rich and powerful men who would have been punished for their connection with the Julian outrage had these men remained in Los Angeles County." The entire chain of events reinforced public conceptions of an ongoing Julian Pete cover-up.

In the final days of April, Shuler, who often imagined conspiracies, became the victim of a real plot. The Los Angeles Bar Association, in hopes of preventing the Federal Radio Commission from renewing Shuler's license, had hired stenographers to record all of Shuler's radio talks. On April 29, the Bar Association filed a sixty-three-page affidavit based on these transcripts charging Shuler with contempt of court for remarks intended to "to interfere with, sway, control, embarrass, and influence the actions of Judge McComb," to "sway and intimidate" potential witnesses and jurors, and in other ways to "interfere with the orderly and due administration of justice." The attorney for the Bar Association demanded, "Not merely a fine, but a fine and a term in jail."

The charges against Shuler rested on a questionable legal pretext. As Shuler himself later noted, the concept of contempt of court was traditionally designed to ensure order in the courtroom itself. Shuler's criticisms had been voiced in the media. At a hearing on May 2, a "healthy, hearty, self confident" Shuler, "big and plain as the old brown farm horse," acted as his own lawyer. He confessed that he had attempted to "spur the judges to action in enforcing the law," to help District Attorney Fitts, and "to see the men guilty of wrecking the Julian Corporation convicted," but insisted that he had intended no contempt. "Let all men know that my offense was that I sought to assist in the punishment of a bunch of millionaires who had stolen over a hundred million dollars from more than 40,000 Julian stockholders," proclaimed Shuler. Asked to substantiate his allegations, the minister characteristically explained, "It was my understanding," or "So I was informed," or invoked the proverbial "man who claimed to know." But Shuler nonetheless vowed "I'll lie in jail until I die before I retract anything."

At the hearing Shuler had pleaded that he not be imprisoned to spare his family the stigma of being "kinfolks of a jailbird." But two days later, when the judge sentenced him to twenty days in jail, Shuler leapt to his feet and announced that he wished to begin serving his sentence immediately. Amidst hundreds of cheering supporters, Shuler made a triumphal march from courtroom to jail cell. At the elevator he shook hands with Mayor Porter; Buron Fitts awaited him at the Hall of Justice with words of support. Shuler, meanwhile, distributed copies of a prepared statement describing himself as a "prisoner of war." "I stand convicted of having spoken in defense of justice," wrote Shuler, "I will gladly surrender not only my freedom but my life rather than swerve one jot or tittle from the course that my God and my conscience have dictated."

If Shuler's persecutors had hoped that imprisonment would weaken or embarrass him, they proved sorely mistaken. Pampered by his jailors (one guard was fired for serving Shuler butter rather than the authorized margarine), Shuler held court in his cell, entertaining reporters and a parade of public officials who came to pay their respects. Pictures of Shuler smiling comfortably from behind bars appeared in the local papers. Meanwhile, Trinity Methodist Church unveiled Shuler's sequel to "Julian Thieves," "Julian Thieves in Politics." The experience also gave Shuler inspiration for his next booklet, "Jailed."

Although his incarceration provided a public-relations bonanza for Shuler, he ordered his attorneys to appeal his sentence in order to determine "whether or not we have such a thing as free speech in California." On May 10, after Shuler had served fifteen days of his sentence, the State Supreme Court ordered the preacher released on bond pending a decision.

While the *Times* and *Examiner* supported the contempt sentence, others in the community expressed dismay over Shuler's plight. "In all the writer's previous life he had never heard so much disrespect expressed for the courts as he has heard expressed in the last few days," reported a columnist for the *Hollywood Daily Citizen*. "One man asks: 'Where has the Los Angeles Bar Association been during this Julian mess—when honest people have been robbed of millions?' " Even the *Record*, Shuler's prime antagonist, opposed his imprisonment, arguing tongue-in-cheek that Shuler was but "a mischievous little boy" and "no good is accomplished by putting minors in jail." The *Fullerton Daily News Tribune* offered perhaps the most accurate assessment:

> Whether Shuler has been the victim of injustice or the perpetrator of injustice against certain judges and individuals in Los Angeles, the fact remains that his twenty-day sentence for contempt of court, with its accompanying publicity, has done more to promote Shuler than he could have achieved by the expenditures of hundreds of thousands of dollars. . . . The Los Angeles Bar Association, much as it despises the radio preacher, has done him a tremendous service and played directly into his hands. . . . Whether Shuler is a hero or a charlatan, he will come out of jail ten times stronger than he went in.

## II

While Shuler luxuriated in prison, the statute of limitations for Julian Petroleum crimes expired. Amidst rumors of impending indictments of 100 people, including C. C. Julian, the grand jury launched a frantic last-minute investigation. On April 29, it subpoenaed financial records from five downtown banks. On May 2, H. L. Carnahan "swamped" the panel with "an avalanche of documents" from the receivers' files. On May 4, the jury reportedly probed stockbroker activities, and on May 5, Arthur Loeb appeared before the panel. But there were no further indictments. On May 6, the grand jury foreman announced, "We are through with the Julian investigation unless we find that illegal acts in this connection with stock selling occurred after May 6, 1927."

"They didn't get the big crooks," lamented Arthur Loeb. Loeb, like many others, blamed Buron Fitts. For two years, said Loeb, he had gone to Fitts with "conclusive evidence," and for a long time he had believed that "Fitts had something up his sleeve." But the district attorney never called him before the grand jury, and when Loeb had voluntarily testified on the final day, the jurors ignored his evidence. "Could the 'Julian Thieves' have more completely escaped punishment [under] Asa Keyes?" asked the "Town Meeting," an unsigned newspaper column. Despite his vaunted staff of detectives, Fitts had allowed the statute of limitations to pass and "the guilty ones to forever [escape] punishment behind a smokescreen of indictments of jury bribers, usurers, etc."

Fitts, naturally, blamed his detractors. "If the district attorney's office and the grand jury had been permitted to conclude their work in this matter without political and other interference," the outcome would have been different, protested Fitts. "We have had inexcusable meddling by those whose objects are entirely political."

The *Los Angeles Times* cited "lack of evidence, not lack of vigilance on the part of the District Attorney" for the failure to secure indictments of stockbrokers. "Rumor and suspicion remained merely rumor and suspicion," stated the *Times*, which concluded, despite overwhelming evidence to the contrary, that had the brokers been aware of the overissue, "they would no more have handled the shares than they would have dealt in counterfeit money—not, perhaps because of ingrained honesty, but for fear of the consequences." The "Town Meeting" not only rejected this argument but implicated the *Times* in a Julian Pete cover-up. "Was Fitts' failure due to lack of ability or something else? Draw your own conclusion," remarked the columnist. "What newspaper in town has always been the loyal friend and protector of the higher ups in the business world, that group which belonged to the Julian wreckers? The Los Angeles *Times*, ever and anon. And who was the *Times*' candidate for District Attorney and now for governor? Buron Fitts."

Attention now shifted back to the various Julian-related cases moving through the courts. The extortion trial of Morris Lavine and Leontine Johnson provided an appropriately unorthodox curtain-raiser. The seemingly open-and-shut case began to unravel before the trial began. Charges against Lavine and Johnson had been based largely on the accounts of Jacob Berman and Charlie Crawford. But when the trial began on April 29, Berman was in the penitentiary and Crawford refused to testify because of his recent indictment in the Friedlander case, which his attorneys claimed was inspired by a desire to "nullify the effect of his testimony" against Lavine. The prosecutors had corroborating evidence about the events in Crawford's office from detectives Leslie White and Frank Meyers. But Meyers died, ironically, of lockjaw, shortly before the trial, leaving White as the sole witness to the blackmail transaction. Defense attorneys attacked White, who moonlighted as the author of hardboiled detective stories, with allegations that his testimony represented "the machinations of his young and aspiring novelist's mind," and most of it was stricken from the record. The trial judge ordered the arrest of another of Fitts's detectives when he admitted that he had used an illegal search warrant to sieze the infamous suitcases and other materials from Leontine Johnson. Attorneys for Johnson and Lavine rested without calling any witnesses.

On May 6, Deputy District Attorney Edward J. Dennison gave his closing argument, demanding convictions of Lavine, his former friend, and Johnson. "It was a tough case," said a visibly fatigued Dennison as he left the courtroom. "I feel like I have traveled a long way. I am near the end of my trail." Dennison then went home and died in his sleep. ("Somebody started the story that the seventy-five thousand dollars was *tainted money*," wrote White. "Frank Meyers had died soon after touching it, now old Ed Dennison.") Two days later jurors reported that they were hopelessly deadlocked and could not reach a verdict, ending the trial with a hung jury.

In early June, Fitts's prosecutors suffered another defeat in the Bankers' Pool No. 1 usury cases. On June 6, after four weeks of legal battles over the admissability of photostatic copies of the original pool agreements, defense attorney Byron Hanna asked Judge William Tell Aggelar, "Does your honor think this prosecution is serving the best interests of society? . . . The District Attorney is trying to make criminals out of good citizens. These are men with impeccable reputations, men of standing in the community, who are brought into court for engaging in an innocent business transaction." Judge Aggelar concurred. The contract in question, he noted, was between the defendants, S. C. Lewis and Wagy & Co, and had nothing to do with the Julian Pete overissue. "These documents do not prove a conspiracy to violate the usury law and are, therefore, immaterial," stated Judge Aggelar, who then acquitted the defendants for lack of evidence. Several days later he also dismissed charges against the members of Bankers' Pool No. 2.

The district attorney's office fared no better in its efforts to convict Lewis of bribing juror Caroline Love. Lewis, as usual acting as his own attorney, posed as the victim of a plot by those who wished to silence him in the stockholders' suit. He broke down in tears as he pleaded with the jury not to add another ten years to his sentence. On June 7, one day after the quashing of the usury charges, the jurors found Lewis not guilty. Two days later Lewis returned to McNeil Island. Fitts announced he would not seek to bring the former Julian Pete president back for trial on either the overissue or Friedlander bribery cases, which were still pending.

Fitts hoped to improve his chances in the remaining Julian Pete cases by pulling his ace from the bottom of the deck, Jacob Berman. Berman voluntarily returned to Los Angeles on June 18 to testify in the Lavine-Johnson retrial; the Krause, Groves, and Grider jury bribery cases; and the final Julian Pete overissue conspiracy trial. In exchange for his cooperation, Fitts promised immunity for Berman and his brother in the Krause case. Berman nonetheless had to stand trial on the embezzlement and forgery charges unearthed by Judge McComb.

Berman's testimony led to a conviction in the second Lavine-Johnson trial. Lavine and Johnson received sentences in the county jail; Lavine for one year and Johnson for nine months. Lavine also had to pay a $5,000 fine. On July 10, jurors took only thirty-five minutes to convict Louis Krause of jury bribery. One week later, John Groves, the juror whom Krause had befriended and compromised, was also convicted. Frank Grider, whose home Jacob Berman had purchased for an inflated amount, escaped prosecution when Berman declined to testify against him.

By mid-July 1930, more than three years after the Julian Pete crash and years of courtroom activity, over half a dozen people had gone to prison for crimes related to the scandal. Asa Keyes and Ed Rosenberg were in San Quentin; Lewis remained at McNeil Island for his Lewis Oil conviction; Berman was in protective custody; Krause and Groves awaited transport to the state penitentiary, and several minor moneylenders nabbed by City Prosecutor Lloyd Nix had served brief sentences in the county jail. Yet the original Julian Petroleum crime remained unpunished. No one had yet been convicted for perpetrating the massive stock overissue. The stockholders' civil suit that had triggered the 1930 Julian Pete *redux*, continued to work its way through the courts with no immediate resolution in sight.

The events of 1930 also exacted a personal toll. In March, plumbing magnate Harry M. Haldeman suffered a heart attack while testifying in a trial unrelated to the Julian case. Haldeman was fifty-eight years old. Among the mourners was his three-year-old grandson, Harry Robbins. Forty years later, H. R. Haldeman, a plumbing expert of a different sort, would become embroiled in his own celebrated scandal. In May, L. J. King, the Associated Oil executive who had served on the California-Eastern board of directors, lost hundreds of thousands of dollars in the Julian Pete crash, worked on the Sunset Pacific reorganization, and helped finance the stockholders' suit, died at age fifty. According to Arthur Loeb, King died of a "broken heart."

Many believed that the strains of the Julian Pete affair had contributed to these deaths. That remains conjectural, but in the case of Motley Flint there can be no doubt. After residing in France for three years, Flint had voluntarily returned to Los Angeles in May 1930 to face charges in both the usury and corporate securities cases. Although the usury charges were dismissed in early June, the pending indictment for violations of the Corporate Securities Act forced Flint to remain in Los Angeles. There, on July 14, he faced his courtroom rendezvous with assassin Frank Keaton's bullets.

### III

On the morning of July 15, 1930, one day after Frank Keaton murdered Motley Flint, the flags at Warner Brothers studios flew at half-mast. "He was the first Los Angeles banker to recognize the value of the fast-growing motion picture industry," stated Jack Warner. "We have lost a great friend and benefactor." Flint's will bequeathed the bulk of his fortune to programs providing summer outings and Christmas gifts for poor children and assistance to the elderly and destitute, reminding the city that it had lost its long-time "Santa Claus."

To many southern Californians, however, the meaning of Flint's death was more ambiguous, salting the wounds opened by the Julian Pete affair and offering further ammunition for the wars that had raged throughout 1930. "The Inglewood stock dabbler isn't the real murderer of Motley Flint," argued the *Record*. "The true murderers are the BIG CROOKS who wrecked Julian Petroleum . . . and [the] inefficient and dilatory officials who have done nothing to put the big crooks in prison."

Foes of "Fighting Bob" Shuler lost no time in capitalizing on the reported discovery of "Julian Thieves" among the assassin's effects. On the evening after the shooting, City Prosecutor Lloyd Nix, whose handling of the usury cases had provoked repeated criticism from Shuler, unleashed an attack on the minister. Nix reiterated and exaggerated charges that Shuler's writings and broadcasts had inspired Keaton. "There is a tension in this city tonight," reported Nix in a radio address, reminding people that Shuler had named many others in his pamphlets, including Nix himself. "Every precaution should be devised for the protection of those men whose names have been bandied about in such a way that public opinion may be inflamed against them." Nix warned that while the Constitution guaranteed free speech, "there is a limit to even this freedom."

Shuler characteristically rose to the bait. "I am delighted to accept my full responsibility for every word I wrote," he responded. "I am delighted that I have had the privilege of painting in all its blackness the tragedy of this gigantic conspiracy and the perfidy of those public officials who have lent themselves to the protection of those who robbed the Julian Petroleum stockholders." But when Shuler took his defense one step further, he came perilously close to justifying Flint's murder. "The picture is far more tragic than some of our newspapers always on the side of the rich would have you believe," he argued. "The man who is killed has a hundred thousand dollars in certified checks, currency and diamonds on his person. The arresting officers find a thin dime in the pockets of the killer." Keaton should hang, stated Shuler, "But God speed the day when the rich will hang just as poor hang."

More significantly, within twenty-four hours of Nix's radio address, Mayor Porter demanded the city prosecutor's resignation. Nix struck back angrily. "There is a species of terrorism abroad in City Hall," he charged. "A department of the city government cannot do its normal job without vacillating interference apparently inspired by Shuler. . . . Now tell me, who is the Mayor of Los Angeles? Where is the seat of government? Is it down here in City Hall or is it in the Trinity Methodist Church?"

A further controversy raged over whether Keaton had actually possessed "Julian Thieves" when he shot Flint or whether the pamphlet had been planted among his effects by police officers opposed to Shuler. Fitts's deputy Robert Stewart denied seeing the pamphlet. Detective Earl Kinette of the sheriff's office, who had close ties to the underworld (he would later be convicted of planting a bomb in a newsman's car), was the leading witness who placed "Julian Thieves" in Keaton's pockets. The booklet first disappeared, then suddenly reappeared, prompting charges that pro-Shuler individuals in either the Porter administration or Fitts's office had attempted to tamper with the evidence.

Frank Keaton, whose mental state had rapidly deteriorated, provided no assistance in resolving the dispute. Keaton alternately denied and admitted possession of the pamphlet. How much Shuler's writing actually influenced Keaton posed yet another point of contention. Despite assertions to the contrary, "Julian Thieves" contained only four brief mentions of Flint. The most damaging passage described him as being "as close to S. C. Lewis and Jack Bennett as triplets in the womb." Investigation revealed that Keaton had never held Julian Petroleum stock but had owned shares of First National Bank stock, which he hurriedly sold when the stock dropped in the aftermath of the Julian Pete crash. Had he held it several months longer, he would have recouped even this loss. Even so, Keaton had profited handsomely from this investment. His most serious losses occurred after the 1929 stock market crash.

Emboldened by the debates over Shuler's responsibility for the Flint assassination, Shuler's enemies took aim at his greatest asset, radio station KGEF. On July 18, Nix called on the Federal Radio Commission to investigate Shuler. Other local figures supported this request, and on July 21, the Federal Radio Commission announced that it would conduct an inquiry into Shuler's broadcasts.

Amidst these controversies, Jacob Berman stood trial for forgery and embezzlement. With the statute of limitations now long past, this case would mark the final opportunity for the State of California to extract retribution from either Lewis or Berman for their crimes. Berman, who had allegedly salted away millions of dollars, announced he was bankrupt and could not pay his legal fees. Judge B. Rey Schauer assigned a public defender to represent the master swindler. Berman waived a jury trial and agreed to have Judge Schauer decide his fate.

Berman's trial marked a reunion of the old gang from the Julian Petroleum transfer department as Fred Packard, Ted Harris, and T. P. Conroy, whom Judge Schauer described as the "white collar galley slaves," all testified against Berman. On July 25, Schauer found Berman not guilty on charges of embezzling $10,000 from Daniel Barnes but guilty on three counts of forging signatures to stock certificates. Schauer gave Berman the maximum sentence of three to forty-two years in the state prison, to be served after Berman had completed his seven-year federal sentence. The verdict, as the *Times* reported, marked the "first conviction of any of the asserted principals in the Julian Petroleum collapse." At long last, noted the

*Times,* "somebody is going to be punished for a conscienceless and nefarious scheme."

To the *Times,* the sentencing of Berman also "dispose[d] of the silly talk" that Fitts did more for Berman than his cooperation had required. In reality Berman's forgery conviction was a great embarrassment for Fitts, who had sought to dismiss the charge and described the case as unwinnable. If not for Judge McComb's persistence, "Immunity Jack" would have escaped the Julian Pete swindle virtually unscathed. With the exception of the Lavine-Johnson extortion trial and the Krause-Groves jury bribery cases, both of which concerned events after the Julian Pete crash, Fitts had failed to secure any convictions from his well-publicized investigation.

The Berman verdict occurred just one month before the August 26 gubernatorial primary. In Los Angeles the campaign had long since ceased to be a referendum on Governor Young's stewardship but had evolved instead into a judgement on the roles that Fitts and Young had played in the Julian Pete affair and what the *Los Angeles Citizen* called "a test of the strength of Reverend Shuler." Throughout the summer, "Fighting Bob" had stepped up his support of the Fitts campaign. "The outstanding fact that Governor Young is forced to concede his relation to the wrecking of the Julian Petroleum Corporation . . . will never down so long as honest men look candidly through the vital issues of their day," wrote Shuler in explaining "Why Fitts Will Win."

The primary gave Los Angeles voters a chance to cast their own verdict on the Julian Pete prosecutions. Although Fitts ran poorly outside of Los Angeles County, he dominated the local balloting. Fitts polled almost 54 percent of the county vote, while Young ran a distant third with less than 18 percent. Young's poor showing in Los Angeles tipped the statewide decision to Rolph, who triumphed by a scant 20,000 votes out of more than one million cast. Rolph went on, as expected, to swamp the Democratic opposition in November.

Several historians have attributed Young's defeat to a split in the "dry" vote with Fitts. Another consideration was southern California's hostility to Young's banking policies. But in an election decided by such a narrow margin, the Julian Pete controversy, which discredited Young and elevated Fitts, had played a decisive and probably greater role than the other issues. Another Julian-tainted candidate, H. L. Carnahan, suffered an embarrassing third-place finish in the race for lieutenant governor. Candidates backed by Shuler, meanwhile, succeeded in forcing four incumbent Superior Court judges into run-off elections. In each instance Shuler's man emerged victorious in November.

As the voters cast their ballots, the trial of Frank Keaton proceeded in a local courtroom. Keaton sat motionless throughout the trial, appearing to sleep at the counsel table, while neighbors and friends described his descent into madness and psychologists attributed his actions to "gross morbid brooding." "He was and is insane," stated one expert. Only once did Keaton show any emotion, jumping to his feet, with "fists clenched about his head and his face flushed with rage," during the prosecution's closing arguments. Neither judge nor jury displayed any sympathy or compassion. On September 5, the jurors found him guilty of first-degree murder. Ten days later Judge Schauer sentenced Keaton to be hanged. Wiser heads ultimately prevailed. Jurors at a 1931 sanity trial took only ten minutes to declare Keaton mentally deranged, sparing him the death penalty.

With the conviction of Keaton, only two of the myriad indictments brought during 1930 remained unresolved: the charges against Lewis, Berman, Flint, and six others for overissuing Julian Petroleum stock, and the Jack Friedlander-Charles Crawford bribery case. On October 14, District Attorney Fitts admitted to Judge Schauer that he had insufficient evidence to merit a prosecution of the overissue defendants and asked for the dismissal of all charges. Schauer accepted the motion but publicly lambasted Fitts:

> The court entertains these motions with chagrin and shame. It is a confession of inadequacy and inefficiency in the administration of justice. If the evidence is now insufficient after months of preparation, it must certainly have been insufficient when the indictments were returned. I cannot see how any grand jury properly advised could have returned such indictments.

Invoking the statute of limitations, Schauer concluded, "These are the last of the Julian cases. I hope that the Julian corpse from the last exhumation of which there has been intensified stench, will remain forever buried."

The trial of former corporation commissioner Friedlander and underworld boss Crawford had been continued, pending the resolution of the overissue case. Fitts vowed to continue this prosecution, but each of his key witnesses had fallen under a cloud of suspicion. Morris Lavine had been convicted of extortion; Frank Cressy had been accused of misappropriating funds; and Jack Roth, Julian's former right-hand man, also faced imminent indictment. On October 20, the district attorney's office admitted the futility of its position and asked to have the charges dropped. Friedlander's attorney hailed the result as a total exoneration of his client.

The state courts had now disposed of all criminal proceedings arising from the 1930 grand jury. The fate of "Fighting Bob" Shuler's license for KGEF, however, remained unresolved. In January 1931, the Federal Radio Commission held hearings into the charges that Shuler had misused the airwaves. A parade of witnesses, including Judge McComb, former mayor Cryer, and Judge William Doran, testified against the minister. An equal contingent, including Fitts, Mayor Porter, Municipal Judge Bogue, and Mrs. Frank Keaton, defended Shuler. "Fighting Bob" testified at great length, admitting that he had based many of his broadcasts on rumor and hearsay evidence. When asked if he could "judge the courts, the lawyers and almost everything else," Shuler invoked the Julian Pete affair. He could, he asserted, judge that something was wrong when jurors were bribed and millions were stolen. During two weeks of testimony the radio commission examiner accumulated 2,300 pages of transcript, the lengthiest single radio commission inquiry to that date. While he pondered the evidence, Shuler and Los Angeles became embroiled in yet another improbable offshoot of the Julian affair.

The new sensation involved gambling boss Charles Crawford. The Julian Pete scandal had given Crawford what an underworld figure could least afford—publicity. Shuler and others had linked him to the mysterious Tia Juana Pool. A suit by the trustees of Wagy & Co. had implicated Crawford in Lewis's October 1926 stock manipulations. The Lavine-Johnson extortion attempt had further propelled him into the headlines and his indictment for conspiring with Lewis and Friedlander to sell Corporation Department permits intensified his predicament. By mid-1930, the Gray Wolf stood dangerously exposed.

Crawford sought shelter in the next-to-last refuge of a scoundrel—religion. "I'm praying for all those mixed up in this terrible Julian case," Crawford had told skeptical reporters at the first Lavine-Johnson trial in June 1930. Several weeks later he appeared in the congregation of St. Paul's Presbyterian Church. As the collection plate passed through his pew, Crawford removed his "huge twin-diamond ring," valued at $3,500, and dropped it in with a note requesting that it be dedicated to the construction of a new Sunday school building. Moments later Crawford marched to the altar, where the Reverend Gustav A. Briegleb baptized him with water brought from the River Jordan.

Reverend Briegleb had been a staunch and unswerving ally of Robert Shuler's, but "Fighting Bob" had no illusions about Brieglieb's latest convert. "I would rather handle a skunk than place my hand on the head of a politician who has so helped to corrupt this city," proclaimed Shuler. Briegleb criticized Shuler's "unwillingness to give a man a chance." "If Christ did not mean his church for sinners, for whom did he mean it?" asked Briegleb in his Sunday sermon. Crawford's conversion caused an irreparable rift between the two popular ministers.

During the next year Crawford lavished donations on Briegleb and St. Paul's. He gave $25,000 to erect the Amelia Crawford Parish House honoring his ninety-five-year old mother; he also funded a radio station to enable Briegleb to compete with Shuler over the airwaves. But Shuler knew his man far better than the more impressionable Briegleb. In early 1931, with the Julian Pete trials at an end and his own indictment dismissed, Crawford attempted a comeback. The vehicle he chose to clear the way for his return was a new weekly publication, the *Critic of Critics*.

The *Critic of Critics* had first appeared on the Los Angeles scene in 1930 with longtime local newspapermen Mike Schindler and Herbert Spencer at its helm. Posing as a liberal reform voice, it repeatedly attacked Shuler, Fitts, and the *Times*. The *Critic* also lashed out at the local underworld, which, since the baptism of Charlie Crawford, had fallen under the aegis of Crawford's one-time ally, former police officer Guy "Stringbean" McAfee. In early 1931, advertisements for Crawford's insurance and real estate business began to appear frequently in the pages of the *Critic*. Rumors circulated that Crawford was underwriting the magazine and using it to discredit McAfee and regain control of the rackets. By May 1931, the competition between Crawford and McAfee had intensified. The rivals held a "peace meeting" at which they reportedly agreed to reunite, but Crawford left visibly disturbed.

Several days later, on May 20, Crawford greeted a visitor to his Hollywood real estate office. Crawford had secured his stucco Sunset Boulevard suite with an elaborate alarm system with buttons he could press to instantly summon his bodyguards, but he took no unusual precautions for this appointment. He spoke for an hour with the visitor, after which *Critic* editor Herbert Spencer joined them. Crawford, seemingly "in very good spirits," emerged and spoke briefly with his confidential secretary before returning to his office. Two minutes later shots rang out. The visitor walked out slowly and then broke into a run. Moments later Spencer staggered out bleeding profusely from a bullet in his heart. He collapsed and died on the spot. Crawford lay near his desk, his finger inches from his warning buzzer. A bullet had ripped through his abdomen, liver, and a kidney. Rushed to the hospital, Crawford regained consciousness long enough to refuse to identify his killer. He died that evening with the Reverend Briegleb at his bedside.

For twenty-four hours the mystery of the Crawford-Spencer killings puzzled Los Angeles. The solution proved even more baffling when David H. Clark, a war hero, former deputy district attorney and candidate for municipal court judge in the upcoming elections, surrendered to his former boss Buron Fitts and confessed to shooting Crawford and Spencer. "Had the chief suddenly accused *me* of the crime, I could not have been more astounded," wrote Clark's former colleague, Leslie White. The tall, handsome thirty-three-year old Clark had worked in the district attorney's office for eight years. He had won his greatest fame for successfully prosecuting Albert Marco, a Crawford-McAfee crony. More recently Clark had worked in the corporate securities division before resigning to run for a judgeship. His campaign had won the support of both the *Times*, which hailed him as "the sort of timber needed for the bench," and McAfee. Crawford and the *Critic of Critics* had endorsed his opponent.

Since Clark steadfastly refused to elaborate on the shootings prior to his trial, rumors about his motives spread through the city. Some speculation, noting Clark's stint with corporate securities, tied the killings to the Julian Petroleum controversies. " 'David H. Clark knows the contents of the suitcase'—dangerous information to be in the hands of a young judge," wrote Lorin Baker in his overwrought 1932 exposé, *That Imperiled Freedom*. Journalist Carey McWilliams later attributed the shooting to a struggle over the spoils of the Julian affair. But no evidence ever tied the killings to the Julian Pete scandal.

Most people seemed inclined to two alternative explanations. Buron Fitts advanced the most commonly held interpretation: the double shooting was an underworld assassination with Clark acting as McAfee's agent. "There were three racketeers in the room where Crawford and Spencer were slain and Dave Clark was one of them," posited Fitts. The other scenario involved what the *Times* described as "a pitiless plan to ruin [Clark] socially, financially and politically and strip him of every last vestige of respect." Police and district attorney's investigators reported that Crawford had lured Clark to a "party . . . with wine, women, and song" and procured, or at least claimed to have procured, compromising photographs of Clark and a woman enlisted to entrap him. When Crawford and Spencer threatened to blackmail Clark if he was elected to the judgeship, Clark had angrily shot them. This sordid version, however, disappeared from the official repertoire almost as quickly as it had surfaced.

The Clark case played itself out in typically bizarre fashion. Clark refused to withdraw from the race for municipal judge and, despite losing, polled 60,000 votes. By the time of his August trial, "Handsome Dave," who had yet to indicate what his defense would be, had become, according to Leslie White, a "matinee idol." The usual star-studded Los Angeles crowd flocked to the courthouse. They heard the charismatic Clark testify that Crawford had outlined a plan to get "back in the saddle" by controlling the 1931 grand jury and using the Reverend Briegleb's pro-posed radio station. Crawford asked Clark's cooperation in framing Police Chief Roy Steckel, Clark's close friend. When "Handsome Dave" refused and threatened to expose Crawford, Crawford pulled a gun on him. Clark responded instanta-neously, firing his own pistol at Crawford and then shooting the oncharging Spen-cer.

Clark's testimony was filled with gaping improbabilities. Police had found no guns in Crawford's office (although Crawford had an empty shoulder holster). Nor did it seem logical that Crawford, who had no personal record of violence, would lure Clark to his office in order to shoot him. Special Prosecutor Joseph Ford warned the jurors that Clark "has a handsome face and a fine physique, but I am afraid he has lost his soul." Eleven of the twelve jurors disagreed. "He is one of our noblest Americans," said the jury forewoman. "We consider him a fine, clean upstanding young man . . . . We think that he was protecting his life against those men." Only the intractable stubbornness of juror William Weller prevented acquittal and forced a hung jury. Two days later Weller's wife found an undetonated bomb on their front lawn. A second jury acquitted Clark in October 1931.

The "Strange Death of Charlie Crawford," as Shuler entitled his newest pam-phlet, did produce one Julian Pete–related twist. At the time of Crawford's assassi-nation, the fate of the "tainted money," the $75,000 he had used to entrap Morris Lavine in March 1930, remained unresolved. Both Crawford and Lavine had made legal claim to these funds. In addition, the trustees for Wagy & Co., arguing that both Lavine and Crawford had illegally profited at the expense of the bankrupt brokerage house, had won an injunction blocking payment to either party. The court ordered the county clerk's office to place the seventy-five $1,000 bills in its vault for safekeeping. On May 19, 1931, the day before his death, Crawford had reached a settlement with the Wagy trustees, clearing the way for him to regain the $75,000.

One year later the Crawford estate demanded payment of the disputed money. Only an empty envelope and photostatic copies of the bills remained in the vault. Police learned that a deputy county clerk with the lyrical name of Liberty Hill had departed for an extended vacation. Hill, a trusted twenty-year veteran of the county clerk's office, had long since appropriated the $75,000 to bankroll his stock market speculations. By May 1932, he had lost it all. Hill was arrested and convicted of six counts of grand theft and sent to San Quentin. A federal appellate court later held that the county could not be held liable for the money, and the heirs of Charlie Crawford lost the funds.

While the Crawford-Spencer murders transfixed the city, the dispute over "Fighting Bob" Shuler's radio license continued. On August 7, 1931, the examiner handed down his decision. Disavowing responsibility for how far a broadcaster might go in "criticizing, comdemning, lauding, praising and supporting" individuals and public officials, he ruled that the commission could only deny a license if the "public disservice" created by Shuler's broadcasts outweighed "all the public service rendered by all other broadcasts made. . . ." He recommended that the Federal Radio Commission renew KGEF's license.

Shuler's victory proved short-lived. His opponents appealed the ruling to the full commission, which held a hearing in Washington, D.C., on September 26. Shuler chose this opportunity to demonstrate several of his least desirable traits. He had told his listening audience that he believed that the commission would rule against him. When the commissioners questioned him about this assertion, Shuler, as usual, cited an unsubstantiated report from an unnamed party. Shuler further offended the panel with a comment about "the Catholic member of the commission." The statement proved as erroneous as it was bigoted. No Catholics sat on the commission. "You didn't take the trouble to check that out, did you?" asked one of his interrogators. Six weeks later, on November 13, the Federal Radio Commission voted unanimously to revoke KGEF's radio license and ordered the station to cease broadcasting.

Shuler's performance at the hearing had hurt his cause, but the Los Angeles minister always believed that the result had been preordained. He charged that the politicians had gone "over [the examiner's] head and got right to President Hoover and had him close me down." Shuler claimed that Henry M. Robinson had personally flown to Washington to consult with the President. Julian pool members Louis B. Mayer and Mendel Silberberg (whom Shuler derided as "my Jewish friends"), both active Hoover re-election supporters, had also applied their influence, according to Shuler. Movie producer Alexander Pantages and newspaper titans Harry Chandler and William Randolph Hearst rounded out the conspiracy. The explanation was classic Shuler—perfectly plausible, but totally unsubstantiated. Shuler appealed the decision to the United States Supreme Court and sought other ways to make his voice heard, but the ruling effectively silenced the radio station that had mesmerized Los Angeles and transformed local politics. Without KGEF, Shuler's power began to wane.

# 18  WHAT PRICE FUGITIVE?

I

As 1931 drew to a close, forty-two-year-old Rosebud Harris found she could no longer endure the pain. More than four years had passed since she had lost her "considerable fortune" in the Julian Pete crash. In the interim Harris had grown despondent, suffered a nervous breakdown, and been institutionalized. On the afternoon of December 11, Harris, recently released from the Glendale Sanitarium, fired eight bullets into the body of her nineteen-year old daughter Helen, then locked herself in the bathroom and turned on a gas jet, taking her own life.

The city and county of Los Angeles had long featured one of the highest suicide rates in the nation. Yet the gruesome murder-suicide of Helen and Rosebud Harris shocked southern California. It linked the Julian Petroleum scandal with the broader calamity that had befallen the nation. Southern California, like the rest of the United States, sat mired among the ruins of the speculative debauch of the 1920s. Suicides had reached epidemic proportions. For all its spectacle, the Julian Petroleum affair now seemed but a dress rehearsal to the opening of a long-running, more profound tragedy.

The Great Depression exposed a morass of corruption underlying the foundations of Los Angeles. "In the early 'thirties," wrote Carey McWilliams, "it would have been possible to have assembled, in the prison yard of San Quentin, a group of former 'civic leaders,' 'empire builders,' 'captains of industry,' and 'outstanding public officials' from Southern California." In December 1930, the Guarantee Building and Loan Association failed, amidst charges that its prestigious chairman had embezzled $8 million. Shortly thereafter the American Mortgage Company collapsed, its president destined for San Quentin. In April 1931, federal officials indicted bankers Thomas Morrissey and Arthur Adkisson, who had figured prominently in Berman and Rosenberg's moneylending schemes, for misappropriating funds. A more crushing blow came in 1932 when Richfield Oil, the state's third-largest distributor, went into receivership. Auditors discovered a widespread pattern of fraudulent bookkeeping, theft, and extravagant spending by corporate officers. Several people who had played roles in the Julian Petroleum scandal, including Alvin Frank, H. J. Barneson, and Joe Toplitsky, also figured in the Richfield Oil affair.

Receiverships, virtually unknown in Los Angeles prior to the Depression, became common. More than 300 enterprises representing over a half a billion dollars in assets were placed into the hands of receivers during the early 1930s. Despite its flaws, the Julian Pete example became the model for these actions. Among the properties entering receivership were the original Julian wells at Santa Fe Springs. In March 1931, the courts placed Julian wells 1, 2, 3, 11, and 12 under the guidance of Charles Allison. Within a year, federal officials charged Allison, a young man with minimal business experience, with creating a receivership "racket," currying favor with municipal court judges through loans, gifts, and campaign contributions in order to acquire lucrative receivership positions. Allison was sentenced to two years in San Quentin.

Sunset Pacific also joined the ranks of companies in receivership, generating new controversies. Anyone carefully analyzing the original reorganization plan could have foreseen the inevitable failure of Sunset Pacific. The corporation had assumed a debt load of almost $900,000 a year, far beyond its revenue-producing capacities. The only possibility for any return to stockholders lay in a merger with a larger company. The Great Depression doomed this solution. H. L. Carnahan, who had remained Sunset Pacific president as "a public service," entered serious negotiations with potential buyers on four occasions, but these efforts failed to produce a deal.

The survival of Sunset Pacific had always depended on the largesse of its largest creditor, the First National Bank. First National had promised Carnahan that it would protect his company from any newly incurred obligations, establishing, in effect, a sustaining line of bank credit. By mid-1931, however, the new Security-First National decided to divest itself of all involvement with the successor to Julian Petroleum. It dumped its $7 million of Sunset Pacific bonds in exchange for 100,000 shares of Associated Oil stock worth only $500,000.

In acquiring these notes at a bargain price, Associated Oil, which had long possessed a financial stake in the Julian Petroleum outcome, became the sole secured creditor of Sunset Pacific. The California petroleum giant seized the opportunity to convert potential losses into clear profit. It moved to foreclose on Sunset Pacific's debts and seize the company's remaining assets.

In July 1931, Associated Oil informed Carnahan that unlike First National, it would no longer underwrite new obligations. Without this guarantee Sunset Pacific could no longer function. On August 27, the Berreyesa Cattle Company, a wholly owned subsidiary of Associated Oil and a minor creditor of Sunset Pacific, filed suit requesting the appointment of a receiver, who would maintain Sunset Pacific until

Associated Oil could institute foreclosure. Under this arrangement, Associated Oil would benefit while all unsecured creditors and stockholders would receive nothing. The suit specifically requested that James Lewis, an Associated Oil employee, be named receiver.

Associated Oil had moved cleverly and carefully to seize the remnants of Julian Petroleum, but they had not reckoned on the wrath of Arthur Loeb. Carnahan had given Loeb assurances that he would be informed prior to any receivership action. But Loeb first learned of the Associated Oil coup in the August 27 newspapers. Loeb immediately filed a bill of intervention. He protested that James Lewis would protect the interests of Associated Oil at the expense of other creditors and stockholders. Loeb demanded the appointment of another receiver.

The task of resolving this latest dispute fell to federal judge Harry Holzer. Holzer created a joint receivership under Lewis and local businessman Frank H. Rolapp. Within two days it became apparent that this arrangement could not stand. Lewis insisted on employing Associated's legal counsel as attorney for the receivership, creating a blatant conflict of interest. Holzer immediately demanded Lewis's resignation and named Rolapp sole receiver.

Loeb thus successfully thwarted Associated Oil's power grab. Whether or not this would ultimately benefit the stockholders now depended not only on Rolapp's skills but on the outcome of the $12 million stockholders' suit still pending in the courts. From the beginning, attorneys representing the stockholders had found themselves in an uphill struggle. "We found the task was so stupendous as to make the physical obstacles of proof almost insuperable," reported Guy Crump. "Between fifteen and twenty of the largest and ablest firms in the state were against us." He estimated that opposition legal fees totaled $200,000, while his firm worked on a contingency basis.

Attorneys for the LASE and stockbrokers filed numerous challenges. Forty-six of the sixty-three firms involved won dismissals in November 1930 due to improper filing of their summonses. The State Appellate Court also approved a writ of prohibition filed on behalf of the LASE on the grounds that the exchange was not a legal business entity and therefore could not be sued. Crump appealed this decision. On July 31, 1931, the California Supreme Court reversed the appellate ruling, but the LASE took the matter to the United States Supreme Court. On January 4, 1932, the high tribunal refused to hear the case, finally paving the way for its prosecution.

By this time the basis of the suit had shifted in several fundamental ways. In May 1930, Sunset Pacific, which had originally refused to participate, had filed an identical cross complaint against the LASE and the stockbrokers. Since the suit had originally been instigated to benefit the oil company, Sunset Pacific now replaced Loeb and his associates as the primary litigants. Crump and his legal team remained as attorneys.

Meanwhile, Loeb had shifted to an adversarial position. In January 1931, Sunset Pacific stockholders elected him to an advisory committee to monitor the affairs of the company and the progress of the suit. Loeb tenaciously battled to get access to all records regarding the litigation. He learned, among other things, of Crump's connection to S. C. Lewis and that both Lewis and Jacob Berman were being paid for their cooperation. Loeb also clashed with Crump over the resolution of claims against A. C. Wagy, one of the primary targets of the suit. Loeb claimed that Wagy had made $4 million in Julian Pete stock speculation. When Wagy died in 1930, Crump feared this "would leave the plaintiffs out in the cold as far as any recovery . . . was concerned." He hastily negotiated a $75,000 settlement with the Wagy estate. Loeb angrily protested the paltry amount.

By this time, Crump had come to believe that the impact of the Depression had crippled an already weak case. Many of the defendants had suffered severe losses in the stock market crash and become judgement-proof, maintained Crump. In April 1932, he told Loeb that the remaining defendants had settled on a total payment of $200,000, most of which would go to legal fees. Loeb blocked this agreement and unsuccessfully attempted to replace Crump with another attorney.

In July 1932, Loeb and the other original plaintiffs received letters informing them that Sunset Pacific had approved a $265,000 settlement recommended by Crump. "This recovery was a brand pulled from the burning," Crump later asserted. "Salvage absolutely." Loeb disagreed, noting that under the proposed resolution, "there will be nothing available to the stockholders." Loeb rejected Crump's descriptions of the defendants as judgement-proof. While some of them might well be bankrupt, argued Loeb, others, like Abe Morris, who Loeb alleged had earned $8 million in Julian Pete stock transactions, retained large fortunes. The LASE was not judgement-proof, and many members were still worth millions of dollars. Loeb and those who had joined him in the initial action immediately sued to block the settlement.

On August 3, 1932, Judge Claire Tappan dismissed the objections of the original plaintiffs and approved Crump's proposal. "I most sincerely sympathize with the stockholders in this case," stated Tappan, but "I have felt all along that proof of this case could not be adequately presented in court." He noted that the $265,000 represented "the only light seen in the whole procedure and the first real money that has been collected." Some of the defendants "are now in the penitentiary and others have folded up their tents and have stolen away into the night," he lamented. Loeb and his fellow stockholders appealed this decision, but the California Supreme Court ultimately rejected their claim.

In the end, the celebrated stockholders' suit, originally filed for $12 million, netted $340,000. The Wagy settlement brought $75,000; the final settlement an additional $265,000. Of this money, legal fees and expenses claimed $167,000. The Sunset Pacific receivership collected the remaining $173,000. Loeb and the other stockholders received nothing. Loeb requested $8,045 to cover his personal expenses (which in actuality exceeded $25,000). The court awarded him $5,500. Even this reward would be denied him. Adolph Ramish placed a lien on these funds. The $2,500 Loeb had received from Ramish in 1930 to initiate the suit turned out to be, not a donation, as Loeb always described it, but a loan secured by a promissory note. The multimillionaire Ramish now demanded that the courts repay him out of the settlement. This was necessary, argued Ramish, because after years of unsuccessful litigation, Arthur Loeb and his wife were insolvent. Loeb eventually received only $3,385 for his four years of expense and effort.

Sunset Pacific also failed to benefit from the long-festering usury suits. In March 1932, the California Supreme Court ruled in favor of the receivers in the Hollingsworth test case, which had challenged their right to pursue the usury cases. By this time, however, the Sunset Pacific receivership had supplanted that of Julian Pete. Rolapp sought to intervene on behalf of the new company, but his efforts proved unsuccessful. The case against Hollingsworth was dismissed in December 1932. Another attorney assumed responsibility for seven of the remaining usury suits, but ultimately concluded that they could not be won. Arthur Loeb made one last effort. In October 1933, he sued receivers Scott and Carnahan for $657,304, claiming that they had allowed the usury cases to "die through neglect and gross negligence." Loeb failed to convince the judge of the validity of his charges.

The failure to extract a substantial settlement from either the stockholders' suit or the usury cases helped shape the Sunset Pacific receivership. In Frank Rolapp the Julian Petroleum properties had finally found an honest and competent administrator. Rolapp assumed the Sunset Pacific receivership under the worst possible conditions. The Depression had crippled the oil industry, destabilizing conditions and triggering price wars, making Rolapp's task even more formidable. Sunset Pacific stood on the brink of a bankruptcy, which would have yielded returns only to Associated Oil. Both the federal and state governments threatened to seize properties for unpaid taxes.

Rolapp, however, proved equal to the challenge. He assumed all administrative tasks, employing no additional managers or supervisors. He settled all tax liabilities and reduced unsecured indebtedness from over $2 million to $500,000. He amassed a $1.9 million cash reserve from which current unbonded obligations could be met. With the passage of the National Industrial Recovery Act (NIRA) in 1933, Rolapp was named to a fifty-two-man National Emergency Committee which drafted a national code for the petroleum industry. Despite the Depression and the reduced production required by NIRA codes, Sunset Pacific managed to increase its monthly sales and register a profit prior to interest payments.

Despite these accomplishments, there remained the daunting task of reducing Sunset Pacific's staggering $12 million bonded indebtedness to Associated Oil. Rolapp negotiated with the petroleum giant for two years. Blocked in its efforts to seize the company, Associated Oil ultimately accepted $2 million in cash and properties and an additional $3 million in bonds in a newly reorganized Sunset Oil Corporation. Annual interest payments dropped from $900,000 to less than $180,000, a sum that could easily be covered by operating profits.

"It has always been my desire to preserve the ownership of the reorganized company in the present stockholders . . . that they might have an opportunity to salvage some of their losses," explained Rolapp. In successfully driving down the company's obligations, he had made this a possibility. For the first time since S. C. Lewis had begun his handiwork, assets would exceed liabilities. All current Sunset Pacific shareholders would receive common stock in a streamlined Sunset Oil. The reorganization, Rolapp reported, "presents a reasonable probability of successful operation."

Rolapp announced the agreement with Associated Oil and the preliminary details of the reorganization in November 1933 during his testimony at an unusual set of United States Senate hearings held in Los Angeles. The Special Senate Committee to Investigate Bankruptcy and Receivership Proceedings convened in the federal courtroom in which Lewis and Berman had been convicted in the Lewis Oil case. The panel spent its first week hearing Rolapp, Loeb, Carnahan, Scott, and others relive the Julian Petroleum affair. The assembled senators reacted with amazement. "Amidst all the wild, reckless, relentless escapades, that is the most remarkable," commented one senator.

The hearings also shed light on the fate of Lewis and Berman. Since 1930, they had resided in federal penitentiaries, first at McNeil Island, Washington; and later at Leavenworth, Kansas; serving seven-year terms for the Lewis Oil convictions. Although Berman faced up to forty-two additional years for his state embezzlement conviction, Lewis would be eligible for parole in late 1934. Unprosecuted federal mail fraud indictments from the Julian Petroleum case, with potential sentences of up to sixty years, still hung over both men. In March 1933, Lewis had demanded dismissal of these charges.

The Lewis-Berman cases resurfaced at an inauspicious time, according to testimony at the hearings by an Assistant United States Attorney. Federal court calendars faced a three-year backlog. The current crop of U.S. Attorneys, who knew little of the Julian Petroleum case, anticipated that a Lewis-Berman trial would last a minimum of six months, further clogging the courts. Lewis's correspondence from prison indicates that he planned to subpoena many of those implicated in the Julian Pete scandal, including Henry M. Robinson and Louis B. Mayer, and reopen the whole affair. In addition, Lewis had discovered irregularities in the grand jury that had indicted him. Four members of the panel had been Julian Petroleum stockholders. The foreman was an official of Pacific Southwest Bank.

Rather than confront Lewis, federal attorneys offered a deal. If Lewis and Berman would plead no contest to the twelve counts of mail fraud and conspiracy, they would receive eighteen-month sentences on each count. The sentences would run concurrently with each other and with the earlier Lewis Oil charges. The effect would be to extend their current federal sentences by only six months, enabling Lewis to be released after only five years in prison and Berman to concentrate on evading his state penalty. Both men readily accepted. Julian Petroleum receivers Scott and Carnahan appeared in court as character witnesses on their behalf.

The plea bargain was completed on October 8, 1933, just weeks before the Senate hearings. It guaranteed that Lewis, rumored to have millions of dollars cached in safe places, would pay minimally for his audacious theft. The saga of crime and punishment, however, had begun, not with Lewis and Berman, but with their predecessor at the helm of Julian Petroleum, C. C. Julian. Fittingly, Julian had returned for a grand finale.

## II

On January 14, 1930, the day after hearings in the Julian Petroleum stockholders' suit had begun, hundreds of former Julian Pete and Western Lead investors had received a letter bearing an unfamiliar postmark but an undeniably familiar message. The letter, emanating from Oklahoma City and emblazoned with capital letters and exclamation points, announced the birth of a great new oil company built "upon the dead ashes of our blasted hopes—A FINANCIAL STRUCTURE THAT SHOULD ULTIMATELY WITHSTAND THE ASSAULTS OF TIME." The author boasted that the signature that adorned the letter represented,

> a name unsullied—I hope unmarked, by the hint of scandal—with the priceless heritage of what I believe to be an unblemished escutcheon—a name which I will to my dying day—God being my helper, keep clean—a name which—(and I say it with all humility) is celebrated in the annals of Petroleum History!

The name was C. C. Julian.

The years between the collapse of Julian Petroleum and Julian's 1930 reappearance at the head of his new venture, the C. C. Julian Oil and Royalties Company, had not treated the flamboyant promoter kindly. Although he had escaped indictment in the Western Lead and Julian Pete scandals, he had suffered almost uninterrupted reverses on other fronts. His various mine mergers had failed to capture the public imagination, and an Arizona judge had stripped him of his interests in his flagship property, the Monte Cristo Mining Company. In 1928 and 1929, Julian had returned to Santa Fe Springs in an attempt to re-create his initial successes. He sold interests in four new wells (Julian Wells no. 14 through 17) on the acreage acquired in his second lease with Globe Petroleum. Two of these efforts struck oil, but neither in amounts sufficient to justify the expenditure, and Julian abandoned the wells. Meanwhile, lawsuits filed by former investors and business partners had generated hundreds of thousands of dollars worth of judgements against him.

In late 1929, after Julian had clearly worn out his welcome in Los Angeles, he relocated to Oklahoma City, the site of the nation's newest "gusher field," and began anew. Julian acquired leases, properties, and several subsidiaries in Oklahoma and Texas and initiated drilling operations. He opened an elaborate office and purchased an expensive, brightly colored red and black airplane. His wardrobe earned him a reputation as the "Beau Brummel" of the Oklahoma oil fields.

Most significantly, Julian launched a mail campaign soliciting investors. His letters, in which he bestowed upon himself the mantle of "empire builder," ran as long as nine pages. Fifteen thousand people, most of them southern Californians, responded to his call and invested $3 million. The federal government described the scale of Julian Oil and Royalties as "the largest ever put over by an oil promotion company in Oklahoma."

In April 1930, Julian flew to Los Angeles to testify before Buron Fitts' grand jury investigation. He detoured to Santa Fe Springs, where the landowners had terminated his lease and awarded drilling rights to United Oil Well Supply. In the early morning hours of April 18, Julian and four comrades routed the guard at the old Brunson lease and forcibly reclaimed the property. That afternoon, John A. Smith, Jr., president of United Oil Well Supply, arrived at the site with a drilling crew. According to Smith, Julian and his cohorts suddenly reappeared armed with pistols. Julian punched Smith in the face and his associates then joined in the attack. Smith emerged with his face battered, nose bleeding, and clothing badly torn.

Julian readily admitted the battery. "I socked him plenty," he told newsmen, obligingly posing for a photograph demonstrating his fighting form. "I don't think they'll ever arrest me. I'll be in the air before they get me," he added. But when mechanical difficulties grounded his plane, he had to appear in court, where he pleaded guilty to assault charges and paid a fine, peeled, according to the *Times* "from a bulging roll of $100 bills."

In John A. Smith, Julian had selected a target even shrewder and more ambitious than himself. "Blackjack" Smith was one of two sons of a fugitive from federal penitentiary. His younger brother, C. Arnholdt Smith, had begun to establish a banking empire in San Diego. John, described in one history as "Arnholdt Smith's smarter brother," had devoted his attention to oil. During the 1920s, Smith had prospered as an independent oil man, whose tactics including illegally pumping Standard Oil reserves into his own hidden underground tank. Using several business associates as proxies, Smith had secured de facto ownership of the Brunson properties Julian's wells stood on. In the aftermath of the April 1930 donnybrook, Smith filed suit against Julian, enjoining him from interfering with work at these sites and demanding that all of Julian's Santa Fe Springs wells, including Nos. 1, 2, and 3, which continued to produce oil and pay dividends, be placed in receivership. The court ruled in Smith's favor, barring Julian from any involvement with the wells that had brought him fame.

Julian seemed to fare better in Oklahoma. In August 1930, C. C. Julian Oil and Royalties delivered its first gusher, a 4,400-barrel-a-day producer. Several days later, a second Julian well burst loose and raged uncontrollably. Nearby residents fled their homes and workers blocked roads and extinguished fires as they "rushed to pit their skill against the strength of nature." Julian immediately announced that he had acquired a refinery and would construct a network of 100 service stations. C. C. Julian seemed to be back in business.

Two familiar obstacles arose to block his path. Like California, Oklahoma had a state corporation commission. In oil-rich mid-continent Oklahoma, the state agency concerned itself less with investor interests and more with protecting the oil industry. The new discoveries and Depression conditions had combined to drive petroleum prices to new lows. The Oklahoma Corporation Commission responded by imposing prorationing. Unlike California in the 1920s, where prorationing was voluntary, Oklahoma demanded that all oil operators limit production to 8½ percent of potential capacity and shut off all wells within twenty-four hours of their initial flow.

Julian found prorationing no more to his liking in 1930 than he had in 1923.

Forced curtailments, he still contended, favored big companies over independents. Julian refused to shut down his wells and won a writ of prohibition from Oklahoma Supreme Court chief justice Charles W. Mason, preventing the corporation commission from enforcing its order and challenging its constitutionality. Sixty-one small companies supported Julian's protest.

On September 11, 1930, the Oklahoma Supreme Court rejected Julian's position. Despite an admission by the state umpire who administered prorationing that he was paid by the large oil companies and never considered consumer interests in his decisions, the court dissolved Julian's writ of prohibition. One month later, in a decision that became a cornerstone for the right of states to impose prorationing, it upheld the constitutionality of these restrictions.

The months following this judicial setback produced one embarrassment after another for Julian. On November 30, he threw a lavish all-night rooftop party at a leading Oklahoma City hotel. The orchestra played into the early morning hours, but, as Julian departed at dawn, the husband of one of his guests accosted him. "So you're the big shot from Los Angeles?" shouted his assailant as he knocked Julian down repeatedly. "Well, how do you like this?"

In February 1931, Mary Olive Julian filed for divorce. Her complaint, dutifully reproduced in the Los Angeles press, documented Julian's "miscellaneous and promiscuous relations." In addition to Maybelle Smith, who had remained Julian's companion since 1919, Mary Julian listed three other women whom Julian had lavished his attention on, including an Oklahoma "girl of tender years" with minor children, and a Cherokee Indian princess whom Julian had flown to Los Angeles on his private airplane, where they had "committed the act of adultery . . . in the Biltmore Hotel." Julian, according to Mary, took delight in flaunting these women before her, asking his wife whether they were more beautiful than she and boasting that he "knew how to pick women, knew how to handle them when picked, and kept them in their place."

Two months later a Texas state highway patrolman arrested Julian at a Laredo hotel when he attempted to remove a guest at gunpoint. Julian and his bodyguard had accosted a former publicity director for Julian Oil and Royalties and demanded that he pay them $50,000. "Come on across or I'll let you have it," Julian allegedly threatened his victim. The police officer removed a sawed-off revolver from Julian's pocket. Texas officials charged Julian with kidnapping, prompting the Los Angeles *Times* to run a lead story wishfully headlined "Texas May Ask the Death Penalty." Julian seemed less concerned and predicted that he would "come through with flying colors." His former employee, he claimed, had extorted $112,000, and he was simply attempting to recover these funds.

Julian ultimately extricated himself from the Laredo indictment by paying several thousand dollars in fines, but John Salsberry, Julian's former partner in Western Lead, took the occasion to press for collection of $200,000 in judgements he had won in California courts. Salsberry secured a court order forcing Julian to reveal his properties and assets. The resulting hearing featured vintage Julian theatrics. Julian, who had allegedly told reporters several days earlier that $50,000 was "like chicken feed" to him, now pleaded poverty. Although Julian Oil and Royalties might be worth as much as $10 million and he was the sole trustee of its funds, stated Julian, he possessed no personal assets. His total wealth amounted to "four one dollar bills, a knife and a bunch of keys," all of which he dramatically removed from his pockets.

Further trouble awaited Julian in Oklahoma. On May 9, 1931, a group of California investors demanded a receivership for C. C. Julian Oil and Royalties. The complaint described Julian as "incapable to manage affairs of the company because of riotous living." It charged that Julian had appropriated three-quarters of the $3 million that he had raised in stock sales for his personal use. An Oklahoma judge placed Julian Oil and Royalties in receivership, but Julian won a restraining order. On May 19, Julian maneuvered around the issue by resigning as trustee. Former Oklahoma chief justice Mason, whose favorable rulings had assisted Julian in the prorationing matter, replaced him.

In July, however, investors filed suit to oust Mason, who, they asserted, "not only failed to take charge of affairs," but allowed Julian to run the company, expend its funds, and expand his "high-powered" stock sales campaign. The plaintiffs included H. A. Penn, a former Julian employee from Burbank, who had named one of his children Courtney Julian Penn before souring on his onetime hero.

Five days later, on July 13, the crushing blow fell. Federal officals arrested Julian for mail fraud. The government alleged that Julian's financial statements had reported $10 million in assets, but "the company had nothing like this and never has had at any time since its operations were started." Bail was set at $50,000. "I couldn't make the bond if it was cut to $5,000," Julian protested. "I haven't got a thing." The "poverty-stricken" Julian spent five days in jail, attended by a valet named Chappy, and bemoaning the absence of "one good pinochle player" among the inmates. On July 17, the judge reduced the bond to $25,000, and three of Julian's friends posted the necessary funds.

Confronted by numerous civil, criminal, state, and federal problems, Julian had, according to a Los Angeles *Times* editorial, "performed outside loops, barrel rolls, side slips and tailspins without number in the legal empyrean without even bending an aileron." But Julian's daredevil flight was running out of gas. On August 18, Mary Julian put their mansion on the auction block. Hundreds of people thronged through the house and grounds examining the lavish tapestries, furniture, books, and art and marveling at the heart-shaped swimming pool, gold-lined bathtub, and luxurious Rolls-Royce. The house reportedly netted Mary Julian $65,000 and the possessions tens of thousands more. She used the money to move back to Winnipeg with her teenaged daughters.

On September 25, the federal government indicted Julian and thirty-two of his associates on mail fraud charges. The total number of defendants ultimately reached fifty, including Julian's brother, his former Los Angeles sidekick Jack Roth, and former chief justice Mason. Federal attorneys contended that Julian had threatened to kill witnesses who cooperated with the prosecution. "The only persons I feel like killing are newspapermen," countered Julian.

More than a year would pass before Julian and his codefendants would be brought to trial. During this time, the oil promoter's life continued to spin out of control. San Francisco police arrested Julian along with "notorious gangster, racketeer" Frank Foster, in February 1932, for questioning in a $100,000 jewel robbery. Julian had no connection to the theft, but he was held for two days on vagrancy charges while police investigated his relationship with Foster. "It would appear that the association is something of a comedown from hobnobbing with the best spenders of Hollywood as was once his wont," commented the ever-vigilant *Times*.

In June 1932, Julian Oil and Royalty stockholders won control of the company. Three days later an Oklahoma judge declared Julian bankrupt and turned his affairs over to a referee. The decree cited over $3 million in liabilities, all but $19,000 accumulated in California. Julian listed among his debts $5,000 to his Los Angeles tailor, $1,253 to the Mackin Shirt Company, and almost $1,000 owed to Rolls-Royce. Julian, meanwhile, was spending increasing amounts of time in his native Canada, prompting his Oklahoma bondsman to revoke his bail guarantee. Julian spent the last two days of 1932 in an Oklahoma City jail until his attorney and business associates posted a new bond.

The federal government had scheduled Julian's trial for February 6, 1933, bringing in close to 200 witnesses from around the country. But on February 3, the managing editor of the Oklahoma City *Times* received a letter from Julian bitterly attacking the methods used by federal investigators and declaring that he could "get a fairer trial in Russia than in the U S of A." The letter was postmarked "Chicago," but dated "Vancouver, British Columbia." C. C. Julian had jumped bail and fled the country.

Federal officials hoped to retrieve Julian and bring him to trial, but the United States extradition treaty with Canada did not cover mail fraud. Julian could, however, be returned for embezzlement. On February 15, the government added an embezzlement charge against Julian and issued a wanted poster. The flyer described the forty-six-year old Julian with "hair turning gray . . . neat appearance, wears expensive clothing; fond of liquor and women, lives at best hotels" and offered a $5,000 reward for his return. But Julian had vanished without a trace.

During the next two months, reward-seekers reported seeing Julian engaged as a toreador in Mexico, an elephant salesman in Siam, and a general in the Chinese army. Those placing Julian in Asia had the wrong occupations but the right locale. In mid-April a California woman visiting the Astor Hotel in Shanghai suddenly stopped short and exclaimed, "Look, there's C. C. Julian."

Julian, who had arrived in Shanghai on March 23, had selected his destination carefully. Shanghai was an international city, controlled jointly by the Chinese, French, and British. Its extraterritorial status and complex extradition laws made Shanghai a haven for fugitives. It was virtually impossible to extradite a Canadian citizen like Julian from the International Settlement for a crime committed in the United States. "You can tell the United States to go to hell," boasted Julian upon discovery. "They can't touch me."

Approximately twenty-five nationalities resided in Shanghai, including 4,500

American citizens. Most non-Chinese lived in two neighborhoods named the "International Settlement" and the "French Concession." Once exposed, Julian made no attempt to hide his identity or lifestyle. Despite the earlier bankruptcy judgement, Julian resided in the lavish Hotel Metropole in the International Settlement, where, according to one report, "he entertains the cabaret girls lavishly with liquids that bubble." Julian blamed his predicament on "crooked politics" in Oklahoma and "Standard Oil's unending opposition" but vowed revenge. "I will develop the oil fields of China and then maybe I will have the last word with the Standard Oil Company," he predicted.

In August 1933, Los Angeles *Times* reporter Harry Carr interviewed Julian. "I have always lived up to the lessons of honesty that I learned at my mother's knee," related Julian. "But sometimes I wonder if it pays to be honest when the world does not appreciate it." Carr described Julian's "degree of sobriety" as "an uncertain condition . . . at best." At one point, according to Carr, Julian made a talking picture to present his version of events. After viewing his drunken performance, Julian ordered the film destroyed.

Shortly after the Carr interview, Julian reportedly ran out of funds. On September 10, Julian bought a friend a drink and reported, "Well, that's my last dollar." Later that day the Hotel Metropole evicted him with his bill two months overdue. Julian withdrew from the limelight and moved to a small hotel in the French Concession, where he lived in relative obscurity for the next several months. One American lawyer recalled that Julian would frequently drop by his office on the pretext of retaining him, "but I am sure that what he wanted was a little companionship, for like all wanted men he was lonely."

In December, Julian proclaimed a new beginning. "Since arriving in China I have basked in the sunbeams of idleness, but I am 100 per cent alive and can't hold my hands for the remainder of my life," he announced. "Shanghai is my home and I am going to work. I rose and fell, but I will come back." A new organization specializing in American securities had named him general manager, reported Julian. Three months later Julian resigned when asked to sell stock in an undertaking firm. "Selling dead ones isn't my line," he quipped. According to one account Julian began borrowing funds from Chinese businessmen whom he promised to enrich through stock speculation. Within a few months he incurred a reported $25,000 in debts.

Julian also conjured up one final moneymaking scheme. He would write an autobiography in which he would reveal his true story and identify the real culprits involved in the collapse of Julian Petroleum. Julian called his book *What Price Fugitive?* and subtitled it *A Refugee from Justice.* He predicted it would sell a million copies. In March 1934, Julian left a copy of the 150-page manuscript with his Shanghai attorney, Oscar Fischer, who he hoped would advance him $5,000 and negotiate its sale. According to Fischer, *What Price Fugitive?* was "indeed an exciting story, written breezily with much slang and would appeal to the public. . . . He named names."

His hard times notwithstanding, Julian retained his way with women. In the early months of 1934, he appeared frequently in the company of "his secretary and confidante," twenty-year-old Leonora Levy, described in one account as " 'bewitchingly pretty' and an 'exquisite young beauty.' " A second woman, Mary Cantorovich, referred to Julian as her "dear friend."

On March 23, 1934, his landlady evicted Julian from his French Concession hotel for nonpayment of his bill. The following day, Julian called attorney Fischer and asked to meet with him but failed to appear. Fischer later learned that Julian was drunk at a local club. That evening Julian returned to the International Settlement. He checked into the fashionable Astor House and reserved a table for two for a late dinner at the hotel restaurant. There Julian hosted Leonora Levy in the extravagant manner that had nurtured his legend. Champagne flowed freely. Inebriated, Julian proposed marriage to Levy, who turned him down. Julian threatened to take his own life, but Levy did not take him seriously.

Shortly after midnight Julian excused himself, leaving Levy seated at "a table expansively littered with champagne glasses." When Julian did not return, Levy went to their room. There she found Julian stretched out on the floor, still breathing, a vial of drugs nearby. Julian was rushed to the hospital. He never regained consciousness. Five hours later, Courtney Chauncey Julian died. He was forty-eight years old. Physicians on the scene labeled his death a suicide.

Even in death, Julian continued to create sensations. Mary Cantorovich appeared at the hospital, where, upon learning of Julian's death, she wept and wailed and demanded admission to the room where his body lay. Meanwhile, Leonora Levy had returned to the Astor House. There she found the remnants of the fatal drugs that Julian had swallowed. She downed the potion and sank into a coma.

Fortunately, Levy had not consumed a fatal dose. The following day she awoke long enough to tell officials, "He said he would do it. I did not believe him, but he did it. He is a very brave man." Several days later, she added, "I will never attempt suicide again, but damn the doctors I wanted to die." Despite Levy's account, many of Julian's friends both in Shanghai and Los Angeles refused to accept that he had committed suicide. They contended that in his inebriated state he had accidentally overdosed on his sleeping potion. A coroner's inquest returned a verdict of "death by narcotic poisoning, self-administered."

Julian's body, reported the *Winnipeg Evening Times* of its native son, lay "upon a slab in a large darkened room, with a cement floor, awaiting burial. About it lay other bodies of men of different races. Chinese, Indian, Burmese, and Malayan were only a few of the nationalities represented in this strange silent company." Penniless at death, Julian seemed destined for a pauper's funeral until a fellow Canadian requested the "privilege" of paying the $46 funeral expenses. On March 28, the pastor of the Shanghai American Church presided over the services. Julian lay in a cheap casket upon which rested a wreath of flowers with a card signed "from Irt." A cablegram arrived later from Los Angeles ordering flowers "from Blackbird." Nine people attended the funeral, including the landlady who had evicted Julian the day before his death, two newspapermen, and "two young twittering girls—apparently curiosity seekers."

Julian's sister Violet, who lived in Los Angeles, attempted to have the body cremated and shipped back to the United States for burial, but her efforts failed. On May 12, Julian was buried in a "cheap unadorned" beggar's coffin in an obscure corner of the foreign cemetery in Shanghai. Leonora Levy, dressed in white, appeared as the sole mourner. She wept bitterly as the coffin entered the earth, then walked away alone. The life and, indeed, the times of C. C. Julian had come to an end.

## III

The death of C. C. Julian effectively ended the Julian Petroleum scandal, but the episode nonetheless haunted southern California for years to come. The veterans of the Julian swindle attempted to restore their lives, with mixed success. Archibald Johnson, who as an investor in a short-lived pooling scheme had been an incidental figure in the scandal, nonetheless called the Julian incident "perhaps the bitterest experience of my life." Dozens shared this sentiment. For some business leaders, most notably Pacific Southwest bankers Charles Barber and Charles Stern, both of whom became utilities executives, the Julian Petroleum scandal lowered their career trajectories. Others retreated to their opulent lairs, licked their wounds, and resumed their positions among the lions of the Los Angeles establishment.

Some victims never recovered. Former Los Angeles district attorney Asa Keyes was released from prison in October 1931. A crowd of 400 supporters greeted a thinner, visibly aged man. Two years later, Keyes, who continued to maintain his innocence, won a full gubernatorial pardon and applied for reinstatement with the state bar. The board of governors unanimously denied his request. His health wrecked, Keyes suffered minor strokes in 1933 and early 1934, before a massive paralytic stroke felled him in his Beverly Hills home on October 17, 1934. He died two days later at the age of fifty-seven. Two months earlier, Jack Friedlander, the controversial state corporation commissioner, had succumbed to a heart attack at age forty-five.

State and federal prisons seemed to maintain a revolving door for those implicated in the Julian scandal. Morris Lavine served ten months for his role in the Charlie Crawford affair. He received a gubernatorial pardon in 1933. His newspaper career behind him, Lavine became one of the city's leading criminal defense attorneys. In 1932, stockbroker C. C. Streeter was convicted of income tax evasion for failure to report his Julian Pete earnings and sentenced to the federal penitentiary. Streeter, the only stockbroker penalized in the scandal, was released in 1933. In June 1934, Jack Roth, Julian's one-time right-hand man, began serving an eleven-to-ninety-five- year sentence at San Quentin for defrauding elderly investors in his brokerage activities. The trial judge called him "a type of criminal that must be eliminated from the community." Roth was joined in San Quentin by Jacob Berman's brother, Louis. Once dismissed as the "dumb one" by S. C. Lewis, Louis

Berman was convicted of grand theft in a real estate swindle in 1935 and received a five-to-fifty-year sentence. Roth was paroled in 1937; Louis Berman in 1939.

S. C. Lewis and Jacob Berman, the prime architects of the swindle, had long since been released from prison. On February 24, 1935, Lewis completed his sentence for the Lewis Oil and Julian Petroleum convictions. A $16,000 fine still hung over him, but Lewis claimed he was insolvent and could not pay. He served an additional thirty-day sentence in lieu of the fine. On March 27, Lewis took a "pauper's oath," certifying his wealth at less than twenty dollars and that he possessed "no property in any way conveyed or concealed, or in any way disposed of for my future use or benefit," and walked out of prison.

Despite accusations by Arthur Loeb and others, federal officials had found no evidence to dispute Lewis's oath of poverty. Lewis made one final offer to assist Julian stockholders in recouping their losses but found little interest in his proposition. He announced plans to re-establish himself in the oil industry and claimed that two Eastern companies had already solicited his services. "Instead of organizing any company, I plan to operate independently in Texas and California in actual drilling and oil development work," he explained. S. C. Lewis subsequently headed for the Texas oilfields, never to be heard from again in southern California.

The case of Jacob Berman proved more complex. Berman had yet to serve time for his 1930 state embezzlement conviction, which carried a prison term of up to forty-two years. On June 21, 1935, one day before his federal sentence ended, the government transferred Berman from a Pennsylvania penitentiary to Alcatraz to enable California officials to seize him upon his release. Berman prolonged his stay in federal custody by refusing either to pay his $16,000 fine or to sign a pauper's oath. His attorneys demanded that he be returned to Pennsylvania. The deadlock persisted for fifteen months before California governor Frank Merriam commuted Berman's state sentence to time served. On October 19, 1936, Berman became a free man. "I have paid my debt and I hope I am out of the spotlight for good," he told reporters.

But the spotlight continued to follow Berman. Within weeks of his release he sailed to Europe, where he embarked on a "spending orgy," confirming longstanding rumors that he had cached millions in European vaults. When he returned to New York in the company of a blonde companion in early January 1937, his wife had him arrested for refusing to pay child support. Berman expressed outrage. He had already given his wife "a million or so," he protested. "Certain human vultures have come between me and my wife and are using her as a cat's paw to try and get that money," stated Berman. But, he quickly added, "As a matter of fact, I'm broke." Shortly thereafter, Berman disappeared.

In 1944, a group of creditors filed suit in New York State Supreme Court against Jacob Berman, seeking to locate the "bright youngster's" hidden fortune. The plaintiffs alleged that Berman had stashed $1.25 million in Barclay's Bank of London and another $1 million in banks in Paris and Rome. According to the action, World War II had blocked Berman's efforts to secure the funds. In the interim he had induced several individuals to finance his business ventures, using the Julian Pete booty as collateral and defrauding them in the process. With the war coming to a close, Berman's more recent victims hoped to seize his European accounts. The *Los Angeles Examiner* revealed that Berman was now living in Washington, D.C. Nothing further seems to have come of the suit. Whether Berman or anyone else ever retrieved the missing millions, if the money indeed existed, remains unknown.

Ed Rosenberg, Jacob Berman's accomplice and the co-defendant in the Keyes bribery trial, also attempted to put his Julian Petroleum years behind him. Rosenberg, like Keyes, had been paroled in 1931. On leaving prison, Rosenberg shook hands with reporters and asked to be forever left out of the newspapers. His wish was largely granted until October 10, 1940, when Rosenberg urged his attorney to rush to his home. The lawyer found Rosenberg's body slumped on a swing in the patio. Like C. C. Julian, Rosenberg had swallowed poison. Rosenberg left two notes. The first asked that his body be removed before "the children will be home from school." The second told his wife, "Dearest: Can't go on any longer. Too bad, depressed and no prospect. Take good care of the children and forget everything as quickly as possible." Rosenberg was forty-eight years old.

For the Reverend Robert Shuler, the Julian Pete years marked the high point of his career. Shuler appealed the decision that silenced radio station KGEF and sought other ways to make his voice heard. In 1932, he ran for the United States Senate in order to force radio stations to sell him air time. He fared remarkably well, polling more votes in the primary election than any other candidate, before losing the general election. Three blows in 1933 further weakened Shuler's influence. In January the United States Supreme Court refused to hear his case against the Federal Radio Commission, cementing the ban on KGEF. In June, Mayor John Porter failed in his re-election bid. Shortly thereafter the minister announced that *Bob Shuler's Magazine*, a victim of the Depression, would take a year's "sabbatical." The year stretched into a decade.

With neither radio station nor journal to plead his case, Shuler retreated to his pulpit. He surfaced periodically, most notably in 1944, when he unsuccessfully attempted to unseat liberal congressman Jerry Voorhies. The themes he raised proved more effective two years later when Richard Nixon launched his political career by defeating Voorhies. Nixon, who had occasionally attended Shuler's church, sought and received "Fighting Bob's" endorsement. Shuler again achieved national recognition in 1960 when he challenged the suitability of a Roman Catholic to run for President. Five years later, Shuler died, a largely forgotten man, his brief obituary buried on the back page of the second section of the *Los Angeles Times*.

Shuler's ally, Buron Fitts, remained at the forefront of Los Angeles politics for a somewhat longer period. Stymied in the 1930 gubernatorial race, Fitts served for the next ten years as an aggressive district attorney. He administered the largest prosecution office in the country, handling 5,000 felony cases a year and, at its peak, winning 82 percent of its cases. But critics challenged Fitts's zeal and even his probity. The 1934 grand jury indicted him for perjury, but Fitts won acquittal. In the late 1930s he played a leading role in anti-union activities. Alleging that Communists seeking to control unions had committed mass murders, Fitts organized a "small army" to combat their "subversion." In 1940, Fitts lost his bid for a fourth term. With the outbreak of World War II, Fitts re-enlisted in the Army, enabling him to re-create his youthful heroism. He served in North Africa, Europe, and the Pacific, participating in twelve battles. Upon his return he practiced law until retiring in 1969. Four years later, dogged by ill health, Fitts put a handgun to his right temple and committed suicide.

Frank Keaton, Motley Flint's assassin, reappeared in the headlines in June 1940. State officials ruled that Keaton, who had been incarcerated at the Mendocino Hospital for the Insane, had regained his sanity and could now be executed. Keaton was moved to San Quentin's death row and his hanging rescheduled for September 1940. The governor commuted the sentence just days before Keaton was to die. David Clark, who escaped conviction for the 1931 slayings of Charlie Crawford and Herbert Spencer, did not get away with murder a second time. In 1954, the former assistant district attorney and judicial candidate, once described by a juror as "one of our noblest Americans," pleaded guilty to second-degree murder after shooting his former law partner's wife.

Echoes of the Julian Pete swindle also were heard in California politics. In the 1938 Democratic gubernatorial primary, candidate William Neblett reminded voters that his opponent, Culbert Olson, had once represented S. C. Lewis and A. C. Wagy & Co. In reality, Olson, the cousin of Utah senator William King, had appeared on all sides of the scandal, representing Lewis in the mid-'twenties, but earning a reputation as a champion of fraud victims in the early 'thirties. The voters no longer seemed to care. Olson overcame Neblett's charges and was elected governor. Following a precedent established by his predecessors, Olson reappointed "Iron Mike" Daugherty, C. C. Julian's nemesis at the Corporations Department, as his corporation commissioner. Daugherty, originally returned to office by Governor James Rolph in 1931, presided over a department greatly strengthened by revisions in the "blue sky law" inspired by the Julian Petroleum and Western Lead escapades. He ultimately became the dean of California state officials, appointed by a total of seven governors, before accepting a mandatory retirement in 1954.

In 1941, the final curtain fell on the Julian Petroleum receivership. H. L. Carnahan had continued to serve as receiver in an unpaid capacity in order to handle unclaimed shares of Sunset Pacific stock and any developments in the remaining usury cases. On February 27, Carnahan filed his final report and resigned, recommending the dismissal of the unresolved usury cases. The court waived the regulation requiring notification of the 30,000 Julian Petroleum stockholders, citing both the expense and fears that "such a notice would probably only cause inquiry to be made about matters that were, more or less, finished and ended some ten years ago."

Ironically, the Julian Petroleum Corporation itself had evolved into a legitimate company. In April 1934, just days after Julian's death, receiver Frank Rolapp incorporated Sunset Oil, heir to the Julian Petroleum, California-Eastern, and Sunset Pacific corporations. Shareholders in the earlier ventures received stock in the new enterprise, offering hope that they might recoup at least some of their losses. In December 1934, Sunset Oil purchased the assets of its predecessor firm at auction. Only one competing bid was heard. Arthur Loeb, in a final protest gesture before fading into obscurity, offered fifteen cents for over one million shares of worthless California-Eastern stock.

The reorganization created a voting trust that included Adolph Ramish, again a major shareholder, and individuals representing Associated Oil and J. A. Smith's Oil Well and Supply Company. Indeed, "Blackjack" Smith, whom Julian had bested at fisticuffs in 1930, now played a major role in virtually all of the California properties that Julian had once controlled. After unsuccessful legal efforts to squeeze Julian's original unitholders out of their interests in the Santa Fe Springs wells, Smith acquired the company that administered the properties. His successful oil ventures reportedly enabled him to bail out brother C. Arnholt Smith's U.S. National Bank on four occasions during the Great Depression. The bank prospered after World War II, propelling C. Arnholt Smith to the forefront of southern California business and politics until the early 1970s, when he went to prison in an offshoot of the Watergate affair.

Rolapp's Sunset Oil handiwork proved surprisingly durable. Functioning primarily as a gasoline marketing company that received most of its supplies from Associated Oil, Sunset Oil prospered during World War II, and in the late 1940s was valued at $10 million, with fifty producing wells and 175 employees. In the early 1950s, Sunset embarked on a major expansion. By 1953, it had become the largest independent petroleum marketer on the Pacific Coast. Three years later, after reporting record earnings, Sunset Oil merged with the International Mining Corporation to form Sunset International Petroleum. Sunset Oil shareholders received three shares of stock in the new company for each share they held. Those who had retained their shares over the three decades since the creation and collapse of Julian Petroleum thus received at least some return on their investments.

Meanwhile, Julian No. 1, the well that had started it all, continued pumping oil. The first and most prolific of Julian's wells remained on production for over forty years. It was abandoned in March 1964, on the thirtieth anniversary of Julian's suicide.

C. C. Julian's wife, Mary, also proved a durable survivor. She had returned with her daughters to her native Winnipeg in 1931, but had never pursued her well-publicized divorce suit. She was grief-stricken upon hearing of her husband's suicide. "It can't be so," she sobbed, remembering him more kindly in death than she had in life. "He never brought any of his troubles to us. . . . It didn't matter whether things were right or wrong, he was always cheerful." The Julian family remained in Winnipeg. Lois Julian became a nurse; Frances worked as a hairdresser. Mary Julian lived with her sister for more than forty years before she died in the 1970s.

Several thousand miles to the south, S. C. Lewis had reconstructed his life in Corpus Christi, Texas, where he had settled in 1936. Lewis developed the local market for natural gas and formed several concerns, including the Southern Community Gas Company. He became a wealthy and respected citizen. Lewis made his greatest mark as a patron of the Corpus Christi Boys Club, which he founded in 1957 and underwrote for many years. He served as a national director of the Boys Club alongside, ironically, Herbert Hoover. In 1970, at the dedication for a Boys Club Neighborhood Center in Corpus Christi, for which Lewis had donated the land and much of the funding, the crowd chanted "Viva Sheridan Lewis!" and named the pool in his honor. But a trace of the confidence artist remained. In relating his life to a reporter, Lewis omitted his Julian Petroleum, Lewis Oil, and McNeil Island experiences. He did, however, take credit for creating Sunset International. Lewis died in 1972, at the age of 82, hailed as a "businessman, lawyer, and philanthropist."

During most of these years, the whereabouts of C. C. Julian's *What Price Fugitive?*, the autobiography that reportedly told all, remained a mystery. In the mid-1960s a Russian woman, possibly Mary Cantorovich, approached actor George Hamilton with a manuscript typed on thin onion-skin paper. She thought that it might make for an interesting film project. Hamilton bought the document, then set it aside until years later when he chanced upon other information about the Julian Pete swindle. In recent years, Hamilton has attempted to interest movie producers in a film based on Julian's version of events.

But, as Hamilton discovered, the Julian Petroleum scandal has long since faded from southern California's collective memory. Periodic retellings in magazine articles, textbooks, and once, in a television documentary failed to stir more than passing interest. These accounts relied heavily on either Guy Finney's 1929 book *The Great Los Angeles Bubble*, or Carey McWilliams' description in his 1946 history *Southern California: An Island on the Land*. Both Finney and McWilliams saw the Julian affair as the outgrowth of conditions unique to Los Angeles. In *Southern California*, unquestionably the most influential work on Los Angeles history, McWilliams used the Julian Pete scandal as a fulcrum on which to balance his central argument about the impact of incessant migration and runaway growth on the region. In the concluding section of his pivotal chapter, "The Sociology of the Boom," McWilliams argued that "mass migration and easy money had corrupted the community." The "Julian fiasco" revealed that California "had never achieved integration as a society, that its government was a hastily improvised affair and that the gambling spirit had bred a reckless social irresponsibility in its residents."

But McWilliams overestimated the uniqueness of Los Angeles and southern California. The Julian Petroleum scandal is not so much the story of a city or a region as it is the saga of an era. The decade of the 1920s, with its idealization of business, get-rich-quick mentality, unwise tax policies, and absence of government regulation and responsibility, bred speculation, corruption, and corporate chaos throughout the nation. More than half a century would pass before we would see its like again.

# NOTES

Frequently used citations and their abbreviations.

| | |
|---|---|
| BSM | *Bob Shuler's Magazine* |
| CSMB | California State Mining Bureau, Long Beach Office |
| CG | *California Graphic* |
| CH | *California History* |
| FBI | Federal Bureau of Investigation, C. C. Julian file. |
| HL | The Huntington Library, San Marino, California |
| HP | John Randolph Haynes Papers, Special Collections, University Research Library, UCLA |
| IDN | *Illustrated Daily News* |
| JP | Hiram W. Johnson Papers (C-B 581), Bancroft Library, University of California, Berkeley |
| LAEX | *Los Angeles Express* |
| LADN | *Los Angeles Daily News* |
| LAR | *Los Angeles Record* |
| LASC | Los Angeles Superior Court |
| LAT | *Los Angeles Times* |
| LAX | *Los Angeles Examiner* |
| NA | National Archives |
| NYT | *New York Times* |
| OA | *Oil Age* |
| SEP | *Saturday Evening Post* |
| SFC | *San Francisco Chronicle* |
| SP | Charles Stern Papers, Special Collections, University Research Library, UCLA |
| USC | Hearst Clippings Collection, Regional History Center, University of Southern California |

Since advertisements appeared in several Los Angeles newspapers, all advertisements other than those which appeared only in specific newspapers are referenced only by date. Where dates are clearly stated in the text, they are not included in the notes.

All stock prices are from the Los Angeles *Times*. All overissue figures are drawn from the Haskins & Sells stock audit included in the 1930 FBI report.

## Chapter 1

### Section I

p. 3   Keaton's biography is drawn from LAT, LAR, LAX, July 15, 1930.
p. 4   "My husband seemed . . ." LAT August 22, 1930.
       "I remember during last November . . ." LAT July 15, 1930.
       "My husband has had a terrible time . . ." LAR July 15, 1930.
       "Frank blamed the Julian crash . . ." LAT July 18, 1930.
       Flint is described as Santa Claus in Guy Finney, *The Great Los Angeles Bubble*
       (New York, 1929), 72–73.
       The biography of the Flint brothers is drawn from LAT July 15, 1930, and *History*
       *of Los Angeles City and County* (n.d.), 19–25.
p. 5   On Motley Flint as booster, see LAT, LAX July 15, 1930, and Finney, 72–73.
       "the best known lodge man . . ." LAT July 15, 1930.
       "Under an exterior . . ." LAX July 15, 1930.
       On Flint and film industry, see Benjamin B. Hampton, *A History of the Movies*
       (New York, 1931), and Neal Gabler, *An Empire of Their Own: How the Jews*
       *Invented Hollywood* (New York, 1988), 134.
       On Flint's hotel suite, see LAT July 15, 1930, and Finney, 133.
       "one of the most important players . . ." Kevin Starr, *Material Dreams: Southern*
       *California Through the 1920s* (New York, 1990), 123.
p. 6   On Frank Flint's business interests, see Frederick C. Jaher, *The Urban Establish-*
       *ment: Upper Strata in Boston, New York, Charleston and Los Angeles* (Urbana,
       1982), 634.
       On Flintridge, see Starr, 199, and LAT July 15, 1930.
       "They have broken my heart . . ." LAX July 16, 1930.
       On Flint in France, see LAR March 27, 1930.
       "He followed carefully . . ." LAT July 18, 1930.
       On Keaton and Shuler, see LAR July 15, 1930, and LAX July 17, 1930.
       "he finally came to the belief . . ." LAT July 18, 1930.
       On Keaton and stock market, see LAT July 15, 1930.
       For Keaton at Julian trials, see LAX July 16, 1930 and LAT July 17, 1930.
       For description of Keaton's gun, see LAT July 15, 1930.
       On Keaton at stock office, see LAT July 17, 1930.
p. 7   For descriptions of Selznick trial, courtroom, and shooting, see LAT, LAX, LAR
       July 15, 1930, and LAX July 16, 1930.

### Section II

p. 8   "a decade of debauchery . . ." NYT *Book Review*, November 17, 1967.
       "displayed an inordinate desire . . ." John Kenneth Galbraith, *The Great Crash,*
       *1929* (Boston, 1961), 8.
       "no one engineered . . ." *ibid.*, 4.
       On speculation in 1920s, see Frederick Lewis Allen, *Only Yesterday* (New York,
       1964), 225–40; Robert Sobel, *The Great Bull Market: Wall Street in the 1920s*

(New York, 1968); and Roger M. Olien and Diana Davids Olien, *Easy Money: Oil Promoters and Investors in the Jazz Age* (Chapel Hill, 1990).

p. 9     "You should have $10,000 ..." and "Everybody Ought to Be Rich" William Leuchtenburg, *The Perils of Prosperity, 1914–1932* (Chicago, 1958), 241.

On Charles Ponzi, see Sobel, 17–21.

"ballyhoo years ..." Allen, 155.

"the 1920s are best ..." Ellis Hawley, *The Great War and The Search for Order in Modern America, 1917–1933: A History of the American People and Their Institutions* (New York, 1979), vi.

"a crucial change ..." Warren I. Susman, *Culture As History: The Transformation of American Society in the Twentieth Century* (New York, 1984), 106–21.

p. 10    For statistics on advertising, see Sobel, 43–44.

On Frederick Cook, see Olien and Olien, 92–103.

On federal tax policies, see Sobel, 52–53.

"every corner bootblack ..." George Soule, *Prosperity Decade* (New York, 1947), 293.

For estimate of stock ownership, see Robert McElvaine, *The Great Depression* (New York, 1984), 43.

p. 11    "the perfectly prodigious business ..." Albert W. Atwood, "Money from Everywhere," SEP, 195 (May 12, 1923), 10.

"The steady speedy growth ..." Bruce Bliven, "Los Angeles: The City That Is Bacchanalian," *The New Republic* 51 (July 13, 1927), 197.

"Los Angeles is not a mere city ..." Morrow Mayo, *Los Angeles* (New York, 1933), 319.

On the "new beginning," see Carey McWilliams, *Southern California: An Island on the Land* (Santa Barbara, 1973), 128–29.

"These square miles ..." Paul D. Augsberg, *Advertising Did It! The Story of Los Angeles* (Los Angeles, 1922), 6.

"From the highest to the humblest ..." Finney, (Los Angeles, 1929), 9–10.

"Super Babbitts ...," Louis Adamic, "Los Angeles! There She Blows," *Outlook and Independent* 155 (August 13, 1930), 594.

p. 12    "the psychology of boost, don't knock ..." James C. Findlay, "The Economic Boom of the Twenties in Los Angeles" (Ph.D. diss., Claremont, 1958), 20.

On the darker side of Los Angeles, see Louis Adamic, *The Truth About Los Angeles* (Girard, Kans., 1927), and the anonymous *Sunshine and Grief in Southern California* (Los Angeles, 1931).

"for the tourists ..." Adamic, *ibid.*

On the "white spot" campaign, see Findlay, 139.

"to exploit the nearby bathing beaches ..." Finney, 16.

On the All-Year club, see Findlay, 131.

"Money from Everywhere," Atwood, 10.

p. 13    "Speculation is in the air ..." Albert W. Atwood, "Mad from Oil," SEP 196 (July 14, 1923), 86.

"particularly ripe ..." Finney, 9–17.

"an insistent element of speculation ..." Atwood, "Money from Everywhere," 147.

"stock market craze of the first magnitude ..." *ibid.*, 144.

On number of new subdivisions, see Bruce Henstell, *Sunshine and Wealth: Los Angeles in the Twenties and Thirties* (San Francisco, 1984), 14.

p. 13    "city size of Albany . . ." Atwood, "Money from Everywhere," 144.
         "One is struck . . ." *ibid.*, 147.
p. 14    On percentage seeking to resell, see Findlay, 176.
         12,000 realtors, *ibid.*, 175.
         "one is disgusted . . ." Atwood, "Money from Everywhere," 147.
         On Palos Verdes promotions and "enterprising philistine," see Robert Glass Cle-
            land, *California in Our Time* (New York, 1947), 128–31.
         "The flame . . ." and "stark, staring . . ." Samuel Blythe, "Gambling with Grief,"
            SEP 197 (July 5, 1924), 2, 72.
p. 15    "Before the effects . . ." Albert W. Atwood, "When the Oil Flood Is On," SEP 196
            (July 7, 1923), 3.
         "greatest outpouring . . ." Fred H. Viehe, "Black Gold Suburbs: The Influence of
            the Extractive Industry on the Suburbanization of Los Angeles, 1890–1930,"
            *Journal of Urban History* (1981), 13.
         "For all the high tides . . ." and subsequent quotes are all from Atwood, "Oil
            Flood."
         On town lots, see OA July, November, 1921, and California Division of Oil and
            Gas, *Summary of Operations, California Oil Field* (May, 1923), 8.
         On encyclopedia lots, see *Los Angeles Times*, April 31, 1923.
         "A wonderful, unprecedented . . ." Atwood, "Oil Flood," 11.
         "Afforded an opportunity . . ." Atwood, "Mad from Oil," 11.


## Chapter 2

### Section I

p. 17    Except where otherwise noted, this account of C. C. Julian's life is based on LAR
            February 11–17, 1924, and FBI report (November 6, 1924).
p. 18    Information on Morris from Lenore Edise, ed., *Furrows in the Valley: A History of
            the Rural Municipality of Norris and Its People* (Intercollegiate Press, 1980), 22–
            23, 61.
         "a lusty, gutsy . . ." Alan F. J. Artibise, *Winnipeg: An Illustrated History* (Toronto,
            1977). See also Gerald Friesen, *The Canadian Prairies: A History* (Toronto,
            1984).
         On Julian's family in Winnipeg, see *Henderson's City of Winnipeg Directory*, 1899–
            1974.
p. 19    Julian as newsboy from ads, February 21, 1926, and August 12, 1923.
         Julian's other jobs, LAR February 12, 1924, and August 12, 1923.
         On Regina, see Earl G. Drake, *Regina: The Queen City* (Canada, 1955), 118–20;
            on Julian in Regina, author's correspondence with Saskatchewan Archives Board,
            July 27, 1987.
p. 20    Return to Regina and Winnipeg, LAR February 15, 1924, and *Henderson's City of
            Winnipeg Directory.*
         "California will have . . ." Walton Bean, *California: An Interpretive History*, 3rd
            edition, (New York, 1978), 310.
         Lack of energy, Harold F. Williamson *et al.*, *The American Petroleum Industry.* Vol.
            2, *The Age of Energy, 1899–1959* (Evanston, 1963), 175.

p. 20      "California geology . . ." Hartzell Spence, *Portrait in Oil: How the Ohio Oil Company Grew to Become Marathon* (New York, 1962), 150.

"Fuel is so scarce . . ." Williamson, 175.

p. 21      On Doheny and Los Angeles City Field, see Bean, 310–12.

On changes in oil industry, see Williamson, 169–71, 190–94.

On Central California fields, see Albert W. Atwood, "When the Oil Flood Is On," SEP 196 (July 7, 1923), 89; and Bean, 312.

p. 22      "the center of the oil universe . . ." William Rintoul, *Spudding In: Recollections of Pioneer Days in the California Oil Fields* (San Francisco, 1976).

Julian in Vancouver, LAR February 15, 1924; FBI report June 9, 1924; LASC Case #D92410, *Mary Olive Julian vs. C. C. Julian* (February 2, 1931); *Henderson's City of Vancouver Directory*, 1911–1914.

On the Vancouver boom, see Margaret A. Ormsby, *British Columbia: A History* (Toronto, 1971), 354–65; Eric Nicol, *Vancouver* (Toronto, 1970); and LAR February 15, 1924.

p. 23      Julian returns to Winnipeg, LAR February 15, 1924, and *Henderson's City of Winnipeg Directory*.

p. 24      On Texas oil schemes in Los Angeles, Atwood, "Oil Flood," 86.

For Julian's return to Los Angeles, see LAR February 17, 1924, and August 28, 1925.

On Maybelle Smith, see LASC Case #D92410.

On Huntington Beach, see Walter A. Tompkins, *Little Giant of Signal Hill: An Adventure in American Enterprise* (Englewood Cliffs, N.J., 1964), 4–5.

Julian's account of the Huntington Beach well appeared in the LAR February 17, 1924, and August 28, 1925.

## Section II

p. 25      For the early history of Santa Fe Springs, see City of Santa Fe Springs, *History in the Making* (1976), 6–8.

"excessively portly . . ." OA, February, 1924.

"to dissolve . . ." *History in the Making*, 7–8.

Unless otherwise noted, the Bell biography is derived from John O. Pohlman, "Alphonso E. Bell: A Biography," *Southern California Quarterly* 46 (1964), 197–217; and Bruce Henstell, "Remembering Mr. Bell," *Los Angeles* (January, 1985), 107–9.

p. 26      Pioneer Oil Company, *History in the Making*, 10.

On Marius Meyer, see Marilyn Jensen, "The Giant That Was Santa Fe Springs," *Pacific Oil World*, 4th Annual Review Issue (January, 1980), 114–15; *History in the Making*, 9; California Division of Oil and Gas, *Summary of Operations, California Oil Field* (May, 1923); and LAT November 26, 1922.

On Meyer wells, *Summary of Operations* (May, 1923); Kendall Beaton, *Enterprise in Oil: A History of Shell in the United States* (New York, 1957), 182; and OA, November, 1921.

On Bell No. 1, see Jensen, 116; Pohlman, 216; Beaton, 182; Atwood, "Oil Flood," 99; and OA, November, 1922.

p. 27    Alexander No. 1, OA, December, 1922; Rintoul, 163; Jensen, 116, Taylor and Welty, 147–48.

Bell No. 2 fire, Jensen, 116; *Whittier Daily News*, May 18, 1980; OA, March, 1922; *Summary of Operations* (May, 1923); LAT, February 19–24, 1922.

"greatest thrill . . ." LAT March 20, 1922; see also Atwood, "Oil Flood," 94–96.

"an ugly place . . ." *History in the Making*, 12.

p. 29    "great carnival type midway . . ." *ibid.*, 29.

Description of Santa Fe Springs is drawn from Rintoul, 54–55 and 156–61, unless otherwise noted.

"California Style buildings . . ." Jensen, 118.

"the right to steal oil . . ." Spence, 180–81.

"the idea is . . ." Atwood, "Oil Flood," 89.

p. 30    On crowding of wells, see OA, July, 1921, and November, 1922; and *Summary of Operations* (May, 1923).

Julian's arrival in Santa Fe Springs, LAR February 17, 1924, and August 28, 1925.

On oil leases, see Atwood, "Oil Flood," 89–93.

On royalties, see Tompkins, 6; OA, August, October, December, 1922; *Summary of Operations* (May, 1923).

"One-well field . . ." LAT April 3, 1922.

"has not proven. . ." LAT July 3, 1922.

On Gettys, see Ralph Hewins, *The Richest American: J. Paul Getty.* (New York, 1960), 83–84; and Robert Lenzner, *The Great Getty: The Life and Loves of J. Paul Getty—Richest Man in the World* (New York, 1985), 32–36.

p. 31    On Brunsons, see Atwood, "Oil Flood," 99; for terms of Brunson lease, see copy of lease between Globe Petroleum and C. C. Julian, FBI report (November 6, 1924).

For Julian's account on fund-raising, LAR August 28–29, 1925.

Details on actual fund-raising, see FBI report (November 6, 1924).

p. 32    "I had to have the money . . ." LAR August 29, 1925.

On common-law trust, see Albert W. Atwood, "Mad from Oil," 196 SEP (July 14, 1920), 98.

"I never saw . . ." *History in the Making*, 29.

p. 33    "Is there anything . . ." LAR March 27, 1923.

"It is difficult . . ." OA, March 7, 1923.

"never run across . . ." Atwood, "Mad from Oil," 98.

"If he is an honest man . . ." LAR March 27, 1923.

Julian sets up office and meets with ad men, LAR August 29, 1925.

p. 34    Writing the ad, LAR February 17, 1924, and August 29, 1925; LAT September 23, 1923.


## Chapter 3

### Section I

p. 36    "thousands of oil lunatics . . ." Samuel Blythe, "Gambling with Grief," SEP 197 (July 5, 1924), 8.

p. 36    "one syllable," Albert W. Atwood, "When the Oil Flood Is On," SEP 196 (July 7, 1923), 4.

On the variety of promotions, see Atwood, "Oil Flood"; OA, March 14, 1923; Paul N. Wilson, "All Within the Law: How the Oil Sharks Trim Their Victims and Still Keep Out of Jail," *Sunset* 50 (May, 1923), 56; and LAR March 28, 1923.

On inventions, see LAR December 28, 1923, and February 10, 1923; Wilson, 56; and OA, August, 1923.

p. 37    The most complete descriptions of the oil tent shows can be found in Albert W. Atwood, "Mad from Oil," SEP 196 (July 14, 1920); and Wilson, "All Within the Law." All quotes and descriptions are from these articles unless otherwise noted.

"SEE THE GUSHERS" ad appeared daily in the Los Angeles newspapers in 1922.

p. 38    "sucker tents . . ." Marilyn Jensen, "The Giant That Was Santa Fe Springs," *Pacific Oil World*, 4th Annual Review Issue (January, 1980), 118.

"On the earlier wells . . ." City of Santa Fe Springs, *History in the Making* (1976), 28.

Norma Talmadge, *Whittier Daily News*, December 19, 1980.

"We should now . . ." *History in the Making*, 29.

p. 39    "the retired farmers . . ." Louis Adamic, "Los Angeles! There She Blows," *Outlook and Independent* 155 (August 13, 1930), 595.

"the supply of morons . . ." LAR December 18, 1926.

"People here are especially . . ." Atwood, "Oil Flood," 86.

"more intelligent people . . ." Atwood, "Mad from Oil," 92.

On doctors, see LAR March 26, 1923; on Jewish and Japanese investors, see chapters 9–12.

p. 40    "king of the sucker tents. . ." Jensen, 118.

On response to Julian's first ad, see LAR August 29, 1925.

"Santa Fe Springs . . ." ad, June 20, 1922.

"Never Before . . ." ad, June 21, 1922.

p. 42    Salesmen listed in ad, June 30, 1922.

"Do you know . . ." ad, June 28, 1922.

"the privilege of participating . . ." and "13 days more . . ." ad, July 2, 1922.

"closed forever . . ." ad, July 13, 1922.

"For the love of Mike . . ." ad, July 4, 1922.

"FIVE DAYS LEFT . . ." ad, July 10, 1922.

"Talk about being snowed . . ." ad, July 11, 12, 1922.

"GOOD NIGHT NURSE" ad, July 14, 1922.

"IT'S A GRAND OLD WORLD . . ." ad, July 17, 1922.

"the Sunday comic page . . ." Robert Glass Cleland, *California in Our Time* (New York, 1947), 154.

"I wrote just as if . . ." LAR August 29, 1925.

"Do you realize . . ." ad, July 14, 1922.

"Kittens Cuffs . . ." ad, August 9, 1922.

"Pigeon's pajamas . . ." ad, September 24, 1922.

"Fuzz on frog's face . . ." ad, August 30, 1922.

Puns appeared in Julian ads on July 9, 1922, September 6, 1922, and October 2, 1922.

"Written by anyone else . . ." and "He demonstrated . . ." Finney, *The Great Los Angeles Bubble* (New York, 1929), 29.

p. 43   Walter V. Woehlke, "The Great Julian Pete Swindle," *Sunset* 54 (September, 1927), 14.

"Pole Horse . . ." ad, June 30, 1922.

"I have no millionaires . . ." ad, June 20, 1922.

"I'm the guts . . ." ad, September 6–7, 1922.

"Every word . . ." ad, September 3, 1922.

"I don't want you to class me . . ." ad, September 6, 1922.

No town lots, ads, July 2, 1922, and August 9, 1922.

"If I can only win . . ." ad, June 28, 1922.

On drilling crew, see ads of June 18, 1922, and June 30, 1922.

"that for the past fifteen years . . ." ad, July 9, 1922.

"ordinary every day driller . . ." ad, August 9, 1922.

"Action . . ." ad, June 20, 1922.

"While you are snuggled up . . ." ad, September 8, 1922.

"Do you know that I have pledged . . ." ad, June 28, 1922.

p. 45   "A promise made . . ." ad, July 27, 1922.

"fall headfirst . . ." ad, July 15, 1922.

"Buzzards pick my bones . . ." ad, September 3, 1922.


## Section II

p. 45   "An oil well is a tricky . . ." Samuel Blythe, "Gambling with Grief," SEP 197 (July 5, 1924), 4.

Unless otherwise noted, descriptions of drilling are from Upton Sinclair, *Oil!* (New York, 1927), 58–65; William Rintoul, *Spudding In: Recollections of Pioneer Days in the California Oil Fields* (San Francisco, 1976), 115–23; and Blythe, 74.

On Julian's schedule, see ads of June 23–30, 1922.

On champagne, author's interview with Henry Hathaway.

p. 47   "cards face up . . ." ad, June 18, 1922.

"When you lay your 'Jack' on the line . . ." ad, September 5, 1922.

"I am here . . ." ad, November 29, 1922.

Julian's difficulties in drilling Well No. 1 are chronicled in his ads and drilling log, CSMB.

"99 per cent . . ." ad, August 24, 1922.

p. 48   "Are you saving . . ." ad, August 18, 1922.

"If I drill . . ." ad, September 26, 1922.

"Fortune knocks . . ." ad, September 10, 1922.

"Faint Heart . . ." ad, September 6, 1922.

"OUR DOUBTS . . ." ad, September 29, 1922.

"one-armed paper-hanger . . ." ad, October 31, 1922.

On Julian's work regimen, see LAR February 17, 1924.

p. 49   On additional fishing jobs, see drilling log, Well No. 1, CSMB.

"Called the roll . . ." ads, September 21 & 28, 1922.

On fishing job and malaria, LAR August 29 & 31, 1925.

p. 50   Well No. 1 strikes oil, ad, October 27, 1922, and drilling records, CSMB; see also LAX October 30, 1922.

"Daddy's Sweet Baby . . ." ad, November 12, 1922, and LAX November 27, 1922.

p. 50    Well No. 2 strikes oil, ad, November 17, 1922, and drilling records, CSMB.
"Julian No. 1 investors . . ." Lucas Stewart ad, October 27, 1922.
"Julian Makes Good . . ." OA, November, 1922.
"experienced operator . . ." LAX December 4, 1922.
On dividend checks, ad, December 17 1922.
"Santa Claus" ad, October 17 and November 4, 1922.

## Section III

p. 50    "Happy family . . ." ad, November 20, 1922.
"A winning Hand . . ." ad, November 21, 1922.
"Honesty's his name . . ." ad, April 20, 1923.

p. 51    "Tho you have tried . . ." ad, November 22, 1922.
"I notice they are trying . . ." ad, December 31, 1923.
"sharks . . ." ad, December 29, 1931
"knockers . . ." ad, January 14, 1922.
On charges of manipulating unit prices, see FBI report, June 12, 1925.
On Santa Fe Springs, see OA, October, 1922.
On discovery of Meyer sands, see LAT October 23, 1923, and California Division
     of Oil and Gas, *Summary of Operations, California Oil Field*, May 23, 1925.
"stands without equal . . ." LAT November 13, 1922.

p. 52    Julian Nos. 3 and 4, LAX December 4, 1922.
"As long as we can pull . . ." ad, December 10, 1922.
On new lease, see LASC, Case #172400 Wilshire Oil *vs.* Citizens Trust and Savings
     Bank *et al.* (June 18, 1925).
"This new issue . . ." ad, February 2, 1923.
"Christmas present" ad, December 18, 1922.
"Don't weaken . . ." ad, December 27, 1922.
"We twisted off . . ." ad, January 11, 1923.
"speared the old fish . . ." ad, January 12, 1923.
"the prodigal son . . ." ad, January 19, 1923.
"all eyes . . ." OA, January 23, 1924.

p. 53    "that darling No. 4," ad, February 19, 1923.
"a real discovery . . ." LAT February 19, 1923.
"tremendous importance . . ." OA, February 21, 1923.
"Where are the birds . . ." ad, February 17, 1923.
"I feel that the results . . ." ad, February 19, 1923.
"I'm not appealing to unintelligent . . ." ad, February 28, 1923.
"to every oil operator . . ." ad, March 6, 1923.

p. 54    $600,000 ad, March 10, 1923.
"It Will Soon Be Curtains . . ." ad, March 16, 1923.
"Struttin Her Stuff . . ." ad, March 27, 1923.
8,100 barrels, ad, March 28, 1923.
"Julian No 1. is today . . ." ad, March 28, 1923.
"cartload . . ." ad, April 10, 1923.
"highest price . . ." LAT March 29, 1923.
On Compton field, see OA, September, 1923.

p. 54        "Play Ball . . ." ad, April 25, 1923.
             "How is Julian . . ." ad, April 15, 1923.
p. 55        "You won't find me . . ." ad, March 7, 1923.
             "one collar . . ." ad, April 15, 1923.
             On Julian's wardrobe, see Norman Carlisle, "Con King of the Golden Coast," *True*
                 46 (October, 1965), 113–14.
             On Pierce-Arrow, ad, April 5, 1923, and LAR February 18, 1924.
             For descriptions of Julian's home, see LAR February 16, 1924, and LAT August 18
                 & 19, 1931.

## Chapter 4

## Section I

p. 56        On overproduction controversy, see Henrietta Larson and Kenneth Porter, *History
                 of Humble Oil and Refining Company* (New York, 1959), 265.
             On oil prices, see James C. Findlay, "The Economic Boom of the Twenties in Los
                 Angeles" (Ph.D. diss., Claremont, 1958), 352.
             On small *v.* large companies, see Albert W. Atwood, "When the Oil Flood Is On,"
                 SEP 196 (July 7, 1923), and LAT June 21, 1922.
p. 57        "There is a lot of propaganda . . ." LAT September 7, 1922.
             "Naturally . . ." ad, May 27, 1923.
             On oil industry meeting, see LAT April 28, 1923; FBI report, November 6, 1924;
                 and LAR August 21, 1925.
             "big boys could honeyfugle . . ." LAR August 31, 1925.
             On General Petroleum, see FBI report, November 6, 1924.
p. 58        "The refining of crude . . ." LAT May 20, 1923.
             "having achieved one or two . . ." Samuel Blythe, "Gambling with Grief," SEP
                 197 (July 5, 1924), 71.
             "My great ambition . . ." ad, March 30, 1923.
             On incorporation, see Julian Petroleum Company Charter, California State
                 Archives, Sacramento, California.
p. 59        "My policy . . ." ad, June 11, 1923.
             "One lonely share . . ." ad, May 22, 1923.
             On Standard Oil profits, ad, June 13, 1923.
             "whispers . . ." ad, May 23, 1923.
             "as a menace . . ." ad, May 27, 1923.
             "It can't be did . . ." ad, June 13, 1923.
             "avalanche . . ." and "were I to live . . ." ad, May 31, 1923.
             Julian No. 3, ad, June 26, 192.
             Dividends, ad, June 14, 1923.

## Section II

p. 60        The biography of Corporation Commissioner Mike Daugherty is drawn from LAR
                 February 20, 1924, and IDN January 13, 1924.

p. 60    On the campaign for the "blue sky" law, see Mansel G. Blackford, *Politics of Business in California, 1890–1920* (Columbus, Ohio, 1977), 117–21; and John Dalton, "The California Corporate Securities Act," *California Law Review* 18 (January, 1930), 115–16.

On the "blue sky" law itself, see Dalton, 130, and Lionel Benas, "The Corporate Securites Act: Recent Cases and Amendments," *California Law Review* 14 (January, 1926), 104.

On campaign to repeal the "blue sky" law, see Blackford, 124, and John Dalton, "The California Corporate Securities Act," *California Law Review* 18 (May, 1930), 397.

p. 61    "degree of sanity and stability . . ." Blackford, 124.

On inherent conflict in the "blue sky" law, see Benas, 101, and Dalton (January, 1926), 115.

On misuse of permits, see Albert W. Atwood, "Mad from Oil," SEP 196 (July 14, 1923), 98, 101.

"hopelessly overcapitalized . . ." LAR October 7, 1922.

"pocketed enormous profits . . ." LAR March 9, 1923.

"the money of one sucker . . ." LAR April 27, 1923.

"were purely promotional . . ." OA, November, 1922, and LAR April 27, 1923.

"One of the mangiest fakes . . ." Blythe, 2.

"sold more percents . . ." City of Santa Fe Springs, *History in the Making* (1976), 30.

p. 62    On declining yields, see Atwood, "Mad from Oil," 102.

On field investigating division, see OA, August, 1922, and Atwood, "Mad from Oil," 101.

Court upholds corporation commissioner, OA, November, 1922.

"handcuff[ed]," see LAR March 12, 1923.

On changes in "blue sky" law, see LAR April 27 & 28, 1923.

"As long as I . . ." LAR February 12, 1923.

"The reduction . . ." LAR April 20, 1923.

p. 63    Calculations on Julian's profits are based on records in FBI report, June 12, 1925, schedules 3 thru 17.

On preliminary conference with Daugherty, see LAR September 1, 1925.

On Daugherty and refineries, see LAT June 1, 1923; LAR August 1, 1923; and LAR February 20, 1924.

"Ruin was just . . ." LAR September 1, 1925.

Julian's June 15 letter to Daugherty was reproduced in ad, July 1, 1923.

Julian seeks Chandler's assistance, FBI report, June 9, 1924.

p. 65    "It might be a week . . ." ad, July 1, 1923.

On writ of mandate, see LAT June 29, 1923.

"Houdini enough . . ." LAR September 1, 1925.

On Goodwin's plan, LAT June 30, 1923.

On trip to Las Vegas, see LAR June 20, 1923, and FBI report, November 6, 1924.

On stock sales, LAR September 1, 1925.

p. 66    "California investors . . ." LAT June 29, 1923.

"bludgeon . . ." LAR June 28, 1923.

Threat, LAR June 30, 1923.

p. 66    Attempt to seize Julian's books, LAT July 1, 1923; LAR June 30, 1923 and November 6, 1923.

Brokerage houses, LAR June 29, 1923, and LAT July 1, 1923.

Kipling quoted, ad, June 30, 1923.

p. 67    Julian's arrest and arraignment, books seized, LAR June 30, 1923 and LAT July 1, 1923.

On Governor Richardson, SFC, July 3, 1923.

p. 68    "Patience is a Virtue . . ." ad, July 5, 1923.

"use them as (he) saw fit . . ." ad, July 8, 1923.

"saved the bacon . . ." LAR September 1, 1925.

On suggestion to sell notes, LAR September 2, 1925.

"Time and Tide . . ." ad, July 26, 1923.

On refinery site and tank farm, ad, July 20, 1923.

p. 69    Defiance Gasoline, ad, July 25, 1923.

"When I resign . . ." LAT, July 29, 1923.

Permit issued, LAT July 31, 1923, and FBI report, November 6, 1924.

## Section III

p. 69    "What More Could you ask . . ." ad, August 2, 1923.

"safely as grocery business . . ." ad, August 29, 1923.

"With 40,000 partners . . ." ad, August 28, 1923.

"Did you ever hear . . ." ad, July 27, 1923.

"Who puts up . . ." ad, August 5, 1923.

p. 70    "$50,000 to $75,000" ad, August 16, 1923.

"I'm busier than a one-eyed man . . ." ad, August 23, 1923.

"Crack Your Whip . . ." ad, October 1, 1923.

Deal with General Petroleum, ad, November 11, 1923.

On orders for gas pumps, see LASC Case #178110, Joy Manufacturing *vs.* Julian Petroleum, September 4, 1925.

"to the masses . . ." ad, October 28, 1923.

On criminal hearings, LAR November 1, 1923.

Accounts of hearing before Daugherty are drawn from LAT and LAR, November 2 & 5–7, 1923.

p. 72    Criminal charges dropped, LAT December 15, 1923.

On Daugherty and FBI, see FBI report, June 9, 1924.

On Well No. 2, ads July 4, 1923, September 10, 1923.

"flowers and soft music . . ." ad, August 26, 1923.

On Wells no. 6 and 7, ad, October 14 & 26, 1923.

On Wells no. 11 and 12, ad, November 25, 1923, and FBI reports.

On Santa Fe Springs, see California Division of Oil and Gas, *Summary of Operations, California Oil Field*, February, 1924, and OA, April, 1924.

On Compton field, see OA, September and December, 1923, and January, 1924.

p. 73    "This is my funeral . . ." ad, October 7, 1923.

For description of Julian stations, see Norman Carlisle, "Con King of the Golden Coast," *True* 46 (October, 1965), 114.

# Chapter 5

## Section I

p. 74    On charges of washing stock, see LAR February 20, 1924.

On oversubscription, see FBI report, November 6, 1924.

On second permit, LAT December 30, 1923.

On property in new oil field, LAR September 2, 1925.

"The public . . ." LAR September 2, 1925.

p. 75    Julian advances money, LAR February 18, 1924.

"pawns of powerful . . ." LAR September 3, 1925.

Julian buys stock, LASC Case #135080, C. C. Julian *v.* Edwin M. Daugherty (January 5, 1924).

Julian applies for release of funds, *ibid.*

"a complete examination . . ." *ibid.*

Daugherty launches investigation, LAT December 30, 1923.

Roy West case, LAR December 31, 1923, and LAT January 1, 1924.

p. 76    "pending the disclosure . . ." and "most cruel . . ." LAT December 30, 1923.

"Powerful unseen forces . . ." and "resent contemptible . . ." IDN January 3, 1924.

Accounts of the publication of *Truth* appeared in LAR September 3, 1925; IDN January 4, 1924; FBI report, November 6, 1924. See also LASC Case #137799, M. H. Harris *v.* C. C. Julian (January 9, 1924).

p. 77    On Vanderbilt, see IDN April 6, 1924, and Robert Gottlieb and Irene Wolt, *Thinking Big: The Story of the Los Angeles Times*, (New York, 1977), 233.

On newspaper circulation wars, see Gottlieb & Wolt 107–8, 233.

p. 78    "permanent feature . . ." IDN January 4, 1924.

"It was cheaper . . ." FBI report, November 6, 1924.

## Section II

p. 78    On death threats and shooting, see IDN and LAR January 4 & 5, 1924, and LAR September 7, 1925.

p. 79    LAR editorials, January 3–5, 1924.

LAT editorial, January 5, 1924.

On LAT advertising policies, see Gottlieb & Wolt 28.

Record estimates of advertising, LAR February 9, 1924.

p. 80    On Chandler, see Gottlieb & Wolt, especially 121–26.

"blackout on truth . . ." William Bonelli, *Billion Bollar Blackjack: The Story of Corruption and the Los Angeles Times* (Beverly Hills, 1954), 34.

On Times rivals, see Gottlieb & Wolt, 233–34.

"Who and what . . ." IDN January 19, 1924.

"sinister, subtle conspiracy . . ." "seeming evil . . ." "No man is free . . ." IDN January 12, 1924.

"stood up against . . ." IDN January 22, 1924.

p. 81    "may not be all . . ." IDN January 14, 1924.

"We are fast losing . . ." IDN January 18, 1924.

p. 81    "Who is this man . . ." IDN January 11, 1924.
         "will not be safe . . ." IDN January 14, 1924.
         "When we need protection . . ." IDN January 19, 1924.
         "What faith . . ." IDN January 17, 1924.
         "a square man . . ." "towers above . . ." IDN January 22, 1924.
         "first water . . ." January 17, 1924.
         "honestly trying to help . . ." January 29, 1924.
         "Julian City . . ." IDN ad, January 6, 1924.
         Accounts of trial are drawn from LAR, LAT, IDN January 17–19, 1924.
p. 82    "Isn't it strange . . ." IDN January 21, 1924.
         On grand jury report, IDN February 14, 1924.


## Section III

p. 83    On Roth-Tower fight, see LAR, SFC, January 25, 1924.
         On Chaplin Fight, see LAR, LAT January 22–25, 1924.
p. 84    "the people of California . . ." IDN January 25, 1924.
         "Forerunner of a series . . ." IDN January 25, 1924.
         Torrance and Whittier wells, IDN January 9, 22, 25, 1924.
p. 85    Wells 8 and 9, IDN ad, January 30, 1924.
         Tanker, IDN ad, February 13 & 28, 1924.
         "Smoke . . ." IDN ad, February 27, 1924.
         "a little hard to find . . ." IDN February 20, 1924.
         On the *Los Angeles Record*, see Reuben W. Borough, *Reuben W. Borough and California Reform Movements* (Oral History Project, UCLA Special Collections, 1968), 48, 54–56.
         Earlier committee suggestions appeared in LAT, IDN January 5, 1924.
p. 86    Julian meets with committee, LAT February 11, 1924, and LAR February 12, 1924.
         For biographies of committee members, see LAR February 12, 1924.
         "Answer to Julian Tangle . . ." LAR February 12, 1924.
         On tour of Julian Petroleum, see LAR, IDN, and LAT February 15, 1924.
p. 87    The biography of C. C. Julian appeared in LAR February 13–20, 1924.
         "The truth about Julian . . ." LAR February 20, 1924.
         FBI investigation, FBI report, June 9, 1924.
p. 88    For Citizen's Committee report, see LAT, IDN April 9, 1924.
p. 89    On stock exchange listing, IDN April 15, 1924.


## Chapter 6

### Section I

p. 90    New permit, FBI report, November 6, 1924.
         New directors, LAT April 11, 1924.
p. 91    Salary demands of new directors, FBI report, November 6, 1924.
         On Jacob Berger, see LAT May 2, 1924, and ad, February 21, 1926.
         Colorado well, ad, April 19, 1924.

p. 91    "the beginning of the end . . ." LAR September 4, 1925.

p. 92    Installment plan, ad, May 16, 1924.

Torrance field, OA, February, 1924.

"nine-foot bird proof . . ." OA, March, 1924.

Pico wildcat shutdown, California Division of Oil and Gas, *Summary of Operations, California Oil Field* (May 1924).

"Biggest Thing . . ." ad, May 1, 1924.

$40 million, ad, May 26, 1924.

For Madeira suit, see LASC Case #144683, C. C. Julian *v.* W. I. Madeira *et al* (May 28, 1924).

"cannot by the widest stretch . . ." LAT May 29, 1924.

Letter never made public, LAT June 3, 1924.

p. 93    "She's Closed to You . . ." ad, June 11, 1924.

On stock sales and "I decided . . ." FBI November 6, 1924.

"went over the top . . ." ad, June 15, 1924.

On Julian in New York, see LAR September 4–5, 1925; FBI November 6, 1924; FBI report on Lewis Oil, 1930; Norman Carlisle, "Con King of the Golden Coast," *True* 46 (October, 1965), 113–14; and Carey McWilliams, *Southern California: An Island on the Land* (Santa Barbara, 1973), 243.

p. 94    Rumors and stock prices, FBI report, November 6, 1924, and ad, August 28, 1924.

## Section II

p. 95    Resignation of Treat *et al.*, FBI report, November 6, 1924.

"tough luck . . ." ad, August 29, 1924.

"adverse propaganda . . ." ad, August 28, 1924.

"the best means . . ." LAR September 5, 1925.

"I am personally . . ." ad, August 31, 1924.

"I will answer . . ." ad, September 3, 1924.

"Everybody and his dog . . ." ad, September 5, 1924.

Hollywood Bowl meeting, LAT September 6, 1924.

Escrow plan, LAT September 6, 1924; ad, September 14, 1924; FBI report, November 6, 1924.

p. 96    Bus trips, ad, September 8, 1924.

Movie, ad, September 26 & 28, 1924.

Fist fight, LAT, and ad, September 30, 1924.

Escrow figures, FBI report, November 6, 1924.

Federal Grand Jury, LAT, IDN October 4, 1924.

p. 97    Julian turns over books, LAT October 11, 1924.

Miller investigation, FBI report, November 6, 1924.

"nothing more than . . ." and Miller activities, FBI report, June 12, 1925.

Athens wells, OA, November, 1924.

Santa Fe Springs wells, FBI report, June 12, 1925.

"I was like an animal . . ." LAR September 5, 1925.

p. 98    Lewis version of meeting Julian, LAR June 16, 1927.

p. 98    Descriptions of Lewis in Lorin L. Baker, *That Imperiled Freedom* (Los Angeles, 1932), 32; LAX July 24, 1927; Walter V. Woehlke, "The Great Julian Pete Swindle," *Sunset* 54 (October, 1927), 18; *Coast Investor and Industrial Review* (April, 1927), 159; LAR June 15, 1927; Robert P. Shuler, *Julian Thieves in Politics* (Los Angeles, 1930), 13; and 1930 FBI report.

Lewis's biography drawn from *Who's Who in Los Angeles* (Los Angeles, 1926), and LAR June 15–17, 1927.

On McAdoo, King, and Lewis, see Guy Finney, *The Great Los Angeles Bubble* (New York, 1929), 36–37, and LAT March 9, 1928.

p. 99    Julian wires Lewis, LAR September 5, 1925, and LAR June 16, 1927.

Negotiations between Julian and Lewis, LAT March 9, 1928.

The contract between Julian and Lewis may be found in NA, Equity Case #M-12-M, Lewis Oil Company Intervention, Second Amended Report (December 17, 1928).

New board, LAT December 23, 1924.

On Sartori, see Bruce Henstell, *Sunshine and Wealth: Los Angeles in the Twenties and Thirties* (San Francisco, 1984), 40.

"without one thin dime . . ." LAR September 5, 1925.

Julian cancels stock in Julian Petroleum Company Minutes, (December 10, 1924), in FBI report, June 12, 1925.

On December 19 meeting, see Julian Petroleum Company Minutes, (December 19, 1924) in FBI report, June 12, 1925.

"Lewis and I decided . . ." LAT February 18, 1928.

p. 100   "It is simply a question . . ." LAT December 23, 1924.

"I feel . . ." LAR December 23, 1924.

"The day I turned over control . . ." LAR September 5, 1925.


## Section III

p. 100   "He has put his name . . ." LAR August 27, 1925.

"A gay cafe figure . . ." LAR August 31, 1925.

"the farmer boy . . ." Woehlke (September, 1927), 13.

"It was the same old story . . ." *Coast Investor and Industrial Review* (April, 1927), 159.

For contemporary commentators, see Finney, 33–35, and McWilliams, 242–45.

p. 101   Unless otherwise noted, calculations on Julian's profits are drawn from FBI reports, June 9, 1924, November 6, 1924, and June 12, 1925.

p. 102   On Julian wells no. 6 and 7, see LASC Case #172400, Wilshire Oil *vs.* Citizens Trust and Savings Bank *et al.* (June 18, 1925).

On Stanley, see C. W. Demond, *Price, Waterhouse & Co. in America: A History of a Public Accounting Firm* (New York, 1951), 39, 76, 312–13; and FBI report, June 12, 1925.

p. 103   "The net financial result . . ." LAR, September 5, 1925.

## Chapter 7

### Section I

p. 107  "This is the one time . . ." ad, January 16, 1925.

"No stockholder's gathering . . ." LAR January 17, 1925.

For descriptions of stockholders' meeting, LAT, LAX January 18, 1925; LAR January 17, 1925.

p. 108  On McAdoo, LAR January 17, 1925.

Oil supplies and drilling reports, ad, March 15, 1925.

Sierra Refining and new wells, OA, January, February, 1925.

New York Curb Exchange, LAT May 6, 1925.

On ethyl gasoline, see OA, March, 1925.

Controversy over Lightning, OA, March, 1925, and LAT, February 11, 1925.

p. 109  Refinery, LAT April 28, 1925, and FBI report, June 12, 1925.

For new FBI report, FBI, June 12, 1925.

p. 110  On Lewis at Biltmore Hotel, for Packard's account, see LAX July 24, 1927. See also Lorin L. Baker, *That Imperiled Freedom* (Los Angeles, 1932), 31–32; and Walter V. Woehlke, "The Great Julian Pete Swindle," *Sunset* 54 (October, 1927), 18.

p. 111  On Johnson letters to FBI, see FBI reports, W. J. Burns to Rush D. Simmons, July 3, 1923, and Arnold Gates Johnson to FBI, August 17, 1923; August 24, 1923; and September 25, 1923.

For revised Lewis biography, FBI report on Lewis Oil, 1930.

p. 112  Unless otherwise noted, the history of Lewis Oil is drawn from FBI reports, April 28, 1928, and 1930.

"You saw these . . ." LAT October 5, 1928.

On Berman's aliases, see 1930 FBI report and LAT, October 8, 1928.

On Meridian mechanic, LAT September 27, 1928.

On Montreal couple, LAT July 13, 1927, and October 13, 1928; and FBI report, April 28, 1928.

p. 113  Baldwin, LAT September 21, 1928.

Burlington, Iowa, LAT September 28, 1928.

Louisiana baker and Sharkey, LAT September 27, 1928.

On overissue, see FBI report, April 28, 1928.

"I trusted him . . ." LAR June 17, 1927.

### Section II

p. 113  On stock transfer office, see Woehlke, (October 1927), 80.

Bennett's desk, LAR June 17, 1927.

On T.P. Conroy, see Guy Finney, *The Great Los Angeles Bubble* (New York, 1929), 142.

p. 114  All overissue figures are drawn from the Haskins & Sells stock audit included in the 1930 FBI report.

p. 114   Harris and Shipp testimony, LAT, March 20, 1928, and Woehlke, (October 1927),
         80.

         Campbell demands, LAX July 24, 1927.

         Lewis and common stock, 1930 FBI report.

p. 116   "I've been working . . ." ad, January 30, 1925.

         "all my efforts . . ." ad, February 8, 1925.

p. 117   "came in so fast . . ." ad, March 9, 1925.

p. 118   Julian stock pools, LAT July 27, 1927, and Robert P. Shuler, "Julian Thieves" (Los
         Angeles, 1930), 16.

         "We can't help saying . . ." Durst ad, March 17, 1925.

         "handed Lewis back . . ." LAT July 27, 1927.

         Julian in Mexico, LAT March 18, 19, 1925.

         "The new money . . ." and "he resented it . . ." Finney, 53.

p. 119   On Board of Directors meeting, see Julian Petroleum Company minutes, FBI
         report, June 12, 1925.

         Reactions to Julian leaving firm, see LAT, LAR April 15, 1925.


## Section III

p. 120   For the history of the LASE, see James C. Findlay, "The Economic Boom of the
         Twenties in Los Angeles" (Ph.D. diss., Claremont, 1958), 328–30.

         "Ten years ago . . ." Albert W. Atwood, "Money from Everywhere," SEP, 195 (May,
         12, 1923), 144.

         Pettingell biography, LAT May 10, 1926.

p. 122   "Lack of funds . . ." Wagy ad, November 25, 1925.

         On bucket shops, see LAR April 6, 1923.

         On switchers, see Sunshine and Grief in Southern California: where good men go
         wrong and wise people lose their money by an old promoter, forty years in the field
         of real estate (Detroit, 1931), 155.

         "uninitiated brokers . . ." Robert Sobel, The Great Bull Market: Wall Street in the
         1920s (New York, 1968), 65.

p. 123   "Sharpshooters" and "The bolshevik element," Finney, 55–56.

         "Since taking over . . ." ad, August 16, 1925.

         Account of stockholders' meeting, LAT August 20, 1925.

p. 124   "We advised you . . ." ad, September 24, 1925.

         "Pandemonium prevailed . . ." Wagy ad, October 1, 1925.

         New York Curb Exchange, LAT October 4, 1925.

         On bears and shorts, see Sobel, 67n, 70.

p. 125   dominates trading, LAT October 11, 1925.

         "THE STORY HAS JUST STARTED . . ." ad, October 11, 1925.

         "It is easy to understand . . ." Jameson ad, October 13, 1925.

         "the greatest buying power . . ." ad, October 16, 1925.

         "avalanche of selling orders . . ." LAT October 18, 1925.

p. 126   Lewis charges and Times, Pettingell responses, LAT and ad, October 17, 1925.

         On Wagy firm, see Finney, 58.

         "$10,500 Reward" ad, October 15, 1925.

         "spirit, promptness . . ." ad, October 19, 1925.

p. 126   "faith and belief . . ." Wagy ad, October 20, 1925.

Auditors' report, Wagy and Lewis ads, October 28, 1925.

p. 127   On Lewis and Chessher and Wagy, see LAT, January 23, 1930, and Finney, 55–56.

Bennett suit against Welch, LAT October 30, 31, 1925; LASC #182129, Jack Bennett *vs*. James M. Welch (October 29, 1925); and LASC #182298, Fred A. Ballin *vs*. James M. Welch (October 30, 1925).

For Welch countersuit, see LAR June 20, 1927, and LASC #204639, James M. Welch *vs*. S. C. Lewis *et al*. (September 2, 1926).

p. 128   "efforts to protect . . ." ad, October 28, 1925.

245,000 shares, Woehlke, November 1927, p. 80.

Drilling activities, LAT November 15, 1925, and California Division of Oil and Gas, *Summary of Operations, California Oil Field* (December, 1925).

Shipp on issuing new stock, LAT July 28, 1927, and March 6, 1928.

Transfer of books, Woehlke, (October, 1927), 80.

p. 129   "This space . . ." LAT, October 31, 1925.

## Chapter 8

### Section I

p. 130   On Julian's Santa Fe Springs wells, LAT September 27, 1925; LASC Case #301676, J. A. Smith *vs*. C. C. Julian (April 25, 1930).

On Mary Julian, LAT March 24, 1928, and LASC Case # D92410, Mary Olive Julian *vs*. C. C. Julian (February 2, 1931).

p. 131   Julian's illness, LAT August 20, 1925, and LAR August 27, 1925.

Julian and IRS, LAR August 27, 1925, and Walter V. Woehlke, "The Great Julian Pete Swindle," *Sunset* 54 (September, 1927), 14.

"Notoriety . . ." and "The public citizen . . ." LAR August 27, 1925.

"never turned a hair . . ." LAR August 29, 1925.

"like a mischievous . . ." LAR August 27, 1925.

"If I had been trying . . ." LAR September 7, 1925.

"The facts and figures . . ." LAR August 27, 1925.

"Is C. C. Julian a dead meteor . . ." LAR August 28, 1925.

"Thousands know me . . ." LAR September 5, 1925.

p. 132   "Organized capital . . ." LAR September 7, 1925.

"smart, sophisticated . . ." LAR August 27, 1925.

"somewhere in the hills . . ." LAR September 7, 1925.

"Desert Rat . . ." ad, February 1, 1926.

On mining in Death Valley, see Carl Burgess Glasscock, *Here's Death Valley* (New York, 1940), 279.

"I don't suppose . . ." ad, February 1, 1926.

"miniature grand canyon . . ." L. Burr Belden, "Wild promotion but little ore from Leadfield," *San Bernardino Sun Telegram*, December 16, 1956.

Chambers, Salsbery, and Berger, ad, February 21, 1926.

p. 133   Western Lead stock agreement, ad, March, 12, 1926, and LASC cases #206766 and 206767, John Salsberry *vs*. C. C. Julian (September 29, 1926).

p. 133    Additional stock purchases, LAT April 7, 1926.
          C. C. Julian and Company Articles of Incorporation, November 23, 1925, Cali-
              fornia State Archives.
          Julian and IRS, Woehlke (September, 1927), 14.
          Julian and FBI, FBI reports, Memorandum to Hoover from C. R. Luhring, March
              20, 1926.
p. 134    "Ze Grand Opening . . ." ad, January 18, 1926.
          "Looks just like a bank . . ." and "shaking hands . . ." ad, January 12, 1926.
          Office opening, LAR January 23, 1926, and ad, January 24, 1926.
p. 135    Julian's January 10, 1926, letter to Daugherty appeared in *Truth* April 9, 1926, HP.
          Proceedings at hearing, LAT January 30, 1926.
          Permit denied, LAT February 3, 1926.

## Section II

p. 135    "I've got a WONDER . . ." ad, January 12, 1926.
          "Just be prepared . . ." ad January 17, 1926.
          "IT IS NOT OIL . . ." ad, January 12, 1926.
          "I'm out of the oil business . . ." ad, January 13, 1926.
p. 136    "It's Black and White . . ." ad, January 20, 1926.
          Western Lead and LASE, LAT January 27, 1926.
          First day of listing, Harold O. Weight, "Leadfield Died of Complications," *Desert*
              (November, 1977), 36–37.
p. 137    "widows and old folks . . ." ad, February 3, 1926.
          "unscupulous, blackhearted . . ." ad, March 8, 1926.
          "all a gamble . . ." ad, January 27, 1926.
          "I'll be tetotally . . ." ad, February 8, 1926.
          "13,000 percent," ad, February 23, 1926.
p. 139    Idaho mine man and "cut down 50 percent . . ." ad, February 7, 1926.
          "Go hire yourself . . ." ad, February 8, 1926.
          Steamship ads, Belden.
          "There ain't a-goin . . . ." ad, February 10, 1926.
          "JAZZ BABY . . ." ad, February 24, 1926.
          "WESTERN LEAD IS costing . . ." ad, March 7, 1926.
          On the Postes, see Weight, 38.
          Julian Avenue, LAT March 18, 1926.
          "highest regard . . ." and turkeys, *Pony Express Courier* 7 (November, 1940).
          "No one came . . ." Weight, 38.
          Chalant report in Belden.
p. 140    "the world along . . ." ad, February 1, 1926.
          "special invitations . . ." ad, March 10, 1926.
          "By no means wear . . ." ad, March 13, 1926.
          Leadfield Chronicle article reprinted in Glasscock, 279–80.
p. 141    Julian for Governor, LAT March 18, 1926.
          "bring more than a thousand . . ." Weight, 34.

## Section III

p. 142    Unless otherwise noted, Julian's charges and the details of the controversy and suit
          against Daugherty and Chandler may be found in LASC #192592, Western Lead
          *vs.* Harry M. Chandler *et al.* (March 25, 1926), reprinted in *Truth*. A copy of
          *Truth* may be found in the Haynes Papers, UCLA Special Collections.
p. 143    "When the claims . . ." LAT, March 28, 1926.
p. 144    "determine whether . . ." "disguised blow . . ." "It is not our business . . ." "We are
          not in a position . . ." LAT March 18, 1926.
p. 145    Trading in Western Lead and meeting, LAT March 19, 1926.
          "brokers strained . . ." LAR March 19, 1926.
          "yelled hoarsely . . ." LAT March 20, 1926.
          "It will come back . . ." LAR March 19, 1926.
          Pettingell and Saturday session, LAR March 20, 1926.
          "biggest half day . . ." LAT March 21, 1926.
p. 146    "for no other purpose . . ." LAT, March 28, 1926.
          First day of hearings, LAT, LAR March 29, 1926.
p. 147    "That's the price . . ." ad, LAR March 26, 1926.
          Julian stock purchases, ad, LAR April 2, 1926.
          McCormick decision, LAT April 4, 1926.

## Section IV

p. 147    On opening of hearings, LAR, LAT April 6, 1926.
p. 148    "a young, inexperienced lad," *Truth*, HP.
          Morris background, LAT April 6, 1926.
          "I think the action . . ." LAT March 20, 1926.
          Morris testimony, LAT April 6 & 7, 1926.
          Insider trading, Robert Sobel, *The Great Bull Market: Wall Street in the 1920s*
          (New York, 1968), 62.
          "time to time . . ." "inasmuch as he could not . . ." LAT, LAR April 7, 1926.
          Julian denies "wash sales," LAT April 8, 1926.
          Abel, Wyman, Gardner testimony, LAT, LAR April 8, 1926.
p. 149    Reese, Wagy, Pettingell testimony, LAT April 9, 1926.
          Julian cross-examines himself, LAR April 9, 1926.
p. 150    A copy of *Truth* may be found in the Haynes Papers, UCLA Special Collections.
p. 151    April 13 session, LAT, LAR April 13 & 14, 1926.
p. 152    April 14 session, LAT, LAR April 15, 1926.
          "Each day . . ." and "Thousands . . ." LAR April 29, 1926.
          "MAMMOTH CAVE . . ." ad, LAR April 21, 1926.
p. 153    Joralemon testimony, LAT April 28, 1926.
          Root and Tucker testimony, LAT April 29, 1926.
          "I am broken . . ." LAR April 29, 1926.
          End of hearings, LAT, LAR April 30, 1926; LAT May 1, 1926.

## Section V

p. 154   On Salsberry, see LASC Cases #206766 and #206767, John Salsberry *vs.* C. C.
           Julian (September 29, 1926).
           On Pettingell death, see CG June 11, 1927, and LAT May 10, 1926.
           "man of honor . . ." LAT May 16, 1926.
           Walther decision, LAT May 28, 1926.
p. 155   "For the life of me . . ." LAR June 1, 1926.
           93,752 shares, LAR June 11, 1926.
           "I will be able . . ." LAR June 1, 1926
           "I will admit . . ." LAR June 11, 1926.
           Doran Decision, LAT June 30, 1926.
           On Leadfield decline, see Weight, 34–46, and Glasscock, 284.

## Chapter 9

### Section I

p. 157   On Adolph Ramish, see Joseph Malamut, *Southwest Jewry: An Account of Jewish
           Progress and Achievement in the Southland* (Los Angeles, 1926); *Press Reference
           Library . . . being the portraits and biographies of progressive men of the Southwest*
           (Los Angeles, 1912), 310; and *Who's Who in the Pacific Southwest* (Los Angeles,
           1913), 468–69.
           On Ramish and Empire Drilling, see Guy Finney, *The Great Los Angeles Bubble*
           (New York, 1929), 111–12.
p. 158   On the Ramish-Lewis loans, see *ibid.*, 62, 111–12 and LAX July 27, 1927.
           On Ramish stock holdings, see Finney, 112, and LAT September 8, 1927.
           "I got brothers . . . ," LAX July 27, 1927.
           On Laemmle, LAT September 8, 1927.
           On Ed Rosenberg, see LAR June 17, 1927; Finney, 97, 138, 144, 148–49; and
           Walter V. Woehlke, "The Great Julian Pete Swindle," *Sunset* 54 (November,
           1927), 66.
           "pinch hitter . . ." Robert P. Shuler, "Julian Thieves" (Los Angeles, 1930), 55.
p. 159   All overissue figures are drawn from the Haskins & Sells stock audit included in
           the 1930 FBI report.
           "Our company has been very successful . . ." ad, December 13, 1925.
           "out of the woods" ads, January 10, 1926, January 21, 1926.
           "This particular meeting . . ." ad, February 11, 1926.
           For stockholders' meeting, see LAR, LAT February 12, 1926.
           "S. C. Lewis has proven himself . . ." OA, February, 1926.
p. 160   "I sincerely believe . . ." ad, January 17, 1926.
           On Julian Pete and Western Lead, see Finney, 63–64.
           "In my judgement . . ." ad, February 25, 1926.
p. 161   Listings on NYSE and SFSE, ad, February 25, 1926.
           Protecting investors against loss, Finney, 146.

p. 161    On Donald Spurlin, see LASC #224208 and #224209, D. Spurlin *vs.* Wagy *et al.*
          (May 18, 1927).
          On alliance with Wagy and Chessher, see Finney, 64–65.
          On Olson role, see Finney, 60, and LAX January 7, 1938.
          On Raymond Reese, see Finney, 60; *Who's Who in Los Angeles, 1926–27* (Los
          Angeles, 1927), 92; LAT March 15, 1928; and LAX July 28, 1927.

p. 162    "temporary reaction . . ." ad, March 13, 1926.
          "those who desired a small profit . . ." ad, March 19, 1926.
          On meeting of board, see ad, March 24, 1926.
          On special accounts, see Finney, 61, and Woehlke (October, 1927), 19.
          "NO MARGIN CALL PLAN," Wagy ad, March 27, 1926.
          "Julian Petroleum Brokers," Finney, 60.

## Section II

p. 163    Lewis announces merger, LAT April 30, 1926.
          On age of consolidation, see John Kenneth Galbraith, *The Great Crash, 1929*
          (Boston, 1961), 49.
          On social theorists, the Supreme Court, and mergers, see Ellis Hawley, *The Great
          War and the Search for Order in Modern America, 1917–1933: A History of the
          American People and Their Institutions* (New York, 1979), 90–91.
          "that he considered it . . . ," C. Joseph Pusateri, *A History of American Business*
          (Arlington Heights, Ill., 1984), 287.
          "Rumors of Julian Pending Merger," Durst ad, April 30, 1926.
          Two Eastern companies, ad, May 13, 1926.
          On New England Oil and Refining, Finney, 68–69.

p. 164    For Arnold report, see Frank Arnold Papers, HL.
          "a drilling campaign . . ." ad, January 15, 1926.

p. 165    On H. J. Barneson, see *Who's Who in Los Angeles, 1925–26,* (Los Angeles, 1926),
          and OA, April, 1927.
          "like poison . . ." Finney, 156.
          "idle rumors . . ." ad, May 13, 1926.
          "If you don't sell me stock . . ." Finney, 57–58.
          May totals on Julian trading, LAT May 30, 1926.
          "same crowd . . ." ad, June 3, 1926.

p. 166    "certain well-known brokers . . ." ad, June 8, 1926.
          "Satisfactory progress . . ." ad, June 10, 1926.

## Section III

p. 166    On the history, assets, and stock of Marine Oil, see OA, July, October, 1926; Finney
          70; and LAT June 30 and July 1, 1926.
          On Marine No. 7, see Wagy ad, April 29, 1927.

p. 167    On shedding Julian name, see Finney, 66.

Lewis announces California-Eastern, LAT June 30, 1926.

On relationship between Associated Oil and First National Bank, see Rockwell Hereford, *A Whole Man, Henry Mauris Robinson, and a Half Century, 1890–1940* (Pacific Grove, Calif., 1985), 89.

Lewis moves offices, LAR June 17, 1927.

"At a time . . ." John McGroarty, *Los Angeles: From the Mountains to the Sea* (Chicago and New York, 1921), 901.

p. 168    The biography of Henry Robinson is drawn from Hereford; Bertie C. Forbes, *Men Who Are Making the West* (New York, 1923), 278–300; McGroarty, 901–2; Ira B. Cross, *Financing an Empire: History of Banking in California* (Chicago and San Francisco, 1927), 424–27; Rockwell D. Hunt, ed., *California and Californians* (Chicago, 1926), 49–51.

On retirees to California, see Kevin Starr, *Material Dreams: Southern California Through the 1920s* (New York, 1990).

p. 169    On John Barber, see *Pan-Pacific Who's Who* (Honolulu, 1941), 41.

p. 170    On A. P. Giannini and branch banking, see James C. Findlay, "The Economic Boom of the Twenties in Los Angeles" (Ph.D. diss., Claremont, 1958), 290–92.

Barber and Robinson drive to work, author's interview with Rockwell Hereford.

On Pacific Southwest expansion, see Albert W. Atwood, "Money from Everywhere," SEP, 195 (May, 12, 1923), 141; Hereford, 86; and Cross, 401–2.

On First Securities, see Hereford, 88.

On territorial banking, see Findlay, 290–92.

Stern joins First National, Hereford, 86.

p. 171    On *de novo* rule, see Findlay, 292.

For numbers of branches in Southern California, see Findlay, 293–94, and Hereford, 87.

On Robinson and the Dawes Plan, see Hereford, 167–68.

On William Rhodes Hervey, see Finney, 73.

For 1924 reorganization of Pacific Southwest, see Cross, 402, and Hereford, 88.

"gift of plausible persuasion . . ." Woehlke (November, 1927), 81.

"Talk to him . . ." LAT June 17, 1927.

p. 172    "I took an interest . . ." Finney, 126.

"Why not let me . . ." Finney, 79.

A copy of the agreement with the Anglo-London-Paris Bank may be found in the Texaco Legal Archives.

Lewis announces terms, LAT August 8, 1926.

750,000 share issue, see NA, Equity Case #M-12-M, Receivers Final Report, VI-2-7.

p. 173    On leaks and bugging, see Hereford, 90, and author's interview with Hereford.


## Chapter 10

### Section I

p. 174    For descriptions of Julian Petroleum offices, see LAR June 15, 1927; LAX, July 28, 1927; and Guy Finney, *The Great Los Angeles Bubble* (New York, 1929), 136.

"bubble mill," Finney, 136.

p. 174   On forged signatures, see LAT March 16, 21, 22, 1928.

On sales in New York, LAR June 10, 1927.

On Rosenberg's role, see Finney, 97, 144–45.

p. 176   On Louis Berman and Jack Rosenberg as messengers, see LAR June 17, 1927.

"He is rather a dumb boy . . ." LAR July 29, 1927.

On Bennett's activities, see LAR, July 27 and August 2, 1927; and Walter V. Woehlke, "The Great Julian Pete Swindle," *Sunset* 54 (November, 1927), 66.

On Cecil B. DeMille, see LAT July 9, 1927.

"Lewis was as big hearted . . ." Robert P. Shuler, "Julian Thieves in Politics" (Los Angeles, 1930), 14.

For a profile of Kent Parrot, see Tom Sitton, "The 'Boss' Without a Machine: Kent K. Parrot and Los Angeles Politics in the 1920s," *Southern California History* 87 (Summer, 1985): 365–87.

"a swaggering . . ." quoted in Robert Gottlieb and Irene Wolt, *Thinking Big: The Story of the Los Angeles Times* (New York, 1977), 198.

"daffy and always heterogeneous . . ." Charles Stoker, *Thicker 'n Thieves* (Santa Monica, Calif. 1951), 8.

p. 177   On 1921 mayoral campaign, Parrot as "power broker," "sinister shadow" and "De Facto Mayor," see Sitton, 368–69, 373.

"was a soft-voiced . . ." Leslie T. White, *Me, Detective* (New York, 1936), 124.

Biography of Charlie Crawford drawn from Dwight F. McKinney and Fred Allhoff, "The Lid Off Los Angeles," *Liberty* (November 11, 1939); Bruce Henstell, *Sunshine and Wealth: Los Angeles in the Twenties and Thirties* (San Francisco, 1984), 45–46; and Robert P. Shuler, "The Strange Death of Charlie Crawford" (Los Angeles, 1931).

On other underworld figures, see Stoker, 11, and Henstell, 46.

p. 178   On Parrot at Biltmore Hotel, see Stoker, 8, and Frank Doherty to Hiram Johnson, April 22, 1930, JP.

On 1923 mayoral campaign and LAT exposé, see Sitton, 372–73.

On Crime Commission, see Gottlieb & Wolt, 197.

On 1925 mayoral campaign, see Sitton, 371.

On Parrot and Crawford in Wagy books, see Robert P. Shuler, "Julian Thieves" (Los Angeles, 1930), 17, and LAT May 24, 1930.

For Gans transactions, see LASC #229200, Julian Petroleum *vs.* R. Gans (January 18, 1927).

"knew that thousands . . ." Robert P. Shuler, "Julian Thieves in Politics", (Los Angeles, 1930), 15.

p. 179   On Allen and Greer, see LASC Case #227785, Julian Petroleum *vs.* Walter B. Allen and Perry Greer (June 30, 1927).

"almost rich enough . . ." Shuler, "Julian Thieves in Politics", 17.

"one newspaper reporter . . ." Shuler, "Julian Thieves", 26.

On 1926 gubernatorial campaign, see Jackson Putnam, "The Persistence of Progressivism in the 1920's: The Case of California," *Pacific Historical Review* 35 (1966), 406.

On A. P. Giannini and the 1926 campaign, see Russell M. Posner, "The Bank of Italy and the 1926 Campaign in California," *California Historical Society Quarterly* 37 (1958), 267–75, 359–68.

p. 179   For C. C. Julian's 1926 campaign ads, see LAR August 20–31, 1926.
         On Parrot as fund-raiser, see Frank Doherty to Hiram Johnson, April 22, 1930, JP.
         On Jacob Farbstein, see LASC Case #229199, Julian Petroleum *vs.* J. Farbstein
             (July 26, 1927).
p. 180   On E. F. Hackel, see LASC cases #227780 and #232183, Julian Petroleum *vs.* E.
             F. Hackel (June 30, 1927, August 23, 1927).
         All overissue figures are drawn from the Haskins & Sells stock audit included in
             the 1930 FBI report.
         Lewis accuses Bennett, LAR June 17, 1927.
         "Well some of it did . . ." LAT January 24, 1930.
         On Lewis's finances, LAT February 19, 1930.
         "Lewis was a big liver . . ." LAX April 5, 1930.
         "roly-poly . . ." Lorin L. Baker, *That Imperiled Freedom* (Los Angeles, 1932), 32.
         For Rosenberg's wealth, see LAT April 10, 1928.
         On Bennett's Rolls-Royce, see Finney, 152–53, and LAT July 30, 1927.

# Section II

p. 181   On acquisition of new wells, see LAR August 28, 1926, December 13, 1926; ad,
             October 8, 1926; and OA, September, 1926.
         On Montana wells, see LAT August 4, 1926; OA, October, 1926; and Baker, 218.
         For Texas properties, see LAR October 23, December 10, 1926.
         Barber hires Kotteman, Finney, 167.
         "gyp loans," *ibid.*, 84.
p. 182   On origins of Million Dollar Pool, see Woehlke (October, 1927), 82, and (Novem-
             ber, 1927), 68; Finney, 84–86; and LAR June 7, 1927.
         On stock pools, see Finney, 82, and John Kenneth Galbraith, *The Great Crash,
             1929* (Boston, 1961), 84–85.
         "The art of pool . . ." Woehlke (October, 1927), 78–79.
         For conditions of Bankers' pool No. 1, see Finney, 85–86; LAR June 7, 1927;
             LAEX, undated clipping HP, Box 88; and LASC #230880, Julian Petroleum *vs.*
             W. I. Hollingsworth (August 8, 1927).
p. 183   "Every millionaire . . ." Shuler, "Julian Thieves in Politics", 18.
         For members of Bankers' Pool No. 1, see Finney, 86, and LAR June 29, 1927.
         "In September last . . ." LAT June 15, 1927.
         For a biography of W. I. Hollingsworth, see Rockwell D. Hunt, ed., *California and
             Californians* (Chicago, 1926), 69–70.
         On Joe Toplitsky, see *Who's Who in Los Angeles County* (Los Angeles, 1928);
             *History of Los Angeles City and County* (n.d.), 612–13; Joseph Malamut, *South-
             west Jewry: An Account of Jewish Progress and Achievement in the Southland* (Los
             Angeles, 1926), 147–48; Hunt, 186; and Finney, 102.
p. 184   "That super-virtuous . . ." Rube Borough, "Law and Order in Los Angeles," *The
             Nation* 125 (July 6, 1927), 12.
         "noted churchman . . ." CG July 23, 1927.
         "most useful citizen," LAT March 11, 1930.

p. 184   "could have lived a thousand years . . ." Shuler, "Julian Thieves", 38.
         On Rouse pool, see Finney, 95–96; LAR, June 8 & 29, 1927, and April 3, 1928.
p. 185   On Bennett, Allen and Greer, see Finney, 145, and Shuler, "Julian Thieves in
               Politics", 50–51.
         On Jewish Pool, see Finney, 97–98.
         On Tia Juana pool see *ibid.*, 92; LAR August 2, 1927; and LAEX June 19, 1927.


## Section III

p. 185   On new A. C. Wagy offices, Wagy ad, August 21, 1926.
         On stock escrow plan, ads, September 1, 2, 1926.
p. 186   On failure to receive stock, see LASC Case #224151, Etta Demange *vs.* A. C. Wagy
               (May 17, 1927), and LASC #224208 and #224209, D. Spurlin *vs.* A. C. Wagy
               *et al.* (May 18, 1927).
         "It is our earnest advice . . ." ad, October 1, 1926.
         "Speculators stood spellbound . . ." LAT October 8, 1926.
         "Stockholders needn't become panicky . . ." LAT October 8, 1926.
         "HOLD YOUR STOCK . . ." ad, October 8, 1926.
         On Etta Demange, see LASC Case #224151.
p. 187   "It was the advertisements . . ." LAR June 8, 1927, and LAT March 28, 1928.
         On Bennett's clientele, see LAT, May 24, 1930.
         For Barber and Stern investments in Julian Petroleum, see Finney, 130–32.
p. 188   On Ethridge discovery, see LAX July 28, 1927.
         For Kottemann report, see LAX July 28, 30, 1927.
         McKay urges sale, LAR June 7, 1927.
p. 189   On extension of bank pool, see LAEX, undated clipping, HP Box 88.
         On Silberberg pool, see Finney, 92.
         On E. J. Miley, see LAT September 16, 1926.
         On C. W. Durbrow, see OA, November, 1926, and LAR August 4, 1927.
         On L. J. King see OA, November, 1926, and Finney, 163–65.
         On Frank Flint and the opening of California-Eastern offices, LAT November 1,
               1926.
         King expresses doubts, Finney, 164–65.
p. 190   "These investigators . . ." Woehlke (November, 1927), 20.
         On additional $100,000, see Finney, 77, and Baker, 218.
         "That is a very usual thing . . ." Finney, 127.
         For the history of the orphan check, see *ibid.*, 78; Woehlke (November, 1927), 68;
               LAX July 30, 1927; and LAT August 10, 1927.
         On pool extensions, see Finney, 95–96, and LAR June 7, 1927.
p. 191   On Toplitsky and Frank, see Finney, 103, 108–9.
         "Frank would take an interest . . ." LAT July 28, 1929.
         "just like shooting fish . . ." Finney, 173.
         "to see the thing through . . ." *ibid.*, 103.
         On Merchants National Bank role, *ibid.*, 103–10, 149.
         On Johnson pool, see *ibid.*, 87–88, and LAT July 27, 1927.
         For Shipp report, see LAX July 24, 1927.

## Chapter 11

### Section I

p. 192      For C. C. Young's career, see H. Brett Melendy and Benjamin F. Gilbert, *The Governors of California: Peter H. Burnett to Edmund G. Brown* (Georgetown, Calif., 1965), 349–51.

         On Young, Giannini, and Wood, see James C. Findlay, "The Economic Boom of the Twenties in Los Angeles" (Ph.D. diss., Claremont, 1958), 198, 300.

p. 193      On Parrot's campaign on behalf of Friedlander, see Frank Doherty to Hiram Johnson (January 21, 1927), JP; and C. C. Young to John Randolph Haynes (May 5, 1930), HP, Box 88.

         "If the governor does not appoint . . ." Frank Doherty to Hiram Johnson (January 21, 1927), JP.

         "It is my understanding . . ." Kent Parrot to Hiram Johnson (February 8, 1927), JP.

         "I have just talked to Kent . . ." and "it will be seized upon . . ." Frank Doherty to Hiram Johnson (February 8, 1927), JP.

p. 194      "Chandler-Dickinson combination . . ." Kent Parrot to Hiram Johnson (February 8, 1927).

         On petition campaign and Young in Los Angeles, see Lorin L. Baker, *That Imperiled Freedom* (Los Angeles, 1932), 87–90.

         On Carnahan, C. C. Young to John Randolph Haynes (May 5, 1930), HP, Box 88.

         On appointment of Friedlander, see LAT February 26, 1927, and Baker, 90.

         On Lewis's obligations, see Guy Finney, *The Great Los Angeles Bubble* (New York, 1929), 148; and LASC Case #223726, Pan American Bank *vs.* Lewis (May 10, 1927).

         "the inability to complete . . ." ad, January 19, 1927.

p. 195      "approximated one-sixth" and "a prearranged drive . . ." LAT, February 10, 1927.

         "wild unfounded rumors . . ." Wagy ad, February 21, 1927.

         "wholesale dumping . . ." Wagy ad, February 17, 1927.

         On Bennett's Valentine's Day activities, LAT February 4, 1928.

         For rumors of overissue, see LAT February 25, 1928.

p. 196      On Johnson pool, see Finney, 87–88, and LAT July 27, 1927.

         On meeting at Frank offices, see Finney, 90, 172–73, and LAX July 27, 1927.

         For Ramish after meeting, see Finney, 91; LAT July 28, 1927; and LAX July 27, 1927.

p. 197      On Frank's activities after meeting, see LAT July 28, 1927.

### Section II

p. 197      On the Kottemann audit, see LAX July 24, 1927, and Walter V. Woehlke, "The Great Julian Pete Swindle," *Sunset* 54 (November, 1927), 20.

         "serious irregularities," LAT March 28, 1928.

         "is the last thing . . ." Rockwell Hereford, *A Whole Man, Henry Mauris Robinson, and a Half Century, 1890–1940* (Pacific Grove, Calif., 1985), 90.

p. 197     Barber agrees to junior financing, Finney, 165–66.

            The biography of Harry Bauer is drawn from *Pan-Pacific Who's Who*, (Honolulu, 1941), 46; and *Who's Who on the Pacific Coast, 1949* (Chicago, 1949), 71.

p. 198     On Bauer's stock transactions and meeting with Lewis, see Finney, 157–60.

            "Mr. Lewis would call . . ." LAX July 26, 1927.

p. 199     On Robinson and Barber, see Hereford, 90.

            On Robinson role, see Finney, 162–63.

            Bauer agrees to head California-Eastern, *ibid.*, 161–62.

            Flint, King, Robinson reaction to Bauer demands, *ibid.*, 122, 155, 161.

            "I was not interested . . ." *ibid.*, 163.

p. 200     For Flintridge meeting, see *ibid.*, 122, 174, and LAR June 20, 1927.

            On February 26 stockholders' meeting, LAT February 27, 1927, and Wagy ad, March 16, 1927.

            On Bertholf and Beery, see Finney, 166; LAEX, May 18 & 19, 1927; and LAT May 19, 1927.

p. 201     On March 25 board meeting, see LAT March 26, 1927.

## Section III

p. 201     "If there was any existing doubt . . ." Wagy ad, March 29, 1927.

            "in line with . . ." LAT April 2, 1927.

            On the dissolution of Bankers' Pool No. 1, see Finney, 93–95.

p. 202     On Louis B. Mayer, see *ibid.*, 116–17, and LAT July 29, 1927.

            On Rosenberg pool, see Finney, 98.

p. 203     "nationwide banking syndicate," LAT April 13, 1927, and Wagy ad, April 18, 1927.

            On stockholders' meeting, see LAR April 15, 1927.

            Lewis explains overissue, Woehlke (November, 1927), 20.

            On reimbursement for Lewis, LAR June 20, 1927.

            Shuler on Barber, Robert P. Shuler, "Julian Thieves" (Los Angeles, 1930), 24.

p. 204     On Barber and Hereford, author's interview with Rockwell Hereford.

            "It was about April 18 . . ." Finney, 176–77.

            "postponement," LAT April 23, 1927.

            On LASE meeting, see LAT April 24, 1927, and March 29, 1928.

            Kottemann receives books, LAR August 8, 1927.

p. 205     "Pat, if anything comes up . . ." Finney, 143.

            On Morris and Bennett, see LAR July 28, 1927.

            Rosenberg raises money, Shuler, "Julian Thieves", 55, Finney, 147–49.

            On April 25 directors' meeting, see LAR August 6, 1927.

p. 206     Flint announces reorganization, LAT April 26, 1927 and Wagy ad April 26, 1927.

            "Lewis came to me . . ." LAT September 12, 1927.

p. 207     "were just waiting . . ." LAT July 30, 1927.

            On Bennett illness, LAR July 28, 1927.

            Hackel loan, LAT July 29, 1927.

            "I asked him . . ." Finney, 39.

            "went into a rage . . ." *ibid.*, 167.

            On Bennett-Barnes transaction, see LAT July 7 & 12, 1930.

p. 207    "absolutely and emphatically . . ." LAT September 12, 1927.
On Bennett-Frank transaction, see LAR August 10, 1927.
Bennett's departure, LAT July 30, 1927.
On Bennett's money belt, see Woehlke (November, 1927), 66.
On May 2 California-Eastern meeting, Finney, 167–68.

p. 208    "to get some money," *ibid.*, 150–51.
"rumors running rife . . ." ad, May 4, 1927.
"orphan check," LAX July 30, 1927.
Kottemann discloses overissue, LAR August 7, 1927.
For May 5 meeting, see Finney, 168–69.

p. 209    "indications of an overissuance . . ." and "Whereas . . ." LAT May 7, 1927.
Friedlander suspends permits, LAR May 7, 1927.

## Chapter 12

### Section I

p. 213    For description of crowds, see LAEX May 7, 1927.
On Julian loan to Lewis, see LAX July 27, 1927.
"I recalled reading . . ." ad, LAR September 20, 1926.

p. 214    For Monte Cristo Mine ads, LAR September 23, 1926; October 4, 1926; and LAT
July 11 & 13, 1928.
On Jacques Van der Berg, see LAT November 15, 1927, October 5, 1928.
"THE POT OF GOLD" ad, LAR October 11, 1926.
On reloading, see Walter V. Woehlke, "The Great Julian Pete Swindle," *Sunset*
54 (September, 1927), 15, and ad, LAR September 20, 1926.
"Folks, today opens . . ." ad, LAR October 4, 1926.
"all the dividends . . ." ad, LAR October 13, 1926.
"Why let the public in . . ." ad, LAR October 12, 1926.

p. 215    For Macmillan decision, see LAT October 17, 1926.
On radio in the 1920s, see Frederick Lewis Allen, *Only Yesterday* (New York, 1964),
26, and Geoffrey Perrett, *American in the Twenties: A History* (New York, 1983),
230.
Julian buys KMTR, LAT November 5, 1926.
"FOLKS WATCH THIS BABY . . ." ad, LAR April 29, 1927.
"with their gums . . ." Woehlke (September, 1927), 15.

p. 216    "My purpose is to protect . . ." LAT May 8, 1927.
"We are absolutely capable . . ." ad, LAR, LAT May 9, 1927.
Flint reassures investors, LAT May 7, 1927.
"a few interesting facts . . ." LAT May 8, 1927.
"My investigators have been . . ." LAT, LAX May 9, 1927.
"great uncertainty . . ." and "altogether solvent . . ." LAT May 10, 1927.
"to devote my exclusive time . . ." LAT May 11, 1927.
"DYNAMITE" ad, LAR May 10, 1927.
On Julian and business leaders, see Woehlke (October, 1927), 17.

p. 217    For Julian broadcast, see *ibid.*; Rockwell Hereford, *A Whole Man, Henry Mauris*

*Robinson, and a Half Century, 1890–1940* (Pacific Grove, Calif., 1985), 91; and LAT, LAR May 11, 1927.

"enables him to strike . . ." CG May 28, 1927.

"Any judgement-proof individual . . ." Woehlke (October, 1927), 18.

"I have had three . . ." Henry M. Robinson to Charles Stern, May 13, 1927, SP.

p. 218    On Lucien Wheeler, see Wheeler's report, May 17, 1927, FBI.

On appointment of receivers, see LAT May 14, 1927, and United States Senate, Special Committee on Bankruptcy and Receivership Proceedings in United States Courts (1933), 297, 298, 352 (henceforth cited as "Hearings").

"had been practically unknown . . ." Hearings, 5.

p. 219    For Reese resignation, see LAT May 14, 1927.

Lewis admits Wagy ownership, LAX May 16, 1927.

On Julian charges and closing of Wagy brokerage, see LAR May 16, 1927.

On Margie Pike and George Powell, see LAT, LAR, LAX May 17–19, 1927.


## Section II

p. 220    "This case is . . ." Lucien Wheeler Report, May 17, 1927, FBI.

On Beery charges, see LAT, LAEX, May 18, 1927.

For overissue totals, see LAT May 19, 1927.

On California-Eastern receivership, see LAT, LAR May 20, 1927, and Equity Case #M-12-M, Perkins Oil Well Cementing Company *vs.* Julian Petroleum, NA.

The hopes of the Julian shareholders . . . LAT May 22, 1927.

"to tell what they know . . ." LAR May 23, 1927.

p. 221    For Lickley probe, see LAT, LAR May 24, 1927.

On usury law, see LAR May 24 & 28, 1927.

"a conviction provides . . ." LAT May 25, 1927.

"big game . . ." LAR May 24, 1927.

"I consider these gentlemen . . ." LAR June 10, 1927.

"Large sums of money . . ." LAR May 25, 1927.

Receivers collect $250,000, LAT May 28, 1927.

On Julian plane crash, see LAT, LAX, LAEX May 24, 1927.

"I believe everything Julian . . ." and "these thieves . . ." LAR May 28, 1927.

p. 222    400 money lenders, LAT June 6, 1927.

June 4 deadline, LAR June 3, 1927.

Lickley exposes bankers' pools, LAEX, LAR June 6, 1927.

"The evidence is concrete . . ." LAT June 7, 1927.

On Haldeman and Ramish refunds, see LAT June 7, 1927.

On receivers' activities, see Woehlke (November, 1927), 68; LAT June 2, 1927; and LAX June 11, 1927.

p. 223    "All morning long . . ." LAR May 19, 1927.

"elderly women . . ." LAT June 2, 1927.

"I am sitting . . ." ad, LAR June 3, 1927.

"effrontery," CG, June 11, 1927.

"colossal impudence," LAT May 28, 1927.

p. 223    On Dekalb Spurlin investments and suit, see LAT May 19, 1927 and LASC
          #224208 and #224209, D. Spurlin *vs.* Wagy *et al.* (May 18, 1927).
          "for the benefit of . . ." LAT June 3, 1927.
          For stockholder stories, see LAR June 8, 1927.
p. 224    The description of stockholders' meeting is drawn from LAR and LAEX June 11,
          1927.

## Section III

p. 224    Haldeman returns check, LAR June 14, 1927.
          "The amount of money . . ." LAT June 22, 1927.
p. 225    For criticism of restitution campaign, see LAEX June 8, 1927.
          On role of Walter Tuller, see LAR June 22, 1927, and Rube Borough, "Law and
          Order in Los Angeles," *The Nation* 125 (July 6, 1927), 12.
          Ramish check returned, LAT June 16, 1927.
          On warrant for Haldeman, see LAR June 15, 1927.
          "We are engaged . . ." LAX June 23, 1927.
          Receivers threaten civil suits, LAR June 22, 1927.
          "I do not think . . ." LAT June 18, 1927.
          "vote my stock . . ." ad, LAR June 18, 1927.
          "Folks don't miss . . ." ad, LAR June 16, 1927.
          On broadcast controversy, see LAR, LAEX June 21, 1927.
p. 226    "the utterly unscrupulous . . ." LAT June 22, 1927.
          "From a careful canvass . . ." *Los Angeles Ledger* June 24, 1927, in HP, Box 88.
          For Horchitz-Spurlin fight, see LAT, LAR June 22, 1927.
          On Loeb and Kimmerle, see LAR June 22, 23, 1927.
          Loeb loses eye, LAR June 25, 1927.
p. 227    "Those people have admitted . . ." LAT June 22, 1927.
          "cock of the walk," LAR June 24, 1927.
          "The greatest swindle . . ." LAEX June 22, 1927.
          "malicious heterodyning," LAEX June 23, 1927.
          For Julian's June 22 broadcast, see LAT, LAR, LAX June 23, 1927.
          On Lewis after crash, see LAR May 20, 1927.
          "While they can indict me . . ." LAX June 23, 1927.
          "No crime was committed . . ." LAR May 20, 1927.
          Lewis's biography appeared in LAR June 9–22, 1927.
          "It was a consistent . . ." LAR June 20, 1927.
p. 228    "I was double-crossed . . ." LAX June 23, 1927.
          "moving heaven and earth . . ." LAT June 24, 1927.
          "produced the desired . . ." Woehlke (October, 1927), 18.
          "About the time . . ." LAEX June 24, 1927.
          On indictments, see LAR, LAT, LAEX June 25, 1927.
          On bail bonds, LAT June 29, 1927.
          Grand Jury to continue probe, LAT June 30, 1927.
          Lickley probe, LAR June 25, 1927.
          Julian's new charges, LAR June 29, 1927.

## Chapter 13

### Section I

p. 229  O'Melveny memos (May 17, 1927 and June 27, 1927), SP.
"The actual damage . . ." LAT June 23, 1927.

p. 230  "The really important thing . . ." Charles Stern to Willis Booth (July 26, 1927), SP.
"Don't get stampeded . . ." *Saturday Night*, July 9, 1927.
"attempt to drag and smear . . ." LAEX June 3, 1927.
For letters to Stern, see SP.
"We are covered . . ." Charles Stern to Willis Booth (July 26, 1927), SP.
"the major parent crime . . ." LAT July 9, 1927.
"The Walrus and the Carpenter," LAT June 21, 1927.

p. 231  "frauds of considerable proportions . . ." LAT June 23, 1927.
"salutary effect . . ." LAT June 30, 1927.
"Then the cry . . ." *Saturday Night*, July 7, 1927.
"a herring . . ." *Los Angeles Ledger*, June 24, 1927, in HP, Box 88.
"score or more . . ." BSM, June, 1927.
"Elmer Gantrys . . ." Charles Stern to Louis Everding (July 29, 1927), SP.
"This nasty affair . . ." *Saturday Night*, July 7, 1927.
"'better America' movements . . ." CG, July 23, 1927.

p. 232  "100 per cent Americanism . . ." undated clipping, HP, Box 88.
"This was wrong . . ." *ibid.*
"there are men . . ." BSM, July, 1927.
"It is whispered . . ." BSM, June, 1927.
"Gosh! What do those reporters . . ." LAEX June 15 & 26, 1927.
"for the express purpose . . ." BSM, July, 1927.

p. 233  "a certain Jew," BSM, June, 1927.
"big Jew money lenders . . ." BSM, August, 1927.
"honest confiding child of Israel . . ." CG July 23, 1927.
On anti-semitism in Los Angeles in the 1920s, see Max Vorspan and Lloyd P. Gartner, *History of the Jews in Los Angeles* (San Marino, Calif., 1970); and Neal Gabler, *An Empire of Their Own: How the Jews Invented Hollywood* (New York, 1988).
Wood exonerates bank, LAR June 27, 1927.
"terrible wrong . . ." LAT June 28, 1927.
On Robinson and Barber, see Rockwell Hereford, *A Whole Man, Henry Mauris Robinson, and a Half Century, 1890–1940* (Pacific Grove, Calif., 1985), 92, and author's interview with Rockwell Hereford.
For Robinson's statement, see LAT June 27, 1927.

p. 234  On bank merger, see Hereford, 92, and LAT July 11, 1927.
For LASE actions, see LAT May 21, 1927, and LAX July 7, 1927.
For Department of Corporations reforms, see LAT June 6, 1927, and LAR July 5, 1927.
On defendant actions, see LAT July 1, 1927.
For new shareholders' groups, see LAR July 5, 1927.
On Cecil B. DeMille, see LAR July 8, 1927.

p. 234    On Lickley's health, see LAR July 9, 1927.
          On shadowing charges, see LAR July 6, 1927.
          For bribe charges, see LAX July 2, 1927.
p. 235    "obtained the promise . . ." LAT July 7, 1927.
          On San Francisco and Tia Juana pools, see LAT July 14, 1927.
          Grand jury ends probe, LAT July 17, 1927.
          On Julian broadcasts, see LAR July 7, 1927, and LAT July 19, 1927.


## Section II

p. 235    "In less than two months . . ." Walter V. Woehlke, "The Great Julian Pete Swin-
          dle," *Sunset* 54 (November, 1927), 68.
          "It was a source of grief . . ." United States Senate, Special Committee on Bank-
          ruptcy and Receivership Proceedings in United States Courts, 352 (henceforth
          cited as Hearings).
          On workload of Scott and Carnahan, see *ibid.*, 380, 352.
          On Wagy receivership, see LAT July 2, 1927.
p. 236    On valuation of properties, see Sunset Pacific Report (November 19, 1929), 8–9,
          in Equity Case #U-100-H, Berreyesa Cattle *vs.* Sunset Pacific, NA.
          On Louisiana properties, see General Report, Box 251, and Lewis Oil Company
          Complaint of Intervention January 21, 1928, in Equity Case #M-12-M, Perkins
          Oil Well Cementing Company *vs.* Julian Petroleum, NA.
          On overissue of California Eastern stock, see General Report VI-2-9, Box 251, in
          Equity Case #M-12-M, NA.
          On liabilities, see Hearings, 313, 320.
p. 237    "the preferred stock alone . . ." LAX June 3, 1927.
          "it will be necessary . . ." FBI report, June 22, 1927.
          On liability of stockholders, Hearings, 384–92.
          Receivers should have sued, *ibid.*, 337–38.
          "I thought the bona fide . . ." *ibid.*, 335.
          "I knew very well . . ." *ibid.*, 384–85.
          "We made our decision . . ." *ibid.*, 376.
          "unsophisticated receivers," *ibid.*, 364.
p. 238    On Lewis's work regimen, see Criminal Case 9191-H, United States *vs.* S. C. Lewis,
          NA.
          For Lewis reorganization plan, see LAR July 8, 1927; hearings, 364; and LAT July
          8, 1927.
          Lewis declared bankrupt, LAT July 6, 1927.
          On July 7 stockholders' meeting, see LAR July 8. 1927.
          On Scott and restitution, see Report of Anderson & Anderson, Box 261, in Equity
          Case #M-12-M, NA.
p. 239    On Bankers' Pool No. 1 response, see James A. Anderson to Robert Zimmerman,
          June 23, 1927, and Joseph Scott to Robert Zimmerman, July 7, 1927, in Equity
          Case #M-12-M, NA.
          On Hollingsworth and Haldeman, see LAT, LAR July 9, 1927.
          On objections to receiver policies, see General Report, IX, 9–10, and Report of
          Anderson & Anderson, 25; in Equity Case #M-12-M, NA, and Sunset Pacific

Report, November 19, 1929, in Equity Case #U-100-H.

On settlement of Receiver suits see General Report, XIV 36–44, Box 251, in Equity Case #M-12-M, NA, and LAT July 1, 1927.

## Section III

p. 239   On Robert Kenney and transcripts, Robert W. Kenney, *My First Forty Years in California Politics, 1922–1962* (Oral History Project, UCLA Special Collections), 46.

p. 240   On size of transcript, see Robert P. Shuler, "Julian Thieves in Politics" (Los Angeles, 1930), 53, and LAR July 9, 1927.

"I have been bunkoed . . ." LAX July 26, 1927.

"Pardon me . . ." LAX July 27, 1927.

"Innocents Abroad . . ." LAT July 29, 1927.

"had a lot of gall . . ." LAR August 10, 1927.

"I want to be fair . . ." LAX July 30, 1927.

For Barber testimony, see LAX July 30, 1927.

p. 241   On Barneson transactions, see LAR August 2, 1927.

For Hahn testimony, see LAR August 2, 1927.

"were so wisely . . ." CG July 23, 1927.

"It still remains . . ." Guy Finney, *The Great Los Angeles Bubble* (New York, 1929), 119.

p. 242   "a very knowing man . . ." Robert P. Shuler, *Julian Thieves* (Los Angeles, 1930), 48–49.

## Chapter 14

### Section I

p. 243   "very bright . . ." LAT October 31, 1927.

On Berman in New York, see LAX May 14, 1927.

"I am not in hiding . . ." LAX May 15, 1927.

"pleasure trip . . ." LAX May 21, 1927.

p. 244   On the *Berengaria*, see LAEX May 22, 1927; LAR May 24 & 27, 1927; and LAEX June 1, 1927.

Bennett located in Paris, LAT, LAR June 13–15, 1927.

"He has the interests . . ." LAX June 15, 1927.

Berman vanishes, LAT June 18, 1927.

On Berman back in New York, see LAEX, LAT June 24, 1927.

For Berman telegrams, see LAT, LAR, LAEX June 24, 1927.

Berman told to go to Russia, LAR June 23, 1927.

On rumors of $10,000,000, see LAT June 18, 1927.

For trunks of gold, see NYT September 18, 1927.

Berman's New York apartment, LAR July 28, 1927, and LAT July 29, 1927.

Berman's autos shipped, LAX July 30, 1927.

p. 244   Ben Cohen sent to New York, LAT August 6 & 7, 1927.

p. 245   "ruefully admitting . . ." LAT August 30, 1927.

"very good reasons . . ." LAT September 8, 1927.

Berman surrenders, LAT, LAR September 10, 1927.

"boy Ponzi . . ." LAT September 18, 1927.

"jaunty and dapper . . ." LAT September 10, 1927.

"nattily dressed . . ." LAT September 18, 1927.

p. 246   "Judas," LAT September 14, 1927.

Charges that Berman wired money, LAR September 11, 1927.

"furthermore he overissued . . ." LAT September 13, 1927.

On Hollingsworth and Haldeman, see LAT August 9, 1927; LASC Cases #230880 Julian Petroleum *vs.* W. I. Hollingsworth (August 8, 1927), and #230879 Julian Petroleum *vs.* H. M. Haldeman (August 8, 1927).

"more or less 'pike' . . ." LAT August 1, 1927.

On usury suits, see LAT October 7, and November 29, 1927.

On Keyes and restitution, see LAT September 20, 1927.

Usury charges dismissed, LAT September 24, 26, 30, 1927.

For receivers' report on restitution, see LAT October 1, 1927.

Berman pledges to recover funds, LAT September 15, 1927.

p. 247   Receivers feud with Berman, LAT October 1, 1927.

End of feud, LAT October 14, 1927.

"We hope to recover . . ." LAT November 2, 1927.

## Section II

p. 247   For description of new courthouse, see CG January 21, 1928.

"on the outcome . . ." Finney, 191–92.

p. 248   On congestion at defense table, see LAT January 6, 1928.

"popular confidence and esteem . . ." Ella A. Ludwig, *A History of the Harbor District of Los Angeles* (Los Angeles, 1926), 726.

"putrid and filthy," BSM, June, 1928.

"marked for vilification," and "stood in high public favor," Guy Finney, *The Great Los Angeles Bubble* (New York, 1929), 188.

"logic, good manners . . ." William W. Robinson, *Lawyers of Los Angeles* (Los Angeles, 1959), 216.

For Keyes's work on the McPherson, Taylor, and Hickman cases, see Thomas McDonald, "History of Los Angeles District Attorney's Office" (unpublished manuscript).

p. 249   On Buddy Davis, see Robert W. Kenney, *My First Forty Years in California Politics, 1922–62* (Oral History Project, UCLA Special Collections), 33, 39.

"gross inefficiency," LAT January 31, 1928.

"The District Attorney will stall . . ." BSM, August, 1927.

100 witnesses, evidence, and "startling," LAT January 5, 1928.

"the box is filled . . ." CG January 21, 1928.

p. 250   "Did you have anything . . ." LAT January 20, 1928.

Keyes asks for dismissal, LAT January 21, 1928.

"advanced with all . . ." CG January 21, 1928.

p. 250     For Gregory testimony and Lewis cross-examination, see LAT January 25–February
               17, 1928.
           "Your questions are entirely . . ." LAT March 8, 1928.
           Doran denies motion, LAT April 13, 1928.
p. 251     "If it was overissued . . ." LAT April 17, 1928.
           Efforts to subpoena brokers, LAT April 13 & 14, 1928.
           Courtney testimony, LAT May 17–18, 1928.
           "not being tried . . ." LAT January 6, 1928.
           "Legal definitions . . ." CG January 21, 1928.
           Attorneys present defense, LAT April 25–26, 1928.
           Doran dismisses charges, LAT April 26, 1928.
           Richardson speech to jury, LAT May 4, 1928.
           For Lewis's argument, see LAT May 9–11, 1928.
p. 252     LeCompte Davis oration, LAX May 22, 1928.
           "I want the record . . ." LAT November 2, 1928.
           Keyes addresses jury, LAT May 22–23, 1928.
           As longest trial, see LAT May 21, 1928.
p. 253     "Not guilty" verdicts, LAX, LAT May 24, 1928.

## Section III

p. 253     On Teapot Dome, see Burl Noggle, *Teapot Dome: Oil and Politics in the 1920s*
               (New York, 1962).
           "This is emphatic evidence . . ." *ibid.*, 201.
           "preposterous" and "travesty," CG May 26, 1928.
           "Is this monstrous crime . . ." LAEX May 25, 1928.
           "I consider the same . . ." BSM, July, 1928.
           "We felt you would be angry . . ." LAX May 24, 1928.
           "We felt that there was a great wrong . . ." LAT May 24, 1928.
           "Only one who sat . . ." LAT November 2, 1928.
           "The way the indictments were drawn . . ." LAT May 24, 1928.
           "After careful consideration . . ." LAT May 25, 1928.
p. 254     "We have just concluded . . ." LAT May 30, 1928.
           For Receivers *vs.* Hollingsworth, see LAT March 6, 1928, and LASC Case #230880.
p. 255     "There is now no hope . . ." NYT June 17, 1928.
           On Julian's New Monte Cristo letter, see LAT December 2, 1927.
           "Open Letter" LAT December 7, 1927.
           Corporation Commissioner seeks investigation, LAT December 2, 1927.
           "undoubtedly ill-advised . . ." LAT December 21, 1927.
           "if the response . . ." LAT December 20, 1927.
           For shareholders' suit, see LAT December 30, 1927.
p. 256     For Jacques Van der Berg suit, see LAT November 25, 1927.
           Julian commits wife, LAT March 24, 1928.
           "She has been on a continual debauch . . ." LAT March 30, 1928.
           "using stimulants . . ." LAT March 24, 1928.
           Mary Julian's public hearing, LAT March 27, 1928.
           "airing of the family skeletons . . ." LAT March 30, 1928.

p. 256    "Liquor is the fountain . . ." BSM, May, 1928.
          On Kimmerle conviction, see LAT September 28, 1928.
          For Kimmerle parole, see LAT, May 22, 23, 1928, and Arthur Loeb to Los Angeles
            City Council (September 15, 1928), Los Angeles City Archives.

p. 257    For Lewis-Berman federal trial, see LAT September 18–October 31, 1928.
          On sentencing of Lewis and Berman, see LAT November 7, 1928.

## Chapter 15

### Section I

p. 258    The life history of Milton Pike is drawn from LAT January 12, 1929.
          The description of the tailor shop is drawn from LAT November 1 & 4, 1928, and
            January 11 & 17, 1929.
          "an illiterate, weasel . . ." Guy Finney, *The Great Los Angeles Bubble* (New York,
            1929), 195.
          "shrunken old gnome," Leslie T. White, *Me, Detective* (New York, 1936), 98.
          On Getzoff's stomach operations and "medicine," see LAT January 24, 1929.
          "secret drinker . . ." Finney, 189.

p. 259    For Pike's diary, see LAX, LAEX, LAT November 1, 1928.
          On Pike and Sherman, see LAT January 17, 1929.
          On attempts to sell diary, see Finney, 195, and LAT January 17, 1929.
          On alleged blackmail, see LAT January 12 & 30, 1929.
          *Times* buys story, LAT January 12, 1929.
          For Arthur Loeb's role, see LAX November 2, 1928.

p. 260    For Robert Shuler's role, see LAT January 17, 1929.
          On grand jury indictments, see LAT, LAX, LAEX, LAR November 1, 1928.
          "The ghost . . ." Finney, 194.
          For excerpts from Pike's diary, see LAT, LAX, LAEX November 1, 1928.
          On grand jury charges, see LAT November 1, 1928.
          "broke like a bombshell . . ." LAR November 1, 1928.
          "A public idol . . ." Finney, 194.
          "The confidence of the public . . ." LAT November 13, 1928.
          "I know my skirts . . ." LAX November 2, 1928.
          For November 17 indictments, see LAT November 18, 1928.

p. 261    For Buddy Davis's charges, see LAT November 21, 1928.
          The biography of Buron Fitts is drawn from Justice B. Detwiler, ed., *Who's Who
            in California: A Biographical Dictionary, 1939–40, 1955–56* (San Francisco,
            1940, 1956); *California Blue Book, Legislative Manual or State Roster* (Sacra-
            mento: 1928); *Pan-Pacific Who's Who*, (Honolulu, 1941); *Men of California*
            (San Francisco, 1926); and *Sunset* (December, 1927). See also telegram from
            Frank Doherty to Hiram Johnson (December 16, 1922), JP, and LAT August
            11, 1923.

p. 262    On Fitts's first month as District Attorney, see LAT December 2, 10, 24, 27, 1928,
            and Frank Doherty to Hiram Johnson, January 2, 1929, JP.

p. 262      On Fitts's detective force, see White, 93–94.

"permits it to be said . . ." Frank Doherty to Hiram Johnson, April 29, 1929, JP.

"is very ambitious . . ." Frank Doherty to Hiram Johnson, January 2, 1929, JP.

## Section II

p. 262      On crowds and fights in hallway, see LAX January 16, 1929.

"a spectacle . . ." White, 99.

Berman charges dismissed, LAT January 8, 1929.

p. 263      For Richardson testimony, see LAT January 8, 1929.

For Pike testimony, see LAT January 11, 1929.

"I nursed him in the daytime . . ." White, 98–99.

p. 264      For Berman testimony, see LAT, LAX January 12 & 13, 1929.

p. 265      On Berman cross-examination, see LAT January 13 & 15, 1929.

"I never saw a witness . . ." LAT February 21, 1929.

p. 266      For Sherman, Rittinger, Louis Berman, and Vianelli testimony, see LAT January 17, 1929.

For Buddy Davis testimony, see LAT January 19, 1929.

For testimony of auto dealers and salesmen, see LAT January 22, 1929.

For Keyes's testimony, see LAT, LAX January 24, 1929.

p. 267      For other defense witnesses, see LAT January 27, 1929.

For Robinson and Rosenberg testimony, see LAT January 30, 1929.

For Simpson and Stewart arguments, see LAT February 2 & 3, 1929.

For Rush argument, LAT February 5, 1929.

"The leopard . . ." LAT February 7, 1929.

"to make peace . . ." "of a nationality . . ." and "Well what of it?" LAX February 6, 1929.

p. 268      For Keyes's address to jury, see LAT February 7, 1929.

For Clark's address to jury, see LAT February 9, 1929.

"He does his work . . ." Frank Doherty to Hiram Johnson, January 18, 1929, JP.

"Buron is in the court . . ." Frank Doherty to Hiram Johnson, January 29, 1929, JP.

p. 269      For Fitts's address and the verdict, see LAT, LAEX, LAX, LAR February 9, 1929.

On Berman and gunman, see LAEX, LAR February 9, 1929, and LAT February 10, 1929.

On Getzoff in jail, see LAT February 10, 1929.

For Getzoff confession, see LAR, LAEX February 11, 1929.

p. 270      "It is unfortunate . . ." LAT February 10, 1929.

"sacrificial goat . . ." LAR February 11, 1929.

"Well Done, Fitts . . ." LAEX February 11, 1929.

February 20 sentencing, LAT February 21, 1929.

"It is quite manifest . . ." LAT April 6, 1929.

Buddy Davis charged, LAT May 2, 1929.

For Buddy Davis trial, see LAT July 3–12, 1929.

On Pike bribery, see LAT June 28, 1929.

p. 271      "personal animosity . . ." LAT July 12, 1929.

## Section III

p. 271    Shareholders approve reorganization, LAT January 20, 1928.

Articles of reincorporation filed, LAT February 27, 1928.

"Lollypop Oil . . ." BSM, June, 1928.

McCormack appproves reorganization, LAT May 25, 1928.

Approval percentages from Final Report of Receivers (1941), Equity Case #M-12-M, Perkins Oil Well Cementing Company vs. Julian Petroleum, NA.

California-Eastern auction, LAT November 19 & 20, 1928.

p. 272    For Ramish claim, see Final Report XIV 36–44, Box 256, Equity Case #M-12-M, NA.

On other usury cases, see ibid., XIV 36–44, and IX 4–7.

On Lewis Oil claims, see Petition of Intervention November 15, 1927; First Amended Complaint, January 21, 1928; and Second Amended Complaint, December 17, 1928; all in Equity Case #M-12-M, NA.

p. 273    For Loeb's criticism of the receivership, see United States Senate, Special Committee on Bankruptcy and Receivership Proceedings in United States Courts (1933), 505–29 (henceforth cited as Hearings).

For Carnahan testimony, ibid., 313–14, 373.

For Loeb protests and those of others, see ibid., 505–06, and various documents in Equity Case #M-12-M, NA.

"got a nickel . . ." Hearings, 369.

On legal fees, ibid., 328–29.

"I wanted to get back . . . ," ibid., 365.

On Relativity in Business Morals, see Henry M. Robinson, Relativity in Business Morals (Boston, 1928); Finney, 7–8; and Rockwell Hereford, A Whole Man, Henry Mauris Robinson, and a Half Century, 1890–1940 (Pacific Grove, Calif., 1985), 220–21.

On Hoover election charges, see Lorin L. Baker, That Imperiled Freedom (Los Angeles, 1932), 32.

"Mr. Robinson's selection . . ." Thomas Barrett to Herbert Hoover, February 28, 1929, Herbert Hoover Presidential Library.

p. 274    For Hoover on Robinson, see Herbert Hoover, The Memoirs of Herbert Hoover, vol. 2, (New York, 1951).

On death of Frank Flint, see LAT February 13, 1929.

"Senator Flint was aged . . ." Hearings, 362.

"the victim of the crushing weight . . ." Finney, 122.

p. 275    On collapse of First National Bank and merger with Security Bank, see Hereford, 93–94, and LAT January 25, 1929.

On LASE in 1928 and 1929, see James C. Findlay, "The Economic Boom of the Twenties in Los Angeles" (Ph.D. diss., Claremont, 1958), 333–35.

## Chapter 16

### Section I

p. 276    "poorest of the poor . . ." Edmund Wilson, "The City of Our Lady the Queen of the Angels," New Republic (December 29, 1931), 89.

For the best account of Shuler's background and theology, see Mark S. Still, " 'Fighting Bob' Shuler: Fundamentalist and Reformer," Ph.D. diss. (Claremont, 1988).

On Shuler's biases and mindset, see *ibid.*, 20–21, and Kevin Starr, *Material Dreams: Southern California Through the 1920s* (New York, 1990), 136–37.

p. 277 "ideals of Christian piety . . ." Still, 15.

"I've found very few . . ." Wilson, 89–90.

On Shuler's career in Texas, see Still, Chapter 3.

On assignment to Los Angeles as punishment, see *ibid.*, 139.

On Trinity Methodist Church, see *ibid.*, 141–42.

"Bob Shuler managed a perfect appeal . . ." Wilson, 89.

"The only Anglo-Saxon city . . ." Starr, 137.

"one of God's watchmen . . ." Duncan Aikman, "Savonarola in Los Angeles," *American Mercury* (December, 1930), 426–29.

p. 278 On Vernon Club sermon and other moral issues, see Still, 154–55.

"the result of conditions . . ." Starr, 137.

On Shuler and the Ku Klux Klan, see Still, 167–73, 189–96.

On "McPhersonism," see *ibid.*, 206–9.

"reveal Shuler at his worst . . ." *ibid.*, 268–69.

p. 279 On Shuler's radio station, see *ibid.*, 248–52.

"Champion aginner . . ." LAT June 1, 1930.

"moral racketeer . . ." Lillian Symes, "Beautiful and Dumb," *Harper's* (June, 1931), 30.

"No politician dares . . ." Aikman, 429.

"They can't find a scratch . . ." Wilson, 90.

"considerably more likable . . ." LAR April 20, 1929.

"I have always been of the opinion . . ." Frank Doherty to Hiram Johnson (November 28, 1923), JP.

p. 280 On 1929 mayoral election, see Robert W. Kenney, *My First Forty Years in California Politics, 1922–62* (Oral History Project, UCLA Special Collections), 56–57; Tom Sitton, "The 'Boss' Without a Machine: Kent K. Parrot and Los Angeles Politics in the 1920s," *Southern California History* 87 (Summer, 1985), 383; and Still, 281–85.

On Porter, see Wilson, 91; Duncan Aikman, "California Sunshine," *Nation* 129 (April 22, 1931): 448–50; and Starr, 139.

Shuler on votes by Julian victims, BSM, May, 1929.

"professional Legionnaire . . ." Frank Doherty to Hiram Johnson, January 18, 1929, JP.

Berman promise to Nix, LAT June 20, 1929.

Arrest of Streeter and Slossberg, LAT April 13, 1929.

p. 281 "The Julian case is a national scandal . . ." LAT June 1, 1929.

"A fine in these Julian cases . . ." LAT May 30, 1929.

For Eberts case, see LAT May 28–30, 1929.

For Perry and Benson cases, see LAT June 4 & 5 1929.

"This case must be decided . . ." LAT June 8, 1929.

For Rothman trial, see LAT June 12–24, 1929.

For Streeter case, see LAT June 20, 1929.

"an unjustifiable waste . . ." LAT June 20, 1929.

For Haldeman case, see LAX June 21, 1929.

p. 282    "To grant this motion . . ." LAT May 25, 1929.
          For Katzev case, see LAT June 28, 1929.
          On appeal to Supreme Court, see LAT July 6, 1929.
          For Supreme Court ruling, see LAT February 26, 1930.

## Section II

p. 282    On Loeb and Panama Canal, see United States Senate, Special Committee on
              Bankruptcy and Receivership Proceedings in United States Courts (1933), 505–
              6 (henceforth cited as Hearings).
          On Loeb's stock holdings, see LAT October 21, 1933.
          "had no other occupation . . ." Loeb Deposition, October 30, 1933, Equity Case
              #U-100-H, Berreyesa Cattle vs. Sunset Pacific, NA.
p. 283    On scrapbooks and other activities, see Loeb Deposition, July 1, 1931, "it was
              necessary . . ." ibid.
          "were the real profiteers . . ." S. C. Lewis to Arthur Loeb, June 28, 1929, Equity
              Case #U-100-H, NA.
          On stockholder court cases, see LAT July 1 & 15, and August 2, 1928, and May 8,
              1929.
          On Loeb's activities in August, 1928, see James Brewer Deposition, October 30,
              1933; James Brewer to Joseph Scott, April 11, 1929; and Joseph Scott to James
              Brewer April 13, 1929, in Equity Case #U-100-H, NA. See also Hearings,
              503.
          On Loeb and Crump, see Hearings, 397–98.
          For Lewis Memorandum, see S. C. Lewis to Guy Crump, June 16, 1929, Equity
              Case #U-100-H, NA.
          "To render any and all . . ." S. C. Lewis to Arthur Loeb, June 28, 1929, Equity
              Case #U-100-H, NA.
          "it was too much dynamite . . ." Hearings, 509–10.
p. 284    On Carnahan's refusal to participate in suit, see Sunset Pacific Report, November
              19, 1929, in Equity Case #U-100-H.
          On funds for lawsuit, see Hearings, 517.
          Stockholders' suit filed, LAT, LAX October 9, 1929.
          For federal indictments of Lewis and Berman, see LAT January 14, 1930.
          For Streeter indictment, see LAT January 17, 1930.
          On Crump and Berman, see LAT January 14 & 15, 1930.
          "perpetual surprise witness . . ." LAX January 18, 1930.
          "flood of sensations . . ." LAX January 15, 1930.
          On Johnson, Chessher and Lewis, see LAT January 15, 1930.
          "I have been kind . . ." Leontyne Johnson to Fletcher Bowron, November 19, 1934,
              Fletcher Bowron Papers, HL.
p. 285    "was plenty of money . . ." LAX January 15, 1930.
          "We knew it was the rankest . . ." ibid.
          "intrigue and doublecross," LAT January 16, 1930.
          "nervous collapse," LAT January 17, 1930.
          "little gray account book," LAT January 18, 1930.
          Johnson on Lewis and California Stock Exchange, LAT January 21, 1930.

p. 285 "waive all my constitutional rights . . ." LAX March 14, 1930.

"My Three Years With SC Lewis," LAX, March 10, 1930.

p. 286 "with his peculiar mincing gait . . ." Leslie T. White, *Me, Detective* (New York, 1936), 163.

"Thou shalt have . . ." Kenney, 27.

"I know a lot . . ." and midnight meeting, LAT, LAX March 17, 1930.

"have $75,000 by noon . . ." LAX March 11, 1930.

"prove they meant business," White, 166.

on arrest of Lavine, see *ibid.*, 166–67, and LAT, LAX March 11, 1930.

p. 287 "I didn't know . . ." LAX March 11, 1930.

Lavine's defense appeared in LAT March 11 & 12, 1930.

"The charge of extortion . . ." Leontyne Johnson to Fletcher Bowron, November 19, 1934, Fletcher Bowron Papers, HL.

"WHAT WERE THE FACTS . . ." LAX March 17, 1930.

"loaded with dynamite . . ." LAEX March 11, 1930.

## Section III

p. 287 "What is the matter . . ." LAX March 17, 1930.

"Mr. Fitts has been in office . . ." LAR March 19, 1930.

p. 288 "beyond the reach . . ." LAR March 13, 1930.

For Vianelli affidavit, see LAX March 19, 1930.

For Weaver/Love charges, see LAT, LAX March 20, 1930.

p. 289 For a life history of Love, see LAR March 21, 1930.

"was but the start . . ." LAX March 20, 1930.

On Viannelli and Fitts, see LAR March 20, 1930.

"Jack Bennett is a big shot . . ." LAR March 22, 1930.

"Berman's new desire to talk . . ." LAT March 22, 1930.

"Julian mess . . ." LAT March 21, 1930.

On McComb and Berman, see LAT February 21, 1930.

p. 290 For Appellate Court decision, see LAT March 2, 1930.

McComb exonerated of prejudice, LAX March 16, 1930.

"I have had difficulties . . ." LAX March 25, 1930.

"Before I'll go to trial . . ." LAT March 26, 1930.

"at a standstill . . ." LAX March 25, 1930.

"We don't want little minnows . . ." LAR March 21, 1930.

"not a single indictment . . ." LAR March 19, 1930.

McComb withdraws, LAT March 28, 1930.

New indictments, LAT March 29, 1930.

"again the district attorney's office . . ." LAX March 29, 1930.

Lewis surrenders, LAT March 29, 1930.

"preferring Uncle Sam's frying pan . . ." LAT April 1, 1930.

"I'm tired of it all . . ." LAR March 29, 1930.

p. 291 "The district attorney is attempting . . ." LAT April 2, 1930.

Securities Act indictments, LAT, LAX April 2, 1930.

"The most sensational booklet . . ." LAR April 4, 1930.

p. 291    There are banker thieves . . ." and subsequent quotes are all from Robert P. Shuler,
          "Julian Thieves" (Los Angeles, 1930).
          On sales of "Julian Thieves", see BSM, May, 1930.


## Chapter 17

### Section I

p. 292    On C. C. Young as governor, see Jackson Putnam, *Modern California Politics*, 2nd
          edition (San Francisco, 1984), 11–13.
          "Unless all signs fail . . ." quoted in BSM, April, 1930.
          Shuler broadcasts charges, LAR March 24, 1930.
p. 293    For Shuler on Berman and Friedlander, see Robert P. Shuler, "Julian Thieves"
          (Los Angeles, 1930), 29–31.
          "would make no trouble . . ." LAT April 7, 1930.
          "an absolute and malicious . . ." and Carr and Carnahan responses, LAX April 5,
          1930.
          "too grotesque . . ." LAEX April 8, 1930.
          "apparently tried to give . . ." LAT April 8, 1930.
          "I am not charging . . ." BSM, April, 1930.
p. 294    "This publication does not mean . . ." quoted in *ibid*.
          "If there had been . . ." *ibid*.
          "who said he would like to see . . ." LAR April 9, 1930.
          "whose relationship to the Julian cases" BSM, April, 1930.
          "too silly for serious . . ." LAT April 8, 1930.
          "dragged in . . ." Leontyne Johnson to Fletcher Bowron, April 19, 1934, Fletcher
          Bowron Papers, HL.
          "an enthusiastic admirer . . ." Ray L. Donley to Fletcher Bowron, February 25,
          1935, Fletcher Bowron Papers, HL.
          "one may imagine the glee . . ." clipping, April 20, 1930, HP.
p. 295    "It is to be hoped . . ." LAT April 8, 1930.
          "I don't relish . . ." LAT April 10, 1930.
          On McComb and Berman, see LAT, LAX April 12, 1930.
          For Athearn report, see LAT April 15, 1930.
          For Julian testimony, see LAT April 16, 1930.
          On death of Robert Bursian, see LAX, LAT April 16, 1930.
          On traces of cyanide, see LAT April 17, 1930.
p. 296    "under the guise of lovemaking . . ." LAX April 18, 1930.
          "young and beautiful brunette . . ." LAX April 16, 1930.
          Blonde secretary, LAT April 20, 1930.
          On theories of suicide or accident, see LAR April 19, 1930, and LAT April 20,
          1930.
          "Fitts produces sensation . . ." LAR April 18, 1930.
          For Crawford and Friedlander indictments, see LAX, LAT April 17, 1930.
          Appellate Court decision, LAT April 18, 1930.
          "Unless you, the public . . ." LAR April 18, 1930.

p. 296    On Berman disappearance, see LAR April 21, 1930.

"We have two instances . . ." Frank Doherty to Hiram Johnson, April 22, 1930, JP.

p. 297    On Berman in Seattle, see LAR April 22, 1930.

On racing documents and "first peace of mind," see LAT April 25, 1930.

Lewis ordered to return, LAT April 24, 1930.

Berman enters prison, LAT April 26, 1930.

"the BIG Crooks . . ." LAR April 17, 1930.

"conspiracy as deep and black . . ." BSM, May, 1930.

On Los Angeles Bar Association, see Mark S. Still, "'Fighting Bob' Shuler: Fundamentalist and Reformer." Ph.D. diss. (Claremont, 1988), 312–14.

"to interfere with . . ." LAT April 30, 1930.

"Not merely a fine . . ." LAT May 3, 1930.

For Shuler on contempt, see Still, 315.

"healthy, hearty . . ." LAR May 3, 1930.

"spur the judges . . ." LAX May 3, 1930.

"Let all men know . . ." Still, 316.

p. 298    "kinfolks of a jailbird . . ." LAX May 3, 1930.

Shuler starts to serve sentence, LAT May 6, 1930.

"prisoner of war . . ." Still, 317.

"I will gladly surrender . . ." LAT May 6, 1930.

On Shuler in jail, see Bruce Henstell, *Sunshine and Wealth: Los Angeles in the Twenties and Thirties* (San Francisco, 1984), 51; and Still, 318.

"whether or not . . ." LAT May 11, 1930.

LAT supports jailing of Shuler, LAT May 5, 1930.

LAX supports jailing of Shuler, LAX May 7, 1930.

"In all the writer's . . ." quoted in BSM, June, 1930.

"was a mischievous little boy . . ." LAR May 7, 1930.

"Whether Shuler . . ." cited in BSM, June, 1930.

## Section II

p. 299    For rumors of indictments, see LAR April 29, 1930.

Financial records subpoenaed, LAT April 30, 1930.

Carnahan "swamped" the panel, LAT May 3, 1930.

Loeb testimony, LAR May 7, 1930.

"We are through . . ." LAX May 7, 1930.

"They didn't get the big crooks . . ." LAR May 7, 1930.

"Could the Julian Thieves . . ." clipping dated May 28, 1930, HP.

"If the district attorney's office . . ." LAR May 6, 1930.

"lack of evidence . . ." LAT, May 8, 1930.

"Was Fitts failure . . ." undated clipping, HP.

p. 300    Crawford refuses to testify, LAT April 30, 1930.

On death of Frank Meyers and Johnson-Lavine trial, see Leslie T. White, *Me, Detective* (New York, 1936), 168–70.

On illegal search warrant, see LAX May 3, 1930.

"It was a tough case . . ." White, 172.

"Feel like I have traveled . . ." LAT May 9, 1930.

p. 300    "Somebody started . . ." White, 173.
          "Does your honor . . ." LAX June 7, 1930.
          For Lewis's jury bribery trial, see LAT June 7 & 8, 1930.
          For Lewis at McNeil Island, LAT June 10, 1930.
p. 301    On return of Berman, see LAT June 17, 1930.
          For list of trials involving Berman, see LAT June 14, 1930.
          On Berman's immunity, see LAT May 9, 1930.
          Johnson, Lavine convicted, LAT July 5, 1930.
          Johnson, Lavine sentenced, LAT July 7, 1930.
          For Krause conviction, see July 11, 1930.
          For Groves conviction, see LAT July 17, 1930.
          On Grider, see LAT August 6, 1930.
          For death of Haldeman, see LAT March 11, 1930.
          On death of King, see LAT May 24, 1930, and United States Senate, Special
              Committee on Bankruptcy and Receivership Proceedings in United States
              Courts (1933), 529.

## Section III

p. 302    "He was the first . . ." LAR July 15, 1930.
          For Motley Flint's will, see LAT July 19, 1930.
          "The Inglewood stock dabbler . . ." LAR July 16, 1930.
          "There's a tension . . ." and "I am delighted . . ." LAX July 16, 1930.
          On Nix resignation, see LAT July 17, 1930.
          "There is a species . . ." LAX July 18, 1930.
p. 303    For the controversy over "Julian Thieves", see LAT, LAX July 15 & 16, 1930.
          On Keaton's bank stock, see LAT July 15 & 18, 1930.
          For Nix and Federal Radio Commission, see LAX July 19, 1930, and LAT July 22,
              1930.
          Berman assigned public defender, LAT July 3, 1930.
          "white collar galley slaves," and Berman verdict, LAT July 26, 1930.
          Berman sentenced, LAT August 9, 1930.
          "first conviction . . ." LAT July 26, 1930.
p. 304    "somebody is going to be punished . . ." and "disposed of the silly talk . . ." LAT
              July 28, 1930.
          "a test of strength . . ." Los Angeles Citizen, July 28, 1930, HP.
          "The outstanding fact . . ." BSM, September, 1930.
          On impact of dry vote, see Royce Delmatier et al., The Rumble of California Pol-
              itics, 1848–1970 (New York, 1970), 217–18, and Putnam, 13.
          Shuler judicial candidates elected, Robert Kenney, My First Forty Years in Cali-
              fornia Politics, 1922–1962 (Oral History Project, UCLA Special Collections),
              63, and Still, 340.
          For trial of Frank Keaton, see LAT September 3–6, 1930.
          "gross morbid brooding . . ." LAT August 23, 1930.
p. 305    Keaton sentenced, LAT September 16, 1930.
          For Keaton insanity verdict, see LAT May 10 & 11, 1931.
          "The court entertains . . ." LAR October 14, 1930.

p. 305    Friedlander charges dropped, LAT October 21, 1930.
          On Shuler radio hearings, see LAT January 20–24, 1931, and Still, 345–49.
p. 306    "I'm praying for all . . ." and Crawford's baptism, LAX July 1 & 2, 1930.
          On the Shuler-Briegleb feud, see Still, 323–24.
          For Crawford's donations, see LAT May 21 & 25, 1931.
          Copies of *Critic of Critics* may be found at the Los Angeles Public Library.
          "peace meeting . . ." LAT May 22, 1931.
p. 307    On Crawford/Spencer murders, see LAT May 21, 1931.
          David Clark surrenders, LAT May 22, 1931.
          "Had the chief . . ." White, 287–89.
          Biography of Dave Clark drawn from LAT May 22, 1931.
          "sort of timber . . ." LAT May 4, 1931.
          "David H. Clark knows . . ." Lorin L. Baker, *That Imperiled Freedom* (Los Angeles,
              1932), 225. See also Carey McWilliams, *Southern California: An Island on the
              Land* (Santa Barbara, 1973), 245.
          "There were three racketeers . . ." *Critic of Critics*, June 1931.
          "a pitiless plan . . ." LAT May 25, 1931.
p. 308    "matinee idol," White, 290.
          "back in the saddle . . ." LAT August 14, 1931.
          "has a handsome face . . ." LAT August 20, 1931.
          "He is one of our noblest . . ." LAT August 24, 1931.
          For bomb at Weller's house, see LAT August 26, 1931.
          For Clark retrial, see LAT October 16–21, 1931.
          For Clark's subsequent murder conviction, see David Clark photo file, USC.
          For claims on $75,000, see LAT July 3, 1930.
          On Wagy-Crawford settlement, see LAT May 28, 1931.
          Crawford estate claims money, LAT May 27, 1932.
          On Liberty Hill, see LAT May 27–28, 1932.
          Hill convicted, LAT July 10, 1932.
p. 309    County absolved, LAT May 17, 1935.
          Federal Radio Commission decision, LAT August 10, 1931, and Still, 350–51.
          For Federal Radio Commission hearing, see Still, 353–55.
          For Shuler's conspiracy theories, see *ibid.*, 256–57, and BSM, December, 1931.

## Chapter 18

### Section I

p. 310    On the death of Rosebud Harris, see LAT December 12, 1931.
          On suicides in Los Angeles, see *Sunshine and Grief in Southern California* (Los
              Angeles, 1931), 53; and Carey McWilliams, *Southern California: An Island on
              the Land* (Santa Barbara, 1973), 246.
          "In the early thirties . . ." McWilliams, 247.
          On Guarantee Building and Loan, see Royce Delmatier *et al.*, *The Rumble of
              California Politics, 1848–1970* (New York, 1970), 223–24.
          On American Mortgage Company, see McWilliams, 246.
          On Morrisey and Adkisson, see LAT April 14, 1931.

p. 311    On Richfield Oil, see Delmatier, 224–25; and Charles S. Jones, *From the Rio Grande to the Atlantic: The Story of the Richfield Oil Company* (Norman, Oklahoma, 1972).

On receiverships in Los Angeles, see United States Senate, Special Committee on Bankruptcy and Receivership Proceedings in United States Courts (1933), 5 (henceforth cited as Hearings).

For Julian wells in receivership, see LAT March 20, 1931, and LASC Case #301676, J. A. Smith *vs.* C. C. Julian (April 25, 1930).

On Charles Allison, see LAT April 15, 1932.

For Sunset Pacific merger negotiations, see Hearings, 336.

On agreement with First National, see *ibid.*, 336, 372.

On Associated Oil and Sunset Pacific, *ibid.*, 372, 571.

For Berreyesa suit, see Equity Case #U-100-H, Berreyesa Cattle *vs.* Sunset Pacific, NA.

p. 312    For Loeb intervention, see Petition of Intervention, October 28, 1931, Equity Case #U-100-H, NA.

On appointment of Sunset Pacific receiver, see Hearings, 410–18, 428–38.

"We found the task . . ." *ibid.*, 400–403.

46 firms win dismissal of suit, LAT November 6, 1930.

For State Supreme Court decison, see LAT August 1, 1931.

For U.S. Supreme Court decision, see LAT January 5, 1932.

For Sunset-Pacific cross-complaint, see LAR May 5, 1930.

On Loeb and Crump, see James Brewer to Guy Crump, February 9, 1933, and Deposition of James Brewer, October 30, 1933, Equity Case #U-100-H, NA.

p. 313    On Wagy settlement, see Arthur Loeb to Judge Thomas Gould, April 2, 1931, and Deposition of Arthur Loeb, October 30, 1933, Equity Case #U-100-H, NA, and Hearings, 401, 515.

On settlement proposal, see Joseph Lewinson to Arthur Loeb, April 18, 1931, Equity Case #U-100-H, NA.

On attempt to substitute attorneys, see Rollin McNitt Affidavit, October 28, 1933, Equity Case #U-100-H, NA.

"This recovery was a brand . . ." Hearings, 407.

On protests of settlement, see affidavits from Rollin McNitt, James T. Best, and Paul F. Fratessa, July 25, 1932, and Arthur Loeb, July 26, 1932, Equity Case #U-100-H, NA.

"there will be nothing . . ." Loeb Affidavit, *ibid.*

"I most sincerely sympathize . . ." LAT August 4, 1932.

For State Supreme Court decision, see LAT April 5, 1933.

On Loeb claim, see his Petition, July 1, 1931, and Report of Special Master, July 18, 1933, Equity Case #U-100-H, NA.

On Ramish lien, see Petition by Arthur Ramish, August 2, 1934, Equity Case #U-100-H, NA.

p. 314    For a summary of usury cases, see Receivers' Final Report, February 26, 1941, Equity Case M-12-M, NA.

For 1933 Loeb suit, see LAT October 21, 1933; Loeb Petition of Intervention, October 15, 1933; and H. L. Carnahan's response, October 30, 1933; all in Equity Case #M-12-M, NA.

p. 314    On Rolapp as receiver, see Hearings, 486–97; and Receivers' Reports 1–6, in Equity
          Case #U-100-H, NA.

          On reduced indebtedness, see Sixth Report of Receiver, Equity Case #U-100-H,
          NA; and Hearings, 422–25.

p. 315    "Amidst all the wild reckless . . ." Hearings, 387.

          On Lewis-Berman federal case, see Hearings, 652–63.

          For Lewis's correspondence and grand jury charges, see Criminal Case #C9888,
          U.S. vs. S. C. Lewis (January 3, 1930), NA.

          On resolution of Lewis-Berman case, see Hearings, 652–63.


## Section II

p. 316    For Julian's January, 1930, letter and Monte Cristo mine, see LAT January 15,
          1930.

          On Santa Fe Springs wells, see LASC Case, #301676, J. A. Smith vs. C. C. Julian
          (April 25, 1930).

          On Julian in Oklahoma, see SFC March 25, 1934.

          "Beau Brummel," LAT July 15, 1931.

          "empire builder," LAT February 6, 1930.

          "The largest ever . . ." LAT July 14, 1931.

p. 317    On Julian at Santa Fe Springs, see LAT, LAR, LAX April 18, 1930, and LASC
          Case #301676.

          On J. A. Smith, see Roberta Ridgeway, "The C. Arnholt Smith Story," San Diego
          28 (October, 1976).

          Julian barred from wells, LAT April 1, 1931.

          Julian's Oklahoma gusher, LAT August 16, 1930.

          Julian's runaway well, LAT August 18, 1930.

          On Oklahoma prorationing, see LAT August 16 & 29, 1930.

p. 318    For Oklahoma Supreme Court decision, see LAT September 3, 1930; Northcut
          Ely, "The Conservation of Oil," Harvard Law Review LI (1938), 1,209–44; and
          Harold F. Williamson et al., The American Petroleum Industry. Vol. 2, The Age
          of Energy, 1899–1959 (Evanston, 1963), 542.

          For Julian's rooftop party, see LAT December 1, 1930.

          For details on Mary Julian's divorce petition, see LASC Case #D92410, Mary Olive
          Julian vs. C. C. Julian (February 2, 1931).

          On Texas kidnapping, see LAT April 3 & 4, 1931.

          For embezzlement charges, see LAT April 7, 1931.

          On Salsberry's collection attempt, see LAT April 10, 1931.

          On Texas hearings, see LAT May 5, 1931 and LASC Case #301676.

p. 319    For stockholder suit and receivership, see LAT May 10, 1931.

          On restraining order, see LAT May 12, 1931.

          Mason replaces Julian, LAT May 19, 1931.

          "not only failed . . ." and H. A. Penn, LAT July 9, 1931.

          On Julian's arrest for mail fraud, see LAT July 14–17, 1931.

          "performed outside loops . . ." LAT June 3, 1931.

          Julian's home auctioned, LAT August 18 & 19, 1931.

p. 319    Julian indicted, LAT September 26, 1931, and February 6, 1933.
          Julian makes death threats, LAT September 29, 1931.
p. 320    Julian arrested in San Francisco, LAT April 11–13, 1932.
          Stockholders win control of Oklahoma company, LAT June 14, 1932.
          Julian found bankrupt, LAT June 17, 1932.
          Julian in jail, LAT December 31, 1932, and January 1, 1933.
          200 witnesses, LAT February 6, 1933.
          Julian jumps bail, LAT, NYT, SFC March 24, 1933.
          Reward posted for Julian, LAT February 15, 1933.
          For Julian's wanted poster, see FBI files.
          Rumors of Julian's whereabouts, LAT March 28, 1933.
          Julian discovered in Shanghai, LAT, NYT April 25–26, 1933.
          For descriptions of Shanghai, see Agnes Smedley, "Shanghai Episode," *The New
              Republic* (June 13, 1934), 122–24; and "Where Races Mingle, But Never
              Merge," *Christian Science Monitor Weekly* (June 24, 1936), 4.
          "You can tell the United States . . ." LAT April 26, 1933.
p. 321    "he entertains the cabaret girls . . ." LAT August 6, 1933.
          "crooked politics . . ." LAT May 8, 1933.
          On Julian and Standard oil, see LAT March 25, 1934.
          "I have always lived . . ." LAT August 6, 1933.
          Julian spends last dollar, LAT September 11, 1933; LAX September 10, 1933.
          "But I am sure . . ." Norwood F. Allman, *Shanghai Lawyer* (New York, 1943), 90.
          "Since arriving in China . . ." LAT December 22, 1934.
          "Selling dead ones . . ." LAX February 1, 1934.
          On borrowing funds from Chinese businessmen, see SFC March 25, 1934.
          On "What Price Fugitive?" see LAT February 5, 1934, and IDN August 16, 1935.
p. 322    On Leonora Levy, see LAT, NYT March 25, 1934, and SFC June 24, 1934.
          On Mary Cantorovich, see LAT March 26, 1934.
          Julian evicted, SFC March 28 & 29, 1934.
          Fischer waits for Julian, IDN August 16, 1935.
          Julian's dinner party described in LAT, NYT, SFC March 25–26, 1934.
          On Julian's suicide and its aftermath, see LAT, NYT, SFC, and *Winnipeg Evening
              Times*, March 25–26, 1934.
          "I will never attempt . . ." LAT March 28, 1934.
          Friends question Julian suicide, LAT March 26, 1934 and IDN August 16, 1935.
          For coroner's inquest, see NYT March 30, 1934.
          "upon a slab . . ." *Winnipeg Evening Times*, March 26, 1934.
          Canadian pays expenses, *ibid.* and SFC March 29, 1934.
          For description of funeral service and attempts to return body, see SFC March 29,
              1934.
p. 323    For Julian's burial, see NYT May 12, 1934.

## Section III

p. 323    "perhaps the bitterest . . ." Archibald Johnson to Hiram Johnson, March 6, 1928,
              JP.

p. 323　For John Barber's career, see *Pan-Pacific Who's Who* (Honolulu, 1941), 41.

For Charles Stern's career, see *Who's Who on the Pacific Coast, 1949* (Chicago, 1949), 887.

On Asa Keyes, see LAT October 14, 1931, and clippings, August 20 & 21, 1933; October 18, 1934, USC.

For death of Friedlander, see clippings, August 13 & 14, 1934, USC.

On Lavine as attorney, see clippings, Morris Lavine folder, USC.

For C. C. Streeter case, see Criminal Case #9906-H, U.S. *vs.* Clarence C. Streeter (January 16, 1930), NA.

On Jack Roth, see LAT October 21, 1932; SFC June 23, 1934; clippings, November 1, 1932, and August 7, 1937, USC.

p. 324　On Louis Berman, see clippings, October 30, 1935, and January 11, 16, 1936, USC.

On Lewis's release, see Criminal Case # C9888, NA; LAT April 2, 1935, and clippings, April 3, 1935, USC.

On Jacob Berman, see clippings in FBI Julian files, and clippings, June 12, 14, 17, and 21, July 24, September 7, and December 19, 1935; June 13, September 5, October 19, 1936; January 4 & 5, 1937, USC.

On World War II suit against Berman, see NYT, LAT August 18, 1944; LAX August 24, 1944.

p. 325　On Rosenberg suicide, see clippings, October 11, 1940, USC; and LAT October 11, 1940.

On Shuler's subsequent career, see Mark S. Still, " 'Fighting Bob' Shuler: Fundamentalist and Reformer." Ph.D. diss. (Claremont, 1988).

On Fitts's subsequent career, see Alice Catt Armstrong, ed. *Who's Who in California, 1955–56* (San Clemente, Calif., 1956).

p. 326　On Fitt's suicide, see LAT March 20, 1973.

For the fate of Keaton, see clippings, March 29, July 10, September 8, 1940, USC.

On Culbert Olson, see LAX January 7, 1938.

For Daugherty, see SFC August 29, 1931; *Fortnight*, August 15, 1947; and *Sacramento Bee*, March 31, 1954.

For changes in the blue sky law, see LAT May 28, 1931, and John Dalton, "The California Corporate Securities Act," *California Law Review* 18 (May, 1930), 382.

On the end of Julian receivership, see Receiver's Final Report February 26, 1941 in Equity Case #M-12-M, NA.

On Suspension of Rule No. 16, see *ibid.*, notice, February 27, 1941.

p. 327　On Sunset-Pacific auction, see LAT November 11, 1934.

On J. A. Smith, see Ridgeway.

For the history of Sunset Oil, see clippings, Sunset Oil file, USC.

For Julian No. 1, see well records, CSMB.

"It can't be so . . ." clipping, March 25, 1934, USC.

For Julian's family in Winnipeg, see *Henderson's City of Winnipeg Directory*, 1934–1974.

p. 328　On Lewis in Texas, see *Corpus-Christi Caller-Times*, February 16, 20, 1970, and January 30, 1972.

p. 328    George Hamilton and *What Price Fugitive?* from author's telephone conversation with George Hamilton.

For Finney's opinions, see Guy Finney, *The Great Los Angeles Bubble* (New York, 1929), especially 6–26.

For McWilliams' description of the Julian affair, see McWilliams, 242–48.

# BIBLIOGRAPHY

## I. Primary Sources

### Archives and Collections

Ralph Arnold Collection. Huntington Library, San Marino,California.
Biography Files. California State Library, Sacramento.
Fletcher Bowron Papers. Huntington Library, San Marino, California.
California Historical Society, San Francisco, California.
California Petroleum Industry Collection. Long Beach Public Library.
Californiana Index. Los Angeles Public Library.
John Randolph Haynes Collection on Politics and Government of Los Angeles, 1890–1937.
    Special Collections, University of California, Los Angeles.
Hathaway Ranch Museum. Santa Fe Springs, California.
Hearst Clippings and Photo Collection. Regional History Center, Department of Special
    Collections, University of Southern California, Los Angeles.
Hiram Johnson Papers. Bancroft Library, University of California, Berkeley.
Los Angeles City Archives.
Los Angeles County Hall of Records.
Los Angeles Superior Court Archives.
Carey McWilliams Collection. Special Collections, University of California, Los Angeles.
National Archives. Southwestern Records Center. Laguna Nigel, California.
Oral History Collection. California State University, Long Beach.
Santa Fe Springs Public Library.
Charles F. Stern Collection. Special Collections, University of California, Los Angeles.
Texaco Oil Company, Legal Archives, Los Angeles, California.

### Newspapers and Periodicals

*Los Angeles Times*
*Los Angeles Examiner*
*Los Angeles Express*

*Los Angeles Record*
*Los Angeles Illustrated Daily News*
*New York Times*
*San Francisco Chronicle*
*Winnipeg Evening Tribune*
*Winnipeg Free Press*

*Bob Shuler's Magazine*
*California Graphic*
*Critic of Critics*
*Oil Age*
*Saturday Night Los Angeles*
*Sunset*


## Government Documents

California State Mining Bureau. *Reports.*
California. Division of Oil and Gas, *Summary of Operations, California Oil Field.* 1915–1930.
California. Division of Oil and Gas. Oil Well Records (Federal Building, Long Beach, California).
United States Department of Justice. Federal Bureau of Investigation. Files on C. C. Julian.
United States Senate. Special Committee on Bankruptcy and Receivership Proceedings in United States Courts (1933): 293–559, 651–663.


## City Directories, Who's Who, and Biographical Sketches

*American Bar.* Minneapolis: J.C. Fifield Co., 1925.
*American Biography.* New York: American Historical Society, 1931.
Armstrong, Alice Catt, ed. *Who's Who in California, 1955–56.* San Clemente, Calif.: Who's Who Historical Society, 1956.
Bates, J. C., ed. *History of the Bench and Bar.* San Francisco: Bench and Bar Publishing, 1912.
*Bench and Bar of California, 1937–38.* Chicago: C. W. Taylor, 1938.
*Bench and Bar of Los Angeles County.* Los Angeles: Los Angeles *Daily Journal,* 1922, 1928–29.
*California Blue Books, Legislative Manual or State Roster.* Sacramento: California Secretary of State, 1907, 1924, 1928.
*California Masonry.* Los Angeles: Masonic History Company, 1936.
Case, Walter H. *History of Long Beach and Vicinity.* Chicago: The American Historical Society, 1927.
Cline, William H. *Twelve Pioneers of Los Angeles.* Los Angeles: Times, Mirror Printing, 1928.
Clover, Samuel. *Constructive Californians: Men of Outstanding Ability Who Have Added Greatly to the State's Prestige.* Los Angeles: Saturday Night Publishing, 1926.
Detwiler, Justice B., ed. *Who's Who in California: A Biographical Dictionary, 1928–29, 1939–40, 1942–43.* San Francisco: Who's Who Publishing, 1929, 1940, 1943.

Forbes, Bertie C. *Men Who Are Making the West*. New York: B.C. Forbes, 1923.

Guinn, James Miller. *History of Biographical Record of Southern California*. Los Angeles: Historic Record Company, 1910.

Harper, Franklin, ed. *Who's Who on the Pacific Coast*. Los Angeles: Harper Publishing Company, 1913.

*Henderson's City of Vancouver Directory*, 1911–1914.

*Henderson's City of Winnipeg Directory*, 1899–1974.

*Henderson's Manitoba and Northwest Territories Gazeteer and Directory*, 1881–1900.

Hunt, Rockwell D., ed. *California and Californians*. Chicago: Lewis Publishing Company, 1926.

Ludwig, Ella A. *A History of the Harbor District of Los Angeles*. Los Angeles: Historic Record Company, 1926.

Malamut, Joseph. *Southwest Jewry: An Account of Jewish Progress and Achievement in the Southland*. Los Angeles: Sunland Publishing Company, 1926.

McGroarty, John. *California of the South: A History*. Chicago: S. J. Clarke Publ. Co., 1933.

———. *History of Los Angeles County*. Chicago: The American Historical Society, 1923.

———. *Los Angeles: From the Mountains to the Sea*. Chicago and New York: The American Historical Society, 1921.

*Men of California*. San Francisco; Los Angeles: Western Press Reports, 1926.

*Pan-Pacific Who's Who*. Honolulu: Honolulu Star-Bulletin, 1941.

*Press Reference Library . . . being the portraits and biographies of progressive men of the Southwest*. Los Angeles: The *Los Angeles Examiner*, 1912.

Spalding, William A. *History and Reminiscences, Los Angeles City and County*. Los Angeles: J. R. Finnell & Sons Publishing Company, 1931.

*Who's Who in Los Angeles*. Los Angeles, C. J. Lang, 1924, 1925, 1926.

*Who's Who in Los Angeles County*. Los Angeles, C. J. Lang, 1926, 1928, 1931, 1933, 1951, 1952.

*Who's Who in the Pacific Southwest*. Los Angeles: Times Mirror Printing, 1913.

*Who's Who on the Pacific Coast, 1949*. Chicago: Larkin, Roosevelt, & Larkin, 1949.

## Contemporary Accounts

Adamic, Louis. "Los Angeles! There She Blows." *Outlook and Independent* 155 (August 13, 1930): 563–65, 594–97.

———. *The Truth About Los Angeles*. Girard, Kansas: Haldeman-Julius, 1927.

Aikman, Duncan. "California Sunshine." *Nation* 129 (April 22, 1931): 448–50.

———. "Savoranola in Los Angeles." *American Mercury* (December, 1930): 426–29.

Atwood, Albert W. "Money from Everywhere." *Saturday Evening Post* 196 (May 12, 1923): 10–11, 134–47.

———. "Mad from Oil." *Saturday Evening Post* 196 (July 14, 1923): 10–11, 92–105.

———."When the Oil Flood Is On." *Saturday Evening Post* 196 (July 7, 1923), 10–11, 86–99.

Augsberg, Paul D. *Advertising Did It! The Story of Los Angeles*. Los Angeles, 1922.

Baker, Lorin L. *That Imperiled Freedom*. Los Angeles: Graphic Press Publishing Company, 1932.

Benas, Lionel. "The Corporate Securites Act: Recent Cases and Amendments." *California Law Review* 14 (January, 1926): 101–25.

Bliven, Bruce. "Los Angeles: The City That Is Bacchanalian." *The New Republic* (July 13, 1927): 197–200.

Blythe, Samuel. "Gambling with Grief." *Saturday Evening Post* 197 (July 5, 1924): 2–3, 71–74.

Borough, Rube. "Law and Order in Los Angeles." *The Nation* 125 (July 6, 1927): 12.

Brown, Stonewall. "A Fortune in Oil—The Promoter Speaks." *Atlantic Monthly* (January, 1928): 96–106.

Comstock, Sarah. "The Great American Mirror: Reflections from Los Angeles." *Harper's Monthly*, 156 (May, 1928).

Dalton, John. "The California Corporate Securities Act." *California Law Review* 18 (January, 1930): 115–36; (March, 1930): 254–66; (May, 1930): 373–99.

Elliot, Edward. "Los Angeles Banks and Their Relationship to the Economic Development of the Southwest." *Los Angeles Saturday Night*. 150th Birthday of Los Angeles: Fiesta and Pre-Olympiad, Special Issue, 1931.

Finney, Guy W. *The Great Los Angeles Bubble: A Present Day Story of Colossal Financial Jugglery and of Penalties Paid*. Los Angeles: The Milton Forbes Company, 1929.

Garrett, Garet. "Los Angeles in Fact and Dream." *Saturday Evening Post* 203 (October 18, 1930): 6–7, 134–44.

Hunter, Sherley. *Why Los Angeles Will Become the World's Greatest City*. Los Angeles: H. J. Mallen, 1923.

Jones, Frederick. "The Los Angeles Stock Exchange." *The Architect and Engineer* 104 (March, 1931): 25–45.

Los Angeles Curb Exchange. *Report of the President*. 1931.

"Los Angeles Stock Exchange to Have New Home." *Los Angeles Realtor* 8 (June, 1929): 13.

Robinson, Henry M. *Relativity in Business Morals*. Boston: Houghton-Mifflin, 1928.

Sanford, J. E. "History of Los Angeles Stock Exchange." *Los Angeles Saturday Night*. 150th Birthday of Los Angeles: Fiesta and Pre-Olympiad Special Issue, 1931.

Shuler, Robert P. "The Strange Death of Charlie Crawford." Los Angeles: Robert Shuler, 1931.

———. "Jailed." Los Angeles: Robert Shuler, 1930.

———. "Julian Thieves." Los Angeles: Robert Shuler, 1930.

———. "Julian Thieves in Politics." Los Angeles: Robert Shuler, 1930.

Sinclair, Upton. *Oil!* New York: Albert and Charles Boni, 1927.

Smedley, Agnes. "Shanghai Episode." *The New Republic* (June 13, 1934): 122–24.

*Sunshine and Grief in Southern California: where good men go wrong and wise people lose their money by an old promoter, forty years in the field of real estate*. Detroit: St. Claire, 1931.

Symes, Lillian. "Beautiful and Dumb." *Harper's* (June, 1931): 22–32.

"Where Races Mingle, But Never Merge." *Christian Science Monitor Weekly* (June 24, 1936): 4.

White, Leslie T. *Me, Detective*. New York: Harcourt, Brace Jovanovich, 1936.

Wilson, Paul N. "All Within the Law: How the Oil Sharks Trim Their Victims and Still Keep Out of Jail." *Sunset* 50 (May, 1923), 32–33, 56.

Wilson, Edmund. "The City of Our Lady the Queen of the Angels: I" *New Republic* (December 29, 1931): 89–93.

Woehlke, Walter V. "The Great Julian Pete Swindle." *Sunset* 54 (September, 1927): 12–15, 69, 80–81; (October, 1927): 16–19, 78–82; (November, 1927): 18–20, 66–68, 81–83.

## Memoirs and Reminiscences

Allman, Norwood F. *Shanghai Lawyer*. New York: Whittlesey House, 1943.

Borough, Reuben W. *Reuben W. Borough and California Reform Movements*. Oral History Project, UCLA Special Collections, 1968.

Hoover, Herbert. *The Memoirs of Herbert Hoover*. 3 vols. New York: Macmillan, 1951–52.

Kenney, Robert W. *My First Forty Years in California Politics, 1922–1962*. Oral History Project, UCLA Special Collections.

McWilliams, Carey. *Honorable in All Things*. Oral History Project, UCLA Special Collections, 1982.

Scott, Joseph. "The Joe Scott Story." Bancroft Library, University of California, Berkeley.

## II. Secondary Sources

Allen, Frederick Lewis. *Only Yesterday: An Informal History of the 1920s*. New York: Harper and Row, 1964.

Artibase, Alan F. J. *Winnipeg: A Social History of Urban Growth, 1874–1914*. Montreal and London: McGill-Queen's University Press, 1975.

———. *Winnipeg: An Illustrated History*. Toronto: James Lorimer & Co., 1977.

Bean, Walton. *California: An Interpretive History*, 3rd edition. New York: McGraw-Hill, 1978.

Beaton, Kendall. *Enterprise in Oil: A History of Shell in the United States*. New York: Appleton-Century-Crofts, 1957.

Belden, L. Burr. "Wild promotion but little ore from Leadfield." *San Bernardino Sun Telegram*. December 16, 1956: p. 46.

Blackford, Mansel G. *Politics of Business in California, 1890–1920*. Columbus: Ohio State University Press, 1977.

Bonelli, William. *Billion Dollar Blackjack: The Story of Corruption and the Los Angeles Times*. Beverly Hills: Civic Research Press, 1954.

Burke, Robert. *Olson's New Deal for California*. Berkeley: University of California, 1953.

Carlisle, Norman. "Con King of the Golden Coast." *True* 46 (October, 1965): 50, 113–18.

Chalfant, W. A. *Death Valley: The Facts*. Palo Alto: Stanford University Press, 1930.

Chaplin, Charles. *My Autobiography*. New York: Simon and Schuster, 1964.

City of Santa Fe Springs. *History in the Making*. 1976.

Cleland, Robert Glass. *California in Our Time: 1900–1940*. New York: Alfred A. Knopf, 1947.

Cleland, Robert G., and Osgood Hardy. *The March of Industry*. San Francisco: Powell Publishing Company, 1929.

Cogan, Sara G. *The Jews of Los Angeles, 1849–1945: An Annotated Bibliography*. Berkeley, Calif.: Western Jewish History Center, 1980.

Cross, Ira B. *Financing an Empire: History of Banking in California*. Chicago, San Francisco: S. J. Clarke Publishing, 1927.

Davis, Mike. *City of Quartz: Excavating the Future in Los Angeles*. New York: Verso, 1990.

Delmatier, Royce, *et al. The Rumble of California Politics, 1848–1970*. New York: John Wiley and Sons, 1970.

DeMille, Cecil. *Autobiography*. Englewood Cliffs, N.J.: Prentice Hall, 1959.

Demond, C. W. *Price, Waterhouse & Co. in America: A History of a Public Accounting Firm*. New York: Comet Press, 1951.

Drake, Earl G. *Regina: The Queen City*. Canada: McClellan & Stewart, 1955.

Dumke, Glenn. *The Boom of the Eighties in Southern California*. San Marino, Calif.: Huntington Library, 1944.

Edise, Lenore, ed. *Furrows in the Valley: A History of the Rural Municipality of Norris and Its People*. Intercollegiate Press, 1980.

Ely, Northcutt. "The Conservation of Oil." *Harvard Law Review* LI (1938): 1,209–44.

Findlay, James C. "The Economic Boom of the Twenties in Los Angeles." Ph.D. dissertation, Claremont, 1958.

Fogelson, Robert. *The Fragmented Metropolis: Los Angeles, 1830–1950*. Cambridge, Mass.: Harvard University Press, 1967.

Friesen, Gerald. *The Canadian Prairies: A History*. Toronto: University of Toronto Press, 1984.

Gabler, Neal. *An Empire of Their Own: How the Jews Invented Hollywood*. New York: Crown Publishers, 1988.

Galbraith, John Kenneth. *The Great Crash of 1929*. Cambridge, Mass.: The Riverside Press, 1961.

Glasscock, Carl Burgess. *Here's Death Valley*. New York: Bobbs Merrill, 1940.

Gottlieb, Robert, and Irene Wolt. *Thinking Big: The Story of the Los Angeles Times*. New York: G.P. Putnam's Sons, 1977.

Halberstam, David. *The Powers That Be*. New York: Alfred A. Knopf, 1979.

Halliday, John. "C. C. Julian." Unpublished seminar paper, July 18, 1966.

Hampton, Benjamin B. *A History of the Movies*. New York: Covici, Friede, 1931.

Hawley, Ellis. *The Great War and the Search for a Modern Order: A History of the American People and Their Institutions*. New York: St. Martin's Press, 1979.

Henstell, Bruce. "How God and Reverend Shuler Took L.A. by the Airwaves." *Los Angeles* (December, 1977).

———. "Remembering Mr. Bell." *Los Angeles* (January, 1985): 107–9.

———. "Southern California's First (And Still Best) Oil Scandal." *Los Angeles* (November, 1977): 92–98.

———. *Sunshine and Wealth: Los Angeles in the Twenties and Thirties*. San Francisco: Chronicle Books, 1984.

Hereford, Rockwell. *A Whole Man, Henry Mauris Robinson, and a Half Century, 1890–1940*. Pacific Grove, Calif.: Boxwood Press, 1985.

Hewins, Ralph. *The Richest American: J. Paul Getty*. New York: E. P. Dutton and Company, 1960.

Jaher, Frederic Cople. *The Urban Establishment: Upper Strata in Boston, New York, Charleston and Los Angeles*. Urbana: University of Illinois Press, 1982.

Jensen, Marilyn. "The Giant That Was Santa Fe Springs." *Pacific Oil World*, 4th annual review issue (January, 1980).

Jones, Charles S. *From the Rio Grande to the Atlantic: The Story of the Richfield Oil Company*. Norman: University of Oklahoma Press, 1972.

Kirkpatrick, Sidney D. *A Cast of Killers*. New York: E. P. Dutton, 1986.

Larson, Henrietta, and Kenneth Porter. *History of Humble Oil and Refining Company*. New York: Harper, 1959.

Layton, Edwin. "The Better America Federation: A Case Study of Superpatriotics." *The Pacific Historical Review* 30 (May, 1961): 137–48.

Leader, Leonard. "Los Angeles and the Great Depression." Ph.D. diss., University of California, Los Angeles, 1972.

Lee, Bourke. *Death Valley*. New York: Macmillan Company, 1930.

Lenzner, Robert. *The Great Getty: The Life and Loves of J. Paul Getty—Richest Man in the World*. New York: Signet Books, 1985.

Leuchtenburg, William. *The Perils of Prosperity, 1914–1932*. Chicago: The University of Chicago Press, 1958.

Lockwood, Charles. "In the Los Angeles Oil Boom Derricks Sprouted Like Trees." *Smithsonian* 11 (1980): 187–206.

McDonald, Thomas. "History of Los Angeles District Attorney's Office." Unpublished manuscript.

Marchand, Roland. *Advertising the American Dream: Making Way for Modernity, 1920–1940*. Berkeley, Calif.: University of California Press, 1986.

McCarthy, John Russell. *Joseph Francis Sartori, 1858–1948*. Los Angeles: Ward Ritchie Press, 1948.

Robert McElvaine. *The Great Depression*. New York: Times Books, 1984.

McKinney, Dwight F. and Fred Allhoff. "The Lid Off Los Angeles," *Liberty* (November 11, 1939).

McWilliams, Carey. *Southern California: An Island on the Land*. Santa Barbara: Peregrine Press, 1973.

Melendy, H. Brett, and Benjamin F. Gilbert. *The Governors of California: Peter H. Burnett to Edmund G. Brown*. Georgetown, Calif.: Talisman Press, 1965.

Mayo, Morrow. *Los Angeles*. New York: Alfred A. Knopf, 1933.

Mullins, William L. *The Depression and the Urban West Coast, 1929–1933*. Bloomington: Indiana University Press, 1991.

Nadeau, Remi. *Los Angeles: From Mission to Modern City*. New York: Longmans, Green, 1960.

Nash, Gerald. "Government and Business: A Case Study of the Regulation of Corporate Securities, 1850–1933." *Business History Review* 38 (Summer, 1964): 144–62.

Nicol, Eric. *Vancouver*. Toronto: Doubleday Canada, 1970.

Noggle, Burl. *Teapot Dome: Oil and Politics in the 1920s*. New York: W. W. Norton, 1962.

Olien, Roger M., and Diana Davids Olien. *Easy Money: Oil Promoters and Investors in the Jazz Age*. Chapel Hill: University of North Carolina Press, 1990.

Ormsby, Margaret A. *British Columbia: A History*. Toronto: Macmillan, 1971.

Perrett, Geoffrey. *American in the Twenties: A History*. New York: Touchstone, 1983.

Pohlman, John O. "Alfonzo E. Bell: A Biography." *Southern California Quarterly* 46 (1964): 197–217.

*Pony Express Courier* 7 (November, 1940): 6.

Posner, Russell M. "The Bank of Italy and the 1926 Campaign in California." *California Historical Society Quarterly* 37 (1958): 267–75, 359–68.

Pusateri, C. Joseph. *A History of American Business*. Arlington Heights, Ill.: Harlan Davidson, 1984.

Putnam, Jackson. "The Persistence of Progressivism in the 1920s: The Case of California." *Pacific Historical Review* 35 (1966): 395–411.

———. *Modern California Politics*, 2nd edition. San Francisco: Boyd and Fraser, 1984.

Ridgeway, Roberta. "The C. Arnholt Smith Story." *San Diego* 28 (July, 1976).

———. "The Adventures of Arnholt Smith's Smarter Brother." *San Diego* 28 (October, 1976).

Rintoul, William. *Spudding In: Recollections of Pioneer Days in the California Oil Fields.* San Francisco: California Historical Society, 1976.

Robinson, David. *Chaplin: His Life and Art.* New York: McGraw Hill, 1985.

Robinson, W. W. "The Southern California Real Estate Boom of the 1920s." *Southern California Quarterly* 44 (1942).

Robinson, William W. *Lawyers of Los Angeles.* Los Angeles: Los Angeles Bar Association, 1959.

Servin, Manuel P., and Iris Higbe Wilson. *Southern California and Its University: A History of USC, 1880–1964.* Los Angeles: Ward Ritchie Press, 1969.

Sitton, Tom. "The 'Boss' Without a Machine: Kent K. Parrot and Los Angeles Politics in the 1920s." *Southern California History* 87 (Summer, 1985): 365–87.

Sobel, Robert. *Panic on Wall Street.* New York: E. P. Dutton, 1988.

———. *The Great Bull Market: Wall Street in the 1920s.* New York: W. W. Norton, 1968.

———. *The Money Manias: The Eras of Great Speculation in America, 1770–1970.* New York: Weybright and Talley, 1973.

George Soule. *Prosperity Decade.* New York: Rinehart, 1947.

Spence, Hartzell. *Portrait in Oil: How the Ohio Oil Company Grew to Become Marathon.* New York: McGraw-Hill, 1962.

Starr, Kevin. *Material Dreams: Southern California Through the 1920s.* New York: Oxford University Press, 1990.

Stern, Norton B. *California Jewish History.* Glendale, Calif.: A. H. Clarke & Co., 1967.

Still, Mark S. "'Fighting Bob' Shuler: Fundamentalist and Reformer." Ph.D. dissertation, Claremont, 1988.

Stoker, Charles. *Thicker 'n Thieves.* Santa Monica, Calif.: Sidereal, 1951.

Susman, Warren I. *Culture as History: The Transformation of American Society in the Twentieth Century.* New York: Pantheon, 1984.

Taylor, Frank J., and Earl M. Welty. *Black Bonanza.* New York: McGraw-Hill, 1950.

Tompkins, Walter A. *Little Giant of Signal Hill: An Adventure in American Enterprise.* Englewood Cliffs, N.J.: Prentice-Hall, 1964.

Viehe, Fred H. "Black Gold Suburbs: The Influence of the Extractive Industry on the Suburbanization of Los Angeles, 1890–1930." *Journal of Urban History* 8 (1981): 3–26.

Vorspan, Max, and Lloyd P. Gartner. *History of the Jews in Los Angeles.* San Marino, Calif.: Huntington Library, 1970.

Weight, Harold O. "Leadfield Died of Complications." *Desert Magazine* (November, 1977): 34–46.

Williamson, Harold F., Ralph L. Andreano, Arnold R. Daum, and Gilbert C. Klose. *The American Petroleum Industry.* Vol. 2, *The Age of Energy, 1899–1959.* Evanston: Northwestern University Press, 1963.

Wolf, Marvin W., and Katherine Mader. *Fallen Angels: Chronicles of L. A. Crime and Mystery.* New York: Facts on File, 1986.

# INDEX

A. C. Wagy and Company, 124, 165, 197, 207, 300, 306, 308, 326; and Wray Berthold, 200–201, 204; dealings with S. C. Lewis, 126–28, 161; after Julian crash, 216, 218–19, 223, 228, 237; Lewis buys A. C. Wagy, 161–62; margin plan, 122, 162; receivership 235; trial of officials, 228, 245–53; under Lewis's management, 178, 181–82, 185–90, 195, 208

Abel, Walter, 143, 146, 148

Adamic, Louis, 11, 39

Adams, Earl, 147–48, 153

Adkisson, Arthur, 159, 310

Advertising, 10–12, 163; A. C. Wagy ads, 122, 162, 185–86, 201; C. C. Julian ads, 33–35, 40–45, 46–50, 179, 213–16, 223, 225–26, 316; Julian Petroleum ads, 58–59, 65–70, 71, 79, 84, 85, 90, 91, 93–95, 96, 101, 107–9, 116–118, 123–26, 128, 159–60, 162, 164–66, 185–86, 195, 208; Western Lead ads, 133–40, 144, 146, 152–53, 155

Aggelar, William Tell, 300

Aikman, Duncan, 279

Allan, Scotty, 139

Allen, Walter B., 178–79, 185, 187

Allison, Charles, 311

All-Year Club, 12

Amalgamated Oil, 29

American Legion, 261, 280

American Mortgage Company, 310

Anglo-London-Paris Bank of San Francisco, 172, 236, 238, 271

Anti-Saloon League, 177

Anti-Semitism, 185, 232–33, 263, 267, 278, 286, 309

Arnold, Frank, 164

Associated Oil, 167, 189, 271, 301, 311–12, 314–15, 327

Athearn, Fred, 295

Athens oil field, 97, 128

Atlantic Refining Company, 59

Atwood, Albert, 12, 13, 14, 29, 33, 37, 38, 39, 120

Augsberg, Paul, 11

Babbitts, 11–12, 231

Baker, Lorin, 98, 307

Baldwin, Willie, 113

Bandhauer, J. W., 136

Bank of America, 275. See also Bank of Italy

Bank of Italy, 170–71, 179, 192, 274–75

Bankers' Pool No. 1, 185, 187, 203–4, 281; disbands, 201–2; after Julian crash 222, 228, 233; members of, 183–84; 1930 indictments, 290–91, 300; operations of, 188–91, 195–96; origins of, 181–82; and reparations 224–25, 239, 246, 254, 272. See also stock pools

Bankers' Pool No. 2 (Rouse Pool), 184, 189, 195–96, 202–203, 228, 254. See also stock pools

Barber, John E.: discovers overissue, 203–4; and financing of California-Eastern, 171–73, 181, 189, 196–99, 201, 206, 217–18; and First Securities, 170–73; grand jury testimony, 240; indictment of, 228, 254; invests in Julian Pete stock, 187; and orphan check, 190, 208, 217; and Henry Robinson, 169–170, 217–18; subsequent career, 323

Barnes, Daniel, 207, 295

Barneson and Company. See H. J. Barneson and Company

Barneson, Harold J., 165, 240, 253, 311

Bauer, Harry, 190, 197–99, 201, 204–6, 209, 240–41

Bears and bear markets, 124–125, 128, 154, 162, 165–66

Beery, Ben, 200–201, 220

Beirne, William, 267

Bell, Alphonzo, 15, 25–27

Bell, Herbert A., 183, 188, 202, 239

Bell, James, 25

Bennett, Jack. See Berman, Jacob

Benson, Samuel, 281

Berger, Jacob "Jake," 19, 91, 132–33, 154

Berman, Jacob, 99, 118, 223, 227, 242, 283, 303; and anti-semitism, 232–33, 263; assassination attempt, 269; convicted

Berman, Jacob (*continued*)
　of forgery, 303–4; convicted of mail fraud,
　257, 259; and Buron Fitts, 262, 289–91, 295–
　97, 301; flees Los Angeles, 207–8; as fugitive,
　243–45; indictment, 228; and jury bribery,
　288–89; Keyes trial, 262–65, 268; and Morris
　Lavine, 286, 299–300; and Lewis Oil, 112–13;
　and Lloyd Nix, 280–82; and overissue, 113–
　14, 122–23, 128–29, 158–159, 174, 181, 190–
　191, 195–97, 206; overissue trial, 251–53; and
　Pike diaries, 259–260; pools and loan
　schemes, 161, 175–76, 179–81, 182, 184–85,
　187, 202, 205–7, 239–241, 246–47, 272, 280,
　311; in prison, 315, 324; returns to Los
　Angeles, 245–46; stock manipulations, 127–29,
　165; and stockholders' suit, 284–85;
　subsequent career, 325; and C. C. Young,
　293–94
Berman, Louis, 176, 247, 250–51, 265–66, 269,
　288, 323–24
Berreyesa Cattle Company, 311
Berthold, Wray, 200–201, 204, 220
Better America Foundation, 177, 184, 232
Bishop, William T., 86
Blank, Ben, 269
Bledsoe, Benjamin, 178
Bliven, Bruce, 11
Blue sky laws. *See* Corporate Securities Act
Blythe, Samuel, 14, 36, 45, 46, 58, 61
*Bob Shuler's Magazine*, 278, 325
Bogue, Charles L., 281, 305
Boosterism, 4, 5, 8, 11, 12, 231
Briegleb, Gustav A., 306–7
Broadway Central Bank, 98–99
Broadway Christian Church, 81
Brown, Eugene, 255
Brown, Meyer, 281
Browne, Peggy, 83
Brunson lease, 31–32, 45, 54, 317
Brunson, Lem and Clara, 31–32
Bryan, William Jennings, 277
"Bubble mill," 174, 180–81, 191, 206
Bulls and bull markets, 9, 124, 127–28, 186, 195,
　275
Burns agency, 93, 244
Burns, William J., 111
Bursian, Robert, 295–96
Bus tours of oil fields, 37–39, 96
Butler, Ben, 265–66, 269–70

C. C. Julian and Company, 133–34, 158
C. C. Julian Oil and Royalties Company, 316–
　320
Cafe Petroushka, 83
California Appellate Court, 290, 296, 312
California Bank, 189
California Club, 184
California Department of Corporations: and
　Beery and Berthold, 200–201; investigated by
　grand jury, 82; and Julian mining companies,
　215, 255; after Julian Pete crash, 219–20, 234;

and Julian Petroleum, 62–67, 70–72, 74–75,
　82, 89–90, 114; origins, 60–61; and Western
　Lead, 133–35, 140, 142–44, 146–48, 150, 153,
　155; and C. C. Young, 193–94. *See also*
　Daugherty, Edward Michael; Friedlander,
　Jacob
California Division of Mines, 156
*California Graphic*, 253
California State Mining Bureau, 54
California State Supreme Court, 62, 282, 312–
　13
California Stock Exchange, 285–86
California-Eastern Oil Corporation, 181–82, 187,
　227, 233, 240, 242; board of directors, 189,
　196–201, 284, 301; collapse of, 208–9;
　creation of, 167, 172; after Julian Pete crash,
　215–17; receivership, 220, 235–39, 246–47,
　271, 273, 326; reorganization of, 203–6
Campbell, H. F., 114, 174, 247–48, 251
Canadian Northern Railroad, 18
Canfield, Charles, 21
Cannon, David H., 133
Cantorovich, Mary, 322
Canyon Oil Company, 236
Caress, Zeke, 177
Carnahan, H. L.: as corporation commissioner,
　61, 194; as lieutenant governor, 271, 294, 304;
　as receiver, 218–22, 224–25, 235–39, 246–47,
　254, 271–73, 283–84, 291, 293, 299, 311,
　314–15, 326
Carr, Harry, 321
Carr, William J., 293–94
Chalant, Bill, 139–40
Chambers, Ben, 132–33
Champion and White, 22
Chandler, Harry, 63, 64, 174, 178, 184, 280,
　309; accusations by Julian, 217, 226–27; and
　*Los Angeles Times*, 77, 78, 80, 194; and
　Western Lead, 143–44, 146, 150, 155. *See
　also Los Angeles Times*
Chaplin, Charlie, 83–84
Chase, Lewis B., 82–83
Chessher, H. B., 126–27, 161–62, 251, 285
Cinema Finance Company, 5
Cities Service, 167
Citizens committee investigation of Julian
　Petroleum, 86–91, 91, 93, 101–2
Clark, Bruce L., 143, 151
Clark, David H., 307–8, 326
Clark, William, 262, 268
Cleland, Robert, 14, 42
Coe, Ira, 143
Cohen, Ben, 244–45
Collier, Frank, 7–8
common law trust, 32–33
Communications revolution, 9–10
Compton oil fields, 54, 63, 72, 73
Conroy, T. P., 113, 114, 174, 207, 247–48, 251–
　52, 303
Cook, Frederick, 10
Coolidge, Calvin, 171

Corporate Securities Act, 61–63, 70, 72, 76, 135, 247, 255, 301. *See also* California Department of Corporations
Corporations Department. *See* California Department of Corporations
Costa Mesa oil field, 128
Council of National Defense, 169
Courtney, Norman, 251
Cowan, R. W., 71
Crawford, Charlie: and Gustav Brieglieb, 305–6; and *Critic of Critics*, 307–8; estate of, 309; indictment of 296, 305; and Morris Lavine 286–87, 292, 299–300, 323; murder of 308–310, 326; and Kent Parrot, 177–78, 292; and Tia Juana pool 185, 187, 232; and A. C. Wagy and Company, 178, 187
Cressy, Frank, 305
Crime Commission, 178
*Critic of Critics*, 306–7
Crump, Guy, 272, 283–85, 312–13
Cryer, George, 60, 177, 178, 193, 305


Dabney, Joseph, 198, 201, 241, 284
Daugherty, Edward Michael, 179, 193, 292, 294; background, 60; and Julian Petroleum, 64–82, 87, 89, 101–2; and oil field swindles, 61–62; subsequent career, 326; and Western Lead, 134–36, 142–46, 150–52, 155. *See also* California Department of Corporations
Davis, Buddy, 240, 260, 266, 270–71
Davis, Clifford, 29
Davis, Harold, 220, 249
Davis, Jim, 256–57
Davis, LeCompte, 252, 266
Dawes, Charles, 171
Dawes plan, 171
*Day of the Locust*, 3
Defiance Gasoline, 69, 73, 85, 92
Demange, Etta, 186–87
DeMille, Cecil B., 176, 231, 234, 280
Dennison, Edward J., 300
Dickson, Edward, 194
Dodge, Jonathan S., 171
Doheny, Edward L., 21, 253
Doherty, Frank, 193–94, 262, 268, 279, 296
Donley, Ray L., 294
Doran, William, 247, 250, 253–54, 264, 290, 305
Dunaev, Nicholas, 83
Durbrow, C. W., 189, 200–201, 205–7
Durst, W. H., 118


Earl, E. T., 80
Eberts, Raburn, 281
Emmens, William, 32, 61
Empire Drilling, 157–58
Encyclopedia lots, 15, 30
Ethridge, John, 188
Ethyl gasoline, 108

Fall, Albert, 253
Farbstein, Jacob, 179–80, 239
Federal Bureau of Investigation, 109–110, 112, 237; and Lucien Wheeler, 187, 218, 220, 261; *See also* U. S. Dept of Justice
Federal Radio Commission, 226, 297–98, 305, 309, 325
Federated Church Brotherhood, 280
Finney, Guy, 11, 13, 100, 174, 196, 209, 217, 223, 231–32, 233, 241, 247, 249–50, 251, 258–59, 274, 328
First Investors, 201
First National Bank, 3–5, 7, 184, 187, 189, 206, 276, 283, 311; and Henry Robinson 167, 169–70, 233–34. *See also* First Securities; First National Trust and Savings Bank; Pacific Southwest Trust and Savings Bank; Security-First National Bank
First National Trust and Savings Bank, 271, 274
First Securities Company, 170, 240; and California-Eastern financing, 172, 181, 187, 189, 196–97, 200–201, 203–6, 233
Fischer, Oscar, 321
Fitts, Buron: background, 261–62; and Getzoff confession, 269–71; and Keyes prosecution, 262, 264, 267–68; and 1930 cases, 286–93, 295–301, 303–4, 316; and 1930 gubernatorial campaign, 288, 290–94, 304; subsequent career, 325–26
Flint and McKay, 6, 216
Flint, Frank, 171; career, 4–5, 6; death, 274; and Julian Petroleum, 184, 189, 199–200, 203, 205–6, 208–9, 216
Flint, Gertrude, 5
Flint, Motley: and Bankers' Pool, No. 1, 181–84, 188–89, 199, 202; career, 4–6; after Julian crash, 217, 231, 239; indicted, 291, 305; murder of, 7–8, 301–3; and Pacific Southwest, 5, 171–72
Flintridge, 6
Ford, Henry, 59
Ford, Joe, 251, 308
Foster, Frank, 320
Frank, Alvin M., 183, 191, 196–97, 202, 207, 240, 311
Friedlander, Jack: appointment as corporation commissioner, 193–94, 292–94; blackmailed, 286; as corporation commissioner, 201, 206, 209, 219–20, 232, 234, 293, 295; death, 323; indictment, 296, 300, 305–6
*Fullerton Daily News Tribune*, 298


Galbraith, John Kenneth, 8
Gans, Bob, 178, 185, 205, 241
Gardner, Frank, 144–45, 149
General Motors, 108
General Petroleum, 29, 31, 33, 54, 57–58, 69, 73, 165
Getty, George, 30
Getty, Jean Paul, 30

Getty Oil, 30, 48, 52
Getzoff, Ben, 258–60, 262–71, 288–90
Getzoff, Dave, 260, 262, 266, 270
Giannini, A. P., 169–70, 179, 192, 274
Gillis, R. C., 86
Glide Memorial Church, 278
Globe Petroleum, 31, 54, 316
Goodwin, Dean, 64
Great Depression, 8, 311, 313–14, 317, 327
*Great Los Angeles Bubble, The*, 13, 328
Greer, Perry, 179, 185, 187, 241
Grider, Frank, 288, 301
Groves, John, 288, 301, 304

H. J. Barneson and Company, 165, 241
H. R. Rule and Company, 285
Hackel, A. W., 184, 207, 239, 241
Hackel, E. F., 180
Hahn, Fred, 241
Haldeman, Harry M.: and Bankers' Pool No. 1,
    183–84, 202; criticized for role, 217, 232;
    death, 301; indicted, 228; and reparations,
    224–25, 239, 246, 254, 280–281
Haldeman, Harry Robbins (H. R.), 184, 301
Hamilton, George, 328
Hanna, Byron, 300
Hanover Trust Co., 9
Harris, C. T., 113–14, 159, 218, 303
Harris, Helen, 310
Harris, Mildred, 83
Harris, Rosebud, 310
Hawley, Ellis, 9
Hays, Will, 232, 284
Hearst, William Randolph, 93–94, 309
Hellman, Irving H., 86
Henry Huntington Art Gallery and Library,
    169
Hereford, Rockwell, 173, 197, 204, 233, 274
Hervey, William Rhodes, 5, 86, 171, 183, 202,
    239
Hickman, William, 249, 266, 270
Hill, Liberty, 308
Hollingsworth, W. I.: and Bankers' Pool No. 1,
    183–84, 202; indicted 228, 233; and
    reparations, 239, 246, 254, 272, 314
Hollywood, 5, 7, 13, 100, 120
*Hollywood Daily Citizen*, 298
Holzer, Harry, 312
Hoover, Herbert, 169, 273–74, 309, 328
Hoover, J. Edgar, 97, 109, 133
Horchitz, Louis, 226
Huntington Art Gallery and Library. *See* Henry
    Huntington Art Gallery and Library
Huntington Beach oil field, 14–15, 24, 29–30,
    37, 57, 85

Imhsen, Max, 80
Insider trading, 148
Insull, Samuel, 9
Inter-Allied Reparations Committee (1924), 171
International Match, 8

International Mining Corporation, 327
Investment Companies Act of 1913, 60

"Jailed," 298
Janes, (Judge), 220
Jardine, John Earl, 154, 204–5, 209
Jewish pool. *See* Rosenberg Pool
Jews, in Los Angeles, 179–180. *See also* anti-
    semitism; Rosenberg pool
Johnson, Alphonso Gales, 111
Johnson, Archibald, 191, 196, 235, 323
Johnson, Hiram, 178–79, 191, 193, 262, 279
Johnson, Leontine, 284–87, 294, 299–301, 304,
    306
Johnson, Parley M., 90–91, 95
Joralemon, Ira B., 153
Julian, C. A., 83
Julian, Courtney Chauncey (C. C.), 4, 8, 15, 36,
    60, 123, 129, 160, 167, 250, 299, 315, 325;
    accuses bankers, 217, 229; accuses
    stockbrokers, 228; advertisements, Julian
    Petroleum 58–59, 65–70, 71, 79, 84, 85, 90,
    91, 93–95, 96, 101, 107–9, 116–118;
    advertisements, mining companies, 214–215;
    advertisements, syndicate wells, 33–35, 40–45,
    46–50, 179, 213–16, 223, 225–26, 316;
    advertisements, Western Lead 133–40, 144,
    146, 152–53, 155; arrested in jewel robbery,
    320; arrested for mail fraud, 319–20; attacks
    Governor Richardson, 179; attempt on life, 78;
    autobiography, 321, 327; and Jacob Berman,
    118, 207; borrowing campaign, 68–69; and
    Brunson lease, 45, 54, 317; and C. C. Julian
    and Company, 133–35, 158; and C. C. Julian
    Oil and Royalties Company, 316–320;
    Calexico incident, 118–19; and California
    Department of Corporations, 62–67, 70–75,
    79, 82, 89–90, 114, 133–135, 140, 142–44,
    146–48, 150, 153, 155, 215, 255; in Canada,
    17–20, 22–23, 320; and Mary Cantorovich,
    322; career as oil promoter assessed, 62–63,
    100–103; charged with securities violations,
    67, 70, 72; and Charlie Chaplin, 83–84; and
    citizens committee, 86–89; at Corporations
    Department hearings, 70–72, 146–50, 152–53;
    creates common law trust, 32–34; creates
    Julian Petroleum, 57–58; creates Western
    Lead, 132–133; and Defiance Gasoline, 69,
    73, 85; divorce, 318; education, 19; eastern
    trip looking for backers, 93–94; flees to
    Shanghai, 320; fundraising, 31–32; funeral,
    322–323; grand jury testimony, 228, 295;
    grandfather, 17; and William R. Hearst, 93–
    94; Hollywood Bowl meetings, 95–96, 107–8;
    Huntington Beach well, 24–25, 30; and
    investors, 39–40, 43–45, 47, 51, 53–54, 63, 68,
    95–96, 223, 255–56; issues stock in Las Vegas,
    65; and Julian Merger Mines, 213–214; and
    Julian Stockholders Committee, 223–225; and
    Leadfield, 139–43, 148, 151–53, 155–56; and
    Leadfield extravaganza, 141–42; learns oil

trade in Bakersfield, 21–22, 23–24; leaves
Julian Petroleum, 119–120; and Leonora
Levy, 322–323; and Lightning gasoline, 108;
loans money to Lewis, 213; and *Los Angeles
Record*, 85–87, 131–32; and *Los Angeles
Times*, 73–75, 79, 85, 144, 146, 177–78, 194,
199, 215, 226–27, 230, 293, 318–19; malaria,
49, 131; mandamus suit against Daugherty,
81–82; mansion, 55, 130–31, 318–19;
marriage to Mary O'Donohue, 22; and New
Monte Cristo Mines, 214–15, 255–56; in
Oklahoma, 316–18; parents, 17–18, 22, 321;
in plane crash, 221; and prorationing
campaign, 57; public support for, 81–82; and
radio, 213, 215, 216–17, 225–26, 235; real
estate speculation in Canada, 19–20, 22–23;
reclaims Julian Petroleum presidency, 75;
resigns as Julian Petroleum president, 69, 100;
and John Salsberry, 132–33, 154, 318; at Santa
Fe Springs oil field, 31, 45, 46–54, 59, 130,
134, 214, 317; sells Julian Petroleum to S. C.
Lewis, 98–100; in Shanghai, 320–23; and
Maybelle Smith, 24, 130–31, 318; and stock
escrow campaign, 95–96, 116–17, 131, 185;
and stock pools, 95, 118; sues Daugherty and
Chandler, 146, 150; suicide, 322; in Texas oil
fields, 24; and Texas kidnapping, 318; and *The
Truth*, 76–78, 150–51; and U. S. Department
of Justice (F.B.I.), 31, 87, 97–98, 101–3, 109–
11, 133; and U. S. Internal Revenue Service,
131–33; and U. S. Post Office, 91–92, 111,
133; visit to Durango oil field, 93; and
Western Lead, 132–33, 135–156; and wife, 20,
22, 82, 130–31, 256, 318–19, 327. *See also
individual Julian wells*
Julian, Frances, 23, 130, 327
Julian, Francis, 17–18
Julian, John, 17
Julian, Lois, 23, 130, 327
Julian, Mary (O'Donohue), 20, 22, 82, 130–31,
256, 318–19, 327
Julian Merger Mines, 214–15
Julian Pete. *See* Julian Petroleum Corporation
Julian Petroleum Corporation (Julian Pete) 3, 4,
6, 7, 8, 82, 132, 134, 146, 152, 198, 200, 227,
301; Alamitos Hills oil field, 220, 222, 236;
appraisal of assets, 107–10, 164, 189–90, 197,
236–37; Athens oil field, 108, 164, 166;
auction of properties, 271; and Bankers' pools,
182–185, 188–89, 191; and citizens committee
investigation, 86–89, 92; Compton oil field,
87; Costa Mesa oil field, 160, 164; Defiance
gasoline, 69, 73, 85, 92; and Department of
Corporations, 62–69, 70–72, 74–76, 82, 89–
90, 114, 135, 209; directors added, 90–91;
directors resign, 95; and First National bank
demise, 274–75; founded, 58; and grand jury
testimony, 240–42; Huntington Beach oil
field, 86, 108, 164; investigated by U. S.
Department of Justice, 96–98, 107–12; Julian
resigns, 119, 130; Las Vegas stock issue, 65;

S.C. Lewis buys company, 98–100; and Lewis
Oil, 98–100, 108, 113, 172, 236; Lightning
gasoline, 108; Lomita oil field, 87; and *Los
Angeles Record*, 179; on Los Angeles Stock
Exchange, 89, 124–29, 136, 158, 160–62,
164–66, 186, 195, 197, 203; Louisiana oil
field, 160, 236; and margin buying, 122, 162;
merger with Marine Oil, 166–67, 172, 180–
81; merger with New England Oil and
Refining, 163–64, 166; Montana oil field, 181,
236; 1923 summer stock sale, 74–75; officials
indicted, 228; officials on trial, 245–48, 250–
55; Pico oil field, 92, 102; pipelines, 123; and
Adolf Ramish, 157–59, 196–97, 271–72;
receivership, 220, 235–39, 246–47, 271, 273,
326; refineries, 59, 64, 68, 87, 96, 102, 123,
236–37; and Henry Robinson, 19, 233–34,
273–75; San Pedro wharf, 87; Santa Fe
Springs oil field, 108; service stations, 73, 85,
92, 108, 163, 167; shareholders and
shareholders' meetings, 95–96, 107–8, 160,
185, 200–201, 203, 221, 238, 271–73, 316;
Signal Hill oil field, 160, 164, 166; stock, 118,
180, 186–88, 195, 197, 200, 208, 227, 229,
231, 236, 241, 247, 282, 284, 287; stock crash
(May, 1927), 208–209, 213, 215; stock
overissue, 101–3, 127–29, 162, 164, 180–81,
184, 188, 195–96, 200, 202–6, 208–9, 220,
227, 231, 236–37, 240, 242, 246, 249, 251,
283, 293, 295; stock escrow scheme, 95–96,
116–17, 131, 185–86; stock transfer office,
113–14, 158–59, 174–75, 180–81, 218; tank
farms, 59, 68, 86; Torrance oil field, 84, 87,
92, 164; Whittier oil field, 84, 86. *See also*
Julian Petroleum scandal
Julian Petroleum scandal, 3, 6, 7, 8, 220, 225,
229, 307, 311, 315, 316, 323, 328
Julian Petroleum Trust Pool, 118, 119, 123
Julian Stockholders' Association, 223–27, 234,
282–83. *See also* Loeb, Arthur
"Julian Thieves," 8, 291–93, 298
"Julian Thieves in Politics," 298
Julian, Violet, 20, 323
Julian Well No. 1, 45–54, 63, 72, 97, 101, 130,
134, 327
Julian Well No. 2, 48–53, 63, 72, 97, 101, 130,
134
Julian Well No. 3, 51, 59, 63, 72, 97, 101, 130,
134
Julian Well No. 4, 48–49, 51–54, 63, 72, 97,
101, 130, 134
Julian Well No. 5, 48, 51, 53–54, 63, 72, 101,
130, 134
Julian Wells Nos. 6–9, 53, 63–64, 72, 101–2
Julian Wells Nos. 11–13, 54, 63–64, 72, 102
Julian Wells Nos. 14–17, 316

Keaton, Frank: background, 3–4; murders
Motley Flint, 6–8, 301–4; subsequent life,
326; trial of, 304–5

Keaton, Miriam, 3, 4, 6, 305
Kenney, Robert, 239–40, 286
Kern County oil field, 20, 21
Keyes, Asa, 223, 224, 255, 280, 283, 287, 289, 299, 301; background, 248–49; bribery trial, 262–70, 276; death, 323; Kimmerle case, 256–57; overissue trial, 250, 252–54; and Pike diaries, 258–61; and search for Jacob Berman, 243–44, 246
Keyes, Elizabeth, 266
Kimmerle, H. J., 226–27, 256–57, 282
Kinette, Earl, 303
King, L. J., 189, 197–201, 204–7, 209, 241, 284, 301
King, William, 98–99, 110, 161, 326
Kottemann, Bertha, 187
Kottemann, William C., 187; indicted, 228; Julian Petroleum and A. C. Wagy audits, 181, 188, 197, 201, 203, 205, 208–9, 237; overissue trial 245, 247, 251–52
Krause, Louis, 288–89, 301, 304
Krueger, Ivar, 8

Lacy, William, 86
Laemmle, Carl, 158
Lane, Albert, 185, 189
Lane, Franklin K., 170
Lansing, Robert, 170
Lasker, A. I., 260
Lavine, Morris, 98, 179, 187, 296; blackmail plot, 286–87, 299–301, 304–6, 308; subsequent career, 323
Leadfield, Calif., 139–43, 148, 151–53, 155–56. *See also* Western Lead
Leasehounds, 30, 52
Lee, Bourke, 155
Leopold and Loeb murder case, 10
Levy, Leonora, 322–23
Lewis, James, 312
Lewis Oil, 116, 163, 181, 257, 283, 290; history of, 111–113, 116; and Julian Pete receivers, 236, 272; and Julian Petroleum, 98–100, 108, 172
Lewis Oil Market-Export Corporation, 111
Lewis, Sheridan C., 4, 103, 120, 213, 280–82, 303; and A. C. Wagy and Company, 126–28, 161, 162, 185–87, 219; acquires Julian Pete, 98–100; background, 98, 111; and H. J. Barneson, 165–66; and Jacob Berman, 112–14, 174, 176, 180, 206–7, 242–46; bribery charges, 288–90, 300–301; and California-Eastern, 169–73, 181, 189–90, 196–209; convicted of mail fraud, 257, 259; descriptions, 98, 237–38; and federal charges, 315; grand jury testimony, 228, 240–41; after Julian Pete crash, 216, 227–28, 241–42; and Lewis Oil, 111–13, 257, 272; and loans, 178–79; and mergers, 162–63, 166–67; and 1930 cases, 288–91, 293–97, 300–301, 303, 305–6; and overissue, 174, 176, 180, 204–209; overissue trial, 245–54; and pools, 181–86;

188–89, 190–91, 196, 199, 202; as president of Julian Petroleum, 107–10, 114, 116, 118–119, 122–29, 131, 135, 159–67, 195; in prison, 295, 297, 315; and Adolph Ramish, 158, 196–97; and receivers, 218, 235–39, 271–72, 313; released from prison, 323–24; and stockholders' suit, 283–86; subsequent life, 327–28; and C. C. Young, 179, 194, 293–94
Liberty Bank of America, 192
Lickley, E. J. "Doc": and usury cases, 220–22, 224–25, 228, 234, 246, 255, 291; as Municipal Court Judge 280, 294
Lightning gasoline, 108
Lincoln–Roosevelt League, 192
Lindberg, Charles, 10, 245
Loeb, Arthur, 234, 299, 301, 324, 327; and Asa Keyes, 256–57, 259; loses eye, 226–27; and receivers, 272–73, 312–15, 327; and stockholders' suit, 282–85, 288, 312–13, 324
Loeb, Mrs. Arthur, 314
Loeb, Edwin, 208–9
Loeb, Walker and Loeb, 206, 208–9, 216
Los Angeles, 11, 27, 277; financial district (Spring Street), 51, 66, 74, 75, 120, 126, 142, 145, 164, 173–74, 176, 181, 195, 204, 206, 213, 219, 227–28, 241–42, 244, 275; land boom of 1880s, 4, 11, 25; newspaper wars, 77–80; oil boom, 14–15; population growth, 12; real estate, 13–14, 15, 37
Los Angeles Bar Association, 283, 297–98
Los Angeles Board of Public Works, 86
Los Angeles Chamber of Commerce, 11, 12, 85, 86
Los Angeles Convention League, 5
Los Angeles County grand jury, investigation of Julian Pete scandal, 219, 228, 234–35, 238; and Buron Fitts, 290–292, 294–96, 299; indictment of Asa Keyes, 259; indictment of Lavine and Johnson, 286; transcripts, 239–42
*Los Angeles Examiner*, 75, 80, 179, 240, 248, 285–88, 290, 298, 325
*Los Angeles Express*, 75, 80, 194, 215, 230, 232, 253, 287, 293
*Los Angeles Herald*, 75, 80, 178, 239–40
*Los Angeles Illustrated Daily News*, 77–80, 84, 85, 87, 232
Los Angeles Investment Corp., 60, 206, 218, 233
*Los Angeles Ledger*, 226, 231
*Los Angeles Record*, 62, 78–79, 81, 178, 179, 214, 224, 240, 248; and citizens committee, 85–86; opposes Buron Fitts and Robert Shuler, 262, 270, 287–91, 296–98; publishes Julian biographies, 85–87, 131–32
*Los Angeles Review*, 292, 294
Los Angeles Stock Exchange (LASE), 6; description of, 120, 122; destroys records, 251; and Julian Petroleum 89, 125–26, 128, 158, 160–62, 164–65, 186, 195, 197, 203; after Julian stock crash 219, 234; speculation in bank stocks, 275; and stock overissue, 204–5, 209, 242; and stockholders' suit, 283–85, 287,

297, 312–13; and Western Lead, 136, 145–49, 152–54

*Los Angeles Times*, 53, 60, 64, 289, 306; editorials on Julian Pete scandal, 223, 229–32, 260, 270, 295; and Buron Fitts, 262, 270, 299; and C. C. Julian, 73–75, 79, 85, 199, 226–27, 318–19; and Julian Petroleum 124, 126, 195, 223; and Asa Keyes 248, 259–60; and politics, 177–78, 194, 293; radio station KHJ, 215, 226; role in Los Angeles, 77–78, 80; and Robert Shuler, 279, 298, 306, 318–19; and Western Lead, 144, 146. *See also* Chandler, Harry

Los Angeles Trust and Savings, 5, 170–71

Love, Caroline, 288–90, 300

Macklin Shirt Company, 320

MacMillan, Clifford, J., 215

Madeira, W. I., 92–93

Mammoth Oil Company, 253

Marco, Albert, 177–78, 187, 307

Margin buying, 6, 122, 162

Marine Oil Company, 166–67, 172, 189

Mason, Charles W., 318–19

Mass marketing, 10

Matthews, Blayney, 263, 269, 286

Mayer, Irene, 7

Mayer, Louis B., 7, 232, 309, 315; and Bankers' Pool No. 1, 183–84, 202; indicted, 228; and reparations 246, 249

Mayo, Morrow, 11, 39

McAfee, Guy "Stringbean," 177, 306

McComb, Marshall, 289–90, 295–96, 301, 304–5

McCormick, Paul J., 218–20, 237, 271, 273

McDonald, Al, 267

McKay, Henry Squarebriggs, Jr.: and Bankers' Pool No. 1, 183–84, 188, 190, 199–200, 202; indicted 291; and reparations, 239

McKay, Henry Squarebriggs, Sr., 6, 184

McMullen, P. L., 183, 202, 239

McNabb, Samuel, 201

McPherson, Aimee Semple, 215, 248, 277–78

McWilliams, Carey, 94, 100, 307, 310, 328

*Me, Detective*, 262

Mellon, Andrew, 10

Mencken, H. L., 39

Merchants National Bank, 159, 190–191, 274

Mergers, 8, 163–64, 166–68, 180–81, 186–87, 189, 192, 196, 200, 242, 271, 274, 293

Merriam, Frank, 324

Metro-Goldwin-Mayer, 7

Metropolitan Bank and Trust Company, 4

Meyer, Marius, 26

Meyer Sands, 51–53, 59

Meyers, Frank, 286, 300

Miley, E. J., 189, 201, 205, 207

Miller, J. H., 97–98, 101–2, 109–10

Million Dollar Pool. *See* Bankers' Pool No. 1

*Mining Journal Press*, 143

Mintner, Mary Miles, 83

Monarch Petroleum Company, 111

Monte Cristo Mines, 214, 255–56, 316

Morris, A. W. "Abe," 147–48, 152–53, 205, 207, 286, 313

Morrisey, Thomas, 159, 310

*Municipal News*, 78

Municipal ownership of power, 178

Murphy, John, 248

"My Three Years with S. C. Lewis," 285

National Industrial Recovery Act (NIRA), 314

National Vigilance Committee of Better Business Bureau, 33

Native Sons of the Golden West, 157

Neblett, William, 248, 326

New England Oil and Refining Company, 163–64, 166

New Monte Cristo Mining Company, 215, 255–56

New York Curb Exchange, 108, 124, 162

New York Stock Exchange, 8, 9, 69, 167, 174

New York Supreme Court, 324

Nix, Lloyd, 280, 282, 291, 301–3

Nixon, Richard, 325

Nordstrom family, 30

Northern Club, 177

Nye, Gerald P., 253

Oakes, Louis, 278

Obenchain, Madalynne, 248

Occidental College, 25

O'Donohue, Mary. *See* Julian, Mary

*Oil Age*, 50, 53

Oil drilling, 45–50

Oil fields. *See* individual fields

Oklahoma Corporation Commission, 317

Oklahoma Supreme Court, 318

Old Colony Exchange Company, 9

Olson, Culbert, 161, 326

O'Melveny, Donald, 229, 231

Orphan check, 190, 208, 217

Otis, Harrison Gray, 77

Owens Valley, 12, 80

Pacific Bond and Share Company, 190, 208

Pacific Finance Company, 197

Pacific Indemnity Company, 202

Pacific Light and Power, 169

Pacific Southwest Trust and Savings Bank, 5, 7, 86, 183, 208, 273, 315; attacked by C. C. Julian, 217, 227; and California-Eastern, 172–73, 181, 187–88, 200–202, 236; history of, 167, 169–70; and S. C. Lewis, 171–73, 209, 240–42; merger with First National 234, 271, 274; officers indicted 233, 252, 254; and receivers, 238, 273–74; withdraws from California–Eastern 204–5. *See also* Barber, John; First National Bank; Robinson, Henry; Stern, Charles.

Packard, Fred, 99, 110, 176, 218, 303

Palomar Observatory, 169

Pan-American Oil, 69

Pantages, Alexander, 309
Parker, Marion, 249
Parrot, Kent: blackmailed, 286; and Friedlander appointment, 193–94, 291–94; as political boss, 176–79, 187, 221, 232, 241, 288
Pauley Oil, 109
Penn, Courtney Chauncey, 320
Penn, H. A., 320
Perry, Arthur, 281
Petroleum industry, 13–15, 157; in California, 20, 21; in Los Angeles, 21; overproduction, 56–57; prices rise, 84; prorationing, 57, 317–18; stock swindles, 36, 61; town lot drilling, 56
Pettingell, Frank, 89, 126, 144–45, 149, 204; background, 120; death of, 154
Pico Canyon oil field, 20
Pike, Margie, 219
Pike, Milton, 258–60, 262–63, 267–68, 270
Pinkerton agency, 77
Pioneer Oil, 26
Platt, Benjamin, 184, 233
Ponzi, Charles, 9
Ponzi schemes, 9, 11, 159
Pope, (Judge), 281–82
Porter, John R., 280, 298, 302–303, 305, 325
Poste, Dave and Anna, 139, 155
Powell, George, Jr., 219
Price-Waterhouse, 87–88, 102–3, 109–10
Prorationing, 57, 317–18
Pyle, Flora, 253

Radio, 215–17, 225–27, 305, 309, 325
Ramish, Adolph: background, 157; and Bankers' Pool No. 1, 182–83, 188, 196–97, 202; grand jury testimony, 240; indicted, 228; loans to Lewis and Berman, 158–59, 179, 241; and orphan check, 190; and receivers, 271–72; reparations, 225, 232; and stockholders suit, 284, 313–14; and Sunset Oil, 327
Raskob, John J., 9
Reese, Charles, 161, 247, 251, 291
Reese, Raymond, 149, 161, 219, 228, 247, 250, 252, 291
Reimer, Charles, 260, 262, 266
Richardson, Friend, 62, 67, 152, 179, 192
Richardson, Julian, 249–52, 262–63
Richfield Oil, 109, 311
Rintoul, William, 46
Rittinger, John, 258, 260, 262, 268
Robinson, Dudley, 264, 267
Robinson, Henry Mauris, 174, 178, 309, 315; attacked by C. C. Julian, 217–18, 230; background, 167–69, 204; and First National Bank, 170–171, 193, 271; impact of Julian scandal on, 273–75; and receivers, 235, 238; recruits Harry Bauer, 199; responds to charges 233–34
Rockefeller, John D., 21
Rolapp, Frank H., 312, 314–15, 326
Rolph, James, 292, 294, 304, 326
Roosevelt, Franklin D., 8

Root, Lloyd, 153
Rosenberg, Barnett, 157
Rosenberg, Edward H.: activities with Jacob Berman, 158–59, 161, 174, 176, 179–81, 185, 232, 239–41, 311; death, 325; indicted, 228, 284, 291; as informant, 218; and Keyes case, 258–67, 269–70, 301; and Nix prosecution 281; overissue trial, 245, 247–48, 250–52; and Rosenberg pool, 185, 189, 202, 233; and S. C. Lewis, 159, 176, 205, 208, 227
Rosenberg, Jack, 176, 260, 266
Rosenberg pool (Jewish pool), 185, 189, 202–3, 233
Roth, Jack, 66, 78, 82, 84, 93, 94, 296, 305, 320, 323–24
Rothman, Louis, 281
Rouse, I. Linden, 184–85, 187–89, 196, 202, 251–52
Rouse Pool. See Banker's Pool No. 2
Rule and Company. See H. R. Rule and Company
Rule, H. R., 285
Rule, O. Rey, 191, 196–97, 201–2, 207
Rule of capture, 29
Rush, Jed, 267
Ruth, Babe, 10

Salsberry, John, 132–33, 154, 318
San Anselmo Theological Seminary, 25
San Francisco Daily News, 232
San Francisco pool. See Johnson, Archibald
San Pedro Harbor, 12
Santa Fe Springs oil field, 25, 57, 61, 72; Agee wells, 31, 33; Alexander No. 1 well blowout, 27; and Alphonzo Bell, 25–27; 57, 60, 61, 103; Bell sands, 51–52; Bell wells, 27, 31, 37; descriptions, 25, 27–28; discovery of oil, 14–15, 26–27; "four corners," 30; and C. C. Julian, 30–31, 33–34, 40–42, 45–54, 59, 97, 101, 130, 133, 311, 316–17, 327; Meyer sands, 51–53, 59; promotional tours, 37–39; syndicate wells, 130; "town lot" drilling boom, 29–32, 52
Sartori, Joseph, 99, 108, 110, 170, 174, 193, 274–75
Saturday Evening Post, 12
Saturday Night, 230–31
Schauer, B. Rey, 303, 305
Schenk, Paul, 268
Schindler, Mike, 306
Scopes Trial, 10
Scott, Joseph, 218–22, 224–25, 235–39, 246–47, 254, 271–74, 283, 291, 314–15
Security Trust and Savings Bank, 99, 110, 274
Security-First National Bank, 274–75, 311
Selznick, David O., 7
Selznick, Florence, 7
Selznick, Lewis, 7
Selznick, Myron, 7
Sharkey, Anna, 113

Shell Oil, 29, 82
Sherman, Joseph, 258–60, 262, 267–68, 270
Shipp, Pat, 114, 128–29, 191, 197, 205, 218
Shuler, Robert Pierce: acquires radio station, 279; background, 276–80; and Federal Radio Commission, 297–98, 303, 309; feud with Gustav Briegleib, 305–6, 308; and Buron Fitts, 290–94, 304; imprisoned, 298–299; on C. C. Julian, 256, 259, 271; on Julian Petroleum scandal, 183, 203, 231–32, 242, 253, 271, 279–80; "Julian Thieves," 8, 292–93, 303; and Asa Keyes, 248–49, 259; and Ku Klux Klan, 278; on S. C. Lewis, 98; and Los Angeles politics, 176–79, 280, 304; and murder of Motley Flint, 6, 8, 302; pamphlets, 8, 278, 292–93, 298, 303, 308; and John Porter, 280; subsequent career, 325
Sierra Refining, 108, 167
Signal Hill oil field, 15, 37, 57, 160
Silberberg, Mendel, 185, 189, 248, 309
Silberberg pool. *See* Tia Juana Pool
Silliman, Benjamin, Jr., 20
Sinclair, Harry, 253
Sinclair, Upton, 45
Slossberg, Max, 280
Smith, C. Arnholdt, 317, 327
Smith, Gladys, 83
Smith, John A., 317, 327
Smith, Maybelle, 24, 130–31, 318
Snyder, Meredith "Pinky," 177
Sobel, Robert, 9, 122
*Southern California: An Island on the Land,* 328
Southern California Edison, 169, 199
Southern Pacific Railroad, 4, 167, 189
Speculation, 8–11, 13, 36, 39, 44, 60, 61, 66, 80
Spence, Hartzell, 29
Spencer, Herbert, 306–9, 326
Spreckles, Claus, 157
Spring Street. *See* Los Angeles, financial district
Spurlin, DeKalb, 161, 187, 223–27, 234, 241, 282
St. Albans, Edward, 24
St. Paul's Presbyterian Church, 306
Standard Oil, 21, 24, 26, 29, 30, 32, 48, 51, 56, 57, 58, 59, 82, 97, 103, 108, 317, 321
Stanley, Hubert, 102–3, 109
Starr, Kevin, 5
Staunton, W. E., 133, 154
Steckel, Roy, 308
Stephens, William D., 60, 90–91, 95
Stern, Charles F.: accused by C. C. Julian, 217–18; correspondence, 230; indicted, 228, 233, 254; joins Pacific Southwest Bank, 171–72; and S. C. Lewis, 172, 187, 190; and overissue 196–97, 204; as State Superintendent of Banks, 170–71; subsequent career, 323
Stewart, Robert, 262, 303
Still, Mark Sumner, 277, 278
Stock markets. *See* bears and bear markets; bulls and bull markets; Los Angeles Stock

Exchange; New York Stock Exchange; New York Curb Exchange; margin buying; speculation; stock pools
Stock market crash of 1929, 4, 6, 275
Stock pools, 6, 9, 95–96, 181–82, 186–91, 201–4, 219, 221, 233–34, 238, 240–41. *See also individual stock pools*
"Strange Death of Charlie Crawford," 308
Streeter, C. C., 165–66, 176, 251, 280–81, 284, 323
"Sucker tents," 38–40
Suicides, 223, 310, 322, 325
Sullivan, Maurice, 141
Sun Oil, 103
Sunset International Petroleum, 327–28
Sunset Oil, 314, 326–27
Sunset Pacific Oil Company, 271–73, 276, 282–84, 301; receivership, 311–15, 326
Supply-side economics, 10
Susman, Warren, 9
Swanson, Axel, 272

Talmadge, Norma, 38
Tappan, Claire, 313
Taylor, William Desmond, 83, 249
Tea Pot Dome, 95, 253
Tecumseh Oil, 23
Texas Holding Co., 24
*That Imperiled Freedom*, 307
Thayer, Ezra, 214
*The Truth*, 76–78, 150–51
Tia Juana pool (Silberberg pool), 185, 189, 231, 235, 306
Titus Canyon, 132, 149, 156
Toplitsky, Joe, 183, 188–91, 196, 202, 228, 246, 249, 311
Torrance oil field, 157
Tourism, 12
Tower, Ira, 83
Treat, Charles H., 85, 90–91, 95
Trinity Methodist Church, 277, 298. *See also* Shuler, Robert Pierce
Tuller, Walter, 225
Tully, A. S., 227, 232

Union Oil, 14, 24, 26, 29, 56, 71, 73
Union Trust and Savings Bank, 205
United Oil Well Supply, 317, 327
U. S. Congress, 10
U. S. Dept. of Justice, 36, 111, 133; FBI investigation, 97–98, 101–3; investigates Julian Pete, 87. *See also* Federal Bureau of Investigation
U. S. Internal Revenue Service, 131–33
United States National Bank, 327
U. S. Post Office, 92, 111, 133
U. S. Senate Special Committee to Investigate Bankruptcy and Receivership Proceedings, 315
U. S. Supreme Court, 163, 325
United Studios, 7

Van Cott, E. P., 249–50
Van Der Berg, Jacques, 214, 256
Van, Jac. *See* Van Der Berg, Jacques
Van Viletzy Brugges De Frelinghuysen, Baron
    Jac Ferdinand. *See* Van Der Berg, Jacques
Vanderbilt, Cornelius, Jr., 77–80
Vernon Country Club, 278
Vianelli, Carl, 266, 288–89
Voorhies, Jerry, 325

Wagy, A. C., 126–27, 149, 161–62, 251, 291,
    313
Wagy and Company. *See* A. C. Wagy and
    Company
Wall Street. *See* New York Stock Exchange and
    New York Curb Exchange
Walther, Elmer, 146, 148–49, 151, 154
Warner Brothers Corporation, 5, 302
Warner, Jack, 5, 302
Watergate scandal, 301, 327
Weaver, Jack, 288–89
Welch, James, 127–28
Weller, William, 308
West, Nathanael, 3, 39
West, Roy, 75, 76, 78
Western Lead Mines Company, 160, 162, 167,
    179, 213–14, 316, 318, 326; advertisements,
    133–40, 144, 146, 152–53, 155; and Harry

Chandler, 143–44, 146, 150, 155; creation of,
    132–33; and Department of Corporations,
    133–35, 140, 142–44, 146–48, 150, 153, 155;
    hearings on, 146–49, 150–53; and Leadfield,
    139–43, 148, 151–53, 155–56; Leadfield
    excursion, 140–42; and Los Angeles Stock
    Exchange, 136, 145–49, 152–54; and *Los
    Angeles Times*, 144, 146
"What Price Fugitive?," 321, 328
Wheeler, Lucien, 187, 218, 220, 261
White, Leslie, 258, 262–63, 269, 286, 300, 307–
    8
Wilshire Oil, 101–2
Wilson, Edmund, 277
Wilson, Paul, 38
Wilson, Woodrow, 169–70
Witherell, O. S., 69
Woehlke, Walter, 100, 190, 215–17, 228, 235
Wood, Walton, 289
Wood, Will C., 192, 233
Woolwine, Thomas, 248
World War I, 3, 10, 23, 26, 169, 261–62
World War II, 327
Wyman, L. C., 143, 148

Yankovitch, Leon, 254
Young, Clement C., 179, 192–93, 262, 271, 280,
    288, 291–99, 304